0 463 218 00 59

THE CHEMICAL PHYSICS OF SOLID SURFACES AND HETEROGENEOUS CATALYSIS

THE CHEMICAL PHYSICS OF SOLID SURFACES AND HETEROGENEOUS CATALYSIS

Volume 1
CLEAN SOLID SURFACES

Volume 2
ADSORPTION AT SOLID SURFACES

Volume 3
CHEMISORPTION SYSTEMS

Volume 4
FUNDAMENTAL STUDIES OF HETEROGENEOUS CATALYSIS

Volume 5
SURFACE PROPERTIES OF ELECTRONIC MATERIALS

THE CHEMICAL PHYSICS OF SOLID SURFACES AND HETEROGENEOUS CATALYSIS

EDITED BY

D.A. KING
B.Sc., Ph.D. (Rand), Sc.D. (East Anglia), M.Inst.P., F.R.S.C.
Professor of Physical Chemistry,
University of Cambridge

AND

D.P. WOODRUFF
B.Sc. (Bristol), Ph.D., D.SC. (Warwick)
Senior Lecturer in Physics,
University of Warwick

VOLUME 3

CHEMISORPTION SYSTEMS

PART A

ELSEVIER

AMSTERDAM — OXFORD — NEW YORK — TOKYO
1990

ELSEVIER SCIENCE PUBLISHERS B.V.
Sara Burgerhartstraat 25
P.O. Box 211, 1000 AE Amsterdam, The Netherlands

Distributors for the United States and Canada

ELSEVIER SCIENCE PUBLISHING COMPANY INC.
655 Avenue of the Americas
New York, NY 10010
U.S.A.

ISBN 0-444-42027-4 (Vol. 3A)

with 166 illustrations and 21 tables

Printed in The Netherlands

Contributors to Volume 3, Part A

J.Q. BROUGHTON — Complex Systems Theory,
Naval Research Laboratories,
Washington, DC 20375, U.S.A.

C.R. BRUNDLE — IBM Research Division,
Almaden Research Center,
San Jose, CA 95120, U.S.A.

J.C. CAMPUZANO — Department of Physics,
University of Illinois at Chicago,
Chicago, IL 60680, U.S.A.

J.W. DAVENPORT — Department of Physics,
Brookhaven National Laboratory,
Upton, NY 11973, U.S.A.

P.J. ESTRUP — Department of Physics and Department of Chemistry,
Brown University,
Providence, RI 02912, U.S.A.

M.A. HARRISON — Department of Chemistry,
University of Cambridge,
Cambridge CB2 1EW, Gt. Britain

D.A. KING — Department of Chemistry,
University of Cambridge,
Cambridge CB2 1EW, Gt. Britain

R. RAVAL — Department of Chemistry,
University of Cambridge,
Cambridge CB2 1EW, Gt. Britain

A wide variety of sophisticated techniques have now been developed for the preparation and characterisation of single crystal solid surfaces and for the study of adsorption phenomena on these surfaces. These techniques have greatly advanced our knowledge of the chemical composition of surfaces, the local structural arrangement both in terms of adsorbate–adsorbent registry and order–disorder phenomena, the electronic structure of clean surfaces and adsorbate layers and thus the nature and strength of surface bonding, kinetics and mechanism of adsorption, desorption and catalytic reactions at surfaces. Already quite a number of books and review collections have been published which aim to describe these techniques and their development. What has been far less extensively reviewed, however, is the very recent consolidation which these techniques have undergone and the shift of emphasis from technique development to application, often of several parallel techniques, thus providing a proper understanding of surface adsorption and reaction processes.

Until recently, there has been a paucity of fundamental studies on well-defined heterogeneous catalytic systems and the link between fundamental adsorption studies and industrial catalytic problems has often seemed remote. In this series, we hope to demonstrate that recent extensive work by chemists, physicists and metallurgists has led to a well-developed science, providing a firm physical basis for the understanding of chemisorption and heterogeneous catalysis and that this link is now established.

The series consists of five volumes, the first two dealing with general principles and phenomena at clean solid surfaces and at adsorbate-covered surfaces, respectively, and laying down the foundation of the physical principles of the many techniques utilised. Volume 3 contains a comprehensive account of chemisorption systems which have been studied under well-controlled conditions. Volume 4 contains a series of accounts of heterogeneous catalytic reactions which have also been studied under well-defined conditions, and also illustrates how this work is related to results obtained under less well-defined conditions. Volume 5 provides an account of fundamental properties of semiconductor surfaces and the capabilities and limitations of device fabrication.

The present volume complements Volume 3B, and contains reviews of the present state of understanding of a range of chemisorption systems, illustrating the depth of understanding developed through application of the techniques and principles outlined in the first two volumes. Each of the authors has made major contributions to the subject under review, and the topics cover hydrogen (Davenport and Estrup), nitrogen (Raval, Harrison and King), oxygen (Brundle and Broughton) and carbon monoxide (Campuzano) adsorption on metal surfaces. These topics have developed so quickly since the conception of this volume that the authors have been faced with a major problem in trying to capture within a confined space and time a comprehensive account of their field. Nevertheless we believe that an invaluable account of adsorption on metals has been compiled.

D.A. King
D.P. Woodruff

July 1990

Chapter 4 (J.C. Campuzano)

Chapter 1

Hydrogen on Metals

J.W. DAVENPORT

Department of Physics, Brookhaven National Laboratory, Upton, NY 11973, (U.S.A.)

P.J. ESTRUP

Department of Physics and Department of Chemistry, Brown University, Providence, RI 02912 (U.S.A.)

1. Introduction

The effects produced by exposing a clean metal surface to hydrogen seem out of proportion to that element's size and simplicity. Hydrogen may adsorb on the surface or may penetrate into the metal and form a hydride. If it adsorbs, it can give rise to many different ordered phases depending on temperature and coverage. Hydrogen adsorption can cause the work function of the metal to increase or decrease, may or may not produce a well-defined change in the photoemission signal, and can cause reconstruction of a surface or can stabilize a surface which would otherwise reconstruct.

Fundamental studies of hydrogen adsorption include work on surface states, the nature of the metal–hydrogen bond, vibrational modes, the statistical mechanics of the various surface phases, the reconstruction of the substrate, surface diffusion, sticking coefficients, accommodation, and desorption. These problems have been addressed theoretically by molecular cluster models, slabs, band theory, quantum chemistry, jellium models, lattice gas models, and Monte Carlo simulation. They have been addressed experimentally by temperature-programmed desorption, LEED, work function measurements, photoemission, reflectivity studies, inelastic electron scattering, secondary ion mass spectroscopy (SIMS), IR absorption, NMR, ESR, and helium atom and neutron scattering. Many parallel investigations of bulk hydrides have been carried out [1], although usually without interaction with the surface studies.

Hydrogen adsorption on metals is important in a number of technological areas. It plays a central role in many catalytic reactions, ranging from methanation and Fischer–Tropsch synthesis to processes in fuel cells. In metallurgy, hydrogen is notorious for causing embrittlement of steel [2]; in fusion technology its effects on reaction chamber walls present major problems. Hydrogen can also change the electrical and magnetic properties of metals; for example, PdH is a superconductor, whereas pure palladium is not.

References pp. 33–37

In this review, we wish to describe the principal information available about hydrogen adsorption on metals, drawing attention to parallel bulk phenomena when appropriate. It is impossible to discuss all the work which has been done on these systems and we will refer to several general reviews which have appeared in recent years [3–11]. We also want to introduce the theoretical ideas which have been proposed in attempts to correlate and rationalize these data. Throughout the discussion, the adsorption on transition metals and noble (*d*-band) metals will be emphasized.

2. Structure

2.1 ATOMIC GEOMETRY

The chemisorption of hydrogen on a surface is conveniently discussed in terms of simplified one-dimensional potential energy curves of the type shown in Fig. 1(a) and (b). There are two potential wells on the vacuum side of the surface M: the shallow well, of depth E_{H_2} relative to the desorbed state ($M + H_2$), represents molecular adsorption; the deeper well, of depth E_H relative to ($M + H$), represents atomic adsorption. Depending on the shape

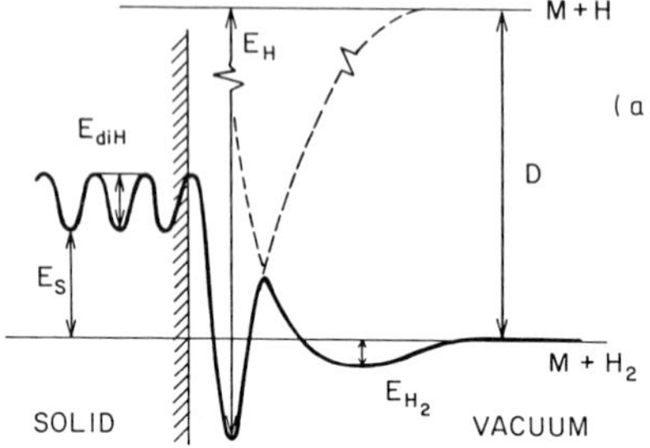

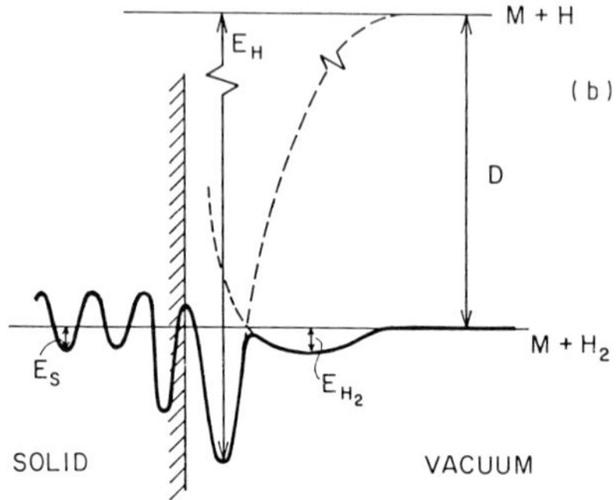

Fig. 1. Schematic potential energy curves for the dissociative chemisorption and bulk sorption of hydrogen.

of the wells, the dissociation may or may not require an activation energy. For many metals, e.g. Cu, W, penetration into the solid is energetically unfavorable as indicated by the position of E_s in Fig. 1(a). For metals like Pd and Nb, which favor sorption by the bulk, the energy of an interstitial H atom is below that of the desorbed state, as illustrated in Fig. 1(b).

Molecular hydrogen may be present on the surface as a precursor state but in detectable amounts only at very low temperature or high pressures. Under most conditions, hydrogen adsorbs in the form of atoms; these will tend to be located at potential minima (not shown in the one-dimensional diagrams in Fig. 1) determined by the crystallography of the metal surface. Unfortunately, it is a difficult experimental task to characterize these sites accurately. X-Ray diffraction, only recently practical for surfaces, has the added complication that hydrogen is the weakest of all scatterers. Neutron diffraction, with its strong sensitivity to hydrogen will not be practical for single-crystal surfaces for some time because of the low fluxes available. However, vibrational modes on large surface area powders such as Raney nickel have been studied in this way [12]. To some degree, low energy electron diffraction (LEED) faces the same problem as X-rays, namely that a proton is a weak scatterer, even for a strongly interacting probe like low-energy electrons. Despite these difficulties, there has been one LEED study done for hydrogen on Ni(111) by Christmann et al. [13]. They find that the H atoms form a honeycomb (2 × 2) structure (at a coverage of 1/2) and occupy the threefold hollows, as illustrated in Fig. 2(a), with a bond length of 1.84 ± 0.06 Å. There are two threefold hollows on the (111) face, corresponding to the stacking sequence abc in the bulk crystal. The results of Christmann et al. indicate that the adatoms alternately occupy the c and b sites (if the outer plane is a) in such a way as to form a graphitic or

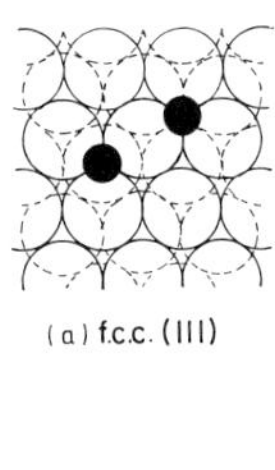

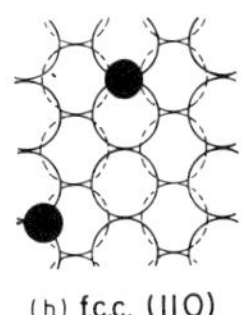

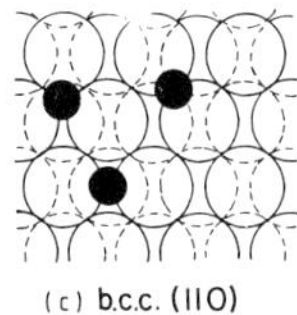

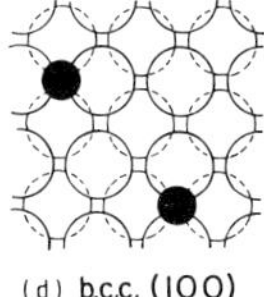

Fig. 2. Models of site geometry for adsorbed hydrogen atoms. The small solid circles represent H atoms. The large circles represent the first layer (full lines) and second layer (broken lines) of the metal substrate.

hexagonal overlayer. The bond length is found to be close to that of bulk NiH, which is 1.86 Å. However, this length is called into question by the vibrational analysis discussed below.

The method of helium diffraction [14–16] has recently been extended to structural studies of adsorbate-covered surfaces. Results have been obtained for hydrogen on Ni(110)[14], indicating that sites of both twofold and threefold coordination can be occupied on this surface [Fig. 2(b)]. Some results are available also for Ni(001) and Pt(111)[15, 16].

Most of the other information concerning sites comes from vibrational spectra of the adsorbed hydrogen measured by means of electron energy loss spectroscopy (EELS)[17]. Vibrational modes are discussed in more detail in Sect. 2.3. By comparing the number and frequencies of the modes with standards, it is often possible to deduce the site. For example, on Ni(001), Andersson [18] has found rather low vibrational frequencies, indicating bonding on the fourfold hollow site. For H on Ni(111)[19] and Pt(111), EELS data [20] indicate that the site is the threefold hollow, in agreement with the LEED study on nickel [13]. However, the bond length deduced from the vibrational frequencies is 1.57 Å for nickel and 1.68 Å for Pt. This is substantially different from the LEED determination and there is no independent assessment of which is correct. Other systems for which vibrational spectroscopy has furnished sites are listed in Table 1. They include Fe(110)[21] and W(110)[22], for which the sites shown in Fig. 2(c) are proposed, and W(100), on which the H atoms occupy bridge sites [Fig. 2(d)][23–27].

TABLE 1

Vibrational energies of H atoms

Surface	Vibrational energies (meV)	Site	Comment	Ref.
Fe(110)	131, 109	Short bridge		21
Ni(100)	74	4-fold hollow		18
Ni(111)	88, 139	3-fold hollow	Two types of site	19
Ru(001)	88, 105, 138	3-fold hollow	Two types of site	29
Pd(100)	64	4-fold hollow		30
W(100)	155 → 130 80, 160	Bridge	Coverage-dependent reconstruction	23–28
W(110)	95, 157	Bridge	Both short and long bridge site proposed	22, 28
W(111)	160	Atop		28
Pt(111)	68, 153	3-fold hollow		20
Bulk hydride				
NiH	70–110	Octahedral		31
PdH	56–66	Octahedral		32

Photoelectron spectroscopy can also be used to infer the geometry of the adsorption sites. In this way, Louie found that, for H on Pd(111), the calculated photoelectron spectrum was inconsistent with an atop site but consistent with either of the threefold hollows [33]. A similar study for H on Ti(0001) by Feibelman et al. [34] revealed an ambiguity in site selection if only photoemission data were used. The total energy for about 25 different sites was calculated and the one with the largest binding energy was selected. This site was again a threefold hollow with a bond length of 1.90 Å, the same as for the compound TiH_2. Other systems for which attempts have been made to calculate the sites include H/Ni (by Upton and Goddard [35]) and H/Cu (by Madhavan and Whitten [36]). They are discussed in Sect. 2.4.

It is useful to compare these results with bulk hydrides where X-ray, neutron scattering, and channeling studies have led to a more definitive picture of the geometry, and with cluster compounds containing metals and hydrogen for which extensive structural data are available [37]. It has been found that H occupies the tetrahedral sites in b.c.c. metals such as Nb and Ta and the octahedral sites in f.c.c. metals such as Pd [1]. While there is no reason to expect a hydrogen atom to have a unique coordination, it is interesting to note that in f.c.c. crystals the octahedral sites correspond respectively to fourfold hollows, bridge, and threefold hollows on the (100), (110), and (111) surfaces. However, the bond lengths are not expected to be the same at the surface, some relaxation would no doubt occur as exemplified by the LEED analysis of H on Ni(111). Studies of hydrides also suggest that the concept of a site may retain its validity even at high temperatures where the hydrogen atoms are diffusing rapidly and have no long-range order. For example, in bulk NbH, neutron scattering measurements indicate that hydrogen atoms spend more than 90% of their time on tetrahedral sites [38]. This implies that some sort of lattice gas model is appropriate in these systems. As discussed below, lattice gas models are also used to treat the analogous two-dimensional (2D) system formed by a layer of chemisorbed hydrogen.

If the adsorbed atoms interact with each other in some way, the surface layer is expected to possess 2D long-range order at certain values of temperature (T) and adsorbate coverage (θ)[39, 40]. The systems for which such 2D phases have been observed (by LEED or He diffraction) include hydrogen on Ni(110)[14, 41, 42], Ni(111)[13], Pd(100)[43], Pd(110)[44], Fe(110)[40, 45], Mo(100)[39], W(100)[40, 49], and W(110)[40, 48]. For some of these, the phase diagram, in the T–θ plane, has been determined and representative examples are shown in Figs. 3–5. For the H/Pd(100) system, only a single ordered phase is observed (Fig. 3), stable only at low temperature and at coverages near half a monolayer of hydrogen. The H/Fe(110) system (Fig. 4) shows both a (2 × 1) and a (3 × 1) superstructure in addition to intermediate regions where antiphase domains are present. The diagram for H/W(100)[40, 49] shows even more phases (Fig. 5); the complexity of this system is caused by the tendency of the W(100) substrate to reconstruct, as discussed below.

Although there have been few direct measurements of the rate of hyd-

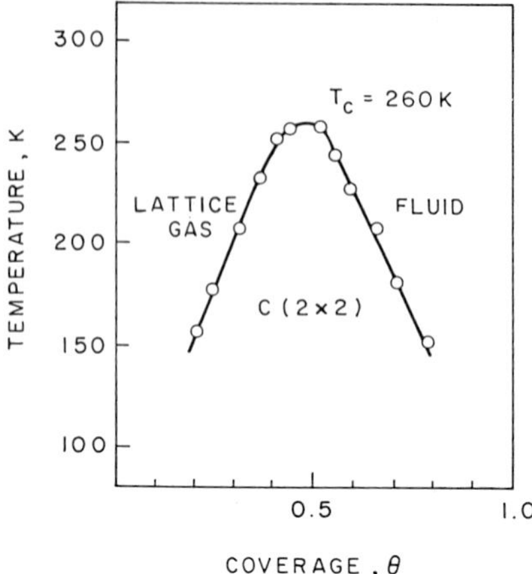

Fig. 3. Phase diagram in the temperature–coverage plane for H/Pd(100). At a coverage of $\theta = 0.5$ the $c(2 \times 2)$ phase disorders at $T_c = 260$ K. Data from Behm et al. [43].

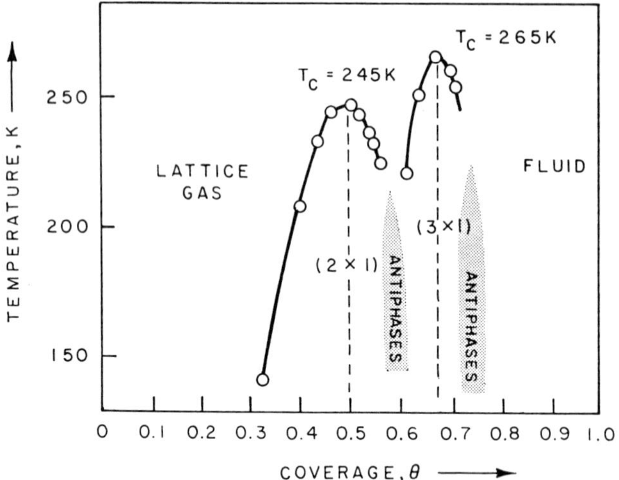

Fig. 4. Phase diagram for H/Fe(110), showing two ordered phases separated by regions with antiphase domain structures. Data from Imbihl et al. [45].

rogen diffusion across a metal surface [50], the adatoms are believed to make frequent jumps from site to site, under most circumstances. It has therefore been suggested that the vibrational modes may be broadened sufficiently to form bands, similar to those formed by the electronic states in the solid [13]. In its extreme form, this corresponds to non-localized adsorption and in this limit it is not appropriate to speak of a site. The adatoms might be thought of as occupying two-dimensional plane wave states with equal probability of being at any point on the surface. Such a picture would be valid for positrons, for example, but whether protons are light enough to produce effects of this type is not clear at present.

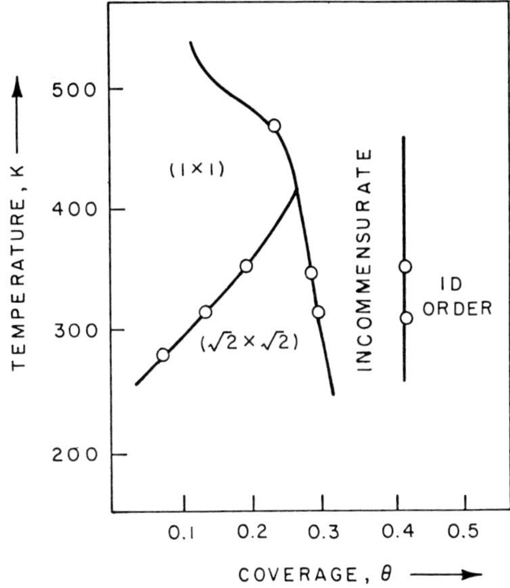

Fig. 5. Phase diagram for H/W(100) in the low coverage region. Data from Horlacher-Smith et al. [49].

Finally, the tendency of many metals to form hydrides suggests that chemisorption into sites below the surface should not be ruled out. Even if a bulk hydride does not exist, the adatoms could distribute themselves over several layers in the surface region. The potential energy curve for such a metal is shown schematically in Fig. 1(b); in addition to the atomic well on the vacuum side, a deep minimum occurs just inside the solid. This has been proposed as the reason for the disappearance of the photoemission signal from hydrogen on the (111) surfaces of Ni, Pd and Pt at room temperature [51], and for the unusual desorption kinetics for H/Pd(110)[44]. On the other hand, nuclear microanalysis has shown [52] that for Pt(111) the absolute coverage (of deuterium) at room temperature is $\lesssim 0.05$ monolayer, and this includes all the adsorbate within $\simeq 1\,\mu$m of the surface. This result would rule out a subsurface site in the case of Pt(111).

2.2 BINDING ENERGY

The binding energy for some surface and bulk hydrides are shown in Fig. 6 [53–55]. By binding energy (BE) is meant the energy required to remove an adsorbed or dissolved hydrogen atom from the system. This number typically lies in the range 2–3 eV (1 eV = 23.06 kcal mol^{-1} = 96.49 kJ mol^{-1}). Experiments can provide the heat of adsorption, q, which measures the energy relative to H_2 molecules at large distances. As is apparent from Fig. 1, since the dissociation energy of H_2 is D = 4.48 eV, for a chemisorbed atom these numbers are related by

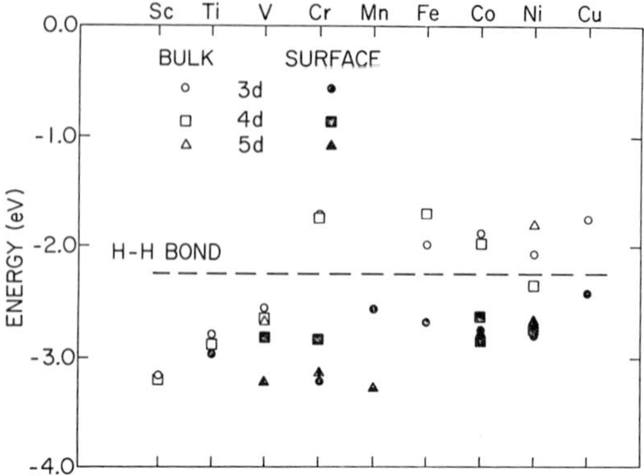

Fig. 6. Binding energy of hydrogen to metals in the 3*d*, 4*d*, and 5*d* series [53]. Data from Flanigan and Oates [54] and Toyashima and Somorjai [55].

$$\mathrm{BE} = E_{\mathrm{H}} = q + 2.24 \text{ eV}$$

(Usually the heat per H_2 is quoted; this is $2q$ in our notation.) Systems with BE values less than 2.24 eV are endothermic adsorbers and those with larger BE values are exothermic. Surfaces of metals appear always to be exothermic even though the same material may be endothermic for bulk solution.

As defined here, the BE is a differential property and is likely to depend on the coverage because of H–H interactions. These will be discussed in more detail in Sect. 2.5.

It is interesting that the binding energy at the surface is larger than in the bulk even though the hydrogen can make fewer bonds there. The origin of this extra energy presumably lies in the fact that the hydrogen atoms are not forced to occupy the small interstices of the metal lattice. Hydrides are expanded compared with clean metals by rather large amounts (~15% in volume) and at the surface they need not be. A theory which relates the difference in bulk and surface energies to the bulk and shear moduli of clean metals has recently been presented by Welch and Pick [57]. These energy differences are typically several tenths of an electron volt. This means that surfaces are always preferred sites and at any given temperature the surface coverage can be orders of magnitude larger than the bulk concentration. This is probably related to the embrittlement and cracking of steels because a crack, once started, will attract other hydrogens [2].

Figure 6 also indicates that the binding energy does not vary by a great deal across a row in the periodic table. On the other hand, the differences which do exist are of great importance to catalysis because they are large compared with thermal energies.

Another interesting feature of Fig. 6 is that the adsorption energies for Cu and Ni are not very different even though one has a filled *d* band and the other does not. Experimentally it is found that Ni readily adsorbs hydrogen when exposed to H_2 gas whereas Cu does not. This difference in adsorption properties is therefore probably a kinetic effect due to a static or dynamic barrier rather than any intrinsic difference in the bonding properties per se. Indeed, molecular beam studies have concluded that a dissociation barrier exists for H_2 on Cu(100)[58], as illustrated in Fig. 1(a).

2.3 VIBRATIONAL MODES

In recent years it has become possible to measure the vibrational frequencies of hydrogen adsorbed on single crystal metal surfaces using both low-energy electrons (1–10 eV) and infrared absorption [17, 59]. Previous work utilized powders or dispersed catalysts for which traditional IR absorption and neutron scattering can be used. Vibrational spectra are important for several reasons. First, in analogy with the gas phase, if the modes could be measured (including all the overtones) the detailed shape of the potential energy wells could be constructed. Second, the low-lying excitations of adsorbates play an important role in thermal processes. Adsorbate modes couple to the substrate and can either overlap phonon bands or be split off as discrete "localized" modes. These mode frequencies enter directly into the partition function and hence partly determine equilibrium concentrations and also influence surface diffusion and desorption. The details of these processes have yet to be worked out. In the interim, vibrational spectroscopy provides several other important pieces of information. As indicated in Sect. 2.1, the number and frequency of the modes can determine the adsorption site, usually by comparison with model compounds for which electron and X-ray diffraction has determined the geometry unambiguously. In contrast to LEED and He diffraction, EELS observations are possible in systems which do not possess long-range order. Furthermore, the vibrational modes are properties of the ground electronic state rather than the ion states which are produced in photoelectron spectroscopy. Experimental results for the vibrational modes of a number of systems are given in Table 1 where they are compared with modes observed in bulk hydrides.

The system which has received the most attention is hydrogen on W(100). It is now known that the clean W surface is reconstructed at low temperatures but that at saturation coverage (corresponding to 2 H atoms per W) the system reverts to the symmetry characteristic of bulk truncation [46, 47, 60–63]. Studies of cluster analogs suggest that a hydrogen atom bridge-bonded to two tungsten atoms should produce symmetric stretch frequencies (i.e. perpendicular to the surface) in the range 135 $\pm$ 35 meV, while linear, "on top" bonding should produce 235 $\pm$ 35 meV [65]. A similar trend has been calculated for H adsorption on Ni clusters as discussed in Sect. 2.4.1. For hydrogen on W(100) at saturation the symmetric stretch frequency has been measured as 130 meV, consistent with bridge bonding [23–27]. Also, no mode corresponding to H_2 stretching has been observed, confirming previous conclusions that the adsorbed state is atomic rather than molecular at room temperature [24]. In addition, two lateral (i.e. parallel to the surface) vibrational modes at 80 and 160 meV are observed as expected for bridge sites if account is taken of next nearest neighbors [23, 26, 27]. High-energy ion-scattering measurements [64] show that at low coverage the tungsten reconstruction reduces the W–W distance in the bridge site by 0.3–0.4 Å; at the same time the symmetric H–W stretch shifts to 155 meV.

It is sometimes possible to calculate the bond lengths from the frequencies using a simplified model. For example, an oscillator consisting of a linear molecule W–H would have a stretch frequency of $\omega_0 = \sqrt{k/m}$ where k is the spring constant. Frequencies for bridge, threefold, and fourfold geometries can be calculated assuming the metal atoms have infinite mass. They are given in Table 2 following Baro et al. [20]. In a bridge geometry with nearest neighbors only, there would be two modes with $\omega = \omega_0\sqrt{2}\sin\alpha$ and $\omega = \omega_0\sqrt{2}\cos\alpha$ where 2α is the W–H–W angle. Measuring the two mode frequencies determines the bond angle and, if the W–W distances are known, the bond length. For the case of H on W(100), since three modes were observed, next nearest neighbor force constants must be considered. A case in which this type of analysis is easier is for the threefold hollows associated with H on Ni(111) and Pt(111). In these systems two frequencies have been observed, corresponding to the symmetric (perpendicular to the surface) and asymmetric stretching modes. Assuming that the lower frequency is the perpendicular mode Tables 1 and 2 yield, for Ni, $\tan\alpha = \sqrt{2}\omega_{\parallel}/\omega_{\perp} = 2.23$. Then, from the bulk Ni–Ni distance of 2.49 Å, the bond length is found to be 1.57 Å [51]. A similar analysis for Pt yields 1.68 Å [20, 51]. The length for Ni is substantially different from the LEED value of 1.84 Å [13]. However, both lengths are in reasonable agreement with the metal–hydrogen bond length in $H_3Ni_4Cp_4$ which is 1.692 Å as determined by an accurate neutron diffraction study (Cp = cyclopentadienyl) [37].

For Ni(100) and Pd(100), only one mode was observed so no such analysis can be made.

Two recent developments seem promising. One is the ability to make high-resolution infrared studies. For H on W(100), Chabal and Sievers [66] have shown that the symmetric stretch mode has a frequency of 129.7 meV

TABLE 2

Stretching frequencies for light adsorbed atoms

Site	Coordination	$l\sin\alpha$	$\omega_{\perp}$	$\omega_{\parallel}$
Atop	1		ω_0	
Bridge	2	$a/2$	$\sqrt{2}\omega_0\cos\alpha$	$\sqrt{2}\omega_0\sin\alpha$
Threefold hollow	3	$a/\sqrt{3}$	$\sqrt{3}\omega_0\cos\alpha$	$\sqrt{3/2}\omega_0\sin\alpha$
Fourfold hollow	4	$a/\sqrt{2}$	$2\omega_0\cos\alpha$	$\sqrt{2}\omega_0\sin\alpha$

The metal atoms are in a plane, a is the metal–metal distance and l is the metal–hydrogen distance. If d is the perpendicular distance from the plane to the hydrogen atom then $\cos\alpha = d/l$. $\omega_0 = \sqrt{k/M_H}$ where k is the spring constant.

and a full width at half maximum of 1.74 meV. The width is an order of magnitude smaller than that obtained with EELS. It was suggested that the width is due to phonon sidebands on the H–M stretch mode. If true, this opens the possibility of measuring the surface phonon density of states. The second result is the observation of rotational transitions for H_2 physisorbed on Cu(100) and polycrystalline Ag using electron scattering [67, 68]. These experiments, incidentally, demonstrate that molecularly adsorbed H_2 can exist at temperatures of 10–15 K because a nearly unshifted H_2 vibrational mode at 516 meV was observed.

2.4 NATURE OF THE BOND

There have been many theoretical investigations in recent years of the electronic structure of metal/hydrogen systems. These include hydrogen adsorbed on metal surfaces as well as bulk hydrides, hydrogen/jellium, and molecular cluster models. The qualitative picture which has emerged from these studies is that hydrogen tends to adsorb dissociatively in high-symmetry sites on the surface. In most systems the main interaction appears to be between hydrogen and metal *s* orbitals with not insignificant metal *d* interactions. The bond is essentially covalent but there is some charge transfer to the hydrogen, as much as $e^-/3$ as measured by a Mulliken population analysis. However, this charge transfer leads to only a small change in the work function, on the order of a few tenths of an electron volt. This implies a net surface dipole which is much smaller than that obtained by multiplying the Mulliken charge by the perpendicular distance of the adsorbate. Indeed, the work function has been found to increase for hydrogen in underlayer sites as well as for overlayers. Calculations of the electronic spectrum indicate that the hydrogen induces a split-off state, below the surface states of the clean substrate which are shifted in energy.

Experimentally, many of these features have been confirmed but the amount of charge transfer and the importance of *d* orbitals have not been determined unambiguously.

2.4.1 Theoretical investigations

The problem of the metal–hydrogen bond has been approached theoretically from many different points of view, ranging from simple electronegativity arguments to sophisticated quantum mechanical calculations.

Perhaps the simplest way of discussing metal–hydrogen interactions is to use Pauling's equation, as was first done by Eley in 1950 [69–71]. Pauling asserts that the bond energy for an AB bond is

$$E_{AB} = \tfrac{1}{2}[E_{AA} + E_{BB}] + \alpha(x_A - x_B)^2 \tag{1}$$

where the x terms are electronegativities and α is a constant which Pauling takes to be 1 eV. Further, the charge transfer should increase with increasing electronegativity difference. The Pauling electronegativity of H is 2.1;

for metals to the left in the periodic table it is substantially less than this value (e.g. 1.5 for Ti) and for metals to the right it is comparable (e.g. 1.9 for Ni). Thus, charge is predicted to flow to the hydrogen for most metals. Important exceptions are the Group VIII metals in the 4*d* and 5*d* rows. While the direction of charge flow can be understood, the actual bond energy is difficult to explain in this way. In practice, the only term in eqn. (1) which varies significantly is the metal–metal bond energy. The gross trend of this quantity is that it peaks in the middle of a transition metal row when the bands are half filled [72]. E_H is therefore predicted to reach its largest value in the middle of the row; however, as seen in Fig. 6, the bond energy decreases smoothly from left to right across the periodic table.

A more complicated phenomenology has been worked out by Bouten and Miedema for binary metal hydrides [73]. All of these approaches are complicated by the difficulty in defining precisely the quantities which enter into them.

In recent years there has been an increased interest in model Hamiltonians where the various parameters have a clear physical meaning. The most widely used is the Anderson model first applied to chemisorption by Newns [74]. Excellent discussions of this model have been given by Schrieffer [75], Lundqvist et al. [76], and others [77]. The model can be derived from an ab initio Hamiltonian in matrix form by retaining only the dominant metal–hydrogen overlap integrals and the hydrogen–hydrogen Coulomb integral. In addition, the metal is modeled by a band rather than a discrete set of states. The model can be solved in the Hartree–Fock approximation [74].

Newns' treatment assumed that the dominant overlap integral is H 1*s*–metal *d*. As indicated below, this is probably not true since the metal *s* overlap is larger. However, since the overlap integrals were obtained by fitting to adsorption energies, the exact source may not matter. Newns then used the fitted parameter to calculate the charge on the hydrogen and obtained values in the range 0.1–0.4e^-. These are consistent with the Mulliken populations from ab initio calculations to be discussed below. It has usually been assumed that charge transfer of this magnitude is ruled out by the small change in work function on adsorption. However, since the more detailed calculations agree with experiment (both for work function and charge transfer) either some sort of screening occurs, which effectively cancels the moment produced by this charge transfer, or the inherent limitations in defining "charge transfer" preclude its use in determining dipole moments.

A theory similar to the Anderson model has been used by Varma and Wilson [78] to explain the trend in bond energies across the periodic table. Although only an *s*–*d* interaction is assumed, the correct trend in the binding energy is obtained.

A number of ab initio calculations have been carried out using simple molecules as models of the surface species. A good example is the study by Bagus and Bjǫrkman [79] of NiH and PdH because a number of questions about the nature of the bond are answered. In the first place, it is found that the configuration of the metal atom in these molecules is close to d^9s. Since the nickel atom ground state is d^8s^2 while Pd is d^{10}, there is a promotion energy to go to d^9 which is regained on bonding. The d shell in these molecules is quite compact relative to the s. Bagus and Björkman give the mean radii of these orbits in the atom and the s orbit is roughly three times as large as the d. The larger s orbital would therefore be expected to interact most strongly with H 1s. To illustrate this, Bagus and Björkman tabulate the overlap integrals for those orbitals and their results are reproduced in Table 3. The ratio of d/s overlap is about 0.3 for Ni and 0.4 for Pd. Therefore the contribution of the d orbitals is expected to be less than the s but both are expected to participate in the bond and neither can be neglected. In addition, somewhat greater d participation should occur for Pd than for Ni. On the whole, these expectations are borne out by the calculations. Orbitals are formed which are bonding and antibonding combinations of metal s and d and hydrogen 1s. There is charge transfer to the hydrogen of approximately 0.3e and the dipole moment μ for the NiH is $-1.6ea_0$. This is larger than the crude estimate obtained by multiplying the charge (0.3e) by the bond length (1.52 Å). These results are substantially in agreement with experiment (for the molecule) though some discrepancies remain. The experimental bond length is $R_e = 1.47$ Å for NiH and 1.53 Å for PdH, while the bond energies are 3.1 eV and 3.3 eV, respectively. As mentioned in Sect. 2.1 for hydrogen adsorbed on Ni(111), the bond length on the surface has been found to be 1.57 Å by vibrational mode analysis and 1.84 Å by LEED.

The second model system to be considered here is hydrogen adsorbed on jellium. In the diatomic molecule model one neglects the effects of band structure of the substrate but retains all of the orbital character necessary to describe the bonding. In the jellium model one neglects the orbital character to some extent (certainly there are no localized d states) but allows for the true semi-infinite geometry. This is reminiscent of the Sommerfeld free electron model of metals. The model has a surface, the jellium edge, where the uniform positive background ends and the electrons self-consistently

TABLE 3

Mean radii of metal orbitals and hydrogen–metal overlap integrals [79]

	Ni	Pd
Metal–hydrogen distance	1.52 Å	1.64 Å
$\langle\phi_d\|r\|\phi_d\rangle$	0.56 Å	0.77 Å
$\langle\phi_s\|r\|\phi_s\rangle$	1.79 Å	1.94 Å
$\langle\phi_d\|\phi_H\rangle$	0.147	0.192
$\langle\phi_s\|\phi_H\rangle$	0.496	0.458

ϕ_d is the atomic 3d orbital centered on nickel (4d for palladium).
ϕ_s is the atomic 4s orbital centered on nickel (5s on palladium).
ϕ_H is the atomic hydrogen orbital centered on hydrogen.

adjust causing some spillout of charge at the surface. Lang has extensively reviewed this model [80]. The parameter which completely specifies the model is the electron density well inside the surface. This is usually characterized by the radius of a sphere which contains one electron. This parameter, universally denoted r_s, is therefore related to the electron density by

$$\frac{4}{3}\pi r_s^3 = \frac{1}{n} \tag{2}$$

For example, Al with an electron density of $6.02 \times 10^{22}\,\text{cm}^{-3}$ has $r_s = 2.07a_0 = 1.10\,\text{Å}$. Lang and Williams [81] have calculated the bond energy and dipole moment for hydrogen on jellium with $r_s = 2a_0$ and find $E_H = 1.5\,\text{eV}$ and $\mu = 0.5\,\text{D} = 0.20ea_0$. Both the bond energy and the dipole moment are smaller than the values found for the molecules discussed previously. However, the direction of charge transfer is such as to make the hydrogen atom negative, as found for the molecules. Gunnarsson et al. [82] have improved this model by including the effect of the substrate using pseudopotentials and calculated the equilibrium position for H on the low index faces of aluminum. They find the same qualitative picture as for jellium but with slightly increased bond energies of 1.8–2.0 eV, depending on the face. The bond lengths are found to be 1.94 Å on Al(100), 1.8 Å on (110), and 1.6 Å on (111). These values are close to those of the AlH diatomic molecule (1.65 Å); however, the bond energy of AlH is 2.9 eV, larger by almost 1 eV.

One advantage of the jellium-plus-pseudopotential model is that it is computationally simple enough so that rather complete potential energy surfaces can be calculated. An example of this, for molecularly adsorbed hydrogen on magnesium, was given by Nørskov et al. [83]. They found a barrier for dissociation which is smaller on more open faces and which contains a molecularly chemisorbed state which is identified as a precursor. This model has recently been extended in another direction, an effective medium theory [84]. In this view, the interaction of hydrogen with transition metals is calculated by assuming that the proton is embedded in a uniform electron gas with a density characteristic of the metal at the H site. Corrections are made so that an average density can be used, and an approximate interaction with the *d* band is included. Nevertheless, the dominant interaction is due to the electron gas which gives a (universal) adsorption energy of 2.45 eV. The variations from metal to metal, and with crystal face are due to the *s*–*d* interactions which are of less importance. This theory has explained the trends in adsorption energy for all of the transition metals [84]. Trends of this type have also been investigated by Muscat [85].

In order to provide a good representation of the surface, cluster calculations have been made with increasing number of metal atoms [35, 36, 86–91]. Early work using the local density method by Ellis et al. [87] and by Messmer et al. [88] showed the resemblance of the hydrogen–metal interaction to bulk hydrides and diatomic molecules in that a hydrogen induced state was found below the states of the isolated cluster. As discussed in Sect. 2.4.2, this state has been observed in photoemission. Bond energies in a Ni_2H_2 cluster were studied by Melius et al. [89, 90] who utilized ab initio quantum chemical techniques, including correlation effects. They obtained reasonable "chemisorption" energies and a 0.22*e* charge transfer to the H atom. They also found that the 3*d* orbitals do not participate significantly in the bond. Proceeding in this way, Upton and Goddard [35] have calculated the adsorption energy and vibrational frequency for hydrogen adsorbed at various sites on the 20 atom nickel cluster shown in Fig. 7. Table 4 gives the results of

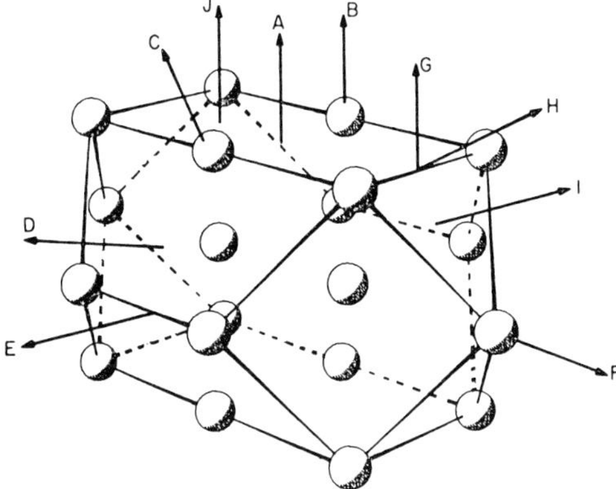

Fig. 7. Hydrogen binding sites on a nickel cluster. Numerical values are given in Table 4. Data from Upton and Goddard [35].

TABLE 4

Calculated bond parameters for H adsorbed on nickel. The various sites are shown in Fig. 7

Site	Surface	Description Ligancy of H[a]	Bond length (Å) $R_\perp^b$	Bond length (Å) R^c	Vibrational frequency (meV)	Chemisorption energy (eV)
B	⟨001⟩	1(7)	1.50	1.50	283	1.56
F	⟨110⟩	1(5)	1.49	1.49	280	1.43
C	⟨112⟩	1(7)	1.49	1.49	(231)	(1.00)
A	⟨001⟩	2(7)	0.99	1.59	177	2.73
G	⟨001⟩	2(5)	0.99	1.59	(173)	(2.17)
E	⟨110⟩	2(6)	0.93	1.55	(161)	(1.56)
H	⟨112⟩	2(5)	0.96	1.57	176	2.43
I	⟨111⟩	3(5, 5) h.c.p.	0.78	1.63	155	3.21
D	⟨111⟩	3(6, 7) f.c.c.	0.79	1.64	(131)	(2.12)
J	⟨001⟩	4(7, 5)	0.30	1.78	73	3.04

[a] The number of nearest neighbors for the Ni atom(s) binding site is given in parentheses. Where non-equivalent surface atoms are present, values are given for each type.
[b] Optimum distance from H to the plane representing the surface.
[c] Distance from H to nearest neighbor Ni atoms.
[d] Values not in parentheses are believed to most closely represent the infinite surface.

Upton and Goddard for the bond length, vibrational frequency and chemisorption energy, all of which are in reasonable agreement with the available experimental data.

Another recent cluster study is by Whitten and co-workers for H, and H_2 on titanium and copper [36, 91]. In these calculations, clusters of about 40 metal atoms were used. Traditional Hartree–Fock and configuration interaction calculations were performed in a local subspace of the full cluster. The conclusions are similar to those reached in the work of Upton and Goddard on nickel. For both Ti(001) and Cu(001), the adsorption energies are in good agreement with experiment. For Ti, the preferred site is the threefold hollow; the calculated bond energy is 3.20 eV and the bond length is 2.00 Å. For Cu, the preferred site is the fourfold hollow when two H atoms are present (as they were in their Ti calculation); the bond energy is 2.6 eV and the metal–hydrogen bond length is 2.01 Å. For Cu, the calculations show the existence of an activation barrier for H_2 to dissociate and adsorb in agreement with experiments but the calculated barrier (1.5–1.7 eV per H_2) is larger than the experimental value of $\sim$0.2 eV [for Cu(100)][58]. Many geometrical configurations were studied which cannot be discussed here. However, some general trends emerge from their work. First, hydrogen is dissociatively adsorbed and the adatoms are hydridic, i.e. they have negative charge in the range $0.3e^-$. Second, the *d* orbitals play a secondary role although their contributions cannot be neglected in the case of titanium.

A number of film calculations have been performed for monolayers of hydrogen adsorbed on Pd(111)[33], Mo(001)[92], Ti(0001)[34], Ni(110)[93], and W(001)[94]. In each case the adsorbate–substrate system was modeled by a thin film of from five to eleven metal layers with hydrogen on each side. All used density functional theory. In contrast to cluster studies, these calculations all reveal bands of hydrogen-like states with substantial dispersion ($\sim$3 eV) in good agreement with angle-resolved photoemission experiments. For example, in Fig. 8 the calculated energy bands for hydrogen on Pd(111) are compared with experiment [95]. One clearly sees the hydrogen-induced state which varies in this case from about -6 to -8 eV relative to the Fermi level. States of this type have been found in various molecules and cluster models as discussed above but, of course, they have no dispersion because of the finite size of the cluster. Similar states were found for H on Mo and Ti; for the latter there is again good agreement with angle-resolved photoemission data.

For Ti(0001)[34], the total energy was evaluated for $\sim$25 different overlayer and underlayer sites. The preferred one, which corresponds to the threefold hollow with no titanium atom directly underneath (the f.c.c. site), has a bond energy of 3.3 eV and approximately the same bond length as in the compound TiH_2 (1.91 Å). It was also found that the total energy of the underlayer sites was higher, corresponding to a bond energy of 2.6 eV, which is close to the heat of solution of H in Ti (2.7 eV). Since the H/Ti calculation gives a good account of the bond energy and of the photoelectron spectrum,

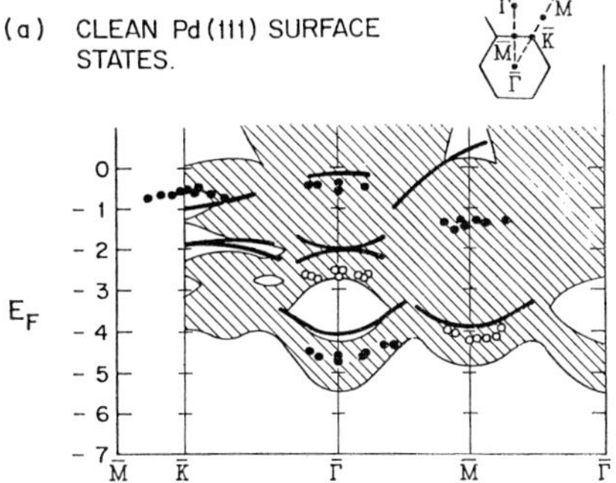

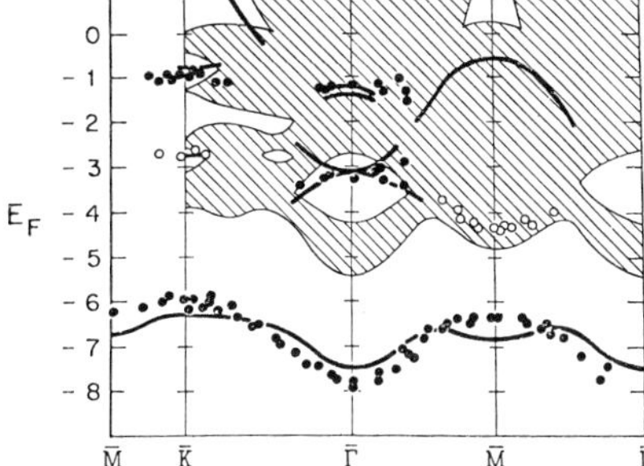

Fig. 8. Surface states on (a) clean and (b) hydrogen-covered Pd(111). The shaded regions are the calculated projection of the bulk bands onto the (111) surface. The solid lines are calculated surface states or resonances; the solid circles are experimental measurements. The open circles represent peaks in the photoemission spectrum for which the assignment is less certain. Data from Eberhardt et al. [95].

it is reasonable to ask how it describes the bond. First, the charge transfer is toward the H atom and a Mulliken-like analysis gives $q \sim 0.2e^-$. This is fully consistent with the other calculations discussed in this section. Second, the change in work function of $+0.2$ eV is in excellent agreement with experiment. This helps to resolve the long-standing puzzle about work function changes induced by hydrogen chemisorption. It shows clearly that charge transfer in a Mulliken sense does *not* imply a change in work function which is q times the bond length. The reason for this is an interesting problem. It could be that Mulliken q values are so arbitrary (basis-set dependent) that it is meaningless to attach significance to them. Alternatively it could be (as suggested implicitly by Cremaschi and Whitten [91]) that there is a compensating Ti–Ti dipole set up in the outer two metal layers, or it could be that other electrons which do not participate in the bond readjust to screen the dipole.

It is interesting to note that the H/Ti(0001) calculation gave an increase in the work function when the H atoms were placed in underlayer sites as well. The population analysis still showed that the adatoms were negative.

This further illustrates the problems with simple interpretations of the work function.

To illustrate that the hydrogen does carry substantial negative charge Fig. 9 shows the difference charge density (H-covered minus clean film) calculated for H adsorbed on Mo(001). There is clearly a charge depletion (represented by the broken contours) around the metal atom. That there is also an enhancement around the H atoms cannot be seen directly from the figure because the hydrogen atom charge density has not been subtracted. Nevertheless, an integration over the negative contributions yields a charge transfer to the H of $0.083e^-$. In view of the inherent difficulty in defining charge transfers this should be taken as consistent with the other calculations discussed previously. The calculated work function, 4.92 eV, is in excellent agreement with experiment. Similar charge density plots have been generated by Hamann for (1 × 1) overlayers of H on Ni(110)[93].

Finally, it should be mentioned that a similar situation exists for bulk hydrides. The electronic structure has been discussed by Switendick [96] who has also performed many calculations on metal–hydride systems. Briefly, the hydrogens occupy interstitial sites which are usually tetrahedral in b.c.c. crystals and octahedral in f.c.c. crystals. They show a hydrogen-induced electronic state, broader than the bands observed for adsorbed H, and such states have recently been observed by X-ray and UV photoelectron spectroscopy [97, 98]. For palladium hydride, a shift to larger binding energy of the Pd 3*d* core level was observed which is most simply interpreted as due to charge transfer to the hydrogen. The dominant effect is that, on addition of hydrogen to the metal lattice, states which are *s*-like about the proton are

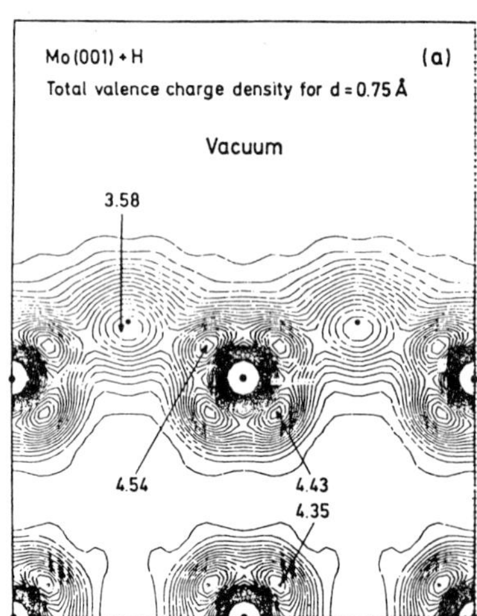

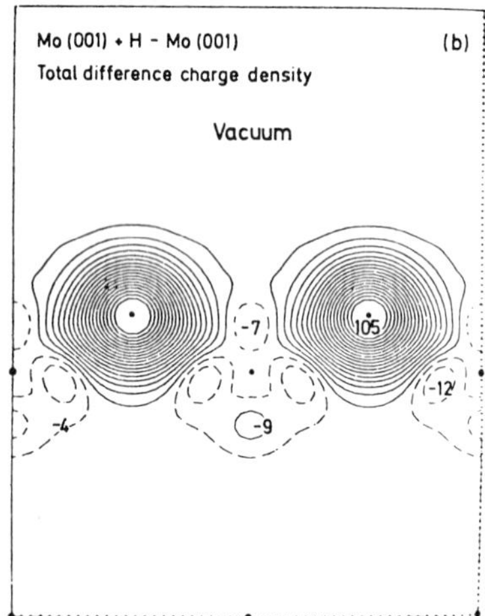

Fig. 9. (a) Calculated charge distribution for one monolayer of hydrogen atoms on Mo(001). The small dots indicate H atoms. The numbers are charge densities in atomic units (e/a_0^3). (b) Difference density between hydrogen-covered and clean surface. The broken contours are negative. Data from Kerker et al. [92].

lowered in energy. These states are predominantly *p*-like about the metal and naively this implies that the *s–p* electrons interact most strongly with the hydrogen. This is consistent with Nørskov's discussion of hydrogen in metals in terms of effective medium theory which was described earlier [83].

2.4.2 Experimental investigations

To what extent does the picture of hydrogen adsorption which has emerged from the calculations agree with experiment?

In the first place, the dissociation of the molecule is confirmed by a variety of experiments. For example, for W(100), the fact that co-adsorbed H_2 and D_2 desorb as H_2, D_2 and HD implies dissociation [6]. However, the most conclusive evidence comes from vibrational spectroscopy, as discussed in Sect. 2.3, which shows modes which are too low in energy to be molecular H_2.

The "molecular orbital" type picture of a hydrogen-induced state has been confirmed by photoelectron spectroscopy. The recent study by Eberhardt et al. [51, 95] for hydrogen adsorbed on the (111) faces of Ni, Pd, and Pt shows clearly that there is a hydrogen-induced state 7–9 eV below the Fermi energy, in excellent agreement with the calculations by Louie for Pd(111)[33]. The comparison between theory and experiment is shown in Fig. 8 and has been discussed in detail in a review article by Plummer and Eberhardt [99]. Results for hydrogen on Ti(0001) also show excellent agreement between the photoemission experiment and the calculations which were discussed in Sect. 2.4.1 [34]. To illustrate the experimental results, Fig. 10 shows the measured photoelectron spectra for clean and hydrogen-covered titanium. There are two hydrogen-induced peaks at about -1.5 and -7 eV. The calculations indicate that these are the bonding and antibonding combinations of H 1*s* orbitals with Ti *s* and *d* orbitals. The antibonding peak happens to lie in a band gap of the Ti so it is a well-defined surface state, not broadened by interactions with the substrate. In the case of palladium, the analogous state overlaps substrate bands and is therefore so broad that it is not readily seen [33].

Incidentally, the good agreement between theory and experiment for H/Pd(111) [95] led the authors to conclude that the bond length must be close to the one chosen by Louie for the calculation. However, as demonstrated by Feibelman et al. for Ti [34] there may be other sites which give equally good agreement with the photoelectron spectrum. They chose a site for the H/Ti system on the basis of the calculated total energy, a procedure which should be generally applicable. In agreement with these findings, calculations [100, 101] have recently shown that the photoelectron spectrum for a H–Pd system in which a monolayer of hydrogen atoms is placed in underlayer sites also agrees with experiment.

Hydrogen-induced features have also been studied for H on W(100) and W(110). These results have been reviewed by Smith and Larsen [102]. Features quite similar to those seen in the H/Ti system have been observed also for H on Nb [103]. As mentioned above, ab initio calculations all predict

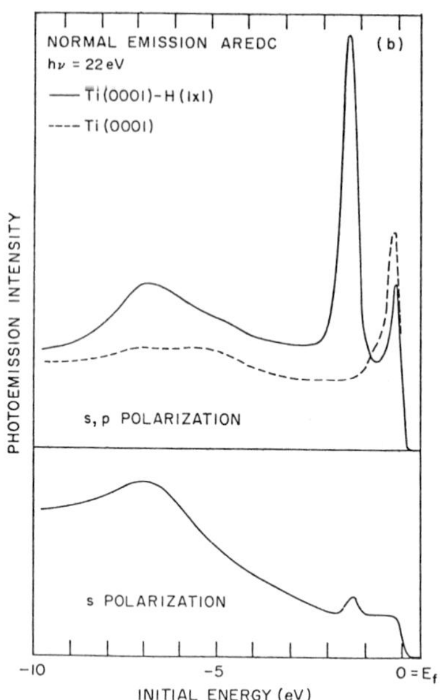

Fig. 10. Photoemission spectrum for clean and hydrogen-covered Ti(0001). Data from Feibelman et al. [34].

some charge flow to the hydrogen, but a precise value cannot be given to the charge transfer because of the ambiguity in assigning charge to a particular site. Experimentally there are two indications of charge flow. One is the observation of surface core level shifts for W(111) and W(100). In particular, the W 4*f* levels (with binding energies of about 30 eV) shift to higher binding energy when hydrogen is adsorbed [104–106]. This is the expected result if electrons were removed from the valence region of the W atoms. (Consider the electrostatic potential q/R inside a spherical shell of charge; if q is reduced, the potential is reduced and the core electrons inside are bound more strongly.) Other factors may contribute to the observed core level shift, however. The charges rearrange themselves in response to the creation of the core hole and changes in "relaxation energy" could, in principle, mimic the charge transfer effect. Furthermore, a hydrogen-induced reconstruction of the substrate may affect the position of the 4*f* levels.

The classical measure of charge transfer is the change in work function. With some exceptions [e.g. H/W(110)], the work function usually increases upon hydrogen adsorption which, if the hydrogen is outside the surface, would imply a negative charge on the adatom. There are several problems with this simple interpretation, however. It has already been noted that the accurate calculations of Feibelman et al. for H on Ti show that the work function can increase when the hydrogen is negative but located below the surface [34]. As another example, in the case of the (111) faces of Ni, Pd, and Pt, which in other respects have very similar behavior, it is found that hydrogen adsorption increases the work function of Ni and Pd but decreases that of Pt [51, 107].

Another problem concerns the magnitude of the change in work function; it is usually quite small, typically $+0.20\,\text{eV}$. The change in electrostatic potential across a dipole layer is

$$\Delta\phi = 4\pi ne\mu \quad (3)$$

where $n\mu$ is the dipole moment per unit area. For $n = 10^{15}\,\text{cm}^{-2}$ and a dipole moment corresponding to a charge transfer of $1e$ over a distance 1 Å, the change is $\Delta\phi \sim 19\,\text{eV}$, about two orders of magnitude too large. Presumably the reason for the discrepancy is that electrons not directly involved in the bond rearrange themselves so that the net dipole is zero. Evidence for this is provided by the slab calculations discussed in the last section which all give realistic changes in work function (typically within 0.1 eV of experiment) yet also indicate charge transfer to the hydrogen (in a Mulliken sense). Unfortunately, no detailed picture of how this cancellation occurs has been presented yet. An alternate possibility is that the Coulomb repulsion on the adatom is sufficiently strong that the ionic contribution to the bond is strongly suppressed [75]. However, this explanation is not in accord with cluster calculations which include substantial amounts of configuration interaction and still indicate significant charge transfer [36, 91].

Finally, regarding the *d* electron participation in the bond, there is only indirect evidence. For example, it is well known that nickel and copper behave differently when exposed to H_2. The latter has, of course, a filled *d* band and the former does not. Note, however, that because of *s*–*d* hybridization, even in copper the atomic configuration is not d^{10}. Nevertheless there are certainly more *d* holes in nickel than copper. The most important difference between the two substrates is that a barrier to the dissociation of H_2 exists on copper, but if this barrier is surmounted then a H–Cu bond is formed which is only a few tenths of an eV weaker than the H–Ni bond. This is consistent with the calculations of Madhavan and Whitten [36] who find that the *d* electrons do not have the dominant effect on the bonding.

An interesting study of the role of the *d* band has been done by Strongin and co-workers who investigated niobium surfaces covered with monolayers of palladium [53, 108]. An ordinary Pd surface readily dissociates H_2 (and so does Nb) but the Pd overlayer deposited on Nb is inert to H_2 in a manner analogous to copper. This behavior has been explained as an alloy effect [109]. The Pd monolayer has an electron configuration similar to that of bulk Pd (at least there are *d* holes at the Fermi level). On interaction with Nb, the Pd states hybridize and form bonding and anti-bonding orbitals. The bonding orbitals are filled, leading to a copper-like configuration and hence a barrier to dissociation. This again suggests that *d* holes are important for hydrogen dissociation.

Photoelectron spectroscopy has also supplied information about *d* participation. The data show clearly that adsorption of hydrogen on transition metals produces changes in the *d* band region [95]. However, it is not clear to what extent this implies that d electrons are directly involved in the bonding.

In conclusion, while a qualitative description of the nature of the bond has been achieved there are still disagreements about the importance of d orbitals and on the amount of charge transfer to the hydrogen. Ab initio calculations from two different perspectives, quantum chemistry and density functional type theories, seem to agree that the role of the d orbitals is secondary in the determination of the binding energy.

2.5 HYDROGEN–HYDROGEN INTERACTIONS

In addition to the adatom–substrate interactions discussed in Sect. 2.4, an adsorbed hydrogen atom is subject to interactions with other H atoms. One manifestation of these adatom–adatom forces is that the binding energy becomes dependent on the coverage [110, 111]. Another is the formation of ordered two-dimensional arrangements of the adatoms of the type illustrated in Figs. 3–5.

Theoretical work on the origin of the hydrogen–hydrogen interactions has been reviewed recently by Einstein [112]. Essentially, the approach involves consideration of the ordinary interactions between two H atoms modified by the proximity to the substrate. For atoms in the gas phase one has, at large distances, Van der Waals-type interactions and, at short range, chemical forces. These are referred to as *direct* interactions. For hydrogen atoms near a surface, other interactions can occur. The two which decay least rapidly with the distance, R, between the atoms are the dipole–dipole and elastic interactions. The dipole–dipole interaction comes about because of charge transfer to the hydrogen, and it leads to the usual repulsive term

$$w = \frac{2\mu^2}{R^3} \tag{4}$$

where μ is the dipole moment. The factor 2 is due to image effects.

The elastic interaction [113] comes about because the metal lattice is deformable and an isolated H atom exerts a force on the metal atoms close to it. This interaction is also repulsive and decays as $1/R^3$. The magnitude depends on the elastic constants of the medium and the metal–hydrogen interaction. It is interesting that the elastic interaction is found to be repulsive on the surface; the same type of interaction is believed to play a large role in H–H interactions in bulk metals [114] but there they are attractive. It is an open question what happens near a surface when one H is inside and the other is outside.

Finally, there is an *indirect* interaction due to the coupling of the hydrogen atoms through the metal substrate. This has the asymptotic form

$$w = A\,\frac{\cos\,(k_{\mathrm{f}}R)}{R^5} \tag{5}$$

Actually, at distances where this form applies the whole interaction is very small or negligible. At distances comparable with the nearest-neighbor spacing, where the interaction is larger, some sort of model for the substrate electron structure must be adopted. An *s* orbital tight binding model, jellium models, or jellium models plus *d* muffin-tins have been considered for this purpose [112].

The ab initio calculations which have been performed have not been aimed at adatom–adatom interaction energies per se. However, the calculation of Madahavan and Whitten [36] for H_2 on copper(100) shows that there are substantial attractive interactions for two hydrogen atoms in adjacent fourfold hollows. By means of a cluster approach, Muscat [115] has recently calculated the interactions between H atoms on iron(110) for which it is necessary also to include a "trio" interaction between three adatoms. These results are referred to again below.

There is little direct experimental evidence with which to compare the theoretical predictions. If the adsorbate tends to form simple superlattices, the sign of the H–H interaction (attractive or repulsive) can be inferred from the dimensions of the new 2D unit cell. For example, if the adsorbate arrangement has a periodicity twice that of the substrate, an effective repulsion must exist between adatoms on nearest-neighbor sites. If "islands" of this superlattice are formed at low coverage, the H–H interaction at longer range must be attractive. The strength of the interaction can then be estimated by comparing the thermal stability of the adsorbed phase with that predicted from statistical mechanical calculations [116]. The H/Pd(100) system provides a simple illustration [43]. The phase diagram is shown in Fig. 3; at a coverage $\theta \simeq 0.5$ the hydrogen atoms form a c(2 × 2) phase which disorders above a critical temperature $T_c = 260\,K$. If it is assumed that the system can be described by a 2D Ising model with only nearest-neighbor repulsion, the magnitude of this repulsion is given by [117, 118]

$$w = 1.76 k_B T_c \tag{6}$$

Thus, $w \simeq 0.9\,kcal\,mol^{-1}$ ($0.04\,eV\,atom^{-1}$) for H on Pd(100). Although w is smaller than the binding energy by more than an order of magnitude, the existence of a H–H repulsion should give rise to a measurable decrease (of magnitude $4w$) in the heat of adsorption q as the coverage is increased beyond $\theta \simeq 0.5$, where the formation of nearest-neighbor pairs can no longer be avoided. However, as seen in Fig. 11, the measured $q(\theta)$ for this system [43] does not show *any* decrease at this coverage. In the case of H/Ni(111)[13] the adsorbate forms a non-primitive (2 × 2) structure which disorders above $T_c \simeq 270\,K$, implying a H–H repulsion of strength similar to that occurring on Pd(100). Structural models indicate that this interaction should affect the heat of adsorption at $\theta \simeq 1/4$. As seen in Fig. 11(b), $q(\theta)$ does drop at this coverage but the decrease is much larger than predicted from the magnitude of w. The reason for the failure of the lattice gas model in this respect is not yet clear.

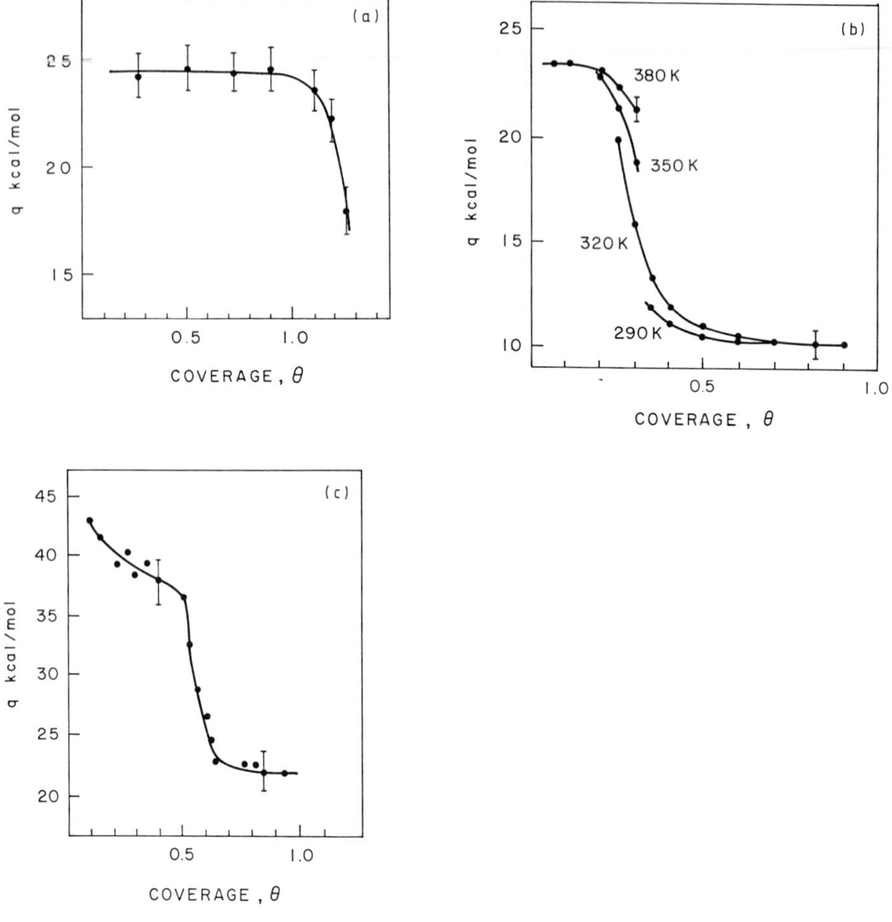

Fig. 11. (a) Isosteric heat of adsorption, q, as a function of the fractional coverage θ for the H/Pd(100) system. Data from Behm et al. [43]. (b) Heat of adsorption as a function of coverage for the H/Ni(111) system. Data from Rendulic and Winkler [56]. (c) Heat of adsorption as a function of coverage for the H/W(100) system. Data from Horlacher-Smith et al. [49].

In order to account for the more complicated phase diagram of H/Fe(110), shown in Fig. 4, a single nearest-neighbor repulsion, w_1, is insufficient [41]. In addition, pair-wise repulsions, w_2 and w_3, between next-nearest and third-nearest neighbors must be postulated. Furthermore, because the phase diagram is asymmetric about $\theta = 0.5$, the existence of a trio interaction w_t is assumed. Using a renormalization group approach combined with Monte Carlo simulations, Kinzel et al. [119] have calculated the phase diagram for different w_i values. The best fit to the experimental data is obtained when $w_1 = 0.3$ kcal, $w_2 = 1.9$ kcal, $w_3 = 0.5$ kcal, and $w_t = 0.8$ kcal. The calculations by Muscat [115] for the same system give $w_1 = 0.08$ kcal, $w_2 = 0.32$ kcal, and $w_3 = 0.05$ kcal, i.e. values which are about five times too low. One possible reason for the discrepancy is that the adsorbate induces a substrate reconstruction [40, 63], in which case the effective interactions are altered by distortions of the metal surface.

It appears that two types of reconstruction can take place [63]. In one, the displacement of substrate atoms occurs as the adsorbate–substrate bond is formed. H/Ni(110) and H/Pd(110) probably fall into this category. In the other type, the substrate has an inherent tendency to reconstruct even when clean, but the adatoms modify its electronic and elastic properties and thereby determine the direction and magnitude of the displacements [120]. H/W(100) and H/Mo(100) are in the latter category. The phase diagram for H/W(100), shown in Fig. 5, therefore cannot be explained in terms of the standard lattice gas models. To reproduce the main features of the diagram it is necessary to include in the Hamiltonian not only the usual H–H interaction but also the substrate distortion energy and the adatom–substrate interaction (which depends on the distortion) [121, 122].

The results for W(100) demonstrate the extent to which reconstruction can dominate the properties of the adsorbed hydrogen. The effect on the vibrational energies was described in Sect. 2.3. The heat of adsorption shown in Fig. 11(c), decreases from a value of 40–50 $kcal\,mol^{-1}$ at low coverage (when the W atom displacements are large) to a value of about 25 $kcal\,mol^{-1}$ at saturation coverage (when the displacements vanish)[49]. The reconstruction of W(100) also causes an effective interaction between the H atoms. At low coverage the free energy per unit area, $f(\theta)$, has positive curvature which means that if an island of c(2 $\times$ 2)H were formed it would tend to expand [123]. The result is an effective H–H repulsion, the magnitude of which is considerably larger than that estimated for the indirect interaction.

3. Dynamics

3.1 OVERVIEW

Dynamics refers to time-dependent phenomena such as adsorption, desorption, and diffusion. To obtain a microscopic understanding of these processes it is necessary not only to know details of the surface structure and adatom bonding but also to have information about the possible excitations of the metal substrate. As an illustration, consider a gas-phase hydrogen molecule incident on the metal surface. If no energy dissipation occurs, the molecule will merely be reflected at the surface as it encounters the repulsive part of the potential depicted in Fig. 1. In order for the sticking probability to be non-zero, the molecule must lose sufficient energy, by excitation of substrate phonons [124] or electron–hole pairs [125, 126], to become trapped in either the molecular or, after dissociation, the atomic well. The adatoms may then transfer additional energy to the substrate to end up in the vibrational ground state. Subsequently the process may re-

References pp. 33–37

verse itself, the adatoms receiving sufficient energy from the substrate to enable them to diffuse across the surface, to penetrate into the bulk, or to desorb back into the gas phase. As might be expected, these processes will occur with different probability on different surfaces. Furthermore, the rates may differ for apparently identical systems due to the influence of defects and impurities; an example is the strong effect of steps on the sticking probability of hydrogen on W(110)[127].

There are several general approaches to problems in dynamics, e.g. Eyring's transition state theory [128, 129], Kramers' treatment based on the Fokker–Planck equation [130], the random walk approach of Montroll and Schuler [131], non-equilibrium thermodynamics [132], and molecular dynamics [133]. Of these the most widely used is the transition state theory (TST). One problem with this theory is that classical mechanics enters in a very fundamental way and it is not clear if it is valid for light particles such as hydrogen. Kramers showed that TST can be derived from a Fokker–Planck equation assuming that the dissipation (energy loss) is neither too small nor too large. This problem has been taken up again by Suhl et al. [134] and by Caroli et al. [135]. Again, the results derived so far assume heavy adsorbates (relative to the substrate) which clearly does not apply to hydrogen.

In the absence of a more complete theoretical foundation, experiments are usually interpreted in terms of ordinary rate equations, as discussed by Menzel [136], King [137], and Ibach et al. [138]. The most important application is to the kinetics of adsorption and desorption. The rate equation is written

$$\frac{\mathrm{d}n_a}{\mathrm{d}t} = 2\Gamma\sigma(n_a) - R(n_a) \tag{7}$$

which gives the rate of change in the number of adsorbed atoms per unit area (n_a) as the difference between the rates of adsorption and desorption. Γ is the flux of particles arriving from the gas phase, σ the sticking probability, and R the desorption rate. The factor 2 comes from the assumption of two adsorbed atoms per incident molecule. In equilibrium we must have

$$R(n_{eq}) = 2\Gamma\sigma(n_{eq}) \tag{8}$$

For an ideal gas, Γ is given by the kinetic theory

$$\Gamma = \frac{1}{4} n_g \langle v_g \rangle = \frac{n_g kT}{\sqrt{2\pi mkT}} \tag{9}$$

Where n_g is the gas density and $\langle v_g \rangle$ the mean velocity in the gas. Furthermore, the chemical potential, μ_g, of the gas is related to the density by

$$\mathrm{e}^{\mu_g/kT} = n_g \frac{h^3}{(2\pi mkT)^{3/2}} \frac{1}{z_{rot}} \tag{10}$$

where z_{rot} is the rotational partition function. It is assumed that vibrations are unimportant since the vibrational energy of H_2 is much larger than kT.

Therefore

$$R(n_{eq}) \;=\; 2\sigma(n_{eq})\,\frac{2\pi m(kT)^2}{h^3}\,z_{rot}\,e^{\mu_g/kT} \tag{11}$$

Also, at equilibrium

$$\mu_g \;=\; 2\mu_a \tag{12}$$

where μ_a is the chemical potential of the adsorbed atoms. This leads to a relation between R and the chemical potential of the adsorbed gas. If it is assumed that this same relation holds away from equilibrium, it follows that

$$R(n_a) \;=\; 2\sigma(n_a)\,\frac{2\pi m(kT)^2}{h^3}\,z_{rot}\,e^{2\mu_a/kT} \tag{13}$$

That R and σ are related has been discussed recently by Iche and Nozières [139] and by Brenig and Schönhammer [132]. An advantage of eqn. (13) is that one can include interactions between the H atoms because they enter via the chemical potential.

In order to apply eqn. (13), $\sigma(n_a)$ must be known. For dissociative adsorption it is often assumed that the sticking probability varies according to

$$\sigma \;=\; \sigma_0(1 - \theta)^2 \tag{14}$$

where $\theta = n_a/n_s$ and n_s is the number of sites on the surface per unit area. If it is also assumed that the hydrogen atoms form a non-interacting lattice gas, so that

$$e^{\mu_a/kT} \;=\; \frac{\theta}{1 - \theta}\,e^{-\varepsilon_a/kT} \tag{15}$$

then

$$R \;=\; 2\sigma_0\,\frac{2\pi m(kT)^2}{h^3}\,z_{rot}\,\theta^2 e^{-2\varepsilon_a/kT} \tag{16}$$

In the opposite limit, that of mobile adsorption, the corresponding expression is

$$e^{\mu_a/kT} \;=\; e^{-\varepsilon_a/kT}\,\frac{n_a h^2}{2\pi mkT} \tag{17}$$

which in the special case of $\sigma = \sigma_0$ becomes

$$R \;=\; \sigma_0\,\frac{h}{\pi m}\,z_{rot}\,n_a^2 e^{-2\varepsilon_a/kT} \tag{18}$$

Instead of eqns. (13), (16), or (18), the desorption rate is often assumed to have the form

$$R \;=\; \nu n_s \theta^2 e^{-\varepsilon/kT} \tag{19}$$

where the pre-exponential factor ν is taken to be equal to kT/h [4]. Equation (19) has the same form as eqns. (16) and (18) but the pre-exponential factors are different. $\nu = kT/h$ is only appropriate for unimolecular desorption when the motion parallel to the surface is the same in the gas phase and the adsorbed phase. The case where ν can be treated as the vibrational frequency occurs for mobile non-dissociative adsorption when $kT \gg h\nu_{vib}$. But, as mentioned previously, the vibrational frequencies are much larger than kT in the temperature range of interest.

It should also be noted that the coverage dependence of the sticking probability can be quite different from eqn. (14). In the case of H/W(100), for example, the measurements described below yield $\sigma \simeq \sigma_0(1 - \theta)$[140]. More serious modifications of the rate expressions are necessary for this system, however, because of the substrate reconstruction. Both the energy and the entropy of the metal surface change upon adsorption [49] and there is no simple relationship between μ_a and ε_a. If the rate data are forced to fit eqn. (19), ν and ε will become functions of both θ and T.

3.2 STICKING PROBABILITY

The absolute value of the sticking probability, needed for the application of rate equations, may be determined by molecular beam techniques. Such techniques can also provide information about the energy and angle dependence of σ. A few examples involving hydrogen on metal surfaces are given in this section.

The departure from the $(1 - \theta)^2$ dependence is illustrated by experimental results for H/W(100). The sticking probability measured by Madey [140] using a molecular beam to give the tungsten surface a calibrated dose of H_2 is shown in Fig. 12. In this case, an approximately linear dependence on θ is clearly indicated, a result which has been confirmed by King and Thomas [46]. The reason for this behavior is not well understood at the present time

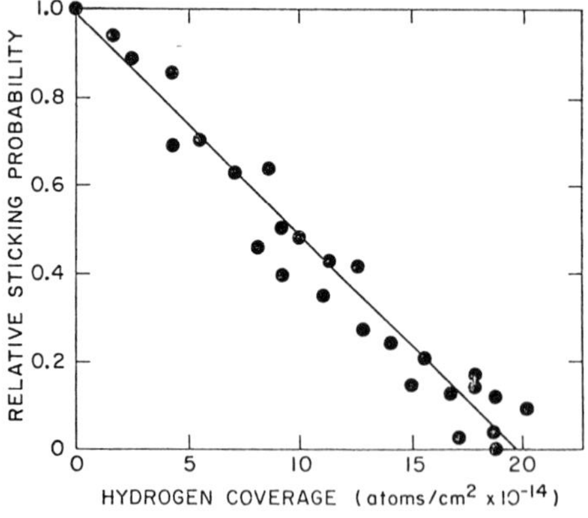

Fig. 12. Relative sticking probability σ/σ_0 for hydrogen on W(100). Data from Madey [140].

and the result cannot be generalized to other surfaces. In the case of H/Pd(100)[43] the coverage dependence of σ is even further from the expected $(1 - \theta)^2$ form, and it has been proposed that initial adsorption into a precursor state is responsible.

As a second example, Fig. 13 shows the energy dependence of the sticking probability for beams of H_2 scattered from (100) and (110) surfaces of copper [58]. The experiments measured the dissociative adsorption probability and, as seen, this probability varies as the "perpendicular energy", i.e. as $E_\perp = E_i \cos^2\theta_i$ where E_i is the incident energy and θ_i is the angle of incidence. These surfaces display a barrier to dissociation (cf. Fig. 1), which is clearly revealed by the increase in the sticking probability for mean energies near 3 kcal mol^{-1} for Cu(110) and 5 kcal mol^{-1} for Cu(100). The desorption rate of H_2, after diffusion through the copper crystals, was also measured. Comparison of the desorption and scattering results indicates that the

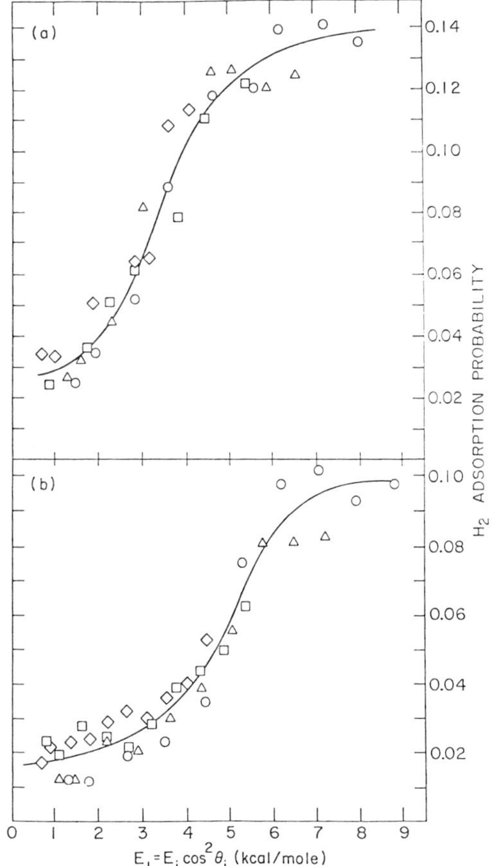

Fig. 13. Energy dependence of the dissociative adsorption probability for hydrogen on (a) Cu(110) and (b) Cu(100). Data from Balooch et al. [58].

principle of detailed balance applies for this system. This is important for the validity of the rate expressions in Sect. 3.1.

Another recent beam experiment monitored H–D exchange on Pd(111) surfaces [141]. In this case, the rate of HD production for incident beams containing a mixture of H_2 and D_2 was studied. The data were analyzed using phenomenological rate equations in two ways. In one the sticking probability was assumed to vary as $(1 - \theta)^2$ and the data were fitted assuming the adsorption energy was a function of θ. In this way, the desorption energy was found to vary from 21.5 kcal mol^{-1} at low coverages, to 18 kcal mol^{-1} at high coverages. This is not unreasonable in view of the adsorbate–adsorbate interactions discussed in Sect. 2.5. In the second analysis it was assumed that the energy barrier was constant at 20.8 kcal mol^{-1} as found by Conrad et al. [142]. The data were fitted to give a sticking coefficient as a function of coverage which was found to vary roughly as $\sigma(\theta) \propto \ln \theta$. A clear choice between these two models cannot be made from the data.

Finally, it should be mentioned that currently there is much interest in beam experiments that permit the observation of the diffraction of hydrogen by the surface [143]. This phenomenon has great potential value for the study of energy exchange in particle–surface collisions as well as for investigations of surface structure.

3.3 DIFFUSION

Surface diffusion is another dynamic process which plays a crucial role in many surface phenomena. The theory of surface diffusion of hydrogen has not been developed to the same degree as for the bulk [1], and only recently has it become possible to obtain experimental measurements for the diffusion rate on well-characterized metal surfaces. The most detailed data, due to DiFoggio and Gomer [50], are for H and D adsorbed on W(110). They yield the diffusion coefficient as a function of both temperature and coverage. The variation with temperature is shown in Fig. 14. A striking feature of this figure is the isotope dependence. Above 140 K both H and D undergo activated diffusion with an activation barrier of 0.21 eV. Below 140 K, the deuterium diffusion constant continues to drop but for hydrogen it is constant down to $T = 28$ K. This is due to tunneling of the protons from site to site rather than thermally activated hopping. There is a strong isotope effect because the zero point motion of the deuterium is much less than that for hydrogen. A similar behavior for H versus D has been observed for bulk diffusion in Nb and Ta [144]. The activation energies are, however, somewhat lower, 0.127 eV for D in Nb and 0.160 eV in Ta.

3.4 THERMAL DESORPTION

Studies of the rate of thermally stimulated desorption have provided a major portion of the empirical information concerning the bonding of ad-

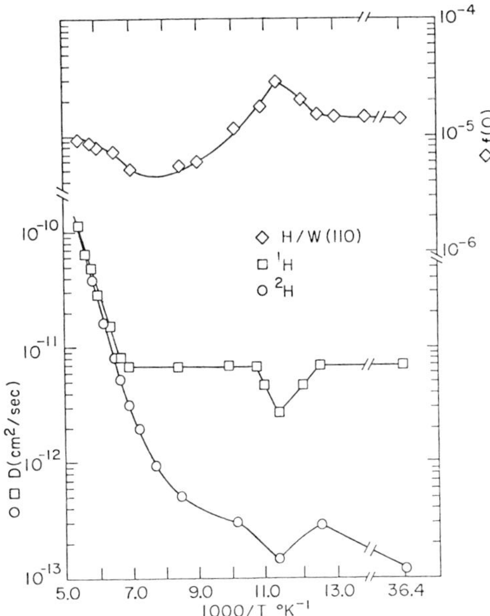

Fig. 14. Temperature dependence of the diffusion coefficient for hydrogen and deuterium atoms on W(110). Also shown is $f(0)$, the correlation function at $t = 0$. Data from DiFoggio and Gomer [50].

sorbed hydrogen. This topic has been discussed at length elsewhere [136, 137]. The typical procedure is to measure the thermal desorption "spectrum" (TDS), i.e. the rate of desorption as the temperature is increased (usually linearly) with time. If the rate is given by

$$-\frac{d\theta}{dt} = \nu\theta^2 e^{-\varepsilon/kT} \tag{19}$$

and the time dependence of the temperature is $T = T_0 + bt$, then a maximum in the TDS will occur at a temperature T_p given by

$$\frac{b\varepsilon_a}{2kT_p^2} = \nu\theta_p e^{-\varepsilon_a/kT_p} \tag{20}$$

where θ_p is the coverage at the maximum. Physically, the maximum occurs because at low temperature the exponential is small and at high temperature the surface is depleted. ε can then be found from eqn. (20) if a value for ν is assumed. However, due to the uncertainty in ν and the shortcomings of eqn. (19) discussed in Sect. 3.1, the value for the binding energy obtained in this manner should be taken with some caution. A comparison with the heat of adsorption (obtained from equilibrium measurements [43, 49, 56]) should be made whenever possible.

For many surfaces the TDS contains more than one peak [6] and this is often used as evidence for more than one binding site. Other interpretations are possible, however; even when all the sites on a surface are identical, multiple peaks in the TDS can occur if the binding energy is coverage-dependent because of adatom–adatom interactions, as for H/Ni(111)[13], or because of substrate reconstruction, as for H/W(100)[49].

As a result the experimental methods used in most desorption studies give a rate which represents an average over momentum, kinetic energy, and molecular excitations. Recent investigations, such as those by Comsa on the desorption of hydrogen from copper and palladium [145], have made it clear that models of the dynamics can be tested by measuring the velocity distribution of the desorbing H_2 molecules. Among the intriguing findings is the fact that on these surfaces the distribution is strongly non-Maxwellian.

3.5 SORPTION

As was discussed in connection with Fig. 1, it is usually assumed that hydrogen atoms in the surface region do not penetrate the bulk to any appreciable extent. On the other hand, reference has been made to the evidence from photoemission, LEED, and kinetics that there is absorbed (as opposed to adsorbed) hydrogen near palladium surfaces, for example. Recently Strongin and co-workers have studied this problem directly by measuring the kinetics of hydrogen sorption by niobium films using the change in resistivity of the film as a monitor of bulk hydrogen concentration [53]. In another study, Kasemo and Törnquist [146] have investigated hydrogen absorption by titanium films using a quartz crystal microbalance.

The simplest model which can describe these effects goes back to Wagner [147]. Assuming a Langmuir-type kinetic equation for the surface but with an added term which accounts for the loss of H atoms to the bulk, the absorption rate can be expressed by

$$\frac{\mathrm{d}\theta}{\mathrm{d}t} = \gamma(1-\theta)^2 - k\theta^2 - N_l\frac{\mathrm{d}x}{\mathrm{d}t} \tag{21}$$

where θ is the surface coverage, x is the bulk concentration (assumed uniform) and N_l is the number of layers. A simple jump model gives

$$N_l\frac{\mathrm{d}x}{\mathrm{d}t} = \nu\theta(1-x) - \beta x(1-\theta) \tag{22}$$

where ν and β contain pre-exponentials and Arrhenius-type exponential terms. At equilibrium

$$\frac{\theta_e}{1-\theta_e} = \beta\Big/\nu\,\frac{x_e}{1-x_e} = \left(\frac{\gamma}{k}\right)^{1/2} \tag{23}$$

Since γ is proportional to the pressure, it can be seen that θ obeys the Langmuir adsorption isotherm. Furthermore, the bulk concentration is proportional to $p^{1/2}$ (for small values of x); this result is known as Sieverts' law.

As noted by Wagner there are two simple limits. The first occurs when surface-to-bulk transfer is the rate-limiting step. Then the adsorbate comes into equilibrium with the gas so that

$$N_1 \frac{dx}{dt} - \nu\theta_e(1 - x) = \beta x(1 - \theta_e) \tag{24}$$

which can be integrated. The second limit corresponds to $\theta = x$ at all times. This yields the usual hyperbolic tangent as the functional form for $x(t)$. Recently, Davenport et al. [53] have found another simple limit; after an initial transient, equilibrium is reached between the adsorbed and absorbed hydrogen but not with the gas. This leads to the rate equation

$$N_1(1 - x + \beta/\nu x)^2 \frac{dx}{dt} = \gamma(1 - x)^2 - k(\beta/\nu)^2 x^2 \tag{25}$$

which can also be integrated. Since, in equilibrium, the left-hand side vanishes, the temperature dependence of the term $k(\beta/\nu)^2$ must be $\exp(-2\varepsilon_s/kT)$ where ε_s is the solution energy. However, the rate can be severely affected by the term on the right-hand side which greatly slows the process of filling.

Acknowledgements

The authors acknowledge support by the Division of Basic Energy Sciences, U.S. Department of Energy under Contract DE-AC02-76CH00016 (for J.W.D.); and by the National Science Foundation under grant DMR 8305802 and through the Brown University Materials Science Laboratory (for P.J.E.).

References

1 G. Alefeld and J. Völkl (Eds.), Hydrogen in Metals, Springer Verlag, New York, 1978.
2 R.M. Latanison and J.R. Pickens (Eds.), Atomistics of Fracture, NATO Conference Series VI, Plenum Press, New York, 1983.
3 T.B. Flanigan and W.A. Oates, Adv. Chem., 167 (1978) 283.
4 D.O. Hayward and B.M.W. Trapnell, Chemisorption, Butterworths, London, 1984.
5 J.R. Anderson (Ed.), Chemisorption and Reactions on Metallic Films, Academic Press, New York, 1971.
6 L.D. Schmidt, in R. Gomer (Ed.), Interactions on Metal Surfaces, Springer Verlag, New York, 1975.
7 R. Gomer, Solid State Phys. 30 (1975) 93.
8 T.N. Rhodin and D.L. Adams, in N.B. Hannay (Ed.), Treatise on Solid State Chemistry, Vol. 6A, Plenum Press, New York, 1976, pp. 343–484.
9 S. Roy Morrison, The Chemical Physics of Surfaces, Plenum Press, New York, 1977.
10 T.N. Rhodin and G. Ertl (Eds.), The Nature of the Surface Chemical Bond, North-Holland, New York, 1979.
11 J.R. Smith (Ed.), Theory of Chemisorption, Topics in Current Physics, Vol. 19, Springer Verlag, Berlin, 1980.

12 R.R. Cavanagh, R.D. Kelley and J.J. Rush, J. Chem. Phys, 77, (1982) 1540.
13 K. Christmann, R.J. Behm, G. Ertl, M.A. Van Hove and W.H. Weinberg, J. Chem. Phys., 70 (1979) 4168.
14 T. Engel and K.H. Rieder, Springer Tracts, Mod. Phys., 91 (1982) 54; Surf. Sci., 109 (1981), 140.
15 K.H. Rieder and H. Wilsch, Surf. Sci., 131 (1983) 245.
16 J. Lee, J.P. Crowin and L. Wharton, Surf. Sci., 130 (1983) 1.
17 H. Ibach and D.L. Mills, Electron Energy Loss Spectroscopy and Surface Vibrations, Academic Press, New York, 1982.
18 S. Andersson, Chem. Phys. Lett., 55 (1978) 185.
19 W. Ho, N.J. DiNardo and E.W. Plummer, J. Vac. Sci. Technol., 17 (1980) 134.
20 A.M. Baró, H. Ibach and H.D. Bruchmann, Surf. Sci., 88 (1979) 384. Note that the factor $1/\sqrt{2}$ missing from the asymmetric stretch mode. This changes the bond length to 1.68 Å as quoted in ref. 51.
21 A.M. Baró and W. Erley, Surf. Sci., 112 (1981) L759.
22 G.B. Blanchette, N.J. DiNardo and E.W. Plummer, Surf. Sci., 118 (1982) 496.
23 W. Ho, R.F. Willis and E.W. Plummer, Phys. Rev. Lett., 40 (1978) 1463. R.F. Willis, W. Ho and E.W. Plummer, Surf. Sci., 80 (1979) 593.
24 A. Adnot and J.D. Carette, Phys. Rev. Lett., 39 (1977) 209.
25 H. Froitzheim, H. Ibach and S. Lehwald, Phys. Rev. Lett., 36 (1976) 1549.
26 M.R. Barnes and R.F. Willis, Phys. Rev. Lett., 41 (1978) 1727.
27 R.F. Willis, Surf. Sci., 89 (1979) 457.
28 C. Backx, B. Feuerbacher, B. Fitton and R.F. Willis, Phys. Lett. A, 60 (1977) 145.
29 M.A. Barteau, J.Q. Broughton and D. Menzel, Surf. Sci., 133 (1983) 443.
30 C. Nyberg and C.G. Tengstål, Solid State Commun., 44 (1982) 251.
31 J. Eckert, C.F. Majkrzak, L. Passell, W.B. Daniels and T.A. Kitchens, in P. Jena and C.B. Satterthwaite (Eds.), Electronic Structure of Hydrogen in Metals, Plenum Press, New York, 1983.
32 T. Springer in G. Alefeld and J. Völkl (Eds.), Hydrogen in Metals, Springer Verlag, New York, 1978.
33 S.G. Louie, Phys. Rev. Lett., 42 (1979) 476.
34 P.J. Feibelman, D.R. Hamann and F.J. Himpsel, Phys. Rev. B, 22 (1980) 1734.
35 T.H. Upton and W.A. Goddard III, CRC Crit. Rev. Solid State Mater. Sci., 10 (1981) 261.
36 P.V. Madhavan and J.L. Whitten, Surf. Sci., 112 (1981) 38; J. Chem. Phys., 77 (1982) 2673.
37 R. Bau, R.G. Teller, S.W. Kirtley and T.F. Koetzle, Acc. Chem. Res., 12 (1979) 176.
38 V. Lottner, U. Buchenau and W.J. Fitzgerald, Z. Phys. B, 35 (1979) 35.
39 P.J. Estrup, J. Vac. Sci. Technol., 16 (1979) 635.
40 L.D. Roelofs and P.J. Estrup, Surf. Sci., 125 (1983) 51.
41 T.N. Taylor and P.J. Estrup, J. Vac. Sci. Technol., 11 (1974) 244.
42 J. Demuth, J. Colloid, Interface Sci., 58 (1977) 184.
43 R.J. Behm, K. Christmann and G. Ertl. Surf. Sci., 99 (1980) 320.
44 M.G. Cattania, V. Penka, R.J. Behm, K. Christmann and G. Ertl, Surf. Sci., 126 (1983) 382. R.J. Behm, V. Penka, M.G. Cattania, K. Christmann and G. Ertl, Surf. Sci., 126 (1983) 382.
45 R. Imbihl, R.J. Behm, K. Christmann, G. Ertl and T. Matsushima, Surf. Sci., 117 (1982) 257.
46 D.A. King and G. Thomas, Surf. Sci., 92 (1980) 201.
47 R.A. Barker and P.J. Estrup, J. Chem. Phys., 74 (1981) 1442.
48 V.V. Gonchar, Yu. M. Kagan, O.V. Kanash, A.G. Naumovets and A.G. Fedorus, Sov. Phys. JETP, 57 (1983) 142.
49 A.M. Horlacher-Smith, R.A. Barker and P.J. Estrup, Surf. Sci., 136 (1984) 327.
50 R. DiFoggio and R. Gomer, Phys. Rev. Lett., 44 (1980) 1258; Phys. Rev. B, 25 (1982) 3490.
51 W. Eberhardt, F. Greuter and E.W. Plummer, Phys. Rev. Lett., 46 (1981) 1085.
52 P.R. Norton, J.A. Davies and T.E. Jackman, Surf. Sci.l, 121 (1982) 103.
53 J.W. Davenport, G.J. Dienes and R.A. Johnson, Phys. Rev. B, 25 (1982) 2165.
54 T.B. Flanagan and W.A. Oates, Ber. Bunsenges. Phys. Chem., 76 (1972) 706.

55 I. Toyoshima and G.A. Somorjai, Catal. Rev., 19 (1979) 105.
56 K.D. Rendulic and A. Winkler, J. Chem. Phys., 79 (1983) 5151.
57 D.O. Welch and M.A. Pick, Phys. Lett. A, 99 (1983) 183.
58 M. Balooch, M.J. Cardillo, D.R. Miller and R.E. Stickney, Surf. Sci., 46 (1974) 358.
59 R. Caudano, J.M. Gilles and A.A. Lucas (Eds.), Vibrations at Surfaces, Plenum Press, New York, 1982.
60 T.E. Felter, R.A. Barker and P.J. Estrup, Phys. Rev. Lett., 38 (1977) 1138.
61 M.K. Debe and D.A. King, J. Phys. C, 10 (1977) L303.
62 R.A. Barker and P.J. Estrup, Phys. Rev. Lett., 41 (1978) 1307.
63 P.J. Estrup, in R. Vanselow and R. Howe (Eds.), Chemistry and Physics of Solid Surfaces, Vol. 5, Springer, Berlin, 1984, pp. 205–230.
64 I. Stensgaard, L.C. Feldman and P.J. Silverman, Phys. Rev. Lett., 42 (1979) 247. L.C. Feldman, P.J. Silverman and I. Stensgaard, Surf. Sci., 87 (1979) 410.
65 U.A. Jayasooriya, M.A. Chesters, M.W. Howard, S.F.A. Kettle, D.B. Powell and N. Sheppard, Surf. Sci., 93 (1980) 526.
66 Y.J. Chabal and A.J. Sievers, Phys. Rev. Lett., 44 (1980) 944.
67 S. Andersson and J. Harris, Phys. Rev. Lett., 48 (1982) 545.
68 Ph. Avouris, D. Schmeisser and J.E. Demuth, Phys. Rev. Lett., 47 (1982) 199.
69 L. Pauling, The Nature of the Chemical Bond, Cornell University Press, Ithaca, NY, 3rd edn., 1960.
70 D.D. Eley, Discuss. Faraday Soc., 8 (1950) 34.
71 E. Miyazaki, Surf. Sci., 71 (1978) 741.
72 J. Friedel, in J.M. Ziman (Ed.) The Physics of Metals, Cambridge University Press, London, 1969.
73 P.C. Bouten and A.R. Miedema, J. Less Common Met. 71 (1980) 147.
74 D.M. Newns, Phys. Rev., 178 (1969) 1123.
75 J.R. Schrieffer, J. Vac. Sci. Technol., 9 (1972) 561.
76 B.I. Lundqvist, H. Hjelmberg and O. Gunnarsson, in B. Feuerbacher, B. Fitton and R.F. Willis (Eds.), Photoemission and the Electronic Properties of Surfaces, Wiley, New York, 1978.
77 T.L. Einstein, J.A. Hertz and J.R. Schrieffer in J.E. Smith (Ed.), Theory of Chemisorption, Topics in Current Physics, Vol. 19, Springer Verlag, Berlin 1980.
78 C.M. Varma and A.J. Wilson, Phys. Rev. B, 22 (1980) 3795.
79 P.S. Bagus and C. Björkman, Phys. Rev. A, 23 (1981) 461.
80 N. Lang, Solid State Physics, 28 (1973) 225.
81 N.D. Lang and A.R. Williams, Phys. Rev. Lett., 34 (1975) 531; Phys. Rev. B, 18 (1978) 616.
82 O. Gunnarsson, H. Hjelmberg and B.I. Lundqvist, Phys. Rev. Lett., 37 (1976) 292; Surf. Sci., 63 (1977) 348.
83 J.K. Nørskov, A., Houmøller, P.K. Johansson and B.I. Lundqvist, Phys. Rev. Lett., 46 (1981) 257.
84 P. Nordlander, S. Holloway and J.K. Nørskov, Surf. Sci., 136 (1984) 59.
85 J.D. Muscat, Surf. Sci., 110 (1981) 389.
86 D.J.M. Fassaert and A. Van der Avoird, Surf. Sci., 55 (1976) 291, 313.
87 D.E. Ellis, H. Adachi and F.W. Averill, Surf. Sci., 58 (1976) 497.
88 R.P. Messmer, D.R. Salahub, K.H. Johnson and C.Y. Yang, Chem. Phys. Lett., 51 (1977) 84.
89 C.F. Melius, J.W. Moskowitz, A.P. Mortola, M.B. Baillie and M.A. Ratner, Surf. Sci., 59 (1976) 279.
90 C.F. Melius, Chem. Phys. Lett., 39 (1976) 287.
91 P. Cremaschi and J.L. Whitten, Phys. Rev. Lett., 46 (1981) 1242; Surf. Sci., 112 (1981) 343.
92 G.P. Kerker, M.T. Yin and M.L. Cohen, Solid State Commun., 32 (1979) 433; Phys. Rev. B, 20 (1979) 4940.
93 D.R. Hamann, Phys. Rev. Lett., 46 (1981) 1227.
94 R. Richter and J.W. Wilkins, Surf. Sci., 128 (1983) L190.
95 W. Eberhardt, S.G. Louie and E.W. Plummer, Phys. Rev. B, 28 (1983) 465.

96 A.C. Switendick, in G. Alefeld and J. Völkl (Eds.), Hydrogen in Metals, Springer Verlag, New York, 1978.
97 L. Schlapbach and J.P. Burger, J. Phys. (Paris), 43 (1982) L273.
98 P.A. Bennett and J.C. Fuggle, Phys. Rev. B, 26 (1982) 6030.
99 E.W. Plummer and W. Eberhardt, Adv. Chem. Phys., 49 (1982) 533.
100 C.T. Chan and S.G. Louie, Solid State Commun., 48 (1983) 417.
101 S.R. Chubb and J.W. Davenport, Phys. Rev. B, 31 (1985) 3278.
102 N.V. Smith and P.K. Larsen, in B. Feuerbacher, B. Fitton and R.F. Willis (Eds.), Photoemission and the Electronic Properties of Surfaces, Wiley, New York, 1978.
103 R.J. Smith, Phys. Rev. B, 21 (1980) 3131; Phys. Rev. Lett., 45 (1980) 1277.
104 J.F. Van der Veen, P. Heimann, F.J. Himpsel and D.E. Eastman, Solid State Commun., 37 (1981) 555.
105 J.F. Van der Veen, F.J. Himpsel and D.E. Eastman, Solid State Commun., 40 (1981) 57.
106 C. Guillot, C. Thuault, Y. Jugnet, D. Chauveau, R. Hoogewijs, J. Lecante, Tran Minh Duc, G. Treglia, M.C. Desjonquères and D. Spanjaard, J. Phys. C, 15 (1982) 4023.
107 K. Christmann, G. Ertl and T. Pignet, Surf. Sci., 54 (1976) 365.
108 M. El Batanouny, M. Strongin, G.P. Williams and J. Colbert, Phys. Rev. Lett., 46 (1981) 269.
109 M. El Batanouny, D.R. Hamann, S.R. Chubb and J.W. Davenport, Phys. Rev. B, 27 (1983) 2575.
110 D.L. Adams, Surf. Sci., 42 (1974) 12.
111 C.G. Goymour and D.A. King, J. Chem. Soc. Faraday Trans., 69 (1973) 749.
112 T.L. Einstein, in R. Vansclow (Ed.), Chemistry and Physics of Solid Surfaces, Vol. II, CRC Press, Boca Raton, FL, 1979, p. 181.
113 K.H. Lau and W. Kohn, Surf. Sci., 65 (1977) 607.
114 H. Wagner, in G. Alefeld and J. Vökl (Eds.), Hydrogen in Metals, Springer Verlag, New York, 1978.
115 J.P. Muscat, Surf. Sci., 118 (1982) 321.
116 C.G. Wang, in D.A. King and D.P. Woodruff (Eds.), The Chemical Physics of Solid Surfaces and Heterogeneous Catalysis, Vol. 2, Elsevier, Amsterdam, 1983.
117 L. Onsager, Phys. Rev., 65 (1944) 117.
118 C.N. Yang and T.D. Lee, Phys. Rev., 87 (1952) 404. T.D. Lee and C.N. Yang, Phys. Rev., 87 (1952) 410.
119 W. Kinzel, W. Selke and K. Binder, Surf. Sci., 121 (1982) 13. W. Selke, K. Binder and W. Kinzel, Surf. Sci., 125 (1983) 74.
120 A. Fasolino, G. Santoro and E. Tosatti, Surf. Sci., 125 (1983) 317.
121 K.H. Lau and S.C. Ying, Phys. Rev. Lett., 44 (1980) 1222.
122 S.C. Ying and L.D. Roelofs, Surf. Sci., 125 (1983) 218.
123 A. Horlacher-Smith, J.W. Chung and P.J. Estrup, J. Vac. Sci. Technol. A, 2 (1984) 877.
124 J.C. Tully, Annu. Rev. Phys. Chem., 31 (1980) 319.
125 K. Schönhammer and O. Gunnarsson, Phys. Rev. B, 22 (1980) 1629.
126 J.W. Gadzuk and H. Metiu, Phys. Rev. B, 22 (1980) 2603. H. Metiu and J.W. Gadzuk, J. Chem. Phys., 74 (1981) 2641.
127 R.S. Polizzotti and G. Ehrlich, J. Chem. Phys., 71 (1979) 259.
128 S. Glasstone, K.J. Laidler and H. Eyring, The Theory of Rate Processes, Mcgraw-Hill, New York, 1941.
129 P. Pechukas, Dynamics of Molecular Collisions, Modern Theoretical Chemistry, Vol. 2, Plenum Press, New York, 1976.
130 H.A. Kramers, Physica, 7 (1940) 284.
131 E. Montroll and K. Schuler, Adv. Chem. Phys. 1 (1958) 361.
132 W. Brenig and K. Schönhammer, Z. Phys. B, 24 (1976) 91.
133 J.H. McCreery and G. Wolken, J. Chem. Phys., 67 (1977) 2551.
134 H. Suhl, J.H. Smith and P. Kumar, Phys. Rev. Lett., 25 (1970) 1442. E.G. d'Agliano, P. Kumar, W.L. Schaich and H. Suhl, Phys. Rev. B, 11 (1975) 2122.
135 C. Caroli, B. Roulet and D. Saint-James, Phys. Rev. B, 18 (1978) 545.

136 D. Menzel, in R. Gomer (Ed.), Interactions on Metal Surfaces, Springer, New York, 1975, p. 102
137 D.A. King, CRC Crit. Rev. Solid State Mater. Sci., 7 (1978) 167.
138 H. Ibach, W. Erley and H. Wagner, Surf. Sci., 92 (1980) 29.
139 G. Iche and P. Nozières, J. Phys. (Paris), 37 (1976) 1313.
140 T.E. Madey, Surf. Sci., 36 (1973) 281.
141 T. Engel and H. Kuipers, Surf. Sci., 90 (1979) 162.
142 H. Conrad, G. Ertl and E. Latta, Surf. Sci., 41 (1974) 435.
143 C. Yu, K.B. Whaley, C.S. Hogg and S.S. Sibener, Phys. Rev. Lett., 51 (1983) 2210.
144 J. Völkl and G. Alefeld, in G. Alefeld and J. Völkl (Eds.), Hydrogen in Metals, Springer Verlag, New York, 1978.
145 G. Comsa, in G. Benedek and U. Volbusa (Eds.), Dynamics of Gas–Surface Interaction, Springer Series in Chemical Physics, Vol. 21, 1982, p. 117.
146 B. Kasemo and E. Törnquist, Appl. Surf. Sci., 3 (1979) 307.
147 C. Wagner, Z. Phys. Chem. Abt. A, 159 (1932) 459.

Note added in proof

Since this review was completed in January 1984, there have been many theoretical and experimental studies of hydrogen in and on metals.

An excellent review has been given by

1 K. Christmann, Surf. Sci. Rep., 9 (1988) 1.

A few examples which illustrate the state of the art are

2 T.E. Felter, S.M. Foiles, M.S. Daw and R.H. Stuhlen, Surf. Sci., 171 (1986) L379.
In this paper an embedded atom calculation was performed to obtain the interaction potential between two hydrogen atoms in the vicinity of a Pd(111) surface. The interaction was then used to predict the phase diagram which was found to be in good agreement with experiment.

3 T.E. Felter, E.C. Sowa and M.A. Van Hove, Phys. Rev. B, 40 (1989) 891.
In this paper an extensive LEED study was carried out for H on Pd(111). After an examination of more than 20 possible geometries it was concluded that there was partial occupancy of surface and subsurface sites.

4 D.R. Hamann, J. Electron Spectrosc. Relat. Phenom., 44 (1987) 1.
This paper re-examines the vibrational mode assignments of H on tungsten (as well as other metals) on the basis of total energy calculations and concludes that many of the previous assignments were incorrect.

Chapter 2

Nitrogen Adsorption on Metals

R. RAVAL, M.A. HARRISON and D.A. KING

Department of Chemistry, University of Cambridge, Cambridge CB2 1EW (Gt. Britain)

1. Introduction

1.1 THE REVIEW IN OUTLINE

Undoubtedly the major and continuing interest in nitrogen adsorption on metals revolves around the catalytic synthesis of ammonia from N_2, a process which has been reviewed in this series by Grunze [1]. The wide-ranging studies which have been performed on nitrogen chemisorption have revealed a wealth of interesting phenomena. On transition metals in Groups IIIB–VIIB of the periodic table adsorption of gaseous dinitrogen leads directly to dissociation and the formation of a very strong M—N bond, while on metals in Groups IB and IIB adsorption of thermal, gaseous dinitrogen does not lead to dissociation. On these metals dinitrogen is adsorbed with a relatively low heat of adsorption, although the N—N bond itself is strongly perturbed, as indicated, for example, by large shifts from the gas-phase N—N stretching frequency. On Group VIII metals dissociation can occur, but the process is slow, as exemplified by Fe surfaces, where the dissociative sticking probability is only about 10^{-7} (compared with 0.6 on W{100}). The chemisorbed dinitrogen state co-exists with the dissociated state at high coverages on Group VIIB and VIII transition metals. The molecular state itself has been characterized in both end-on and sideways bonded modes.

This review is structured in the following way. For each group of transition metals described above only the most extensively studied metals are chosen to exemplify the group, and the strength of the nitrogen–metal interaction increases (roughly) as the review proceeds. Thus, we start with palladium, ruthenium and copper and then deal with nickel, where the heat of adsorption for dinitrogen is greatest. The dinitrogen–copper interaction is very weak, but we deal with copper to review the data obtained by the adsorption of gaseous N atoms or excited-state N_2. The Cu—N bond formed by chemisorbed adatoms is strong, showing that dissociative adsorption of gaseous N_2 is prohibited by kinetics, and not by thermodynamics. Nitrogen on iron is reviewed because of its importance to the ammonia synthesis reaction, and also because it exemplifies a system where dinitrogen is the stable state at low temperatures, with a slow conversion to the dissociated state, which is only relatively weakly bound. Strong dissociative chemisorption, together with dinitrogen at low temperatures, is exemplified by the extensively studied N_2/W system.

As background material, in the remaining sections of this introduction we outline the relevant properties of dinitrogen, some organometallic complexes, and inorganic metal nitrides.

1.2 THE NITROGEN MOLECULE

There are many points of comparison to be made between the adsorbed states of dinitrogen and of the isoelectronic molecule carbon monoxide, and it is therefore useful to set out the orbital characteristics of each molecule. The different molecular orbital notations for homonuclear and heteronuclear diatomics can be a source of confusion. The higher symmetry ($D_{\infty h}$) of the former leads to the *g*, *u* classification of orbitals based on inversion through the centre of symmetry. Thus for N_2 the orbitals are classified as $1\sigma_g$, $1\sigma_u$, $2\sigma_g$, $2\sigma_u$, etc. For CO, however, where there is no *g*–*u* symmetry ($C_{\infty v}$), these orbitals are classified simply as 1σ, 2σ, 3σ, etc. An alternative (unifying) scheme is to label the orbitals by the atomic orbitals from which they are formed: $\sigma_g 1s$, $\sigma_u 1s$, $\sigma_g 2s$, etc. This notation is less confusing, but in the chemisorption literature the heteronuclear diatomic notation has become dominant, and this will also be used here.

Wave functions contour plots for the valence orbitals of N_2 are compared in Fig. 1 with the orbitals from CO [2]. While there are similarities, there are also very significant differences. In particular, the $\sigma_g 2p$ (5σ) orbital is heavily weighted to the C end of CO, and is the major factor in determining the bonding of molecular CO to metal atoms through the C atom at surfaces or in organometallic compounds, whereas in N_2 the $\sigma_g 2p$ (5σ) orbital is evenly weighted at both ends. Conversely, the $\sigma_u 2s$ (4σ) orbital on CO is weighted heavily to the O end of the molecule, while on N_2 it is evenly weighted and even has a somewhat larger spatial extent along the N—N axis than the $\sigma_g 2p$ orbital. The antibonding, unoccupied $\pi_g 2p$ (2π) orbital on CO, which is partially populated when bonded to metal atom(s), is heavily weighted to the C end, through which bonding to the metal occurs, whereas the spatial extent of this orbital on N_2 is somewhat less, with the weight evenly distributed.

This review will demonstrate that molecular N_2 is invariably less strongly adsorbed to metal surfaces than is CO. As pointed out by Horn et al. [2], the reason can be surmised from the comparison in Fig. 1.

For CO, the Blyholder bonding model [3] involves donation from the $\sigma_g 2p$ (5σ) CO orbital to the metal and backbonding into the $\pi_g 2p$ (2π) level. The extent of overlap with the corresponding N_2 orbitals is clearly less. In fact, since the $\sigma_g 2p$ (5σ) and $\sigma_u 2s$ (4σ) orbitals have about the same spatial extent along the N—N axis, when the molecule is terminally bonded to a metal atom (or atoms) overlap between metal orbitals and the $\sigma_u 2s$ and $\sigma_g 2p$ would be roughly equal. As pointed out by Bagus et al. [4, 5] these interactions

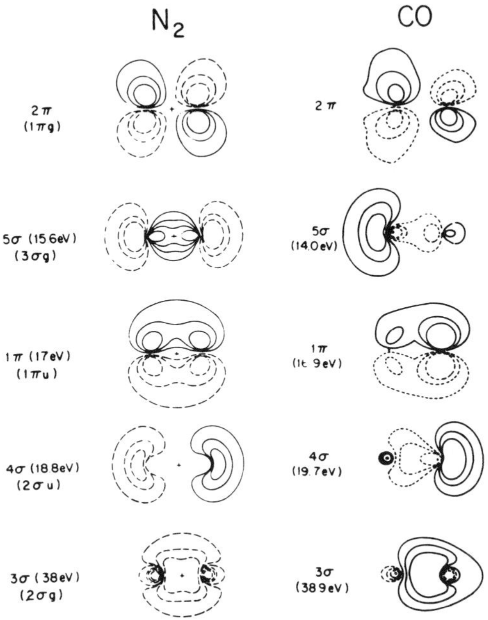

Fig. 1. Wave functions for N_2 and CO with contour values of 0.3, 0.2 and 0.1. The binding energies are given in parentheses. (From Horn et al. [2].)

break the symmetry of the N_2 molecule, forming two new σ states from the $\sigma_u 2s$ (4σ) and $\sigma_g 2p$ (5σ) states which resemble the lone pair 4σ (on the O end) and 5σ (on the C end) states on CO. The extent of the mixing is determined by the strength of the N_2—metal bond. Thus, as has been experimentally determined, *both* the $\sigma_g 2p$ (5σ) and the $\sigma_u 2s$ (4σ) orbitals should be shifted in energy on adsorption compared with gas-phase adsorption due to bonding. In contrast, with CO only the $\sigma_g 2p$ (5σ) orbital is shifted. On the other hand, the spatial extent of the 1π orbital in the direction orthogonal to the N—N axis is somewhat greater than that of the same orbital with respect to the C—O axis, and we might therefore anticipate cases where the side-on π bonded N_2 molecule would be more stable than the terminally bonded species.

According to Doyen and Ertl [6], terminal σ bonding of N_2 is dominated by the $\sigma_u 2s$ (4σ) orbital. Since this orbital is antibonding with respect to N_2, σ donation to the metal should cause an increase in the strength of the N—N bond. However, following Bagus, the mixing of the 4σ and 5σ states will tend to cancel this effect since the $\sigma_g 2p$ (5σ) is an N—N bonding orbital. We might expect, therefore, a relatively weak effect on the N—N bond strength when bonding occurs to a metal. The domination of σ bonding over 2π backbonding is supported strongly by recent studies of N_2 adsorption on Ru{001}, to be described in Sec. 4 of this review, and provides a major difference between CO and N_2.

We note here that the N—N bond length is 1.095 Å, its van der Waals radius is about 1.5 Å, and its dissociation energy is 9.764 eV (942 kJ mol^{-1}).

1.3 ORGANOMETALLIC DINITROGEN COMPLEXES

Metal—N_2 bonds in inorganic complexes are similar to those in metal carbonyls, although the bond strength is consistently weaker. Considerable interest has centred on these complexes in relation to their potential importance in nitrogen fixation, parallel to the biological conversion of N_2 to NH_3 by N-fixing bacteria. Examples [7, 8] of linearly bonded mononuclear complexes, with their observed N—N stretching frequencies are: $Ru(NH_3)_5N_2^{2+}$ (2105–2169 cm^{-1}, depending on anion); $Os(NH_3)_5N_2^{2+}$ (2010–2060 cm^{-1}, depending on anion); and $Ir(PPh_3)_2N_2Cl$ (2190 cm^{-1}). Since the gas-phase N—N stretch is centred at 2331 cm^{-1}, the observed frequency shifts in these complexes is very large (140–220 cm^{-1}) compared with linearly bonded carbonyl complexes where shifts of 50–150 cm^{-1} are usually observed.

There are also several examples of dinitrogen complexes in which the N_2 molecule bridges between two metal atoms, as, for example, in $[(NH_3)_5RuN_2Ru(NH_3)_5]^{4+}$. The bond lengths (in Å) shown for this structure may be

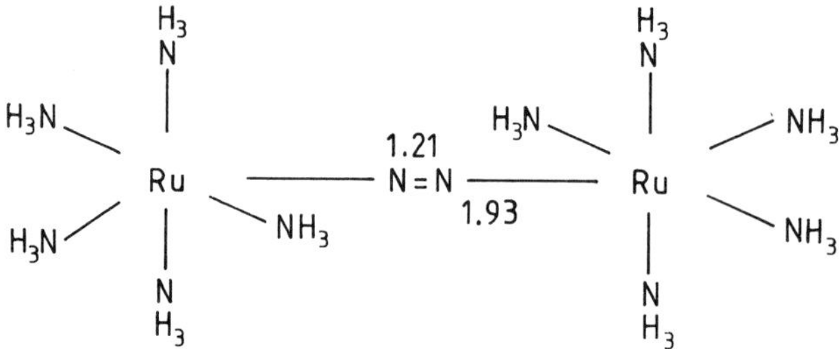

compared, for example, with the values of 1.10 Å for N—N and 1.78 Å for Co—N in $CoH(N_2)(PPh_3)_3$; the N—N bond length is almost identical to that in the free molecule.

Dinitrogen metal complexes have also been formed by matrix isolation through co-condensation of metal atoms with N_2, or N_2 and Ar, at 10 K [9–11]. In this way, the compounds $Ni(N_2)_n$, $Pd(N_2)_m$, and $Pt(N_2)_m$, where $n = 1$–4 and $m = 1$–3, have been formed and identified using infrared spectroscopy. The structures are analogous to metal carbonyls. For example, of the $Ni(N_2)_n$ complexes, $Ni(N_2)_4$ is the most stable, and, in an Ar matrix, it is a regular tetrahedral molecule with end-on bonded N_2. NiN_2 and $Ni(N_2)_2$ are linear, and $Ni(N_2)_3$ is trigonal planar (D_{3h} symmetry). The observed N—N stretching frequencies are listed in Table 1. In all cases the assignment involves end-on bonding, with the exception of $Pt(N_2)_2$ which was assigned as sideways-on bonded, i.e.

N⋯Pt(N≡N), with the second N₂ bonded sideways-on

TABLE 1

N—N stretching frequencies (IR active) in the compounds $Ni(N_2)_n$, $Pd(N_2)_m$ and $Pt(N_2)_m$ [9–11]

	ν_{N-N} (cm^{-1})		
	Ni	Pd	Pt
$M(N_2)$	2088	2213	2173, 2065
$M(N_2)_2$	2106	2234	2104, 2095, 2079
$M(N_2)_3$	2134	2241	2217, 2206
$M(N_2)_4$	2174		

1.4 INORGANIC METAL NITRIDES

The term "refractory hard metals" is used to describe collectively the borides, nitrides and silicides of the transition metals of the 4th to 6th groups of the periodic table. These compounds are characterised by high melting points, hardness and brittleness as well as by metallic, electrical and thermal conductivities (of the same order as the transition metals). This combination of properties is unusual, the former usually being associated with covalent semiconducting compounds such as SiC. However, the conductivity of the nitrides and carbides of the transition metals is not very different from that of the pure metal; in the case of Ti, the conductivity of the metal ($21.7 \times 10^6\, ohm^{-1} m^{-1}$) is actually less than that of the nitride ($46 \times 10^6\, ohm^{-1} m^{-1}$) [12]. The metal–metal distance in the nitrides shows an expansion over the parent metal [13] (e.g. for Ti and V from 2.93 and 2.63 Å to 2.99 and 2.92 Å, respectively). Interestingly, the lattice expansion is least for the oxides (TiO, 2.95 Å; VO, 2.86 Å), whereas the conductivity is generally lower for the oxides. In contrast to the parent metals, which have positive Hall constants and therefore exhibit predominantly hole conduction, the hard metals all have negative Hall constants, with the exception of VN which is slightly positive [14]. We note also that very high melting points [15, 16] have been recorded for these compounds (e.g. those for TiN, TaN, HfN and ZN are all around 3500 K; CrN is 1900 K). The band structure for the hard metals with rock salt lattices has been calculated by Bilz [17] using the tight binding method. Compared with the parent metal, M—M bonding is apparently increased at the expense of M—X bonding, which may arise from metalloid *p* electrons in the metallic *d* band. A minimum in the density of states occurs at a position corresponding to 6 *d* electrons per atom between TiN and TiC.

The available enthalpies of nitride formation for all metals are collected together in Table 2 [18]. While the transition metal nitrides in the 4th to 6th groups of the periodic table are very stable, those in Groups IB and IIB are very unstable, and Group VIIIB metals lie in between. The relatively low bond energy in FeN is believed to be the major factor in the activity of Fe as an ammonia synthesis catalyst.

References pp. 124–129

TABLE 2

Standard heats of formation of bulk nitrides (kJ/mol^{-1} N, 298 K), with approximate decomposition temperatures in parentheses [18] (Ex = explosive)

IA	IIA	IIIB	IVB	VB	VIB	VIIB	VIII	VIII	VIII	IB	IIB	IIIA	IVA	VA	VIA	VIIA
H_3N − 46																
Li_3N − 197	Be_3N_2 − 285 (> 600 K)											BN − 134	$(CN)_2$ + 155	N	ON + 92	F_3N − 109
Na_3N − 151 (420 K)	Mg_3N_2 − 230 (1000 K)											AlN − 243	Si_3N_4 − 188	PN − 84	S_4N_4 + 134 Ex	Cl_3N + 230 Ex
K_3N 84	Ca_3N_2 − 218	ScN − 285	TiN − 305	VN − 172 (> 2600 K)	CrN − 121	Mn_5N_2 − 117 (> 1500 K)	Fe_4N − 12 (710 K)	Co_3N	Ni_3N 0	Cu_3N + 75 (820 K)	Zn_3N_2 − 12	GaN − 105	Ge_3N_4 − 17 (720 K)	(As)	Se_4N_4 + 176	Br_3N + 335
Rb_3N + 180	Sr_3N_2 − 197	YN − 301	ZrN − 343 (> 3300 K)	NbN − 247 (> 2600 K)	Mo_2N − 71	TcN	(Ru)	(Rh)	(Pd)	Ag_3N + 285 Ex	Cd_3N_2 − 79	InN − 21	Sn_3N_4 (< 630 K)	(Sb)	Te_3N_4	I_3N + 272
Cs_3N + 314	Ba_3N_2 − 184	LaN − 301	HfN − 326	TaN − 243 (> 3300 K)	WN − 71	Re_2N	(Os)	(Ir)	(Pt)	Au_3N Ex	Hg_3N_2 + 8	Tl_3N + 84	(Pb)	BiN		
					UN − 335											

2. Nitrogen on palladium

2.1 ADSORPTION OF DINITROGEN

The earliest N_2/Pd study was that by Van Hardeveld and Van Montfoort [19] who used electron microscopy and transmission infrared to study nitrogen adsorption on alumina-supported palladium (for comparison with nickel and platinum) to investigate particle size effects. Various Pd/Al_2O_3 samples giving a range of metallic particle sizes were studied and revealed an infrared band at $2260\,cm^{-1}$ for a limited particle size distribution. By modelling the geometry of the metallic particles and proposing that the large frequency shift of the band from the gas-phase value of $2330\,cm^{-1}$ would require a "high field" adsorbate site, Van Hardeveld and Van Montfoort concluded that the dinitrogen species was adsorbed on B_5 sites occurring at the ridges of their "incomplete cubo-octahedron" models. These sites are characteristic of, for example, sites on a {110} f.c.c. surface. The adsorption behaviour described above was observed to be similar for Pd, Pt and Ni systems, and on the basis of its non-specificity it was concluded that the dinitrogen species was physisorbed.

King [20] studied thermal desorption of nitrogen from deposited Pd films and reported two distinct binding energy states, γ_1 and γ_2 ($-\Delta H^{\gamma_1} \leqslant 30\,kJ\,mol^{-1}$; $-\Delta H^{\gamma_2} \approx 25$–$40\,kJ\,mol^{-1}$). The initial sticking probability at 78 K was 0.67. By comparison of the s vs. θ behaviour for Pd with Ni films it was deduced that the adsorbed N_2 layer was mobile in the case of Pd at 78 K, and, in agreement with van Hardeveld and van Montfoort, it was concluded that the nitrogen species is physisorbed under the experimental conditions.

Kunimori et al. [21] used AES to study NH_x species on a Pd ribbon and found, in agreement with the previous work, that N_2 does not adsorb at room temperature. In the presence of an electron beam, however, a nitrogen species *was* adsorbed which was attributed to the dissociatively adsorbed species, $N_{(ad)}$. On the adsorption of, and subsequent reaction with, hydrogen this yields $NH_{x(ad)}$.

2.2 ADSORPTION OF N_2 ACTIVATED IN THE GAS PHASE

In the most recent study of the N_2/Pd system on a polycrystalline surface, Miyazaki et al. [22] used isotopic nitrogen and thermal desorption spectroscopy (TDS) to further demonstrate the phenomena occurring with activated nitrogen. Activation of an N_2 ambient with a W filament at 2000 K gave an adsorbed layer with a three-peak TDS spectrum (T_p at 493, 633, and 973 K). Isotopic studies revealed the two highest temperature states to be dissociative, while the state desorbing at 493 K was attributed to a molecularly adsorbed species. When the film was heated to desorb all except the highest T_p state and then this surface exposed to N_2, the adsorption state at 493 K

References pp. 124–129

was recreated. Thus the atomic N state with $T_p = 973$ K appears to promote non-activated molecular nitrogen adsorption. Further evidence to support this was drawn from the fact that the coverage of molecular nitrogen adsorbed in this way was always in proportion to the coverage of the atomic state.

Horn et al. [2] have studied the adsorption of N_2 on Pd{111} by angle-resolved ultraviolet photoelectron spectroscopy. The results shown in Fig. 2 for an adsorption temperature of 45 K are particularly significant in demonstrating that the molecule is physisorbed to the surface. The three peaks in the spectra are labelled using the gas-phase (CO) notation. They have the relative intensities seen in the gas phase, and their initial state energies are rigidly shifted upward by 1.3 eV compared with gas-phase N_2. The relative intensity of the three peaks is apparently independent of measurement geometry, indicating a nearly random molecular orientation. Clearly, the N_2 orbitals are not involved in chemical bond formation with the surface.

The vibrational high-resolution electron energy loss spectroscopy (HREELS) technique, has been applied in two studies of nitrogen on Pd{110} by Kuwahara et al. [23, 24]. At low temperatures (120 K) a molecular N_2 species giving a sharp (2 × 1) LEED pattern was observed, with $\nu_{Pd-N_2} \approx 242\ cm^{-1}$ and $\nu_{N-N} \approx 2242\ cm^{-1}$. A band at 467 cm^{-1} was assigned to the first overtone of ν_{Pd-N_2}. The bonding here appears to be stronger than on Pd{111}, and an ARUPS study would prove fruitful as a check on the nature of the metal—surface bond. The dipole nature of the modes observed was confirmed by off-specular measurements and comparison of the intensity dependence on primary beam energy against that predicted by dipole theory calculations. The adsorbed N_2 species, which is fully desorbed by 150 K, was assigned to a linear on-top configuration. Using AES it was also shown that, by activation of the gas phase, the surface nitrogen coverage could be increased fourfold over the saturation adsorption of thermal, gaseous dinitrogen.

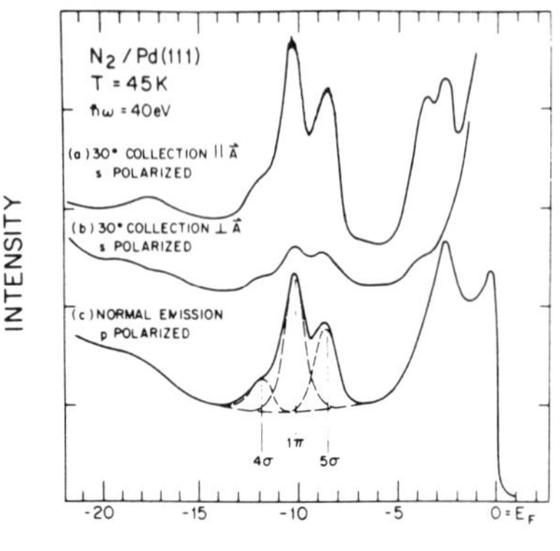

Fig. 2. Angle-resolved photoelectron spectra of N_2 on Pd{111} after an exposure of 2×10^{-6} Torr s. (From Horn et al. [2].)

In their second study Kuwahara et al. [24] followed up on the activated adsorption process mentioned above, using an ion gun to deposit nitrogen (N_2 and N both excited and ionised) species on Pd{110}. HREELS was then used to observe and assign the various adsorbed nitrogen species, both atomic *and* molecular in nature. Various surface preparations were made by varying (N_2 + N) doses, annealing temperatures, etc. The results of this work were interpreted as evidence for the promotion of non-activated N_2 adsorption by atomic nitrogen. It should be noted, however, that at no stage was an $N_{(ad)}$ precovered surface prepared and then ambient N_2 allowed to adsorb as proof of this idea. Instead, the (N_2 + N) adsorption process was carried out using an ion gun. This procedure raises questions about the validity of their proposal since the ion gun beam could contain many excited species (i.e. in an activated form) which could adsorb in a molecular form.

3. Nitrogen on copper

3.1 ADSORPTION OF DINITROGEN

Adsorption from gaseous N_2 does not occur on copper surfaces at room temperature [25]. The observation by Lee and Farnsworth [26] of a layer of "molecular gas" on Cu which could be decomposed by an electron beam has been more recently attributed [27] to gas-phase activation in the electron beam followed by adsorption of the activated N_2.

Schmeisser et al. [28] evaporated Cu into a nitrogen matrix at $T < 40$ K to form N_2—Cu clusters. Using photoemission, they found that the features due to N_2, at 9.7, 11.0, and 12.8 eV corresponding to the $\sigma_g 2p$ (5σ), $\pi_u 2p$ (1π), and $\sigma_u 2s$ (4σ) molecular orbitals are all rigidly shifted due to final-state screening by Cu electrons, but showed no bonding shift. As with N_2/Pd{111}, the species is physisorbed to the Cu cluster. The authors tentatively conclude that the configuration is not entirely random, with some preference being demonstrated for a side-on geometry.

3.2 ADSORPTION OF N_2 ACTIVATED IN THE GAS PHASE

Activation of gas-phase nitrogen using a hot filament [29, 30] or an ion gun [26, 27, 30–35] leads to the formation of a strongly bound adlayer composed of N adatoms. Tibbetts [27] concluded that the sticking probability for gaseous atomic nitrogen is unity on a clean copper surface.

The c(2 × 2) LEED pattern observed by Lee and Farnsworth [26] on Cu{100} is attributed to N adatoms. The structure is well defined after annealing to 500 K, and is removed by desorption after heating to 800–900 K. Several LEED intensity analyses have been performed on this structure, and all agree that the N adatoms occupy the fourfold hollow sites. Earlier quasi-kinematical calculations [36, 37] indicated a vertical interlayer spacing d_z between adsorbate and the first Cu layer of 1.45 Å, but a more recent multiple scattering analysis [32] indicates that the two layers are virtually coplanar. The N adatoms are effectively 5-fold coordinated, and relaxations are reported in the first and second Cu layers. Franchy et al. [38] studied the phonon dispersion of this structure and evaluated d_z as 0.6 Å by comparing their data to a slab calculation. The Cu layer relaxations introduced by Zeng et al. [32] probably explain the discrepancy, and at this stage the coplanar structure is favoured. The system has been subjected to a further HREELS study by Mohamed and Kesmodel [31] who report bands at 155 and 324 cm^{-1} in the specular direction and at 207 and 750 cm^{-1} off-specular. The peaks at 324 and 750 cm^{-1} are attributed, respectively, to the perpendicular and parallel modes of the Cu—N stretching vibration. The peaks at 155 and 207 cm^{-1} are tentatively attributed to surface resonances.

X-Ray photoelectron spectroscopy studies variously place the N_{1s} binding energy for N on a Cu film at 397.0 eV [39] and for Cu{100} c(2 × 2)–N at 397.3 eV [36, 40] and 396.3 eV [29]. The latter is probably the more accurate evaluation. UPS spectra from this structure yielded two adsorbate-induced features [36, 40], a small feature at 1.3 eV and a larger, broad peak at 5.8 eV below the Fermi level: one feature appears above the Cu *d* band and one below. The Auger spectra from this surface, previously attributed [34] to interatomic transitions involving electrons localized on the substrate, were satisfactorily accounted for as intra-atomic transitions. Several theoretical attempts have been made to provide an interpretation of the UPS data. Anderson [41], using an approximate molecular orbital calculation based on metal clusters of up to 32 atoms, was able to reproduce the two experimentally observed adsorbate peaks. Smith et al. [42] reproduced the data using a self-consistent electronic structure calculation, and later Yu and Whiting [43] used an SCF-Xα scattered wave method on a Cu_5N cluster to reproduce the results. The structure above the Cu *d* band arises from the 2*p* electrons on the N adatoms, while the structure below the *d* band is primarily attributed to a spreading of the Cu *d* band local density of states in the surface Cu plane, shifted downward by interaction with the N electrons. The broad peak below the *d* band thus has a contribution from both the N 2*p* band and the alteration of the Cu 3*d* band. The calculations do, however, need to be treated with caution in view of the more recent LEED structural study [32] described above, since the position of the N adatom in the fourfold hollow sites and interlayer relaxations were not accounted for.

The adsorption of activated nitrogen on Cu{111} produces a complex LEED pattern, corresponding to a slightly distorted Cu{100} c(2 × 2)-N structure [33]. Thus it was concluded by Higgs et al. [33] that the N adsorbate induced a restructuring of the Cu{111} surface. It is pointed out that the {100} c(2 × 2) structure is similar to the {100} plane of the bulk nitride Cu_3N. At higher coverages of N adatoms, the surface becomes disordered. However, Mohamed and Kesmodel [31] subsequently concluded that the vibrational spectrum of N/Cu{111} was quite different from that of N/Cu{100}, which does not support the idea of a N/Cu{111} facetted structure. The HREELS data of Higgs et al. [33] showed bands at 266 and 403 cm^{-1}; the latter is attributed to the perpendicular Cu—N mode, and the former to a surface phonon. By contrast, the perpendicular mode is reported at 324 cm^{-1} on Cu{100} [31].

Ferrer and Rojo [44] studied the N/Cu{110} system by analysing changes in Auger lineshapes induced on stepwise heating of the surface. They identified three distinct phases. At 300 K, following plasma deposition of activated N_2, a disordered surface is produced. Heating to 600 K was believed to induce migration of N from subsurface implantation sites to an ordered surface layer. On heating to 650–700 K, further changes are controversially attributed to a surface recombination of N adatoms to produce a strongly bound dinitrogen adsorbed species. Heating to temperatures above 700 K resulted in complete desorption of the nitrogen adlayer. Heskett et al. [45] have used a variety of techniques to investigate the N/Cu{110} system. Activated nitrogen adsorption at 300 K produces a disordered overlayer, with a perpendicular Cu—N stretching mode at 429 cm^{-1} and a phonon mode at 160 cm^{-1}. Disordering is evident from the LEED pattern and the width of the bands in the vibrational spectra. Desorption spectra show a broad feature at 500 K and a second broad feature at 750 K followed by a sharp spike at 760 K (Fig. 3). Annealing the surface to 650 K to remove the first broad peak in the desorption spectrum yields a well-ordered (2 × 3) LEED pattern with a Cu—N stretching mode at 405 cm^{-1} and a phonon mode at

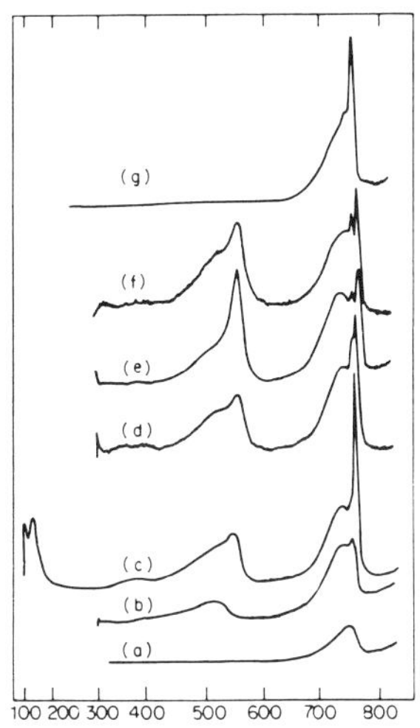

Fig. 3. Thermal desorption spectra (a.m.u = 14) of N/Cu{110}. Curves (a), (b), (d), (g): adsorption at room temperature; curve (c): adsorption at 95 K. Coverages quoted in integrated current on the crystal during nitrogen deposition by sputtering: (a) 114 μC; (b) 432 μC; (c) 1470 μC; (d, e) 2400 μC; (f) 4134 μC; (g) 1470 μC, previously annealed to 720 K. (From Heskett et al. [45].)

184 cm^{-1}. The interpretation of the data is similar to that of Higgs et al. [33] for N/Cu{111} and Kirby et al. [30] for N/Cu{210}: annealing produces a N-induced reconstruction of the Cu surface, with formation of a bulk-like (i.e. Cu_3N) surface copper nitride. The reversion of the surface to (1 × 1) structure as the N_2 is desorbed at 760 K produces the explosively sharp desorption feature (Fig. 3). It is pointed out that bulk Cu_3N decomposes at 720 K.

An early study of adsorption on terrace–step crystal surfaces was conducted by Perdereau and Rhead [35] who studied activated nitrogen adsorption on several vicinal surfaces of Cu from the {100} pole along [001] and [01$\bar{1}$] zones. Stepped surfaces with a terrace width of about 6 atoms gave rise to a simple overlayer structure with N adatoms adsorbed in a c(2 × 2) array on the {100} terraces. On surfaces with a higher step density, the patterns were attributed to a modified form of the c(2 × 2) layer. Kirby et al. [30] subsequently examined the adsorption of activated nitrogen on Cu{210}, and reported a complex series of LEED patterns as a function of coverage and annealing temperature. The analysis implied the formation of {100} facets with a c($11\sqrt{2} \times \sqrt{2}$) R45°–N overlayer at 290 K. Annealing at 520 K produced {110} facets with a p(2 × 3) overlayer, whilst the clean surface was restored after heating to 600 K. (This temperature is anomalously low compared with all other studies reported on Cu surfaces.)

4. Nitrogen on ruthenium

Several studies have been made of dinitrogen adsorption on Ru{001} utilizing thermal desorption spectroscopy, HREELS, photoemission and reflection–absorption IR spectroscopy [46–49, 55]. During adsorption at 85 K an infrared-active band is first observed at 2209 cm^{-1}, shifting downwards with increasing coverage to 2189 cm^{-1} after 4 Langmuirs (L) exposure [47, 49]. This is in the opposite direction to that expected from dipole coupling (as also observed for CO on Cu{111}, for example [50]), implying a relatively large downward shift due to electronic interactions (a chemical bonding effect) between neighbouring molecules in the crowded adlayer. Desorption from this state produces a single first-order thermal desorption peak (E_{des} = 42 kJ mol^{-1}, ν = 10^{16} s^{-1}) and a saturated coverage of θ_{N_2} = 0.35 [49]. At lower adsorption temperatures (78 K), a second state is populated (E_{des} = 28 kJ mol^{-1}) giving a total saturation coverage of θ_{N_2} = 0.58 [46, 47]. LEED results obtained at 85 K demonstrate the onset of a weak ($\sqrt{3} \times \sqrt{3}$) R30° pattern at 2.0 L exposure to N_2, which sharpens up at 2.5 L. Subsequent exposure reduces the intensity of the fractional order beams, and the overlayer becomes disordered. At lower adsorption temperatures a ($2\sqrt{3} \times 2\sqrt{3}$) pattern is observed at higher coverages, corresponding to adsorption of the second N_2 state. The work function is decreased by 550 mV after 2.5 L exposure at 85 K [49], in contrast to CO on Ru{001}, where the work function increases by 500 mV on formation of the ($\sqrt{3} \times \sqrt{3}$) R30° structure. This is attributed to a net charge transfer from the molecule to the surface on adsorption, characterized by σ donation and the *absence* of significant $\pi_g 2p$ ($2\pi^*$) backbonding. This work strongly emphasises the difference between N_2 and CO bonding on Ru.

Considerable clarification of the bonding of dinitrogen to Ru{001} has been obtained by co-adsorption studies [49]. Exposure of the surface to produce low coverages of oxygen adatoms prior to N_2 adsorption shifts the N_2 desorption peak from 130 K on the clean surface to 160 K on the surface with $\theta_0 = 0.25$, corresponding to an increase in adsorption heat of $\sim 6\,\mathrm{kJ\,mol^{-1}}$, and the N_2 coverage at 85 K is increased twofold. There is a corresponding shift in ν_{N-N} from $2188\,\mathrm{cm^{-1}}$ on clean Ru{001} to $2239\,\mathrm{cm^{-1}}$ in the presence of O adatoms, with a decrease in band intensity. With co-adsorbed K, on the other hand, the trend is reversed. Thus, with $\theta_K = 0.05$, the desorption peak maximum is reduced to 115 K, corresponding to a heat of adsorption decrease of $\sim 4\,\mathrm{kJ\,mol^{-1}}$, and the coverage at 85 K is reduced. At this coverage, the N—N stretching frequency is reduced to $2150\,\mathrm{cm^{-1}}$. Nitrogen therefore interacts repulsively with K adatoms and attractively with O adatoms. Consistent with the observed work function change, this is exactly opposite to the effect of K and O adatoms on CO adsorption. Thus, these results strongly support the notion that N_2 chemisorption is dominated by σ donation to the metal on Ru, in contrast to the synergistic model of σ donation and π^* back-donation for CO adsorption. The observed N—N frequency shifts induced by the co-adsorbates are entirely consistent with the effect of the electrostatic field produced by positively (K) and negatively (O) charged adatoms on neighbouring N_2 (upright) adsorbed species.

The conclusion that the adsorption of end-on dinitrogen on Ru is dominated by σ donation with negligible contribution from π^* backbonding contrasts with theoretical models [51–54]; in particular, the calculations of Bagus and co-workers [51, 52] suggest that π backbonding is dominant in these systems, and that σ bonding may even produce a net repulsion. However, this conclusion is not entirely satisfactory. The observed shift in the N—N stretching frequency upon adsorption into an end-on, on-top configuration, with respect to the gas-phase frequency, is generally larger than the shifts observed for CO despite the theoretical conclusion that the $\sigma_u 2s$ orbital is even antibonding with respect to N_2, as discussed in Sect. 1.2 above.

5. Nitrogen on nickel

5.1 INTRODUCTION

Nitrogen adsorbed on nickel represents a weakly and molecularly chemisorbed system. Some of the earliest work on nitrogen adsorption on nickel

References pp. 124–129

was carried out on supported nickel by Eischens and Jacknow [56] who reported infrared spectra exhibiting a strong bond at 2202 cm^{-1} which was assigned to a weakly chemisorbed molecular species, Ni—N≡N^{+} with the N—N bond perpendicular to the surface. Subsequently, Van Hardeveld and Van Montfoort [19, 57] claimed to demonstrate, using electron microscopy and transmission infrared spectroscopy, that the IR-active nitrogen existed as a *physisorbed* species adsorbed only on particles ranging in size from about 15 to 70 Å. On the basis of geometric modelling of small nickel particles, Van Hardeveld and Van Montfoort concluded that the infrared-active N_2 species was adsorbed in special five-coordinate B_5 sites which are more prevalent in catalyst particles of 15–70 Å diameter. Such B_5 sites are similar to those found on f.c.c. {110} surfaces. Using TDS measurements, King [20] refuted the idea of physisorbed nitrogen species on B_5 sites and instead identified three distinct binding sites for nitrogen on vacuum-deposited nickel films; a γ_1 state with an adsorption heat below 29.5 kJ mol^{-1}, a γ_2 state with a heat of adsorption ranging between 25 and 42 kJ mol^{-1}, and finally a γ_3 state ranging from 37 to 59 kJ mol^{-1}, the range of energies for each state being attributed to lateral interactions. By annealing the films to different temperatures, King assigned the γ_1 state to physisorbed N_2 on the {111} planes, the γ_2 state to physisorbed N_2 on {100} planes and the γ_3 state to the infrared-active N_2 species *chemisorbed* on the {110} and higher index planes. It should be noted that the assignments to specific crystal planes remain somewhat tentative. However, transmission IR experiments [58] on CO and N_2 adsorption on vacuum-deposited films also suggest that the γ_3 infrared-active N_2 species is a weakly chemisorbed state which only exists on those Ni films giving rise to νCO frequencies down to 1750 cm^{-1}, i.e. films possessing high coordination sites.

An initial sticking probability of 0.56 was deduced at 78 K on smooth Ni films, and, unlike nitrogen on Pd films, the sticking probability behaviour suggested that the adsorbed N_2 layer on Ni was mobile [20]. Nieuwenhuys [59] has surveyed work function changes associated with N_2 adsorption on Group VIII metals and reports a $\Delta\phi$ of -0.17 eV upon adsorption of N_2 at 77 K on room temperature-annealed Ni films.

Brundle [60, 61] has reported XPS and UPS data, summarised in Table 3, for an evaporated polycrystalline Ni film saturated with N_2 at 77 K, from which he, too, concludes that the molecule is weakly and molecularly adsorbed at this temperature. It should be noted at this point that the majority of the N_2/Ni literature adopts the molecular orbital classification used for CO (Sect. 1.2). Therefore, for clarity, this section on N_2 on nickel will also use the CO MO classification rather than the longer notation used elsewhere in this chapter. The assignments listed in Table 3 will be discussed in greater general detail in the following sections describing N_2 adsorption on single-crystal surfaces. However, it should be noted that the two peaks in the N_{1s} spectrum result from screened and unscreened final state effects. Ranga Rao et al. [62] report that the ratio of the unscreened (US) peak to the screened

TABLE 3

UPS (valence level) and XPS (core level) binding energies (eV) for free N_2 and N_2 adsorbed on Ni

	$(5\sigma)^s$	$(1\pi)^s$	$(4\sigma)^s$	$(5\sigma)^u$	$(1\pi)^u$	$(4\sigma)^u$	$(N1s)^s$	$(N1s)^u$
Free N_2 [72, 73]				15.5	16.8	18.6		409.9
N_2/polycr. Ni [60–62][a]	8.3	8.3	12.7				400.6	405.7
N_2/Ni{100} [74][b]	7.6	7.6	12.4				400.2	405.3
N_2/Ni{100} [67][c]	8.3	8.1	12.8	11.4	12.0			
N_2/Ni{111} [87][b]	8–9	8–9	13				401	405.5
N_2/Ni{110} [2, 96][c]	8.1	7.8	11.8				401.7	406.7
Ni_3N_2, cluster calc. [86]	13.6	12.6	18.1	16.3	17.2	20.1		
NiN_2 ΔSCF calc. [74]	10.4	10.5	15.7	17.1	16.8	21.9		

s = screened final state.
u = unscreened final state.
[a] Referenced to E_F.
[b] Referenced to E_F with work function correction of 5.5 eV.
[c] Referenced to E_F with work function correction of 5.0 eV.

peak (S) decreases as the Ni film is warmed from 95 K. Whereas the US peak is much more intense than the S peak at 95 K, at 150 K the US peak intensity decreases to below that for the S peak. These data are interpreted in terms of two types of nitrogen species existing on the surface, a weakly bound species, possibly existing on {111} planes, which desorbs at $T \geqslant 140$ K and a more strongly bound species which persists to higher temperatures. Such an assignment complies with that previously put forward by King [20].

Just as demonstrated in Sect. 4 for Ru, the bonding of nitrogen to Ni has been clarified by co-adsorption studies with Ba, Al, O and Cl [62]. Co-adsorption with the electropositive species results in increased coupling between the Ni and the N_2 $2\pi^*$ orbitals, leading to a decrease in the νN—N frequency from 2225 to 1945 cm^{-1}, while the electronegative species leads to a decreased coupling between the Ni and the $N_2 2\pi^*$ orbitals. Thus, in the former case the US to S peak intensity ratio suffers a dramatic reduction while the latter situation leads to an increase in the US to S intensity ratio. The US to S intensity ratio is a sensitive indicator of the coupling between a metal and the $N_2 2\pi^*$ orbitals. This point will be discussed at greater length in the following sections.

Finally, Schmeisser et al. [28] have obtained UPS data for nitrogen adsorption on evaporated Ni films and on copper clusters at 7 K. In addition to a condensed phase of N_2 which desorbs at $\leqslant 35$ K, another species is thought to exist on the surface up to a temperature of 42 K, giving rise to peaks which are similar to those obtained for condensed phase N_2 but rigidly shifted down by $\simeq 1$ eV. Comparing the relative intensities and halfwidths of the 5σ, 1π and 4σ bands obtained for this species with those obtained for condensed-phase N_2, the authors assign these bands to a new side-on N_2 species which may have a specific interaction with the surface. Although, the UPS data of Schmeisser et al. [28] provide scant evidence for such an assignment, it is interesting, nevertheless, to note such side-on species have been identified at low temperatures on iron.

5.2 NITROGEN ADSORPTION ON Ni{100}

A range of techniques, e.g. LEED, UPS, XPS and NEXAFS, have been deployed in investigating the N_2/Ni{100} system. As a result a number of structural, thermodynamic and electronic characteristics of this system have been determined, each of which will be discussed in turn in the following sections.

5.2.1 Structural studies

Grunze et al. [63] have reported that at 84 K, the adlayer orders between $\theta = 0.25$ and 0.5, producing a diffuse c(2 × 2) LEED pattern with estimated domain sizes of 17 Å. By annealing the surface in a N_2 ambient, the c(2 × 2) pattern sharpens between 120 and 128 K, leading to larger domain sizes of $\simeq$ 34 Å, this conversion from the diffuse state being kinetically hindered. The ordered c(2 × 2) pattern is assumed to correspond to $\theta = 0.5 \equiv 8.05 \times 10^{14}$ molecules cm^{-2}. XPS measurements [64–66], ARUPS [67] and nitrogen K-edge NEXAFS [68] studies clearly indicate that the dinitrogen molecule is adsorbed with its molecular axis perpendicular to the Ni{100} surface. For such an adsorption geometry, the inter-molecular spacing between N_2 molecules in a c(2 × 2) structure would be 3.52 Å, which corresponds closely to the potential minimum distance of 3.5 Å for two parallel-oriented N_2 molecules [69].

5.2.2 Thermodynamic measurements

Using secondary electron emission measurements, Grunze et al. [63] have recorded the adsorption curve for N_2/N{100} at 82 K in a N_2 pressure of 6×10^{-8} torr. This adsorption curve suggests a high sticking coefficient of $s \simeq 1$ up to temperatures of 117 K, above which desorption becomes a competing process. Below 117 K the sticking coefficient is constant with θ, implying the existence of a possible precursor state to chemisorption. However, XPS investigations carried out at 20 K [70] discovered no evidence for such a precursor, suggesting no appreciable barrier to adsorption exists for the linearly bonded configuration.

The isosteric heats of adsorption as a function of θ are shown in Fig. 4 [63], and reveal that the initial heat of adsorption of 44 kJ mol^{-1} quickly decreases to $\simeq$ 38 kJ mol^{-1} after $\theta \simeq 0.1$–0.2, thereafter remaining constant until $\theta = 0.49$, when a further sharp drop to 25 kJ mol^{-1} is observed. The initial drop in q_{st} is attributed to preferential adsorption on defect sites with a higher heat of adsorption, while adsorption on the ordered {100} plane only proceeds after $\theta \geqslant 0.1$. As the c(2 × 2) structure develops, the heat of ad-

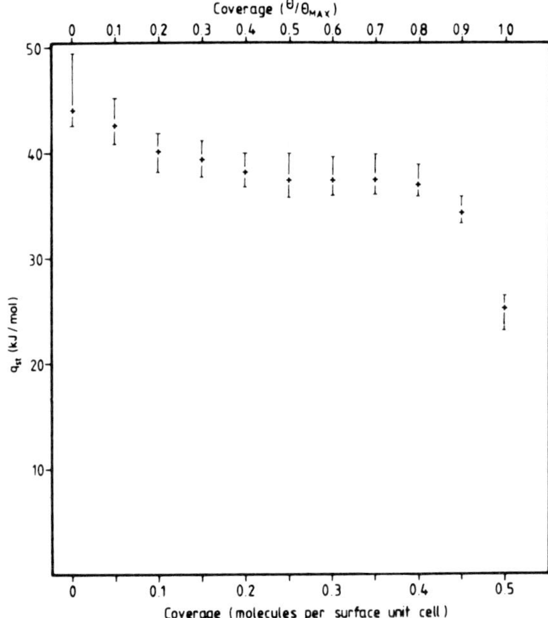

Fig. 4. The isosteric heat of adsorption of N_2 on Ni{100}. (From Grunze et al. [63].)

sorption remains approximately constant, but drops substantially at saturation, presumably due to the presence of domain boundaries [63].

Figure 5 [63] compares the integral experimental entropy, S_{ad}, of the adsorbed layer with calculated integral entropies derived from statistical

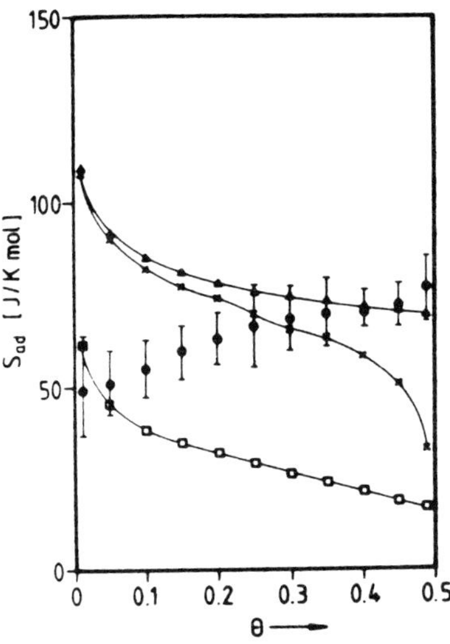

Fig. 5. Comparison of the integral molar entropy S_{ad} of the adsorbed layer (circles with error bars) with calculated entropy values: (□), S_{loc}; (▲), S_{mob} for an ideal 2D gas; (×), S_{mob} for a Volmer gas. (From Grunze et al. [63].)

models. At low coverages ($\theta < 0.1$), when adsorption is considered to occur on defect sites, good agreement is obtained between the experimental entropies and those calculated for localised adsorption. However, with increasing coverage, the experimental values deviate steadily away from the calculated localised values until around $\theta \simeq 0.3$, when ordered domains of 35 Å diameter are being established on the surface, the experimental entropy approaches values calculated for a non-localised Volmer gas. For $\theta > 0.35$, the observed entropy values now approach those calculated for a non-localised ideal gas (where the eigenvolume of the gas is neglected), which obviously represents a physically unrealistic situation. Grunze et al. [63] prefer instead to attribute the steady increase in experimental entropy to the softening or addition of vibrational or rotational modes in the adsorbed layer, for example, arising from antiphase domain boundary regions in which the compressed molecules fluctuate with low frequency and large amplitude perpendicular to the surface in order to relieve local steric repulsions. Such an explanation has been proposed previously to explain anamolously high entropy values [71], but as yet remains speculative since no direct evidence for the existence of such soft vibrational modes exists. Furthermore, as Grunze et al. point out [63], other possibilities for such an entropy increase exist, such as minor reconstruction of the surface upon adsorption.

5.2.3 Electronic properties

The experimentally determined valence and core level I.Ps for nitrogen adsorbed on Ni{100} [74] are shown in Table 3 and will be discussed in turn.

(a) Core level structure

The characteristic two-peak nitrogen 1*s* spectrum observed for N_2/Ni{100} [65, 70, 74, 75] is depicted in Fig. 6. This type of spectrum has previously been interpreted in terms of two different adsorption sites of molecular nitrogen [60] or in terms of two inequivalent nitrogen atoms of a vertically adsorbed dinitrogen species [60, 61]. It is now, however, generally accepted [65, 68, 74, 75] that the N_{1s} double peak structure arises from final-state screening effects for a single, weakly bound, chemisorbed species, which give rise to a main line feature and a higher I.P. satellite feature possessing unusually high intensity. This interpretation was first put forward by Fuggle et al. [65], applying a model proposed by Schönhammer and Gunnarsson [76, 77]. The Schönhammer and Gunnarsson model uses an Anderson-type Hamiltonian to represent the interaction between a surface and a molecule possessing an empty MO just above the Fermi level. Figure 7 illustrates the essential features of the mechanism, proposed on the basis of this model [65, 74–78], for the creation of the two-peak N_{1s} structure. Initially, the substrate bands are occupied up to the Fermi level, just above which lies the essentially unoccupied molecular $2\pi^*$ orbital. The creation of a core hole pulls the empty $2\pi^*$ molecular orbital below E_f by several eV. If the $2\pi^*$ orbital remains empty and there is no charge transfer from the

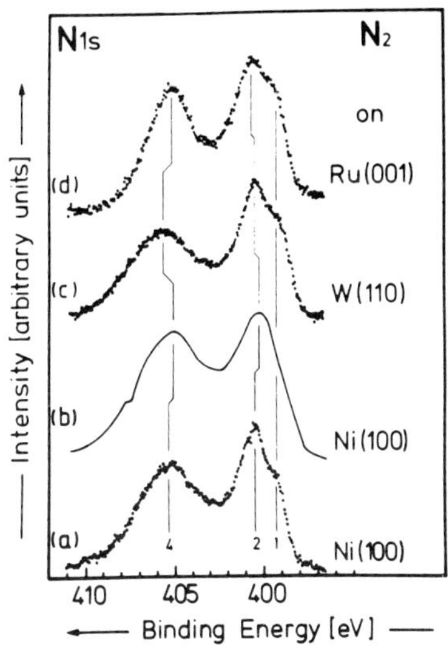

Fig. 6. XPS N_{1s} spectra of weakly chemisorbed N_2 on Ni{100}, W{110} and Ru{001}. Binding energies given with respect to the Fermi level. (From Umbach [75].)

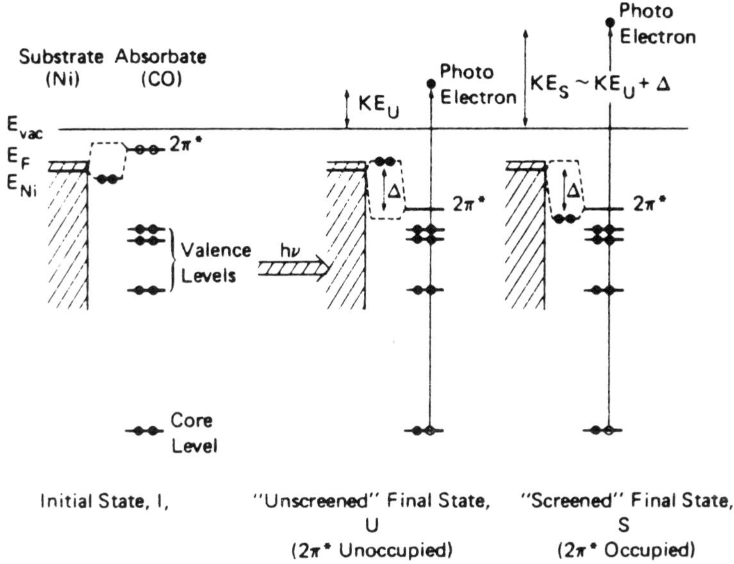

Fig. 7. Schematic representation of the photoemission process taking the adsorbate–substrate system from an initial state I to screened, S, or unscreened, US, adsorbate core-hole final states. (From Brundle et al. [74].)

valence electron MOs into the $2\pi^*$ MO, the *unscreened* final state is created. If, on the other hand, the system relaxes by a "shake-down" process in which partial occupation of the $2\pi^*$ MO by the Ni valence electrons occurs, then a lower-energy *screened* final state is created. Consequently, the high-IP "satellite" peak arises from the unscreened final state and the lower IP "main line" from the screened final state. The Schönhammer–Gunnarsson model predicts that, for a weak coupling between the substrate and the adsorbate $2\pi^*$ MO, the "satellite" peak will dominate the XPS N_{1s} spectrum while for a strong coupling case, the "main line" peak will be dominant. The N_{1s} spectrum for N_2/Ni{100} (Fig. 6) shows approximately equal intensity in both peaks, suggesting an intermediate coupling between the $2\pi^*$ and the substrate MOs exists in this case.

Ab initio Hartree–Fock-LCAO calculations [4, 51, 78] on a linear NiN_2 cluster confirm the general picture presented above and provide good agreement with the experimentally observed N_{1s} core level spectrum, reproducing both the energy separation and the intensity ratio of the two N_{1s} peaks. Brundle et al. [74] have conducted a critical comparison of the N_2/Ni experimental results with various theoretical calculations [4, 51, 66, 78] and emerge with the following conclusions. First, the HF calculations on linear NiN_2 clusters explain well the general core-level photoemission spectra of N_2/Ni{100}, with the main line and satellite structure consistently explained in terms analogous to the Schönhammer–Gunnarsson model. However, good agreement is only achieved with the HF calculations if an electronic structure with significant Ni 4*p* to adsorbate $2\pi^*$ backbonding is employed. For example, the relative intensities of the "main line" and "satellite line" are related to the coupling of the $2\pi^*$ orbital to Ni orbitals. In the ground state of the linear clusters, symmetry restrictions preclude interaction between the Ni 4*sp* valence electrons of σ symmetry and the N_2 $2\pi^*$ level, leading to very weak coupling. In order to simulate such an interaction for an extended surface, an initial state for the linear cluster is constructed from an excited configuration of the Ni atom in which the $4p\pi$ orbital is occupied. Thus a $^3\Phi$ state is used with the configuration $11\sigma^2 5\pi^1 1\delta^3$ as the initial state for the process rather than the $^3\Delta$ ground state of the NiN_2 cluster which has the configuration $11\sigma^2 12\sigma^1 1\delta^3$. In the $^3\Phi$ state, the Ni $4sp\sigma$ valence electron of the $^3\Delta$ $12\sigma^1$ orbital is promoted to Ni $4p\pi$ and contributes to the $5\pi^1$ orbital which has a strong adsorbate $2\pi^*$ component because of the good overlap between the adsorbate $2\pi^*$ and the Ni $4p\pi$. This interaction forms the major part of the metal–adsorbate backbonding, with the metal *d* interaction being of lesser importance. It should be noted that, despite the general success of calculations using the $^3\Phi$ initial state, such models overestimate the backbonding and predict too large a Ni—adsorbate bond distance. This point has been dealt with in greater detail by Brundle et al. [74]. Finally, calculations also show that, for N_2, the σ bonding involves almost equally both the 5σ and 4σ orbitals which are delocalized over the whole molecule.

Umbach [75] has also applied the Schönhammer–Gunnarsson picture to model the interaction of N_2 with an extended Ni surface by using projected DOS of proper symmetry taken from a symmetry-resolved slab calculation for Ni{100} [79]. Again, the experimental results are well mimicked in terms of energy separation and intensity ratios of the peaks. In particular, the calculations reproduce a 3-peak structure observed at high resolution for the "main line" feature. Umbach correlates this fine structure to the band structure of the substrate, whereas Egelhoff [80, 81], on the basis of angle-resolved XPS, attributes the peak splitting to the two inequivalent nitrogen atoms of a vertically adsorbed N_2 molecule. These assignments, however, remain a matter for controversy. For example, the former assignment is supported by Hartree–Fock calculations for a linear NiN_2 cluster [51, 78] which predict that the binding energies of the inner and outer nitrogen 1*s* levels would differ by less than 0.3 eV, which is considerably smaller than the 1–1.4 eV splitting observed experimentally in the screened "main line" peak. On the other hand, Freund et al. [85], on the basis of ab initio generalized valence bond configuration interaction calculations, predict the energy difference between the two inequivalent nitrogen 1*s* ionizations to be about 1.2–0.9 eV, thus supporting the Egelhoff assignment. Freund et al. [85], argue that the Hartree–Fock calculations [51, 78] only predict a small splitting because important electronic correlation effects are neglected. They further point out that the fine structure in the N_{1s} main line peak is very similar for N_2 on Ni{100}, W{110}, and Ru{0001} [64, 75], all of which possess differing densities of states, suggesting the splitting arises from the local metal–molecule interaction which results in two inequivalent nitrogen atoms.

Egelhoff assigns the lower bonding component in the main line peak to the nitrogen atom further away from the surface and the higher bonding component to the nitrogen atom nearest the surface. This assignment, again, is also supported by the calculations of Freund et al. Finally, the main line peak possesses a third component which was interpreted by Umbach [75] as arising from substrate band structure. Freund et al. [85] argue that this component arises due to different spin coupling between the screening electron and the electron in the core of the molecule in the final ion state of the nitrogen atoms.

(b) Valence level structure

Figure 8 shows the He(II) spectrum of a saturation coverage of N_2/Ni{100} [4] alongside the generally accepted assignments [4, 67, 74] also listed in Table 3. By comparison with the CO/Ni system and calculations on $N_2Ni_{(n)}$ clusters [2, 4, 5, 51, 53, 66, 74, 82–84], the peak at $\simeq 8$ eV is assigned to the 5σ and 1π states while the peak at $\simeq 12$ eV represents the 4σ orbitals. The N_2/Ni system differs significantly from the CO/Ni system in a number of ways. Whereas, for CO, bonding arises from the 5σ level only, for N_2/Ni bonding involves both the 4σ and 5σ levels, which are respectively the in-phase and out-of-phase combinations of the same atomic orbitals. This difference in bonding characteristics for the CO and N_2 leads to substantially different behaviour for the relative I.Ps for the 5σ, 4σ and 1π orbitals upon interaction with Ni. For CO, the separation between the 5σ and 4σ states is decreased upon interaction with Ni while it is increased for N_2. These differences arise

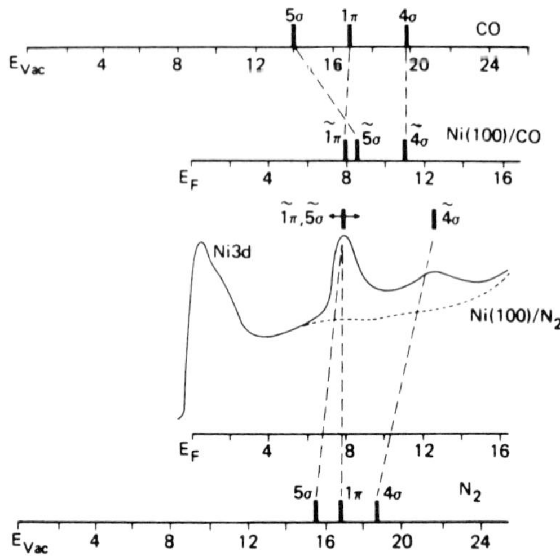

Fig. 8. He(II) UPS data of N_2/Ni{100}. UPS spectra of N_2, CO and CO/Ni{100} are also shown for comparison. (From Bagus et al. [4].)

as a result of both initial state and final state effects. For CO, the overall effect of the interaction with Ni is dominated by the initial state effect of the 5σ CO orbital forming a sigma bond with the Ni surface. Thus, the 5σ level is moved deeper relative to the 1π and 4π levels. On the other hand, the raised CO $5\sigma^{-1}$ and $4\sigma^{-1}$ final state orbitals are similar in character to initial state orbitals and so suffer similar relaxation effects. The total effect is then a decrease in the 5σ and 4σ energy separation.

For N_2/Ni, the bonding arises from delocalized 5σ and 4σ orbitals which therefore suffer similar initial state effects, thus not affecting their energy separation. However, the singly occupied ionized $5\sigma^{-1}$ final state for N_2/Ni becomes localized on the N atom further away from Ni and becomes non-bonding, while the $4\sigma^{-1}$ final state localizes between the Ni and the nearest N atom and constitutes all the σ bonding. The non-bonding character of the $5\sigma^{-1}$ final state means the 5σ I.P. is lower than it would otherwise be, therefore the separation between the 4σ and 5σ levels increases for N_2/Ni compared with the free molecule. Dowben et al. [67] have conducted an ARUPS study of N_2/Ni{100} and have resolved experimentally a number of valence-band satellite structures representing unscreened final states which had been predicted by calculations [5, 53, 66, 86]. For example, N_2 adsorbed on Ni{100} can give rise to six features in photoemission spectra, namely the screened $(5\sigma)^s$, $(1\pi)^s$ and $(4\sigma)^s$ states and the unscreened $(5\sigma)^u$, $(1\pi)^u$ and $(4\sigma)^u$ states. Table 3 shows the assignments of the UPS data to these states. The band dispersion of the N_2 molecular orbitals was determined from $\bar{\Gamma}$ to $\bar{X}'$ of the reduced Surface Brillouin Zone (Fig. 9), and showed little dispersion for the $(5\sigma)^s$ and $(1\pi_x)^s$ orbitals. The $(1\pi_x)^u$ state, however, does show significant dispersion but this could arise from possible $(5\sigma)^u$ and $(4\sigma)^s$ contributions in

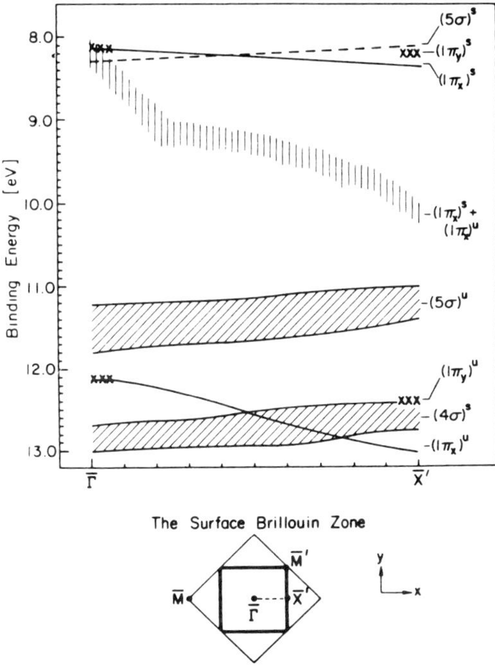

Fig. 9. The band dispersion of N_2 molecular orbitals from $\bar{\Gamma}$ to $\bar{X}'$ of the reduced Surface Brillouin Zone. The character of the bands at $\bar{\Gamma}$ is indicated on the right margin. The broad bands indicate features with considerable uncertainty in binding energies. (From Dowben et al. [67].)

this even geometry. This work suggests that the molecular orbitals of N_2 adsorbed in an ordered c(2 × 2) overlayer form a two-dimensional band structure, probably as a result of direct overlap of 1π orbitals and indirect overlap of the 5σ orbitals.

5.3 NITROGEN ADSORPTION ON Ni{111}

To date, few studies have been reported on the adsorption of dinitrogen on Ni{111}. Nitrogen is very weakly adsorbed on Ni{111}, adsorbing only at very low temperatures. Both physisorbed and chemisorbed states of N_2 on Ni{111} have been investigated and will be examined in turn.

5.3.1 Physisorbed state

Using TPD, UPS and XPS measurements at 20 K, Breitschafter et al. [87] have identified that one weakly chemisorbed and three physisorbed states of nitrogen exist on Ni{111}. The three physisorbed states, ps1, ps2 and ps3, were assigned as 1st layer N_2 interspersed between chemisorbed species, 2nd layer N_2 and multilayer N_2, respectively. These physisorbed species give rise to peaks at about 9.2–9.7, 10.5–11.0, and 12.3–12.8 eV in the He(II) photoemission spectra, which bear a close resemblance to gas-phase values. The N_{1s} core level spectrum for the physisorbed state shows a single peak at 403–404 eV, thus differing significantly from the double-peak structure associated with the chemisorbed species.

Work function changes of −80 meV and −10 meV have been attributed to the ps1 and ps2 species, respectively [87]. The exchange behaviour and the lateral interactions of the physisorbed species have been determined in co-adsorption studies with argon [87]. The first-layer ps1 state, which desorbs at 45 K, is found to be a laterally and orientationally hindered intrinsic precursor whose conversion to the chemisorbed state is kinetically hindered. The second-layer ps2 state, which desorbs at ≃25 K, is an extrinsic precursor and shows a faster conversion to the chemisorbed state.

Recently, Feng et al. [88] have conducted a new study on the physisorbed states and have concluded, on the basis of work carried out below 20 K, that the ps1 intrinsic precursor state can be trapped at low coverage on a clean surface and desorbed directly from it, suggesting that the barrier to chemisorption lies above the barrier to desorption. Using EELS, Feng et al. [88] also conclude that the ps1 state is physisorbed parallel to the surface. However, lack of off-specular data means this conclusion is necessarily speculative.

5.3.2 Chemisorbed states

A single, vertically bound chemisorbed state of N_2 exists on Ni{111} [87–89] and is found to be infrared-active [89], clearly showing that special B_5 sites are not necessary to produce such a species. A saturation layer of the chemisorbed species results in a work function change of −560 meV.

Core level N_{1s} spectra [87, 88] exhibit the characteristic 2-band structure which arises as a result of final state effects, analogous to those discussed for N_2/Ni{100}. The He(II) photoemission spectrum [87], with peaks at 9 and 13 eV, can be similarly assigned, with the 5σ and 1π states contributing to the former feature and the 4σ level representing the latter peak, Table 3.

A recent infrared and LEED investigation [89] reveals that, at 83 K, adsorption leads to a diffuse p(2 × 2) LEED pattern, corresponding to a coverage of 0.25. The infrared spectra show a single band which grows in intensity with increasing exposure, but exhibiting little frequency shift from its initial value of 2186 cm^{-1}. This band is assigned to the N—N stretching vibration and its frequency identifies the chemisorbed N_2 as a vertically bound, on-top species. The corresponding Ni—N_2 vibrational band has been recorded at 350 cm^{-1} by EELS [88]. The invariance of the N—N stretching frequency with coverage has been shown, on the basis of isotope decoupling experiments, to arise from opposing dipole and chemical shifts of +18 cm^{-1} and −18 cm^{-1} respectively. Unlike CO/Ni, a downward chemical shift occurs for N_2 because the bonding arises principally as a result of charge donation from the antibonding 4σ orbital of N_2 into the metal [49] with increasing coverage. This charge donation is reduced, thus decreasing the νN—N frequency.

This chemisorbed N_2 species desorbs at 82–88 K with first-order kinetics and an activation energy of 20 kJ mol^{-1} [87].

5.4 NITROGEN ADSORPTION ON Ni{110}

The adsorption of nitrogen on Ni{110} is the most widely studied of all the N_2/Ni adsorption systems and consequently a number of physical quantities of this system have been determined (see Table 4) [90, 95]. As for N_2/Ni{100} and N_2/Ni{111}, nitrogen is also weakly chemisorbed on Ni{110} with RAIRS [90–94], EELS [2, 95], UPS [2] and ISS [90] data clearly revealing an on-top dinitrogen species, adsorbed on the {110} surface Ni rows with the N_2 molecular axis normal to the surface. It is interesting to note that the short-bridge twofold site is preferentially occupied, despite the presence of B_5 sites on this surface.

5.4.1 Electronic properties

(a) Core level features

The N_{1s} core level spectrum of N_2 on Ni{110} at 130 K shows, at all coverages, the characteristic double peak structure [96, 97] observed for N_2 adsorbed on Ni{100} and Ni{111}. The two peaks at 401.7 and 406.7 eV can, on the basis of the arguments presented for N_2/Ni{100}, be attributed to the screened and unscreened final states respectively (Table 3). This interpretation of the XPS spectrum assumes that the core ionization occurs within the sudden approximation limit leading to the appearance of the higher energy shake-up feature in the N_{1s} excitation spectrum. Such satellite features should not be present in the adiabatic excitation limit which would prevail at low excitation energies. This energy dependence of the satellite intensity had previously been predicted [98] to follow a smooth time-dependent cross-over from the adiabatic to the sudden approximation, the sudden limit being reached at energies > 200 eV above the excitation threshold. Stohr et

TABLE 4

Physical quantities determined for nitrogen adsorbed on Ni{110} [90–95]

Isosteric heat of adsorption, q_{st}, $\theta < 0.5$ [90]	40 (± 1.5) + 2.5 (± 0.3)θ kJ mol^{-1}
Isosteric heat of adsorption, q_{st}, $\theta > 0.5$ [90]	~ 20 kJ mol^{-1}
Sticking coefficient, $T < 165$ K, $\theta < 0.45$ [90]	0.85
Integral entropy, S_{ad}, $\theta < 0.5$ [90]	112 (± 5) J K^{-1}mol^{-1}
Maximum N_2 coverage, $T < 125$ K [90, 93]	0.72
ν_{N-N}, $\theta = 0.5$ [93, 95]	2191 cm^{-1}
ν_{Ni-N_2} [95]	323 cm^{-1}
Chemical shift in ν_{N-N}, $\theta = 0.72$ [94]	− 42 cm^{-1}
Dipole coupling shift in ν_{N-N}, $\theta = 0.72$ [94]	+ 36 cm^{-1}
Vibrational polarizability, α_v, $\theta = 0.72$ [94]	0.26 Å^3
Electronic polarizability, α_e, $\theta = 0.72$ [94]	4.4 Å^3

al. [97] have investigated the energy dependence of the N_{1s} core-level shake-up feature for the N_2/Ni{110} system and report that the transition from the adiabatic to sudden excitation occurs abruptly in the energy region 5–15 eV above the core-level excitation limit. The authors interpret this result in terms of a simple exchange–interaction model, based on the Hartree–Fock approximation, which demonstrates that the two main contributions to the shake-up intensity arise from a direct term and an exchange term process, involving the deep core level K, the valence level a (i.e. the bonding π state) and particles in a photoelectron state, k' and a shake-up level b' (i.e. the antibonding π state, Fig. 10). Thus, the energy dependence of the shake-up intensity is given by

$$\mu^{s} = \mu^{s}_{\infty} \left| 1 - \frac{\langle K|\mathrm{d}|b'\rangle\langle a|k'\rangle}{\langle K|\mathrm{d}|k'\rangle\langle a|b'\rangle} \right|^{2}$$

where μ^s is the shake-up intensity, μ^s_∞ represents all the other overlap integrals contributing to μ^s, apart from the direct term and the exchange term depicted in depicted in Fig. 10, and d is the dipole operator. The exchange term becomes negligible at high energies and the sudden approxi.˙.ation situation results. Importantly, such a treatment can account for the fast turn-on of the shake-up satellite peak. Additionally, it is predicted that the sudden limit is reached faster if the main multielectron excitations are of low energy.

(b) Valence level features

UPS spectra for N_2/Ni{110} [2] also follow the same pattern exhibited by those from the N_2/Ni{100} and N_2/Ni{111} systems, with peaks at 12 and 8 eV, the first representing the 4σ level and the second representing the 5σ and 1π levels (Table 3). Again, the nature of the surface–N_2 bond on Ni{110}

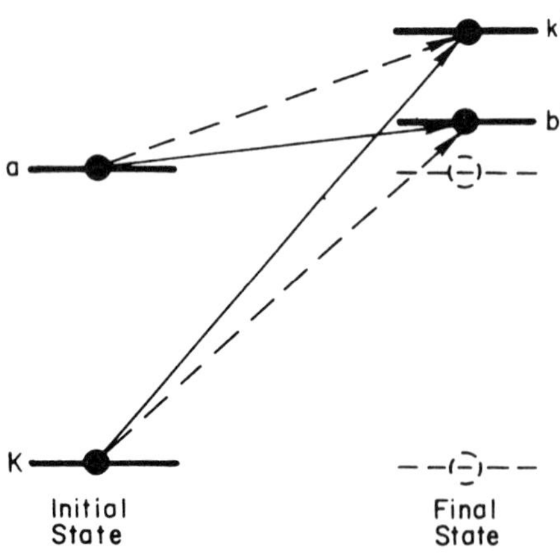

Fig. 10. Direct (solid line) and exchange (broken line) processes contributing to the shake-up intensity. (From Stohr et al. [97].)

can be described in a manner similar to that adopted for the N_2/Ni{100} system with the 1π level shifting up towards E_f by 4 eV due to final state relaxation while the energy separation between the 4σ and 5σ levels increases by 0.5 eV. As for N_2/Ni{100}, this is due to the fact that both the 4σ and 5σ levels are involved in bonding, thus suffering similar initial state effects but quite different final state effects, the $5\sigma^{-1}$ final state becoming non-bonding but the $4\sigma^{-1}$ becoming bonding, leading to an increase in separation of the two levels.

Horn et al. [2] also point out that the 12 eV peak is very broad, which probably indicates the presence of a number of shake-up processes in the valence band excitation contributing in this region alongside the 4σ emission. The broad 12 eV peak can be deconvoluted into three separate bands, but the assignment of one particular peak to the 4σ primary peak was not possible. However, by following the resonance behaviour of the 4σ level, Horn et al. suggest that the 4σ component gives rise to the peak with a binding energy of 11.8 eV. The other contributions in the complicated 12 eV region is attributed to shake-up satellite structures which arise from the interaction of the surface complex with the photo-induced hole.

By comparing their extensive angular and polarization-dependent data with theoretical calculations of the NiN_2 complex, Horn et al. conclude that N_2/Ni bonding occurs mainly via a $N_2 4\sigma$—metal bond with some Nid—$N_2 2\pi$ backbonding.

(c) Work function measurements

Whereas UPS and XPS data for N_2/Ni{110} show little variation with coverage, suggesting that only one type of N_2 species is adsorbed, data from TDS, LEED and vibrational spectroscopy (discussed later) show clear evidence for nitrogen adsorption proceeding in a number of stages involving different types of chemisorbed species. This is also reflected in the work function changes recorded by Golze et al. [90, 96]. In the following discussion, the work function data will be considered within the framework of the N_2/Ni{110} temperature vs. θ phase diagram constructed on the basis of LEED, AES, TDS and UPS measurements (Fig. 11) [90, 99].

Figure 12 illustrates the work function changes observed with increasing coverage at 81 and 127 K [90]. At 81 K, increasing coverage to just below 0.5 leads to a work function decrease of -105 ± 10 mV, consistent with the view that N_2/Ni bonding is dominated by σ donation into the metal with insignificant backbonding. Further increases of coverage to the saturation value of $\theta = 0.67$ lead to an increase in $\Delta\phi$ to -8 mV. Similarly, adsorption at 127 K results in an initial decrease in work function to a minimum of $\simeq -110$ mV. Subsequent increase in coverage, however, only results in an increase in $\Delta\phi$ to $\simeq -90$ mV. $\Delta\phi$ adsorption isobar plots [90] reveal distinct breaks which are associated with phase changes in the adsorbate layer. Thus, initial disordered adsorption to $\theta = 0.3$ causes the work function to decrease approximately linearly with coverage. At $\theta \simeq 0.3$, ordering into the

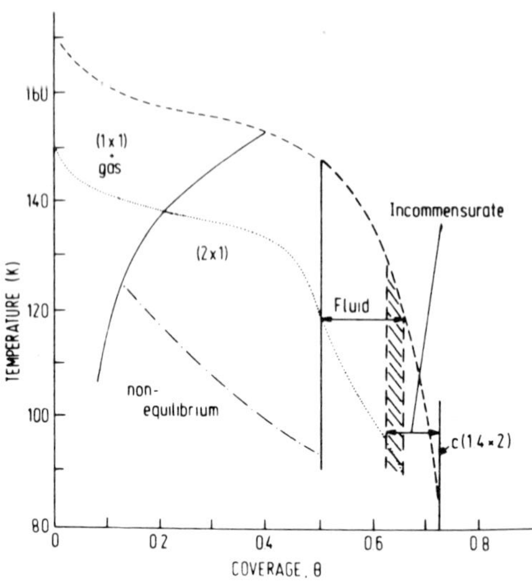

Fig. 11. Schematic phase diagram for N_2/Ni{110}. (From Grunze et al. [99].)

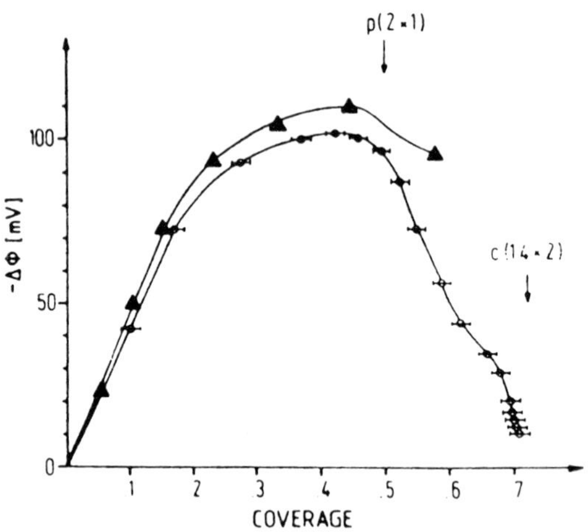

Fig. 12. The work function change with coverage for N_2/Ni{110}. ▲, 127 K; ○, 81 K. (From Golze et al. [90].)

(2 × 1) phase starts to occur and the decrease in $\Delta\phi$ continues but at a slower rate, presumably due to depolarization effects. As the fluid "banana phase" (Fig. 11) comes into existence at $\theta > 0.5$, the work function begins to increase with further change in the rate of increase of work function occurring at around 0.55 and 0.65 as the $\begin{pmatrix} 1 & -1 \\ 2 & 1 \end{pmatrix}$ pattern and the incommensurate c(1.4 × 2) structures form, respectively. The change in $\Delta\phi$ for $\theta > 0.5$ reflects the change in the type of adsorbed species present on the surface. The nature of these species and the LEED structures they give rise to will be discussed in greater detail later.

5.4.2 Kinetic and thermodynamic data

By employing TDS, UPS, XPS and ISS techniques, Golze et al. [90] have determined a number of thermodynamic and kinetic parameters for the N_2/Ni{110} system.

Figure 13 displays the isosteric heats of adsorption measured by XPS isobars [90]. These data illustrate that the heat of adsorption remains constant at about $40\,kJ\,mol^{-1}$ until $\theta = 0.5$ and then decreases sharply to $\simeq 20\,kJ\,mol^{-1}$, presumably due to the strong repulsive interactions which come into existence as the commensurate–fluid boundary is crossed (Fig. 11). At very low coverages, a higher heat of adsorption is observed and is associated with adsorption into defect sites.

Golze et al. [90] have also investigated the entropy changes in the adsorbed layer as the coverage is increased and report fairly constant values for the partial and integral molar entropies in the coverage range $0.1 < \theta < 0.5$. However, as θ increases above 0.5 and the commensurate–fluid boundary is crossed, a significant increase in the integral molar entropy of $\Delta S \simeq 20 \pm 10\,J\,K^{-1}$ is observed. The experimental entropies deviate strongly from those calculated for localized adsorption for all coverages above 0.05 whereas some agreement between the experimental entropies and those calculated for a Volmer gas is seen for coverages less than 0.2, although for higher coverages a large inconsistency is observed. Although calculated entropies for an ideal 2-D gas correspond fairly closely to the experimental entropy values up to $\theta = 0.5$, it is unrealistic to model (2×1) island formation in terms of an ideal 2-D gas with no eigenvolume. Golze et al. therefore invoke the same arguments applied to the N_2/Ni{100} case and attribute the high entropy values in the adsorbed layer to new soft vibrational and rotational modes which, for example, might come into play in fluctuating antiphase domain boundary regions. Again, as for N_2/Ni{100}, the possibility of minor surface reconstruction leading to larger entropy values needs to be considered.

Thermal desorption spectra for various coverages of nitrogen on Ni{110}

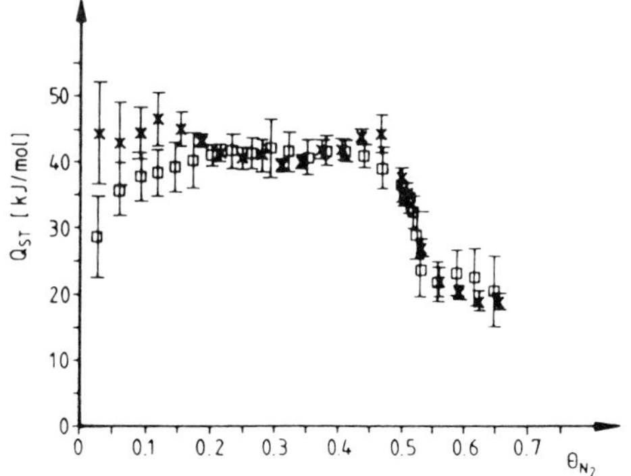

Fig. 13. The isosteric heat of adsorption for N_2/Ni{110}. (From Golze et al. [90].)

have been reported by Golze et al. [90, 100] and show a γ_2 peak and a lower-temperature γ_1 peak at $\simeq 120$ K, which grows in at coverages greater than 0.5. The γ_1 peak appears at a lower temperature due to repulsive interactions present for $\theta > 0.5$. The behaviour of the γ_1 peak closely resembles straight first-order desorption. However, Golze et al. report that the FWHM and FWQM of this peak are not constant with coverage as expected for simple first-order desorption. Instead, the favoured description of the TDS data of the γ_2 peak is in terms of a simple precursor model with attractive pairwise forces in the adsorbed phase for $\theta < 0.5$, represented by the equation

$$r_{\mathrm{des}} = \frac{\theta^*}{1 - \theta^*} \nu N_{\mathrm{s}} \exp\left[-\left(\frac{E_d - w\theta^*}{RT} \right) \right]$$

where θ^* is the relative coverage with $\theta = 0.5$ corresponding to $\theta^* = 1$, ν is the pre-exponential factor for desorption (best fit value of $9.9 \times 10^{13}\,\mathrm{s}^{-1}$), N_{s} is the number of adsorption sites and w is the pairwise interaction energy (best fit for attractive lateral interactions of $0.85\,\mathrm{kJ\,mol^{-1}}$). Further support for precursor-mediated desorption and adsorption processes for $\theta < 0.5$ comes from sticking probability measurements [90] which show that the sticking coefficient has a constant value close to unity up to a coverage of 0.5. For $\theta > 0.5$, the sticking coefficient drops sharply.

5.4.3 Structural studies

(a) LEED measurements

Grunze and co-workers [90, 99] report initial disordered adsorption of N_2/Ni{110} until the coverage range $0.1 < \theta < 0.2$ is reached when (2 × 1) islands nucleate and then grow, leading to a sharp (2 × 1) diffraction pattern at $\theta = 0.5$. The (2 × 1) phase, corresponding to the maximum heat of adsorption of $47 \pm 7\,\mathrm{kJ\,mol^{-1}}$, is the most stable phase for N_2/Ni{110}. As the coverage is increased beyond 0.5, "banana-like" features are observed in the LEED pattern in which curved diffuse streaks connect the 0.5-order beams [Fig. 14(c)]. This structure forms rapidly (< 0.1 s) and reversibly from the (2 × 1) phase. In contrast, the (2 × 1) phase exhibits slower kinetics and heating and cooling rates of $< 0.05\,\mathrm{K\,s^{-1}}$ are required to attain reversible change. The "banana phase" is interpreted as a fluid phase with N_2 molecules restricted to movement along the ⟨110⟩ rows. However, some inter-row correlation does exist, giving rise to curved rather than straight streaks. Further increases in coverage leads to a smooth evolution from the banana pattern into a $\left(\begin{smallmatrix} 1 & -1 \\ 2 & 1 \end{smallmatrix}\right)$ incommensurate structure at $\theta \simeq 0.67$ which then compresses further to produce a c(1.4 × 2) pattern at the saturation coverage of 0.72.

Golze et al. [90] have proposed a number of overlayer structures corresponding to the LEED patterns described above. The (2 × 1) structure is most easily reconciled with rows of dinitrogen species adsorbed in alternate on-top sites [Fig. 14(a)] consistent with the observation of a single infrared

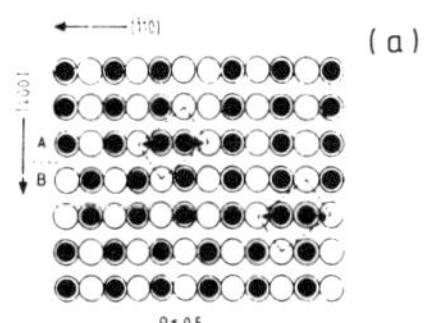

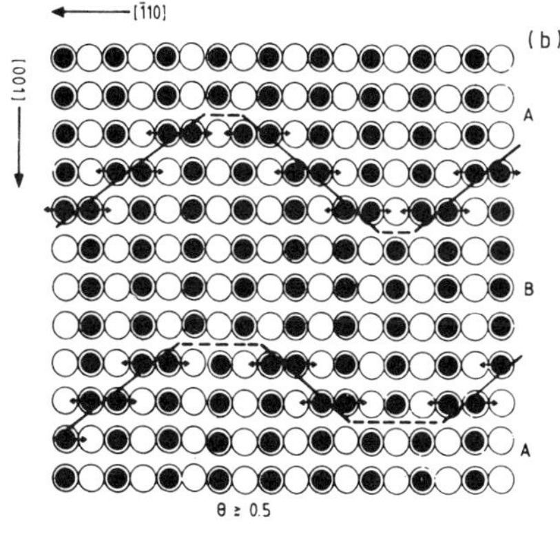

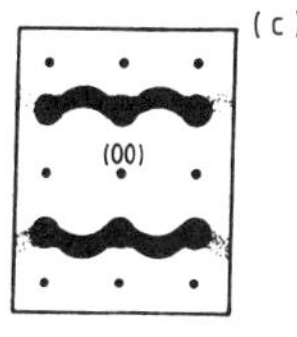

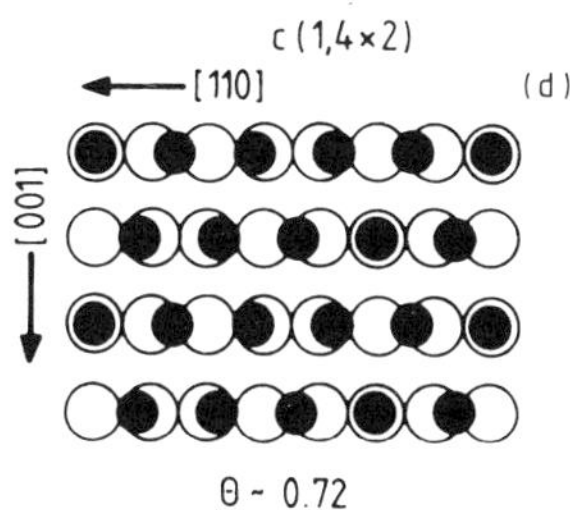

Fig. 14. (a) Ordered domains of (2×1) structure with possible anti-phase boundaries between ordered regions A and B for coverages slightly below 0.5. (b) Model for highly mobile domain walls at $\theta > 0.5$. (c) Schematic drawing of the "banana" pattern. (d) Model for the $c(1.4 \times 2)$ phase. (From Golze et al. [90].)

band at $2192\,\mathrm{cm^{-1}}$ at $\theta = 0.5$ [93]. It is thought that the formation of the (2×1) structure exhibits slow kinetics with an activation energy of 10–20 $\mathrm{kJ\,mol^{-1}}$ because of the difficulty of annealing out antiphase domain boundaries of the type shown in Fig. 14(a). The formation of the "banana phase" is accompanied by a sharp decrease in the isosteric heat of adsorption from 47 to 20 $\mathrm{kJ\,mol^{-1}}$ and a large increase in the integral entropy of $\Delta S \simeq 20 \pm 10\,\mathrm{J\,K^{-1}}$. On the basis of these changes and the fact that this phase exhibits rapid kinetics of formation (< 0.1 s) the "banana phase" has been assigned as a fluid phase [90, 99] rather than a solid with extensive frozen-in disorder. Grunze and co-workers [90, 99] point out it has been predicted [101] that, theoretically, a $(p \times 1)$ commensurate phase (with $p^2 < 8$) on a rectangular substrate can melt to form an intermediate fluid phase prior to transition into an incommensurate phase. For N_2 on Ni{110}, such a fluid phase is envisaged to consist of highly mobile, high-density antiphase domain walls which separate commensurate areas of (2 × 1) structure [Fig. 14(b)]. Essentially, such an adlayer may be described as a "soliton liquid". The fluid phase arises because, presumably, over a certain coverage range greater than 0.5, an overlayer structure with commensurate domains bound by high-density walls is more stable with respect to a weakly incommensurate phase. The high entropy values observed for $\theta > 0.5$ can now be attributed to large density fluctuations in the overlayer arising from the presence of such domain walls. The entropy increase results both from an increased configurational contribution and from additional softened vibrational modes associated with the fluctuations of the domain wall. The N_2 molecules in the domain boundaries are also bonded in on-top sites, consistent with vibrational data which shows no indication of any other type of dinitrogen species being present. However, it is possible, due to steric repulsions, that some relaxation of the adsorbed molecules away from ideal on-top positions does occur, either by lateral movement or by tilting.

No specific overlayer models have been proposed for the $\binom{1\ -1}{2\ \ \ 1}$ incommensurate structure at $\theta = 0.67$. It is, however, proposed [90] that, at this coverage, (2 × 1) domains are denuded from the surface as the density of domain walls increases, leading finally to the formation of the incommensurate c(1.4 × 2) phase at $\theta = 0.72$. The proposed overlayer for this phase is depicted in Fig. 14(d), and shows that a number of different adsorbed dinitrogen species are now present on the surface, ranging from on-top bonded to twofold bridged molecules. However, such an overlayer structure is not supported by vibrational data, discussed below, which show no evidence for the presence of bridged N_2 species.

(b) Vibrational studies

A number of vibrational investigations have been conducted on the N_2/Ni{110} system [2, 90–95]. These indicate that an on-top dinitrogen species is chemisorbed on the surface at all coverages, giving rise to a strong νN—N band between 2200 and 2192 cm^{-1} [91, 93] and a νNi—N_2 band at $\simeq 340$ cm^{-1} [2, 95]. Off-specular EELS studies [95] indicate that both modes are dipole-scattered. No evidence for a bridged species has been found [93, 95], thus conflicting with the interpretation of the c(1.4 × 2) LEED pattern observed at $\theta = 0.72$.

Adsorption at 81 K [93] produces, at low coverage, an infrared band at 2200 cm^{-1} that moves steadily downwards in frequency to 2192 cm^{-1} at $\theta = 0.5$, which coincides with the maximum surface potential change. Increasing coverage to the saturation value of 0.72 causes the IR band to move up slightly to 2194 cm^{-1}. In addition to this main band, high-frequency shoulders at 2204 and 2220 cm^{-1} are also reported [93].

RAIRS studies [92, 93] indicate that the adsorption behaviour varies somewhat at 125 K. First, the saturation coverage only reaches 0.5. More importantly, the IR band shapes exhibit marked differences at 125 and 81 K, the higher temperature data clearly showing multicomponent peaks at coverages up to 0.36 (Fig. 15). According to the phase diagram for N_2/Ni{110} [90, 99] (Fig. 11), non-equilibrium structures exist at 81 K while above 125 K equilibrium overlayer structures involving ordered (2 × 1) islands are formed. Thus, Brubaker and Trenary [93] rationalize the differences in the infrared spectrum at different temperatures in the following way. At 125 K, small ordered (2 × 1) islands are formed at low coverages. The multicomponents in the IR data for coverages up to 0.36 arise from molecules adsorbed in different environments, for example in the centre of the islands, on the island perimeters or at the corners of the islands, each type of species possibly giving rise to a separate band. Further bands can also arise from molecules adsorbed in the disordered "sea" between islands. As island sizes grow, the island edge, corner and disordered "sea" species diminish and a single band dominates the infrared spectrum. At 81 K, a frozen-in disordered layer is adsorbed and the high-frequency shoulders at 2204 and 2220 cm^{-1} are attributed to N_2 molecules adsorbed in high-density antiphase domain boundaries.

A few papers [2, 102] have reported EELS spectra with an extra vibrational band at 575–600 cm^{-1} alongside the νNi–N_2 and νN–N bands. This band has been variously assigned as a νNi–N vibration from dissociated species [2] or an overtone of the Ni–N_2 stretching vibration [102]. However, Schenk et al. [103] have recently reported EELS data from co-adsorption studies of N_2 with

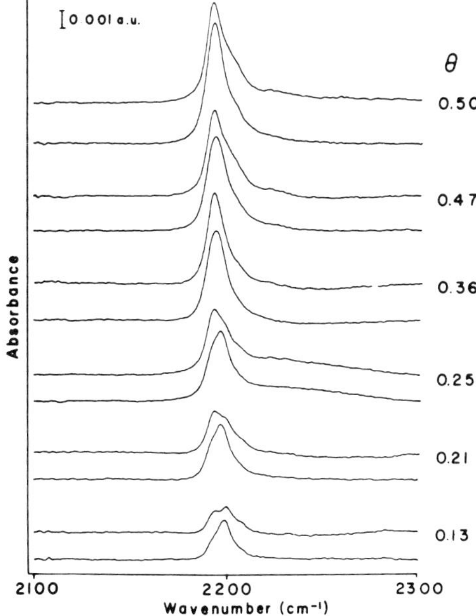

Fig. 15. Pairs of IR spectra of N_2/Ni{110} before and after annealing to 125 K as a function of coverage; the lower spectrum of each pair obtained after adsorption at 81 K; the upper spectrum obtained after subsequent heating to 125 K for 60 s and recooling to 81 K. (From Brubaker and Trenary [93].)

H on Ni{110}. This work provides strong evidence that this extra band is due to residual hydrogen "pick-up".

Bonding in weakly chemisorbed systems such as CO/Cu and N_2/Ni is often described as dominated by the donation of electron density from a slightly antibonding σ orbital of the adsorbate [6, 49]. As the coverage increases, the σ donation decreases, resulting in the decrease of the νCO frequency. It is thought that the insignificance of the $2\pi^*$ bonding in such systems is probably due to the relatively large energy difference in the $2\pi^*$ level and the Fermi level of the metal [55, 104]. Therefore, a negative chemical shift is expected for N_2 adsorbed on nickel.

Ueba [104] provides a somewhat different description for the negative chemical shift in this system. Whereas, the interaction of the adsorbate with the substrate is dominated by σ donation, Ueba proposes that the negative chemical shift arises from increased $2\pi^*$ donation as the coverage is increased. Ueba's model involves electronic band formation from the interaction of the $2\pi^*$ orbitals of the adsorbed molecules, the energy of the electronic band shifting with coverage due to depolarization effects. Band formation results in the broadening of the $2\pi^*$ level which consequently increases the density of adsorbate states at the Fermi level thus increasing the degree of backdonation into the $2\pi^*$ level. In addition to this effect, for a weakly chemisorbed system such as N_2/Ni, the dominance of σ donation in bonding means there is a net positive charge on the molecule. This also leads to a decrease in the energy of the $2\pi^*$ level with coverage, thus reinforcing the negative chemical shift caused by band formation. This model also provides a more satisfactory explanation for the large observed shift in the N—N stretching frequency upon adsorption.

The small frequency change in the νN—N vibration with increased coverage can now be rationalized in terms of opposing chemical (negative) and dipole (positive) shifts. Brubaker and Trenary [94] have carried out isotopic decoupling studies on the c(1.4×2) structure of N_2/Ni{110} at 81 K and report a positive dipole coupling shift of 36 cm^{-1} which is almost nullified by a negative chemical shift of 42 cm^{-1}. These experimental results were modelled well by dipole–dipole coupling theories using a vibrational polarizability of 0.26 $Å^3$ [94].

5.5 ACTIVATED MOLECULAR AND ATOMIC NITROGEN ADSORPTION ON NICKEL

The adsorption of activated gas-phase nitrogen on a number of nickel surfaces has been reported [30, 105–112]. Gas-phase activation can be achieved in a number of ways [26, 27, 29–35] which include using a hot filament, an ion bombardment gun, electron bombardment or a plasma discharge. Alternatively, an atomic nitrogen adlayer can be formed by the decomposition of nitrogen-containing molecules such as ammonia or hydrazine on the surface.

Kirby et al. [30] have reported that the adsorption of activated gas-phase

N_2 on Ni{110} leads to the facetting of that surface. As observed for Cu{210}, exposure to the activated nitrogen at 77 K and warming to $\gtrsim$ 290 K results in the reconstruction of the {210} surface to form {100} facets with a c($11\sqrt{2} \times \sqrt{2}$) R45°–N overlayer, while warming further to $\gtrsim$ 520 K, creates {110} facets with a p(2 × 3)–N overlayer. At 673 K, the clean Ni{210} surface is regenerated due to N_2 desorption and possibly simultaneous diffusion of adsorbate into the bulk.

Benziger and Preston [105] have investigated the dissociative adsorption and recombination of N_2 on Ni{111} using LEED, AES and TPD. Nitrogen was adsorbed by the decomposition of hydrazine on the surface at 500 K. Heating to 850 K resulted in the desorption of molecular N_2. An analysis of the TPD data revealed that the N + N recombination reaction follows second-order kinetics with an energy of recombination of 210 kJ mol^{-1}.

Adsorption of half a monolayer of N atoms on Ni{100} causes a reconstruction of the surface, giving rise to a p4g(2 × 2)N LEED structure. This reconstruction consists of a c(2 × 2) overlayer of nitrogen with a clockwise and counterclockwise rotation of the nickel surface atoms around the two nearest-neighbour nitrogen atoms [111, 112] (Fig. 16). The p4g(2 × 2)N structure possesses mutually perpendicular glide planes along the [110] direction leading to a systematic absence in the LEED structure at ($h + \frac{1}{2}$, 0) and (0, $k + \frac{1}{2}$) positions. On the basis of SEXAFS data [112], the nitrogen is placed in the fourfold hollow site with a bond length of 1.89 Å. The nearest Ni atoms are rotated by a tangential displacement of $\zeta = 0.68$ Å leading to a $d_\perp$ of 0.11 Å. Thus the N atom is nearly equidistantly bonded to the four Ni atoms in the surface and the Ni atom in the layer underneath. This reconstruction is thought to be driven by internal stresses arising from a charge redistribution induced by the adsorbate, which leads to repulsive interactions between the first layer Ni atoms. The dispersion of surface phonons and resonances of the reconstructed Ni{110} p4g(2 × 2)N structure has been measured by Daum et al. [111] using EELS.

Grunze et al. [91] have reported the dissociative adsorption of nitrogen on Ni{110} under high pressure and high temperature conditions. Adsorption at

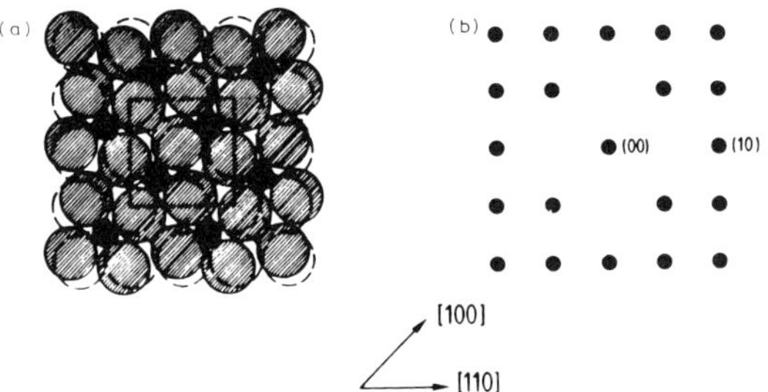

Fig. 16. (a) Real space model of the p4g(2 × 2) reconstructed N/Ni{100} surface. (b) Diffraction pattern of the p4g(2 × 2) structure for normal incidence of beam. (From Daum et al. [111].)

190 K followed by heating to 500 K results in a weak but sharp p(2 × 3) pattern which is intensified by further exposure to N_2 above 500 K. This dissociative state gives rise to TDS peaks at 470 and 800 K. The p(2 × 3) structure has been reported to form from the dissociative adsorption of NO on Ni{110} [106] and on the {110} facets of Ni{210} after exposure to activated nitrogen [30] as described above. Roman and Riwan [107] have investigated the ionic implantation of N_2^+ in Ni{110} using SES, AES, LEED and ELS. Exposing the crystal to high nitrogen doses at 300 K produces a sharp p(2 × 3) pattern with split (h, k/3) and (h/2, k) reflections, attributed to antiphase boundaries [106], while heating to > 750 K produces a simple p(2 × 3) pattern. Roman and Riwan assign the p(2 × 3) pattern to half a monolayer of nitrogen atoms adsorbed in B_5 sites located in the troughs on the Ni{110} surface. They further propose that such a structure could be a precursor state for Ni_4N.

Scattering of N_2^+ ions from clean and nitrogen covered Ni{110} surfaces [108] reveals that the yield of the scattered N_2^+ ions ($N_2^+/(N_2^+ + N^+)$) increases with increasing N coverage. Although it is not clear why this should be so, one possibility is that the molecular dissociation process, thought to proceed via antibonding states created after electron pick-up from the surface [108–110] is less favourable on the N—Ni surface due to work function changes which make electron transfer to N_2^+ less likely. Nitridation studies on Ni{111} and polycrystalline Al, Cu, Mo and Ag [109, 110], using N_2^+ beams, draw a similar picture of the N_2^+/surface interaction. Thus, Shamir et al. [109] report that the nitridation process exhibits a strong energy dependence below ≃ 30 eV and that the threshold impact energy for nitride formation varies for different metals, e.g. it is ≈ 0 for Al and Ni, ≈ 4 eV for Mo and does not occur for E < 200 eV for Ag. Whereas the nitridation of Mo by N^+ is energy-independent in the range 5–80 eV, the behaviour with N_2^+ indicates that the energy dependence is intimately linked to the dissociative process. Darko et al. [110] have developed a model for the nitridation of metallic surfaces and identify four steps; (i) the neutralization of N_2^+ by resonance (RN) and/or Auger electron transfer (AN) from the metal to the cation; (ii) impact dissociation of the N_2 molecules; (iii) collisional de-excitation and thermalisation of the N atoms on the surface and (iv) the actual nitridation chemical reaction. It is concluded that the AN neutralization process is dominant for Mo, Ni and Cu while the AN and RN neutralization channels are equally important for Al. This difference is attributed to the broad valence band of Al which is resonant with a number of molecular N_2 states while Mo, Ni and Cu possess sharper valence bands resonant with only a limited number of molecular N_2 states. This model provides a reasonable framework for analysing the nitridation processes and provides an explanation for the experimental observation that the electronic structure of the metal, the N_2^+ ion and the resultant neutralized species critically determines the low-energy nitridation of a metal surface by N_2^+.

6. Nitrogen on iron

6.1 INTRODUCTION

Nitrogen on iron has been widely studied due to its importance in the elucidation of the mechanism and species occurring in the ammonia synthesis reaction [113]. Catalytic and surface spectroscopic experiments [1] have shown that the dissociative adsorption of nitrogen is the rate-determining step in the catalytic formation of ammonia from nitrogen and hydrogen.

Of the three low-index planes of b.c.c. iron, attention has focused, increasingly in recent years, on the {111} face, the most open of the three. Brill et al. [114] concluded from FEM experiments that nitrogen preferentially adsorbed on Fe{111} and that adsorption of nitrogen on the iron tip caused faceting to increase the {111} surface area of the tip. Using Mössbauer spectroscopy, Dumiesic et al. [115, 116] concluded that a particular coordination site on the iron surface, the seven coordinate (C_7) site, was the most active in ammonia synthesis (and hence in nitrogen adsorption). Of the three low-index Fe surfaces, the C_7 site occurs with the highest density on the {111} plane.

As discussed by Spencer et al. [117], both of the above conclusions were based on indirect evidence, but they have been confirmed more recently by the groups of Ertl and of Somorjai. Under ultrahigh vacuum conditions [118, 119], it was found that the initial rate of dissociative adsorption occurred in the ratio 60:3:1 for Fe{111}:Fe{100}:Fe{110} at crystal temperatures of 500–600 K. Ammonia synthesis on iron single crystals at high pressures (20 atm) gave an activity ratio 415:25:1 for Fe{111}:Fe{100}:Fe{110}.

6.2 NITROGEN ON POLYCRYSTALLINE IRON

Wedler et al. [120] studied N_2 adsorption on iron films at 77 and 90 K, and found three different weakly bound adsorption states, which would be designated as δ, γ, and α, by thermal desorption spectroscopy; from isotopic exchange experiments these were all shown to be molecularly adsorbed. The heat of adsorption was recorded calorimetrically to be $\approx 21\,\text{kJ mol}^{-1}$ at low exposure, agreeing quite well with values subsequently reported for γ and α N_2 phases on Fe{111} of 31 and $24\,\text{kJ mol}^{-1}$, respectively. The annealing temperature used in film preparation was found to have a significant effect on the population ratio of the three adsorbed states. Based on earlier work, which showed the annealing temperature to govern the distribution of crystal phases present in a film, the shift in adsorption state population with annealing temperature was attributed to face-specific adsorption, an argument that is realistic in view of present knowledge about the relative reactivities of the low index phases.

Using XPS, Kishi and Roberts [121] studied nitrogen on iron films at 85 and 190 K. At the lower temperature two bands were seen, at binding energies of 405.3 and 400.2 eV. These were associated with two distinct states, linearly bonded and bridge-bonded, although in the light of more recent XPS data on Fe{111} in combination with HREELS data the XPS data has been reattributed to a single linearly bound species, the α phase. At the higher temperature only a single feature at E_B = 397.0 eV was observed, and assigned to atomic nitrogen. Using XPS, Johnson and Roberts [122] observed molecular nitrogen at 80 K converting directly to atomic nitrogen. Their conclusion that the molecular adsorption was not site-specific has since, however, been disputed [123].

In a kinetic study of N_2/polycrystalline iron between 60 and 273 K, Wedler et al. [120] found the dissociative adsorption sticking coefficient to be in the range 10^{-8} to 10^{-7} and the adsorption activation energy, E_a, ranging from -1 kJ mol^{-1} to $+5$ kJ mol^{-1}. Both results were sensitive to the sample preparation history. Wedler et al. also comment on their failure to observe isotopic exchange in a state of nitrogen on iron now known to be atomic.

Work function changes were reported by Gundry [124] who found $\Delta\phi = -0.28$ eV at 90 K and -0.4 eV at room temperature for N_2 on an iron film. This is in fair agreement with the data of Ponec and Knor [125] who reported $\Delta\phi = -0.1$ eV at 77 K and -0.4 eV at 288 K. Both sets of data indicate molecular (α) adsorption at low temperature [126], and of course dissociative adsorption at the higher temperature.

6.3 NITROGEN ON Fe{111}

Four states of nitrogen on the {111} plane have been identified. The main experimental data are summarised in Table 5.

6.3.1 β-Nitrogen

From isotopic mixing experiments on Fe{111}, Bozso et al. [118] concluded that above room temperature only atomic nitrogen was present, with a desorption temperature of ~850 K. Two independent XPS studies of the N_{1s} emission gave reasonably consistent values of 397.1 eV [130] and 396.9 eV [123] which were observed at surface temperatures of up to 800 K; intensity fluctuations were apparent due to bulk diffusion at higher temperatures [130].

The work function behaviour was observed to be complicated because surface reconstruction and bulk diffusion occur, making the work function change, $\Delta\phi$, vs. exposure temperature-dependent. A maximum value of $+0.25$ eV was produced for nitrogen adsorption in the range 318–643 K [118].

UPS gave a single new nitrogen-induced feature at ~5.5 eV, which increased in intensity with nitrogen coverage. More recently, HREELS has been employed to study the atomic adsorbate, yielding for $^{14}N_2$, a stretching Fe—N frequency of 435 cm^{-1}, whilst for $^{15}N_2$, νFe—N is 460 cm^{-1}. A LEED study of the β species [118] revealed a complex series of diffraction patterns. Up to 410 K, the clean surface spots were seen to decrease in intensity as the

TABLE 5

Physical properties of atomic (β) and dinitrogen (α, γ, δ) states of nitrogen on Fe{111}

Phase	T_p/K	$\Delta\phi$/meV	UPS peaks/eV	XPS peaks/eV	Vibrational data/cm^{-1}		Bonding
β	850 [118]	+ 250 [118]	5.5 [123]	396.9 [123]	$(^{15}N)\nu_{N-Ni}$	460 [128] 435 [129]	
					$(^{14}N_2)\nu_{N-N}$	1555[129]	
α	160 [127]	+ 150 [126]	7.3 [123]	398.8 [123]	$(^{15}N_2)\nu_{N-N}$	1490 [129] 1415 [128]	π (side-on)
γ	95 [126]	− 300 [126]	8, 11.7 [123]	401.2, 405.9 [123]	$(^{15}N_2)\nu_{N-N}$	2100 [128]	End-on
δ	84 [126]						End-on

background became brighter, indicating the formation of a disordered adsorbate overlayer. Between 410 and 470 K, with low nitrogen coverages, a (3 × 3) pattern, attributed to a simple overlayer structure, was seen which disappeared with increasing β-N. Above 510 K, depending on annealing time and temperature, and adsorbate coverage, the surface structures (19.5 × 19.5)R23.4°, (21.5 × 21.5)R10.9°, $(3\sqrt{3} \times 3\sqrt{3})$R30° and (2 × 2) were observed. The strong dependence of the formation of these various structures upon surface coverage, annealing temperature and concentration of bulk dissolved nitrogen led to the proposal that the first few atomic layers were reconstructed. That the reconstruction was confined to the near-surface layers only was inferred by the continued observation of bulk metal characteristics in UPS and LEED following reconstruction. LEED patterns from these structures could often be observed to be composites of several of the forms mentioned above, suggesting that the surface layer contained islands of ~100 Å width. In a detailed discussion of the above work it was confirmed, from the various LEED pattern symmetries, that surface reconstruction was due to the formation of surface nitrides of hexagonal symmetry, closely related to Fe_4N. Love and Emmett [131] found that the thermal decomposition of iron nitrides started in vacuo at around 750 K. This value compares well with the recombinative desorption temperature of nitrogen from the bulk.

In early studies on high-area iron catalysts using nitrogen isotopes it was deduced that dissociative adsorption occurred only at $T \geqslant 700$ K. The work described above clearly shows that this conclusion was incorrect. Bozso et al. [118] have suggested that the early observations, leading to an incorrect conclusion, are attributable to nitrogen diffusion between bulk and surface.

6.3.2 *α-, γ- and δ-nitrogen on Fe{111}*

The first work on molecular nitrogen adsorption on Fe{111} was by Bozso et al. [118]. At 140 K, with a static nitrogen pressure of 4×10^{-4} torr, a continuous increase in $\Delta\phi$ was observed. On evacuation and readmission of nitrogen, the work function was seen to fluctuate, indicating rapid and reversible adsorption which was attributed to molecular nitrogen on iron with an adsorption enthalpy estimated to be in the range 20–40 kJ mol^{-1}.

On adsorbing nitrogen on Fe{111} at 120 K, Ertl et al. [127] observed a nitrogen species giving a continuous $\Delta\phi$ increase of up to 140 meV, which desorbed at 160 K (estimated $E_d \leqslant 42$ kJ mol^{-1}). Since atomic nitrogen was known to undergo recombinative desorption at $T > 700$ K, they concluded that the species desorbing at 160 K must be molecularly adsorbed. This was demonstrated by experiments showing (i) a lack of isotopic exchange upon adsorption of $^{14}N_2$ and $^{15}N_2$ at 120 K and (ii) similar desorption temperatures as found for molecular nitrogen adsorption on Ir{110} and Ni{110}. This molecular N_2 species is denoted as α-N_2.

Ertl et al. [127] speculated from the behaviour of other molecular nitrogen–metal adsorption systems that α-N_2 was a terminally bonded species.

In later papers, however, the α-N_2 state was studied using XPS [132] and HREELS [129, 132], and it was concluded that α-N_2 is a π-bonded side-on molecular species. The vibrational frequency, $\nu_{N-N} = 1555\,cm^{-1}$ ($^{14}N_2$) and $\nu_{N-N} = 1490\,cm^{-1}$ ($^{15}N_2$), recorded for the species was much lower than that for which terminal bonding had been established. By comparison of this vibrational frequency with those of various simple nitrogen-containing molecules, other nitrogen adsorption systems and nitrogen organic metal clusters it was concluded that α-N_2 was a π-bonded molecule with a bond order of ≈ 2, both N atoms being coordinated to Fe surface atoms. The low N—N stretching frequency compared favourably with complexes where the

$$M{-}N{=}N{-}M$$

structure exists. In a structure with a π-bonded N_2-metal interaction, the lowering of ν_{N-N} was attributed to the weakening of the N_2 bond as a result of charge donation into the $N_2\pi_u 2p(1\pi)$ antibonding levels, analogous to olefin/acetylene–metal complexes. This electron acceptor idea was supported by the $\Delta\phi$ increase seen for α-N_2 adsorption [127]. In addition, the N_{1s} spectra for α-N_2 were compared with theoretical core hole calculations for linearly and triangularly bonded N_2—Ni complexes; again the results favoured α-N_2 as being π-bonded to Fe{111}. Finally, Freund et al. [133], from an ARUPS study, conclude that the α phase exists with a molecular axis strongly inclined away from the surface normal. In particular, a σ shape resonance, found in other N_2–metal systems, was absent for α-N_2. The interaction of both nitrogen atoms with the surface was considered to quench this resonance. In conclusion, on the basis of an ab initio valence bond calculation, Freund et al. suggest that an excited state of $N_2(^3\Sigma_u^+)$ stabilised by a non-symmetric high-coordination site (Fig. 17), was responsible for the α phase on Fe{111}.

In an effort [134] to identify possible adsorption sites for α-N_2, ethylene was used to displace α-N_2, illustrating its weakly bound nature, while oxygen adsorption (with its known preferential adsorption sites on Fe{111}), was used to infer the adsorption of α-N_2 into the shallow-hole and deep-hollow sites of Fe{111}. Several possible π-bonding conformations on the surface were proposed for the α phase, one in particular [(a) in Fig. 18] finding support in a theoretical calculation of the molecular nitrogen–Fe{111} system by Tomanek and Bennemann [135].

Kinetic data obtained by Ertl et al. [127] led Grunze et al. [123] to search at lower temperatures for an *additional* molecular adsorption state. Using XPS, a three-peak spectrum was obtained for N_2 adsorption under a static nitrogen pressure of 5×10^{-7} mbar at 85 K. The signal intensity at a binding energy $E_B = 398.8$ eV was seen to increase with time and was associated with the α-N_2 state. A doublet at binding energies of 405.9 and 401.2 eV was assigned to a second molecular state (denoted γ) produced by reversibly adsorbing nitrogen, the XPS doublet intensity correlating directly with the

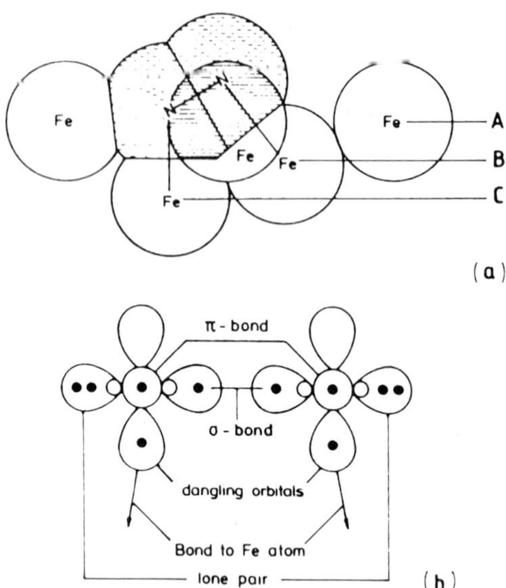

Fig. 17. (a) Schematic representation of a possible non-symmetric bonding site of α-N_2 on Fe{111}. (b) Schematic representation of the $^3\Sigma_u^+$ excited state of N_2 which can form two covalent bonds with the metal atoms in the geometry shown in (a). (From Freund et al. [133].)

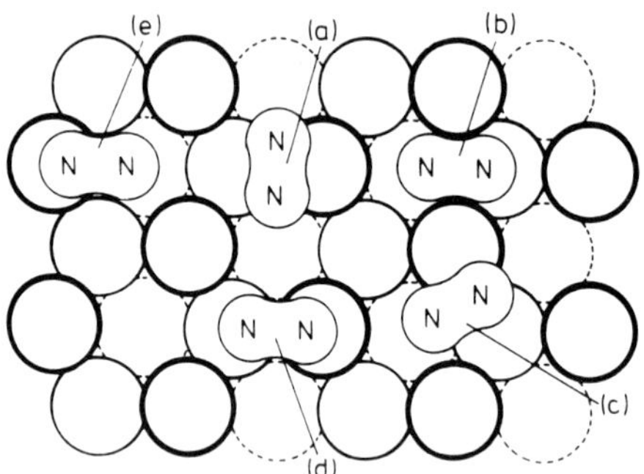

Fig. 18. Possible models for the side-on geometry of the α_1-N_2 state on Fe{111}. For (a), (b), (d) and (e), both N atoms are coordinated to two Fe atoms while for (c), they are coordinated to three Fe atoms. First-layer Fe atoms are shown in thick circles, the second layer with thin circles and the third layer with dotted circles. (From Tsai et al. [129].)

nitrogen pressure. By heating to 97 K the γ-N_2 state could be desorbed ($\Delta H_{ad} \simeq -24\,\mathrm{kJ\,mol^{-1}}$) leaving only the $E_B = 398.8\,\mathrm{eV}$ signal due to α-N_2. Heating of this α-N_2 phase to 126 K produced a shift of the N_{1s} singlet to a final value of $E_B = 397.9$ eV, corresponding to β (atomic) nitrogen. He(II)UPS experiments in the same work gave two bands at 11.7 and 8 eV due to γ-N_2 and one feature at 7.3 eV for α-N_2.

The vibrational spectrum of the γ-N_2 state of Fe{111} was first reported by Whitman et al. [128]. A value of $\nu_{N-N} = 2100\,\mathrm{cm^{-1}}$ was recorded at 74 K after adsorbing a monolayer of $^{15}N_2$ ($\theta = 1$ as defined by Grunze et al. [123] from XPS intensity calibration). On the basis of the ν_{N-N} frequency, γ-N_2 was assigned to a terminally bonded species. Recent ARUPS experiments by Freund et al. [133] have clearly distinguished the γ phase to consist of nitrogen with its axis oriented along the surface normal.

Continuing the search to lower temperature for more information on molecular adsorption, a further molecular phase was reported by Strasser et al. [126] in a thermal desorption experiment after nitrogen adsorption at 68 K. They discovered with increasing exposure (> 3 L) a third molecular desorption peak at 84 K in addition to those of the γ and α phase at 95–101 and 155 K, respectively. By sequential adsorption and desorption of isotopes, it was demonstrated that the low-temperature peak was in fact a distinct state (designated δ-N_2) and not a result of lateral interactions in the γ/α phase layer. The δ state was proposed in this work to be a physisorbed second layer above the γ phase, which acts as a precursor to γ state adsorption. In later work, however, Grunze et al. [134] suggest that δ-N_2 is a physisorbed first-layer state which acts as an "extrinsic" precursor to γ-N_2 occupying distinct but energetically less favourable sites on the surface.

Strasser et al. [126] also measured the work function change and clearly demonstrated the adsorption and desorption of various molecular states under changing temperature and pressure conditions. No distinct LEED patterns were observed for molecular nitrogen adsorption on Fe{111} by Tsai et al. [129] and more recent work has given no indication to the contrary. The properties of the α, γ and δ states are summarised in Table 5.

6.3.3 Kinetics and thermodynamics of N_2 on Fe{111}

Renewing previous studies on iron, Ertl et al. [127] identified the α phase from thermal desorption and work function measurements as molecular nitrogen and proposed a precursor mechanism for dissociative adsorption.

$$N_{2(g)} \longrightarrow N_{2(ad)} \longrightarrow 2\,N_{(ad)}$$

The following data was derived for the reaction sequence from their measurements.

(i) Desorption spectra from $N_{2(ad)}$ gave $T_{max} = 160\,K$ and hence $E_{des} \approx 42\,kJ\,mol^{-1}$.

(ii) The isosteric heat of adsorption for $N_{2(ad)}$ is $30\,kJ\,mol^{-1}$ at low surface coverages ($\sim 1 \times 10^{14}$ molecules cm^{-2}).

(iii) The initial sticking coefficient into the molecular state is $10^{-2\pm0.5}$. (The uncertainty is considered to be inherent since desorption from the molecular state would proceed even as adsorption was occurring.)

(iv) By calibrating the coverage of $N_{(ad)}$ using an AES method [118] the initial sticking coefficient for dissociative adsorption was found to be 5×10^{-6}.

References pp. 124–129

(v) The initial dissociative sticking coefficient s_0 was seen to decrease slightly with surface temperature (earlier work [118] over a smaller temperature range had reported that it is approximately independent of temperature). An analysis gave $s_0^{\mathrm{d}} = \nu_{\mathrm{d}} \exp(-E_{\mathrm{a}}/RT)$ with $\nu_{\mathrm{d}} = 2.2 \times 10^{-6}\,\mathrm{s}^{-1}$ and $E_{\mathrm{a}} = 3.4\,\mathrm{kJ\,mol^{-1}}$. It must be noted, however, that T in this expression refers to *substrate* temperature, and that ν_{d} may well vary with *gas* temperature. The term E_{a} cannot properly be defined as the activation energy for adsorption.

The following facts [127] support the idea that the dissociative adsorption of nitrogen proceeds via a molecular precursor state and not via direct dissociative adsorption.

(i) In an XPS study, Johnson and Roberts [122] reported the presence of $N_{2(\mathrm{ad})}$ at 80 K on polycrystalline iron without the presence of gaseous N_2.

(ii) The observation [118] that $(s_0)_{\mathrm{mol}} \gg (s_0)_{\mathrm{diss}}$.

(iii) A study by Thorman and Bernasek [136] showed that the vibrational and rotational energies of nitrogen molecules formed by recombinative desorption was always at a temperature just slightly above that of the surface. Invoking the principle of microscopic reversibility, this result suggested that desorption proceeded through a molecular, thermally equilibrated state.

From a theoretical study, based on the precursor model [127], Böheim et al. [137] concluded that, if the translational energy of the incoming nitrogen molecules was of the order of the activation barrier between $N_{2(\mathrm{ad})}$ and $N_{(\mathrm{ad})}$, then energy transfer from the $N_{2(\mathrm{g})}$ to the surface would be greatly reduced. On the basis of this conclusion and by assuming that the $N_{2(\mathrm{ad})}$ would be rotationally frozen, a very low value for $(s_0)_{\mathrm{diss}}$ was predicted in agreement with experiment [127]. Furthermore, Böheim et al. [137], in a molecular beam experiment, found that the angular distribution of inelastically scattered N_2 from Fe{110} was very narrow. This observation was interpreted as a sign of the unusually low energy transfer found theoretically.

Some further support for the precursor model has come more recently from XPS [123, 132] and HREELS [126, 129] studies of the kinetics and interstate conversions of the low-temperature molecular adsorption phases. This is also illustrated by the work of Whitman et al. [128], shown in Fig. 19, demonstrating the sequential appearance of the vibrational bands of γ-, α- and β-nitrogen as the temperature of a N_2-precovered surface was increased.

Grunze et al. [123] used XPS to follow the N_{1s} intensities of the α and γ phases, obtaining data for the process

$$N_{2(\mathrm{g})} \rightleftarrows N^{\alpha}_{2(\mathrm{ad})} \rightleftarrows N^{\gamma}_{2(\mathrm{ad})} (\longrightarrow 2\,N_{(\mathrm{ad})})$$

and an empirical potential energy diagram was derived. The initial sticking coefficient for the α phase was found to be $(s_0)^{\alpha}$ (91 K) $\simeq 10^{-2}$ and $(s_0)^{\alpha}$ (128 K) $\simeq 10^{-3}$. The decrease in $(s_0)^{\alpha}$ with temperature was accounted for by a decrease in the surface coverage of N_2^{γ}, the precursor to α-N_2, with increasing temperature. (In fact, a decrease in sticking probability would not

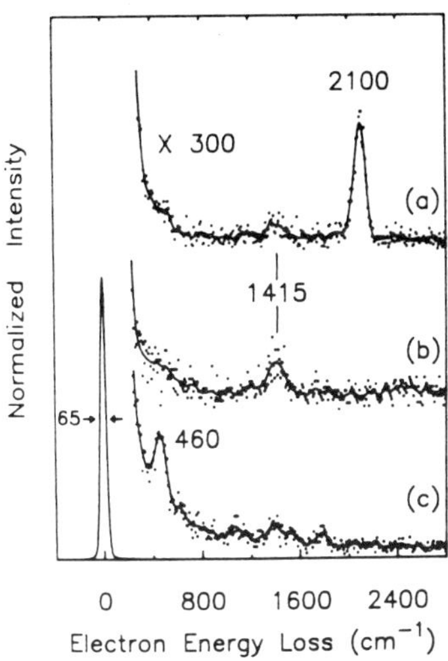

Fig. 19. Electron energy loss spectra of $^{15}N_2/Fe\{111\}$. (a) 74 K showing the γ-N_2 species. (b) After heating to 110 K, showing the α-N_2 species. (c) After heating to 160 K only the Fe—N stretch of atomic nitrogen is observed. (From Whitman et al. [128].)

necessarily follow from a decrease in precursor population: the relevant factor is the partitioning between desorption and passage from precursor to chemisorbed state.) The value for E^{γ}_{ad} was evaluated as $\leqslant 24\,kJ\,mol^{-1}$. Tomanek and Bennemann [135] have evaluated theoretically the total energy for the $N_2^{\gamma} \rightleftarrows N_2^{\alpha}$ and $N_2^{\gamma} \rightarrow 2N^{\beta}$ transitions using a LCAO formalism. The two surfaces are shown in Fig. 20.

Strasser et al. [126] have recorded adsorption isobars for the γ state and calculated the isosteric heat of adsorption as $q^{\gamma}_{st} = 37$–$25\,kJ\,mol^{-1}$, the energy range being due to the coverage dependence of the γ and α phases (since γ is considered to be a precursor for α). Whitman et al. [128] found a maximum coverage of the α-N_2 phase on Fe{111} of 0.10 ML (where the monolayer ML is as defined for the $N_2/Fe\{111\}$ system by Grunze et al. [129] from integrated XPS intensity measurements). This is unusually low compared with other molecular nitrogen systems, and it was suggested that saturation of this phase at such a low coverage was indicative of strong repulsive interactions between adsorbed species, mediated by the metal.

Recent work on the kinetics of $N_2/Fe\{111\}$ combines thermal desorption with model simulation [134]. The full adsorption system is defined as

$$N_{2(g)} \rightleftarrows N^{\delta}_{2(ad)} \rightleftarrows N^{\gamma}_{2(ad)} \rightleftarrows N^{\alpha}_{2(ad)} \longrightarrow 2N^{\beta}_{(ad)}$$

The δ phase was seen as a precursor to γ-N_2 at $T < 100\,K$ with a direct channel $N_{2(g)} \rightleftarrows N^{\gamma}_{2(ad)}$ opening up at higher temperatures. Thermal desorption data for $^{15}N_2$ adsorption at 70 and 83 K were obtained along with their computer-modelled counterparts. Analysis of the data revealed that desorption from the γ-N_2 phase could not be described by first-order kinetics, but

References pp. 124–129

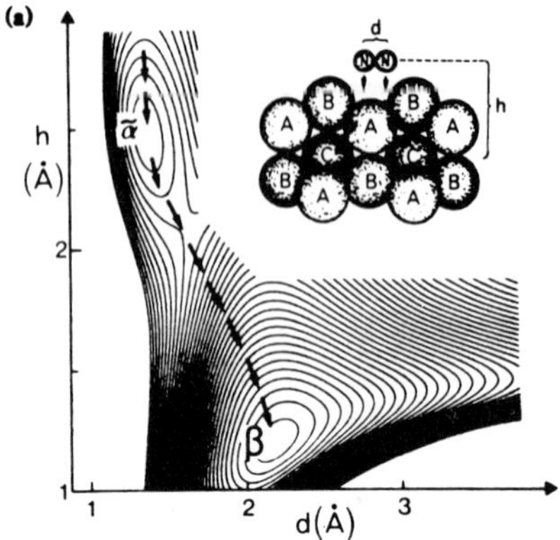

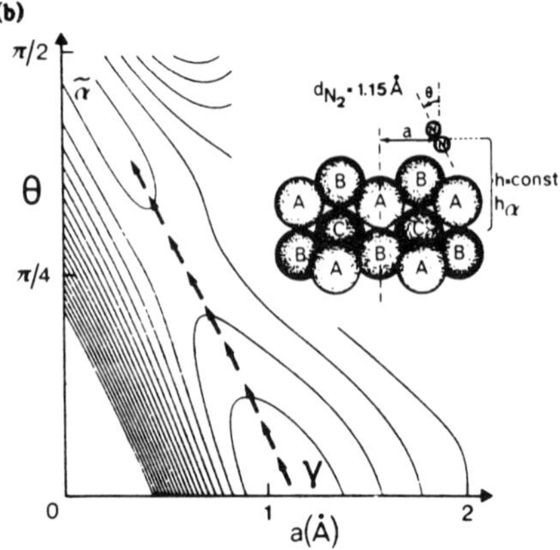

Fig. 20. Total energy contour plot for N_2 adsorption on Fe{111}. (a) $E_{tot}(d, h)$ for N_2 dissociation in the α geometry. d is the intramolecular distance and h is the height with respect to third layer (C) atoms. (b) $E_{tot}(a, \theta)$ for the transition γ-N_2 to α-N_2 in the γ geometry. a is the horizontal distance from the AB "edge" and θ is the tilt angle. The equidistant contours are separated by 0.2 eV. The preferential transition path is marked by arrows. (From Tomanek and Bennemann [135].)

rather by a desorption rate coverage dependence of $\theta/(1 - \theta)$. The activation energy for γ-N_2 desorption was evaluated as $E_d^\gamma = 25 \pm 3\,\text{kJ}\,\text{mol}^{-1}$ with $\nu_d^\gamma = 10^{13\pm2}$ for $0 < \theta^\gamma \leqslant 0.3$. Integration of the desorption spectra, knowing $\theta^\gamma = 1$ when $[N_2^\gamma] = 3.5 \times 10^{14}$ molecules cm^{-2}, yields $s_0^\gamma = 0.7$ for $T \leqslant 83\,\text{K}$.

The coverage dependence of the γ phase desorption rate, the T_p and half-width variation of the desorption spectra along with the constant value of s_0^γ at low θ all supported the idea that adsorption into and desorption from the γ phase proceeded via the δ state as precursor. To test their picture of the total adsorption system, Grunze et al. simulated their desorption spectra based on the model that adsorption of γ-N_2 could proceed via a direct adsorption channel from the gas phase or via δ-N_2. The γ-$N_2 \rightarrow \alpha$-N_2 conversion was ignored due to its slow rate at the temperature used in the simulation. Results of the simulation showed, in agreement with the sequential isotopic

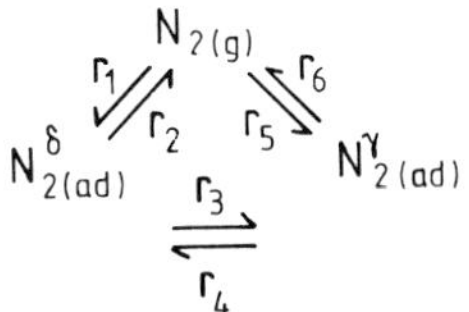

dosing experiments performed previously [126], that adsorption proceeded initially into the γ phase until near saturation when population of the δ phase commenced.

Algebraic expressions were derived for r_1 to r_6 and using experimentally derived data such as ν^{γ}_{des}, $E^{\gamma}_{des} E^{\delta}_{des}$, desorption spectra were simulated and found to be in reasonable agreement with the actual data (Fig. 21), demonstrating a change in the γ-N_2 adsorption mechanism with temperature. Values for E^{γ}_{des} were noted to be systematically lower than q^{γ}_{st} by $\approx 10\%$; the authors discuss this discrepancy in terms of (a) transfer effects between the γ and δ phases and (b) the limited energy exchange phenomenon discussed by Böheim et al. [137], although no convincing conclusions were drawn. A large increase in q^{γ}_{st} in the presence of pre-adsorbed α-N_2 is explained in terms of a mutual attraction between the γ and δ phases.

The evaluation of activation energies for adsorption processes is inherently ambiguous when the data are obtained by adsorption from random gas in a large UHV chamber: only the crystal temperature can be varied, the gas being essentially accommodated to the walls of the reaction vessel.

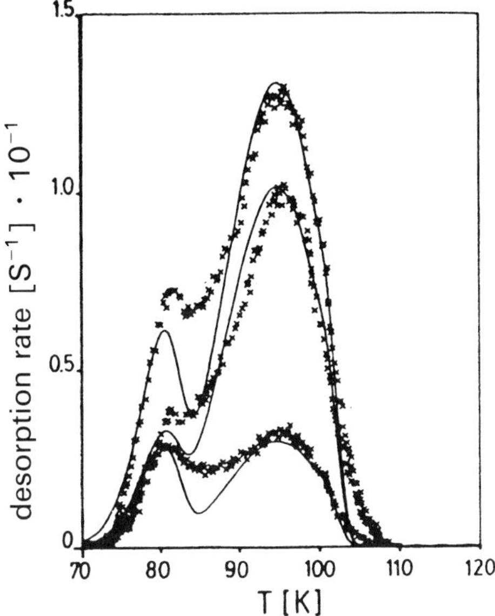

Fig. 21. Thermal desorption data of $^{15}N_2$ ($\times$) and their simulation (solid line) for sequential dosing of $^{14}N_2$ and $^{15}N_2$ on Fe{111} at 68 K. Upper curve, 30 L $^{15}N_2$; middle curve, 15 L $^{15}N_2$ followed by 15 L $^{14}N_2$; lower curve, 15 L $^{14}N_2$ followed by 15 L $^{15}N_2$. (From Grunze et al. [134].)

Rettner and Stein [138] have examined the dissociative chemisorption of N_2 on Fe using a supersonic molecular beam source, with variable beam energy between 0.09 and 4.3 eV. (The higher energy range was achieved by a combination of a tungsten nozzle resistively heated to 2000 K and the use of H_2 as a seeding gas.) The dependence of the zero-coverage sticking probability, s_0, on beam energy at normal incidence and a surface temperature of 510 K is shown in Fig. 22. At the lowest energy $s_0 = 10^{-6}$, in agreement with measurements using random thermal gas dosing. However, s_0 rises sharply with increasing beam energy, reaching a limiting value of about 0.1. This result is consistent with a range of activation energies for the dissociative process deriving from the variety of collision parameters for molecules striking the surface, arising from both rotational orientation with respect to the surface plane and azimuth, and surface coordinates within the surface unit mesh. The largest barrier for the most unfavourable collision parameters would appear to be about 2 eV (190 kJ mol^{-1}). However, this implies a *direct* dissociative chemisorption of the molecule on impact with the surface, without prior trapping into a precursor state. Clearly, if some energy was dissipated on collision with the surface, with capture into the precursor state, the subsequent process should be independent of beam energy. Similarly, the probability of trapping into a precursor state would *decrease* with increasing beam energy, and not increase as would be required to generate an increase in s_0. Thus, precursor trapping would appear to be ruled out. However, Rettner and Stein [138] also report that s_0 is dependent on *surface* temperature. The dependence is not large, with s_0 increasing by about 50% in going from 600 to 300 K. Thus the substrate temperature dependence reported is qualitatively similar to that previously reported from random, thermal gas adsorption measurements. In Fig. 23, the dependence of s_0 on crystal temperature is presented as an Arrherrius plot for a beam energy of 1.05 eV. This plot suggests that precursor trapping does in fact occur, and

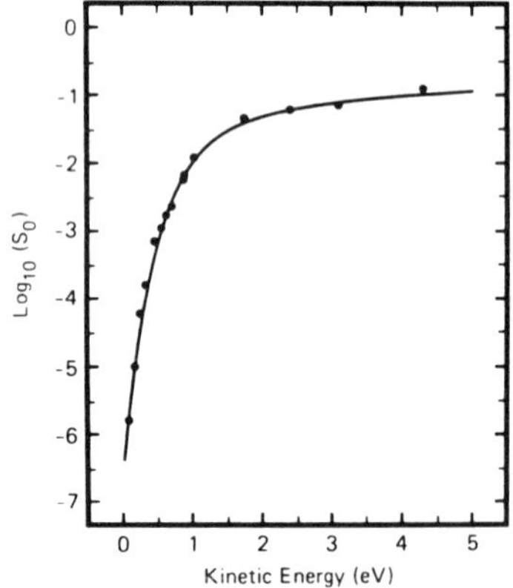

Fig. 22. Effect of incident energy on the dissociative chemisorption probability of N_2 on Fe{111} at normal incidence and a surface temperature of 520 K. (From Rettner and Stein [138].)

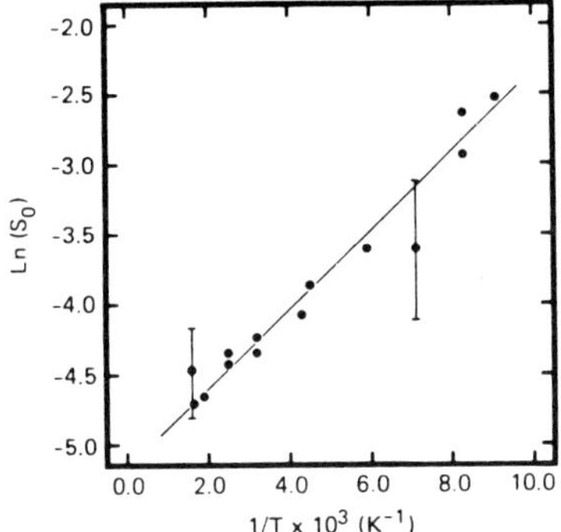

Fig. 23. Effect of surface temperature on the dissociative chemisorption probability of N_2 on Fe{111} at normal incidence and at a kinetic energy of 1.05 eV. (From Rettner and Stein [138].)

from the slope a difference in activation energy of 2.2 kJ mol^{-1} between dissociation and desorption paths from the precursor state is obtained. This interpretation of the data in Fig. 23 would appear to contradict the observation shown in Fig. 22, i.e. the sensitivity of s_0 to beam temperature. The authors propose a precursor model in which access to the precursor state itself is blocked by a distribution of activation barriers. Thus, for example, according to this model the largest barrier height, of ~190 kJ mol^{-1}, corresponds to the barrier between gaseous N_2 and the precursor state. They suggest that the most likely candidate for this precursor state is α-N_2. What is perhaps most surprising, if this model is correct, is that the trapping into the α-N_2 state itself is not attenuated at higher energies. Thus, we might have expected the s_0 vs. beam energy plot to pass through the maximum instead of approaching an asymptote. It is quite possible that the small substrate temperature dependence observed is attributable to subtle variations in, for example, the potential energy surface itself with crystal temperature, rather than to trapping into a precursor state. The situation is not fully resolved at this time, and we feel that this work does bring in to question the modelling based on precursor state trapping.

As is evident from the work reviewed here, a large amount of data has been gathered recently on the N_2/Fe{111} system. Its potential usefulness in catalysis has been demonstrated by the work Bowker et al. [139, 140] and of Stoltze and Norskøv [141], who used surface science data to model high-pressure ammonia synthesis where the dissociative adsorption of nitrogen plays such an important role (as discussed earlier). These analyses do, however, emphasize a very important point relating to the analysis of thermal desorption spectra. Using the desorption energy value obtained from experimental desorption spectra on the assumption of a normal pre-exponential factor, the computed rate of ammonia synthesis was found to be many orders of magnitude too low [139]. However, Stoltze and Norskøv [141] argue that account should be taken of the unusually low sticking probability for β-N formation;

applying "microscopic reversibility" they obtain a low pre-exponential factor for desorption, i.e. $\sim 10^7\,s^{-1}$ instead of $10^{13}\,s^{-1}$. This, in turn, yields a much lower desorption energy, and hence Fe—N bond energy: using this value, excellent agreement is obtained between computed and observed ammonia synthesis rates [141]. Bowker et al. [140] have criticised the evaluation of this energy on the grounds that a good fit to the experimental desorption spectra is still not generated. Undoubtedly, an accurate evaluation of the isosteric (or calorimetric) heat of adsorption for β-N is called for: this is the single most important factor in determining the computed ammonia synthesis rate. This is, however, a very demanding experiment, particularly due to the absorption of N into the bulk of the crystal and Fe_4N formation. We would point out that the microscopic reversibility argument put forward by Stoltze and Norskøv is not necessarily correct. For example, it is found that, for H_2 on W{100}, the sticking probability falls linearly with coverage, showing no remarkable variation at a coverage of 0.46 monolayer where the desorption pre-exponential falls by a factor of 10^7 [142]. Clearly, the microscopic reversibility argument does not apply in this case.

6.4 NITROGEN ON Fe{100}

From a LEED study, Brill [143] deduced that nitrogen was not chemisorbed on Fe{110} and Fe{100}. This result for Fe{100} was disproved by the work of Ertl et al. [144], as shown by the graph of surface atomic nitrogen coverage vs. exposure at various temperatures (Fig. 24). At higher temperatures, θ_N vs. exposure was seen to pass through a maximum before tailing off due to the effects of diffusion into the bulk. From LEED, a c(2 × 2) pattern was seen from all but the lowest coverages up to $\theta = 0.5$. The structure shown in Fig. 25 was proposed with N atoms adsorbed in four-fold sites. Using the LEED result, the initial sticking coefficient, s_0, was calculated as 1×10^{-7} (383 K) and 4×10^{-7} (508 K). The activation energy for adsorption, E_a, was given as $\approx 21\,kJ\,mol^{-1}$ for adsorption in the range

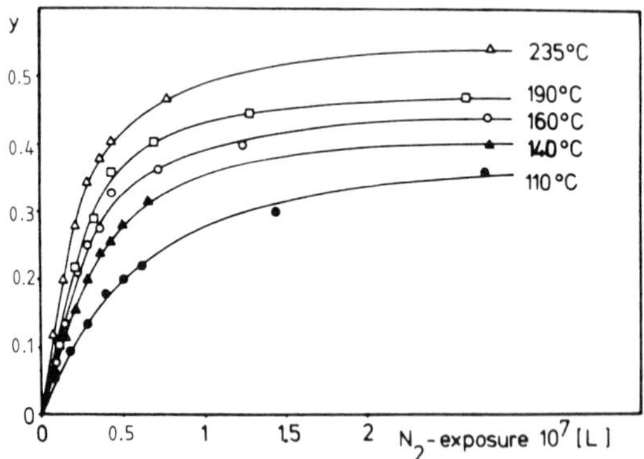

Fig. 24. Variation of the nitrogen surface concentration with N_2 exposure at Fe temperatures between 110 and 235°C. (From Ertl et al. [144].)

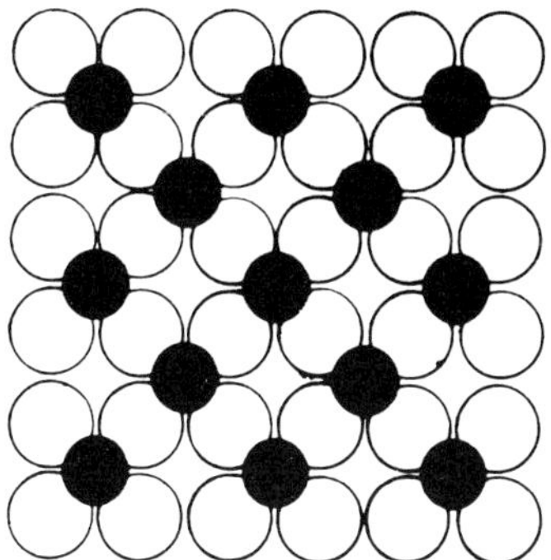

Fig. 25. Structure model for the arrangement of the adsorbed nitrogen atoms on an Fe{100} surface. (From Ertl et al. [144].)

693–783 K. However, this value is based entirely on a variation of the sticking probability with *crystal* temperature. A proper evaluation would involve an analysis with variable crystal and gas temperature.

Bozso et al. [118], using thermal desorption, UPS and $\Delta\phi$ measurements, reported that the dissociative adsorption activation energy, E_a, varied with coverage (Fig. 26); the variation of s_0 with coverage and temperature is shown in Fig. 27. A continuous increase in $\Delta\phi$ was found with surface nitrogen coverage, which was independent of substrate temperature or coverage. The constancy of the surface dipole moment derived from these measurements implies the formation of a simple ordered overlayer rather than surface reconstruction. Thermal desorption data as a function of surface coverage showed T_p shifting about 50 K downwards from ≈ 1000 K with surface nitrogen coverage, the small temperature shift being interpreted as first-order desorption kinetics with a decreasing ΔH_{ad}^{diss} value with coverage.

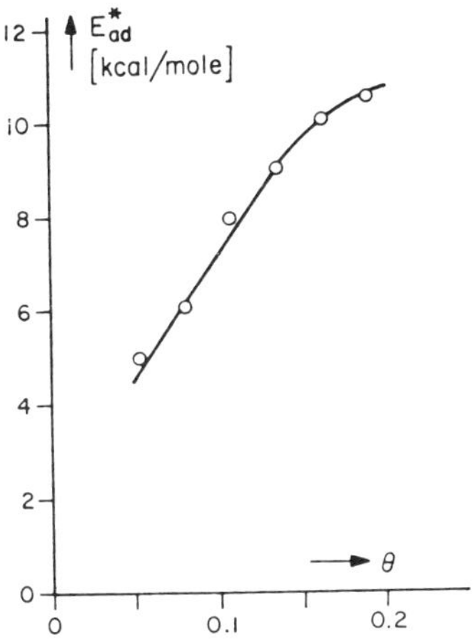

Fig. 26. Variation of the activation energy for N_2 adsorption, E^*_{ad}, on Fe{100} with coverage θ. (From Bozso et al. [118].)

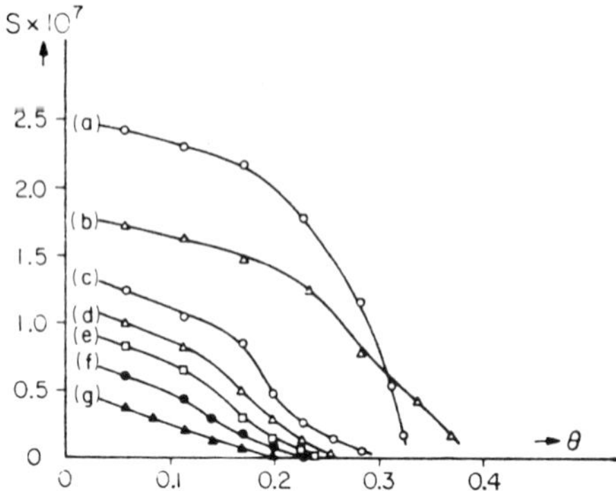

Fig. 27. Variation of the sticking coefficients for N_2 adsorption on Fe{100} with coverage at different temperatures. (a) 743 K; (b) 693 K; (c) 508 K; (d) 463 K; (e) 433 K; (f) 413 K; (g) 383 K. (From Bozso et al. [118].)

Assuming a "normal" value of $\nu_d = 10^{13}\ s^{-1}$, E_d was estimated as ≈ 244 kJ mol^{-1}; this value may need to be revised downwards as a better value for ν might be about $10^6\ s^{-1}$ due to the low sticking coefficient on this system [141].

The major feature from UPS data of N_2/Fe{100} is a band at about 5.0 eV below E_f, derived from N_{2p} states. In a more recent study, Imbihl et al. [145] conducted a LEED I–V analysis of the c(2 × 2) structure on this surface, and confirmed that the adsorbed N atoms occupy fourfold hollow sites.

No molecular nitrogen adsorption was noted in the above studies, in contrast to the {111} plane.

6.5 NITROGEN ON Fe{110}

Gafner and Feder [146] studied a "nearly clean" Fe{110} sample by LEED (patterns and intensities) with surface nitrogen produced by annealing and sputtering from bulk impurities. The LEED data were attributed to contributions from two domains combining to give a complex LEED pattern, the unit cell of which was said to contain six nitrogen molecules close packed with their molecular axes perpendicular to the surface. The assumption that molecular nitrogen would be present on the surface following the pretreatments used is questionable and was to be proved incorrect by later work [147]. Feder and Gafner [146] also performed a LEED I–V analysis on clean Fe{110} and concluded that the surface was unreconstructed. Broden et al. [147] used LEED/AES and UPS to study N_2/Fe{110}. No molecular adsorption was reported even at the highest exposure used at or above room temperature.

As with Fe{100} and {111}, bulk diffusion of nitrogen was observed to occur; in this case it was seen in the temperature range 620–820 K. The UPS spectra show a feature at −5 eV for atomic nitrogen (produced by NO decomposition) similar to the N_{2p} band for Fe{100} and {111} reported by Bozso et al. [118]. In LEED, a pattern similar to that of Gafner and Feder [146], designated the "carbon ring" structure, was observed.

The most recent study of N_2/Fe{110} is that of Bozso et al. [119]. Due to the low reactivity of Fe{110} relative to the other two low-index planes of Fe, the N_2/Fe{110} system suffers from problems arising from bulk diffusion of nitrogen at the high temperatures required to dissociatively adsorb nitrogen, and from the relatively high partial pressures of gas-phase impurities present due to the high total pressures that are required.

The high desorption temperature of nitrogen ($\approx$930 K) necessitated the use of isotopic nitrogen to distinguish surface N from the bulk-dissolved species. This work also confirmed the nature of the adsorbed species as atomic. In LEED, two ordered structures were observed. At low exposure, a (2 × 3) pattern resulted (Fig. 28) whereas at higher exposure, a more complex pattern was seen. Both patterns arise from domain formation, and from the LEED data it was concluded that formation of such large unit cells is indicative of complex structures not reconcilable with simple overlayers.

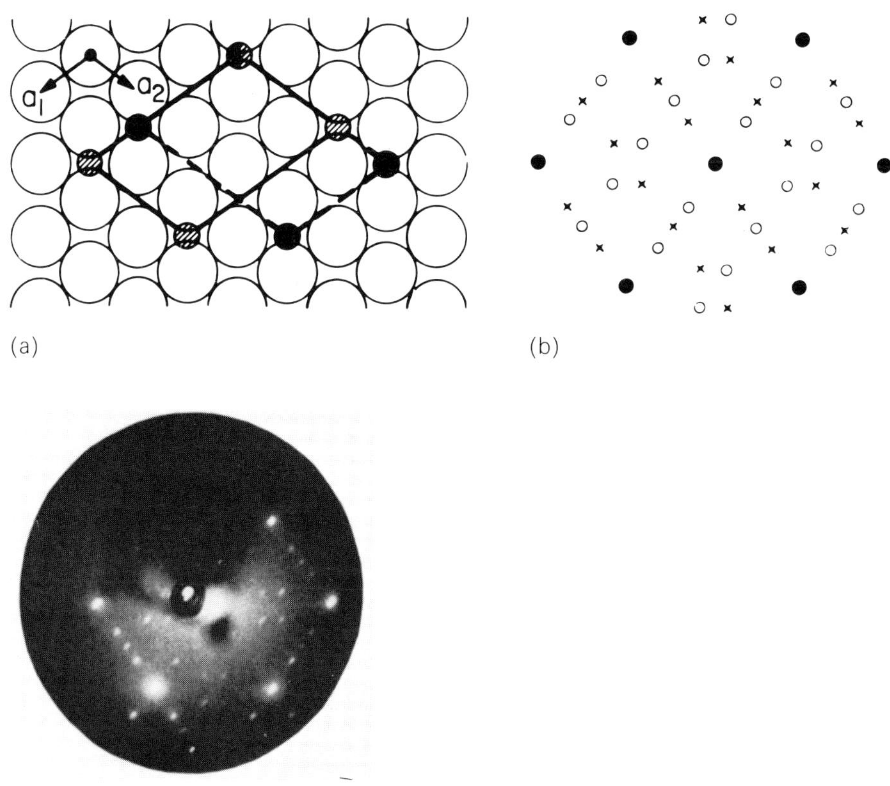

Fig. 28. Structure of (2 × 3)N—Fe{110} LEED pattern. (a) Unit cells of both domain orientation. (b) Constructed reciprocal lattice: (●), substrate lattice (unreconstructed Fe{110}; (○ and ×), reciprocal lattice points from both domain orientations from the surface structure. (c) LEED pattern, 120 eV. (From Bozso et al. [119].)

This, and the fact that LEED pattern formation was subject to the bulk nitrogen concentration, led to the proposition of reconstructed surfaces as seen with Fe{111} [118]. Reconstructed surface structures consistent with the data were proposed, bearing similarities to the {111} plane of Fe_4N, as seen previously [118].

6.6 NITROGEN ON Fe{1210}

In the only paper dealing with nitrogen adsorption on a well-defined stepped iron surface, Dowben et al. [149] adsorbed nitrogen at "high pressures" (7–36 mbar) and 750 K on Fe{1210}. They observed, using LEED to follow spot streaking, that the step density was increased, eventually leading to faceting of the sample. As with previous work, the observed transformation was associated with the formation of iron nitrides, producing surface reconstruction.

6.7 BULK DIFFUSION OF NITROGEN IN IRON

In their work on the {111} and {100} planes of iron Bozso et al. [118] outlined the effects of diffusion of dissociatively adsorbed nitrogen into the bulk of iron. Measurements of the adsorption activation energy, E_a, were only determined from a variation of *crystal* temperatures up to 470 K. The establishment of an equilibrium bulk concentration was described from previous work, where for

$$\frac{1}{2}\, N_{2(g)} \rightleftharpoons N_b$$

the bulk equilibrium concentration N_b in wt.% was

$$(N_b) = 0.098 \exp(-\Delta H_s/RT)$$

with

$$\Delta H_s = 30.2\,\mathrm{kJ\,mol^{-1}}$$

Due to bulk-dissolved nitrogen, thermal desorption spectra were found to exhibit a superposition of peaks from both the surface and bulk species, as shown in Fig. 29. The two contributions to the spectra were demonstrated by the use of isotopes. Various effects are reported as arising from the methods used to "standardise" iron samples by nitrogen pretreatment.

In a study of N_2/Fe{100} Ertl et al. [144] followed N uptake on the crystal at 508 K as a function of sample pretreatment. The effect of annealing time on the bulk nitrogen concentration is clearly shown in Fig. 30. A short annealing time (5 min 640 K) produced limited isotopic exchange between surface ^{15}N and bulk ^{14}N, as evidenced by their distinct appearance in TDS peaks. Annealing to 640 K for 60 min produced a much greater mixed isotope signal with correspondingly reduced ^{15}N and ^{14}N signals. The standardisation of Fe samples by nitrogen pretreatment raises many questions about the detailed interpretation of published data which are not readily answered.

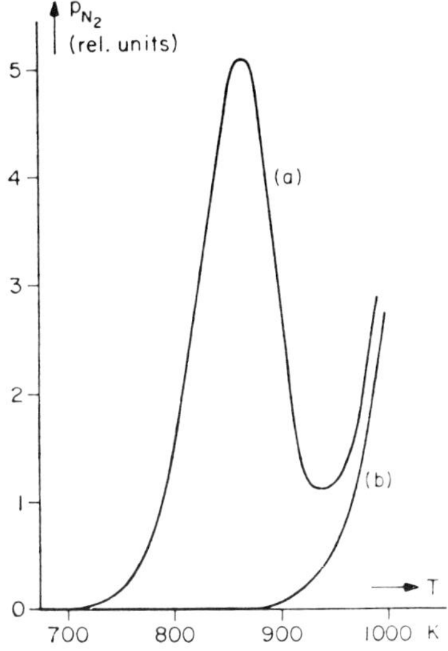

Fig. 29. Thermal desorption spectra from Fe{111}. (a) First run; (b) second run. (From Bozso et al. [118].)

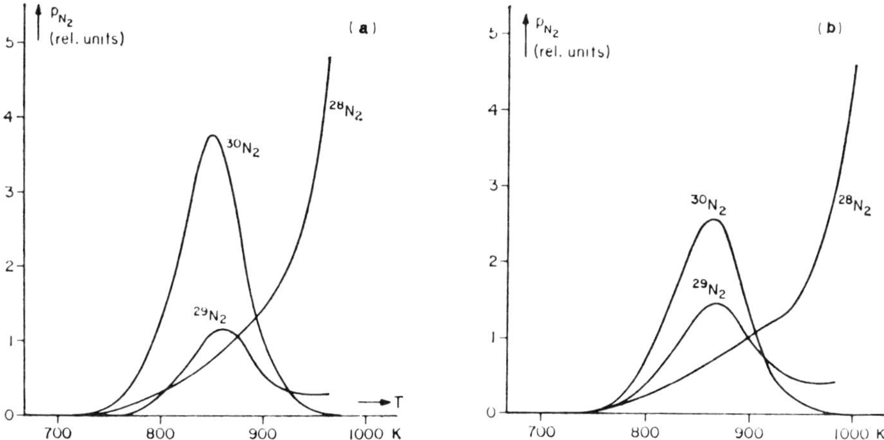

Fig. 30. Thermal desorption spectra for $^{28}N_2$, $^{29}N_2$ and $^{30}N_2$ desorbing from Fe{111}. (a) Five and (b) 60 minutes of annealing at 640 K prior to desorption. (From Bozso et al. [118].)

7. Nitrogen adsorption on tungsten

7.1 INTRODUCTION

The adsorption of nitrogen on both polycrystalline and single-crystal tungsten surfaces has been studied by many investigators using a variety of experimental techniques. Early reviews have summarised most of the polycrystalline data [148, 150–154], and some of the single-crystal work [151, 155].

The major characteristics of this adsorption system can be readily outlined by considering the work carried out on polycrystalline samples. Briefly, Ehrlich [156, 157] identified three states of binding denoted as γ, α and β with respective desorption energies of approximately 40, 80 and 320 kJ mol^{-1}. Two of these, the α and β phases, occur during room-temperature adsorption and the other, the γ phase, exists only at much lower temperatures (below 115 K).

7.1.1 The β state

Adsorption at room temperature leads to the formation of strongly bound β states which desorb above 1000 K. The heat of adsorption of nitrogen into these β states has been reported to be around 360 kJ mol^{-1} [158–161] and to vary little with coverage [157]. This suggested, to a first approximation, that the surface sites available for β-nitrogen were energetically equivalent. However, subsequent experiments revealed greater complexity. First, desorption from the β-state was resolved into two components [162–165], the β_1 and β_2 states. The two β peaks could not simply be the result of heterogeneity present in a polycrystalline sample, since this behaviour is also exhibited by single-crystal planes [164, 166].

The β_2 state appears in desorption spectra immediately on exposure to nitrogen and saturates at about 60% of the total coverage, while the β_1 state starts to populate only after the fractional coverage reaches ≈ 0.2 and continues to grow after the β state has saturated.

The β_1 desorption peak exhibits first-order kinetics with an activation energy for desorption of between 305 kJ mol^{-1} [163] and 343 kJ mol^{-1} [167]. On the other hand, the β_2 state has an activation energy of desorption of about 360 kJ mol^{-1} [157, 162, 163, 167] and was originally assumed to desorb with second-order kinetics. Additional studies by Madey and Yates [165] on the rate of desorption as a function of coverage at a particular temperature, showed that the apparent desorption order actually decreases with increasing temperatures. For example, a desorption order of 1.48 ± 0.03 was observed at 1400 K while at 1225 K, the order was 3.65 ± 0.06.

Isotopic exchange experiments [163, 167, 168] of nitrogen in the β states showed evidence of mixing, indicating both the β_1 and β_2 states to be atomic. A number of models have been put forward to reconcile this fact with the differing desorption kinetics of the two states. Of these, the most satisfying is the model which proposes that essentially only one β adsorption phase exists and the detection of the β_1 and β_2 states is simply a reflection of the differences in desorption mechanism. For example, Robins et al. [169] propose that adsorption sites can be occupied by either adatoms or molecular complexes (comprised of pairs of atoms) which are in dynamic equilibrium with each other. Some molecular complexes desorb with first-order kinetics in the β_1 peak, while both molecular complexes and adatoms desorb in the β_2 peak with apparent non-integral desorption order.

7.1.2 The α state

The second state, of intermediate binding energy, present at room temperature is the α state. Its heat of adsorption has been variously estimated at 76, 84 and 105 kJ mol^{-1} [162, 163, 170]. On tungsten films there is evidence that the α state possesses a range of binding energies [171].

The α state forms simultaneously with the β state during the initial stages of adsorption at room temperature, reaching a peak value of about 15% of the β state and then decreasing until at saturation it constitutes only a small percentage of the total. The decrease in coverage as adsorption proceeds suggests that this state is partly formed on sites which are subsequently displaced by β atoms [163]. Exactly how adsorption in the α state is dependent on the presence of β nitrogen has not been established. However, it has been found that the desorption energy of the α state decreases as β-state coverage increases [148, 172]. This has led to the suggestion [172] that, at high coverages, adsorption of β-nitrogen occurs in nearest-neighbour sites, causing sufficient weakening of the surface—α-nitrogen bond to cause the desorption of the species.

The coverage of the α state and its behaviour with increasing coverage varies on different substrates and seems to be strongly related to the presence of certain crystallographic orientations. For example, work on rolled W ribbons [158, 164] suggests that sites for the α states are concentrated around the $\{111\}$ plane.

This view is further confirmed by single-crystal studies discussed later. The detailed desorption kinetics of this state have been difficult to establish precisely. However, Rigby [163], who studied this state in detail, did not find any evidence for isotopic exchange suggesting α to be a molecular state.

7.1.3 The γ state

When nitrogen adsorbs on polycrystalline tungsten at temperatures below $\sim$150 K, the saturation coverage attained is more than twice that obtained at 300 K. This is due to the adsorption of a third state, γ, which was identified by Ehrlich in flash desorption spectra [156, 170]. The heat of desorption of this state from filaments is about 38 kJ mol^{-1} [154]. On tungsten films, however, Hayward et al. [171] observe a range of binding energies from 25 to 80 kJ mol^{-1}, of which the latter is in the region of the α state. XPS studies on polycrystalline W ribbons [173] also showed a broad distribution of N_{1s} electron binding energies associated with adsorption states present at low temperatures (100 K) only. This range of binding energies could arise from a number of factors such as the polycrystalline nature of the substrate or strong adsorbate–adsorbate interactions. The desorption of the γ state exhibits first-order kinetics [170]. In addition, Yates and Madey [167] have showed that isotopic mixing does not occur, suggesting this state to consist of weakly bound, molecular nitrogen.

Interestingly, the coverage in this state depends on how the adsorption occurs [170]. If adsorption is carried out at low temperatures (115 K), a greater proportion of the γ state forms than when the surface is first saturated at room temperature and then at 115 K. The additional γ adsorption at low temperatures occurs at the expense of β adsorption, showing there is competition between the two types of adsorption with the molecular γ state probably occupying similar sites to the atomic β state. This idea is supported by XPS data which shows that extensive $\gamma \rightarrow \beta$ interconversion occurs on polycrystalline tungsten upon heating [173].

7.2 NITROGEN ADSORBED ON W{110}

The {110} plane of tungsten is the most closely packed and is the least reactive towards nitrogen. In fact, early studies using field emission and field ion microscopy [174, 175] revealed no evidence for the chemisorption of nitrogen at room temperature. This view was further supported by contact potential [164], molecular beam [176] and LEED [177] measurements. However, later experiments showed room-temperature adsorption to take place. For example, Madey and Yates [178] and Hopkins and Usami [179] showed that a β state is formed on this plane but with a very low initial sticking probability (1–5×10^{-3}) and a very small increase in work function (≈ 0.03–0.13 eV). Flash desorption studies [178, 180] confirmed the presence of β nitrogen, which has a very low saturation density (2–3×10^{14} atom cm^{-2}) and gives rise to a high-temperature (> 1100 K) desorption peak. The saturation coverage can be increased by thermal cycling between 300 and 900 K, and the adatoms are believed to occupy underlayer sites [181].

In addition to β nitrogen, another state, γ, occurs for adsorption at temperatures below 150 K [182–184].

7.2.1 γ-Nitrogen

This is a weakly bound state desorbing between 150 and 195 K [182, 185] and giving rise to a work function change of about -0.19 ± 0.003 eV at saturation [185]. This work function change is in fairly good agreement with a previous reporting of ≈ -0.15 eV [164].

Isotope exchange studies [185] clearly reveal γ-nitrogen to be molecular. This is in agreement with the desorption data which essentially shows first-order kinetics. Furthermore, XPS and UPS studies [183] have concluded that this species sits upright on the surface.

The sticking probability curve obtained for γ-N_2 adsorption shows interesting features. Whereas Yates et al. [185] and Fuggle and Menzel [183] conclude that γ adsorption occurs with a constant sticking probability which is close to unity, more sensitive experiments by Bowker and King [182] using a molecular beam method show a pronounced maximum in the

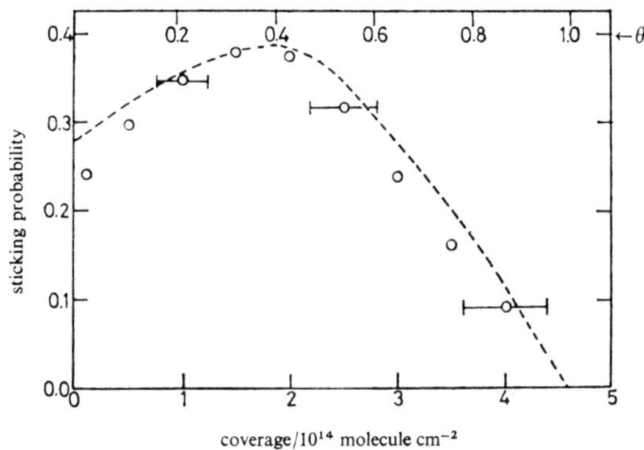

Fig. 31. Experimental sticking probability curve for N_2/W{110} (open circles) and comparison with results from the quantitative model described in the text (broken line). (From Bowker and King [182].)

sticking probability profile (Fig. 31). Bowker and King [182] report an initial sticking probability of ≈ 0.22 which first *rises* with increasing exposure to reach 0.37 at a coverage of $\approx 1.7 \times 10^{14}$ molecules cm^{-2}. After this, the sticking probability falls to less than 0.1 at a saturation coverage of 4.5×10^{14} molecules cm^{-2}. This variation of sticking probability with coverage seems to be unique to the {110} plane. A possible explanation for this behaviour could be based on island formation with preferential adsorption at the island boundaries. For such a growth mechanism, the environment of molecules within the island would be independent of the total coverage, so the work function change should vary linearly with coverage. However, Yates et al. [185] report a pronounced curvature in this plot and interpret it using a Topping model with a randomly distributed, immobile adsorbate. This refutes the island growth mechanism.

Bowker and King [182] have put forward a kinetic model which tries to explain both the low initial sticking probability (s_0) and the unusual sticking probability profile observed for this surface. This model basically invokes the King and Wells kinetic model [186, 187] for a trapping-dominated adsorption mechanism (discussed in greater detail in the N_2/W{100} section below). In addition, the authors propose that transfer from precursor to γ state is improved when a nearest-neighbour site is occupied by a γ species, i.e. a cooperative effect is introduced. This is a reasonable assumption in view of the fact that s_0 for adsorption into the γ state on W{320} is enhanced by pre-adsorption of the β state [187]. The Bowker and King model gives a theoretical curve which shows a reasonably good fit to the experimentally observed sticking probability curve (Fig. 31). It also suggests that trapping into a precursor state is relatively efficient (≈ 0.9) but the low value of s_0 results from an inefficient transfer from the precursor to the γ state. It is further proposed that this transfer is enhanced by the presence of N adatoms or other γ molecules. This would explain why adsorption into the γ state proceeds efficiently on crystal planes where β formation is efficient (e.g. W{100}) and slowly on this plane where only a small number of β adatoms is present.

7.2.2 β-Nitrogen

The β state is atomic and is the major state on the {110} plane for crystal temperatures > 195 K [180]. The saturation density at 300 K and with nitrogen pressures $< 3 \times 10^{-5}$ torr, is about 2×10^{14} atom cm^{-2} [180] which compares well with a previous value of 3×10^{14} atoms cm^{-2} [178]. Saturation of the β state is accompanied by a very small increase in work function ($\approx$0.03–0.13 eV) [178, 179].

Desorption data obtained after adsorption at 300 K show a single peak. The peak temperature at low coverages is $\approx 1450 \pm 50$ K [178, 180, 181, 188] which shifts down in temperature with increasing coverage. However, this shift is not as great as expected for a single state involving second-order desorption kinetics, and the data has been explained by the presence of a second state with a saturation coverage of 10% of the major state [180]. Tamm and Schmidt [180] estimate an activation energy of 332 ± 13 kJ mol^{-1} for the major state. This value is comparable with those observed for β states on other W single-crystal planes.

The sticking probability curve at 300 K has been obtained (Fig. 32), with a consistent profile reported by various workers [178, 180, 181]. In agreement with the desorption data, the sticking probability curve, too, is indicative of two states [180], a minor one having an s_0 value of ≈ 0.05 and saturating at $\approx 2 \times 10^{13}$ atoms cm^{-2} and a major one possessing an s_0 value of 0.004 and saturating at $\approx 2 \times 10^{14}$ atoms cm^{-2}. For the major nitrogen state, the sticking probability data can be fitted by the expression [180]

$$s \approx s_0(1 - \theta)^2$$

This would appear to suggest that the incoming molecule requires two empty sites for adsorption. However, this does not account for the fact that the saturation coverage is only one quarter of a monolayer, with adatoms occupying underlayer sites [181].

The low initial sticking probability along with the low saturation coverages attained on this crystal plane has led to a number of theories about the adsorption mechanism. For example, Adams and Germer [189, 190] have proposed that the β-N adatoms on W surfaces can only be adsorbed into the fourfold symmetric, fivefold coordinate sites characteristic of the {100} plane. Thus the reactivity of any W plane is expected to be proportional to the number of {100} sites in that plane. Singh-Boparai et al. [187], on the basis of experimental data, modified this model somewhat. They concluded that the dissociation of the nitrogen molecule in the β state occurs only at {100} site *pairs* but the chemisorbed atoms thus formed may subsequently migrate out to {110} sites and terraces, thus populating other areas of the crystal. Therefore, dissociative adsorption on W{110} is thought to arise

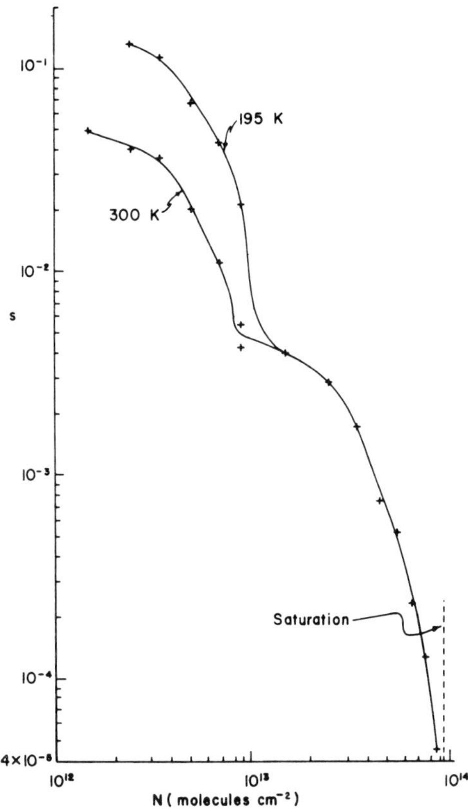

Fig. 32. The dependence of sticking coefficient on coverage for nitrogen on W{110}. (From Tamm and Schmidt [180].)

from defect sites of the required {100} type. These sites would arise, for example, if the crystal surface was slightly misoriented towards the (100) pole, giving rise to steps with the {100} configuration. The low initial sticking probability on W{110} is then correlated to the small number of suitable dissociation sites available. Additionally, the low saturation coverage at 300 K may be due to the fact that occupation of the {110} terraces at 300 K may be limited by the finite number of hops that "hot" N adatoms can make from the defect sites and steps at which they are formed. Ehrlich and Hudda [174] have reported that thermal migration of β-N adatoms on tungsten occurs at temperatures above 700 K. It can, therefore, be anticipated that thermal cycling above this temperature would enable chemisorbed adatoms to migrate away from defect sites and steps at which further dissociative adsorption could then proceed. In accordance with this, Somerton and King [181] and Besocke and Wagner [188] reported increased saturation coverages after the crystal had been thermally cycled to 820–900 K.

Besocke and Wagner [188] have also shown, by studying a multi-faceted

W sample exposing the {110} plane and low vicinal surfaces, that step edges do act as N_2 dissociation centres. However, they concluded that not only {100} step sites but also other step sites devoid of the {100} configuration are able to dissociate N_2 molecules, albeit less efficiently.

It is interesting to consider why direct dissociation of N_2 does not occur on W{110}. Cosser et al. [191] have reported angle-resolved thermal desorption data of N_2/W{110} which exhibits a non-cosine angular distribution, fitting a $\cos^n\theta$ ($3 < n < 4$) expression. The desorption flux distribution was explained in terms of a one-dimensional barrier (1DB) model [192] and yielded an activation energy of adsorption of 17.4 kJ mol^{-1}. From this, the authors predicted s_0 to increase with increasing temperature of the impinging gas. This prediction was verified by molecular beam studies which investigated the effect of varying kinetic energy on the activated adsorption of N_2 on W{110} [193–195]. Although a much higher activation barrier of about 41 kJ mol^{-1} in one dimension normal to the surface was evaluated, the lowest barrier, corresponding to the N_2 velocity at the onset of the observed rise in the sticking probability, was $\sim$ 20 kJ mol^{-1}, in reasonably good agreement with the analysis of the desorption data; dissociative adsorption was concluded to be translationally activated and insensitive to vibrational activation. Increase in translational energy was thought to drive the molecule deeper into the repulsive barrier at the surface, thus facilitating W—N bond formation, leading to dissociation.

However, a number of experimental observations [193–195] conflict with the simple 1DB model. First, the key prediction of a 1DB model is that the dissociative adsorption probability would scale with the normal component of the incident kinetic energy. However, s_0 was found to be independent of incidence angle [194, 195] and to scale approximately with the total energy. Second, a 1DB model predicts a step function change in s_0 with incident energy, but experimental data [193, 195] shows that an increase in s_0 occurs over a wide range of E_i ($>$ 20 kJ mol^{-1}), suggesting the existence of a correspondingly wide distribution of barrier heights.

At present, the favoured explanation for the insensitivity of s_0 to θ_i is that a long-lived surface intermediate or precursor state exists which leads to the "scrambling" of the initial parallel and perpendicular velocity components [194–196]. Kara and De Pristo [196] have theoretically modelled the N_2/W{110} system to determine a potential energy surface (PES) which would yield $s_0(E_i, \theta)$ mimicking the experimental molecular beam data. They attribute the total energy scaling of s_0 to the presence of a molecular well in the entrance channel and to the narrowness of the PES near the activation barrier in the exit channel. The presence of a narrow PES greatly restricts the allowed geometries of N_2 which can lead to dissociative chemisorption. As a result, an incident molecule may "bounce" many times on the surface before obtaining the required geometry, leading to the "scrambling" of the kinetic energy components.

Finally, the presence of an intermediate state which scrambles the parallel and perpendicular velocity components should produce broadened, diffuse scattering data, in contradiction with the experimental data of Cosser et al. [191]. However, it has been suggested [195, 196] that this discrepancy can be explained if two distinct possibilities exist for the incoming molecule, one in which the molecule does not surmount the entrance barrier and the second in which the molecule overcomes this barrier. For the first case, the molecules would be scattered in a peaked distribution while the second class of molecule would either chemisorb dissociatively or desorb, giving rise to diffuse scattering.

Few structural studies of β-N on W{110} have been reported [181, 197]. Somerton and King [181], using ion scattering spectroscopy and angle-dependent AES, surprisingly demonstrated, that the chemisorbed nitrogen adatoms, forming a p(2 × 2) LEED pattern, do not occupy conventional overlayer sites on the undisturbed substrate, but are sandwiched between the top two layers of tungsten atoms. This underlayer model has been further substantiated by recent surface core level shift studies [197]. This phase saturates at a quarter of a monolayer and desorbs as molecular nitrogen. LEED shows a well-defined (2 × 2) pattern, the 0.5-order diffraction beams presumably arising from scattering from the rumpled tungsten surface layer (Fig. 33), the periodicity of which is determined by the arrangement of N adatoms in the underlayer. Pfnür et al. [195] have reported that coverage can

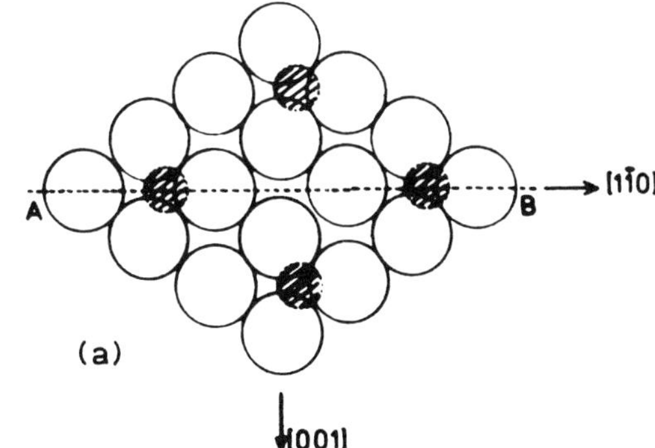

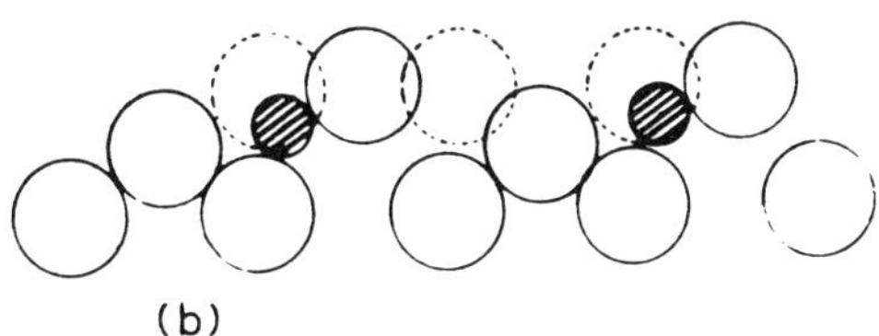

Fig. 33. Possible structure for W{110}-(2 × 2) − N. (a) Plan view. (b) Section along AB through the surface illustrating rumpling of the top W layer. (Broken circles represent W atoms in the row in front of AB; large circles, W atoms, small circles, N atoms.) (From Somerton and King [181].)

be increased further by using high-energy molecular beams. They report a saturation coverage of 0.52. At the half-monolayer stage, a c(4 × 2) LEED pattern is observed and again attributed to underlayer nitrogen adsorption.

7.3 NITROGEN ADSORBED ON W{111}

The sticking probability curve for N_2 on W{111} has been obtained at 300 K using the molecular beam technique [176] and shows a rapid fall in sticking probability with increasing coverage (Fig. 34). This work also showed the initial sticking probability, s_0, to be 0.08 and the saturation coverage as 1.25×10^{14} molecules cm^{-2}. Flash desorption spectra [164] obtained after adsorption at room temperature reveal a low-temperature peak, which evolves as soon as the temperature is raised, and a high-temperature peak, which dominates at temperatures greater than 1000 K. The former peak is assigned to the molecular α-nitrogen and the high-temperature peak to atomic β_2-nitrogen. It is interesting to note that the α state has not been observed on any other single-crystal plane apart from the {111}, implying that it is extremely sensitive to surface structure.

After adsorption at room temperature, the work function at saturation shows an increase of 0.14–0.17 eV [164, 166, 175]. Of this increase ≈ 0.13 eV is thought to be associated with the β state [164]. At lower coverages, the β_2 state shows second order desorption kinetics with an activation energy of ≈ 340 kJ mol^{-1} [164]. As the coverage is increased, the kinetics become more complicated and indicate the presence of a second β-state with a desorption energy ≈ 34 kJ mol^{-1} higher than that of the β_2 state. It is estimated this β

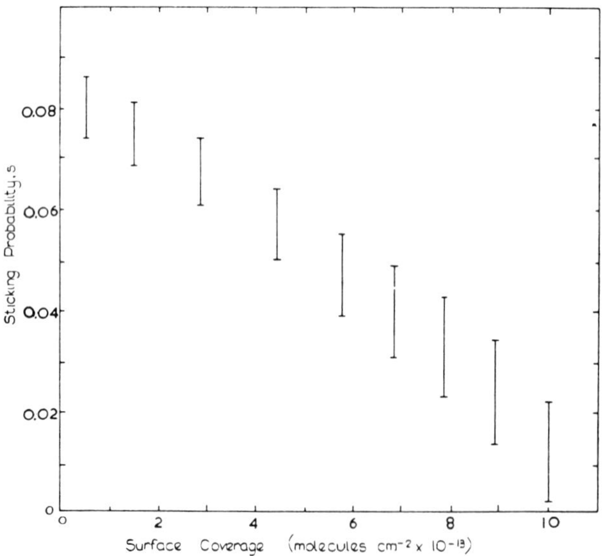

Fig. 34. Sticking probability profile for nitrogen on W{111}. (From King and Wells [176].)

state has one quarter the concentration of the β_2 state. The onset of complication in the desorption kinetics occurs at the same coverage as the first discontinuity observed in the work function curve.

As discussed in the previous section, there is some question about whether nitrogen can dissociatively adsorb on crystal planes such as W{111} which do not contain adsorption sites of fourfold symmetry such as those found in W{100}. It is suggested [176] that on this surface, too, the observed β state may be due to defect sites of the correct geometry existing on the surface.

Adsorption at temperatures below 300 K leads to an increased concentration in the α state [164]. At sufficiently low temperatures (< 130 K) the molecular, weakly bound γ state is also found.

7.4 ADSORPTION OF N_2 ON W{100}

The chemisorption of N_2 occurs readily on W{100}. At room temperature a single, strongly bound state, the β state, is observed, whilst adsorption at low temperature produces an additional weakly bound γ state [164]. Most of the literature concentrates on the β state but some information is also available about the γ binding site.

7.4.1 The γ state

The γ state adsorbs at low temperature only [164, 198]. Flash desorption spectra of nitrogen adsorbed on W{100} at 78 K reveal two peaks due to this state (Fig. 35), at 170 and 190 K. Using the nomenclature adopted by Delchar and Ehrlich [164], these desorption peaks can be assigned to the γ^+ and γ^- states which are estimated to have activation energies of desorption of ≈ 38.5 and 44 kJ mol^{-1}, respectively [198]. Both peaks show first-order desorption kinetics. At saturation, the amounts adsorbed in each sub-state are approximately equal. The total amount adsorbed in the γ state is about twice that in the β_2 state.

Evidence for the two sub-states had first been seen in the early work function studies by Delchar and Ehrlich [164]. The work function change observed on adsorption and desorption could only be explained by the presence of two low-temperature states, one which increased the work function (γ^+) and the other which reduced it (γ^-).

Auger and ELS spectra have been obtained by Chesters et al. [199] which seem to indicate that the γ state (their α state) is molecular with a vertical bonding configuration. This has been confirmed by vibrational EELS studies [200] which show a vibrational loss peak at 2137 cm^{-1} (265 meV), assigned to the N—N stretching vibration of a linearly bound, vertical N_2 molecule. In addition, the vibrational spectra revealed a peak at 1452 cm^{-1} (180 meV) which was assigned to the N—N stretching vibration of a twofold bridge-bonded N_2 molecule. However, this assignment is questionable and a more likely alternative is the N—N vibration of a molecular species lying flat on

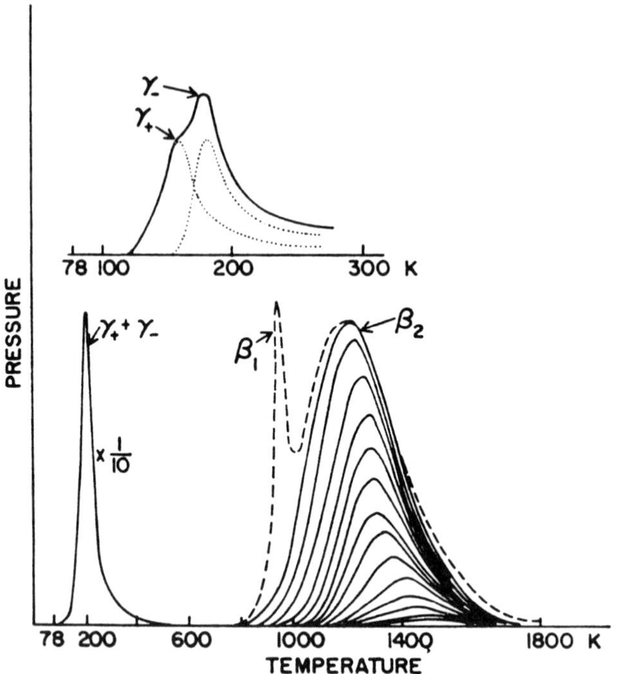

Fig. 35. Flash desorption traces for N_2/W{100}. Solid lines represent traces obtained by moderate exposure ($< 10^{-4}$ torr s) with the crystal at 78 K. The insert shows desorption from the saturated γ states at a lower heating rate. The trace indicated by the broken line is that obtained by high exposure to N_2 at 78 K with intermittent heating to 300 K. The areas under the saturation curves indicate equal amounts in the γ_+, γ_- and β_2 states. (From Clavenna and Schmidt [198].)

the surface, similar to that observed for N_2 on iron. In fact, King and Somerton [201] have concluded from their photoemission studies that the γ^+ species lies flat on the surface while the γ^- species is linearly bonding with a vertical configuration. Proposed configurations [201] for the two species are illustrated in Fig. 36.

7.4.2 The β state

For adsorption at $T > 200$ K, a single strongly bound state is seen for nitrogen on W{100} [164, 178, 179, 190, 198, 202, 203]. This is the atomic β_2 state which exhibits second-order desorption kinetics and has an estimated activation energy of desorption of 309 kJ mol^{-1} [198]. At 300 K, adsorption into this state proceeds with an initial sticking probability, s_0, of 0.59 ± 0.01 [186] and a saturation coverage of 5–6 $\times$ 10^{14} atoms cm^{-2} [186, 198]. Adsorption into this state produces a *decrease* in the work function and at $\theta = 0.5$ the change in work function is -0.6 eV [190].

Both the structure and the kinetics of β-nitrogen on W{100} have been examined in some detail and will be discussed separately.

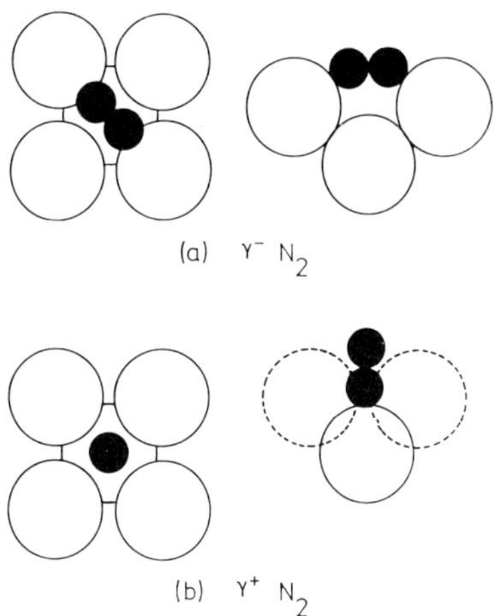

Fig. 36. Possible configuration for the γ_+ and γ_- species on W{100}. (From King and Somerton [201].)

(a) Structural studies

Extensive structural investigations of this phase have been carried out [190, 202–206] and a fairly detailed picture of the overlayer arrangement has emerged.

Estrup and Anderson [202] first found that room-temperature adsorption gives rise to a c(2 × 2) LEED pattern, suggesting a double-spaced structure at half-monolayer coverage. They proposed a structural model in which the N adatom occupies the fivefold coordinate site at the centre of four surface W atoms. Further detailed studies by Adams and Germer [190, 203] supported this model. At low coverages (< 0.2 monolayer), they observed LEED patterns with weak, diffuse crosses centred in the half-order positions [Fig. 37(a)]. These crosses were interpreted as due to antiphase domains of nitrogen adatoms, each of which has a c(2 × 2) structure and is elongated in a particular ⟨11⟩ direction [Fig. 37(b)]. Furthermore, since the integral-order beams showed no sign of streaking, it was concluded that adsorption must occur in the fourfold symmetric hollow sites [207]. In support of this hypothesis, a LEED I–V analysis by Griffiths et al. [205] of the c(2 × 2) structure with N in this adsorption site gave reasonable values for the W—N interlayer spacing and the W—W bulk spacing. In addition, ion scattering studies also confirm the fourfold hollow as the adsorption site [206].

At a coverage of about 0.2 monolayer the streaked half-order features alter to an approximate square shape, indicating that island growth now no longer occurs in a preferred direction. For coverages between 0.3 and 0.4, the half-order features split into a quartet which could be sharpened by annealling [Fig. 38(a)]. This was interpreted as a periodic array of out-of-phase islands each containing 16 N atoms and having a constant unit cell size of (4 × 4) [Fig. 38(b)].

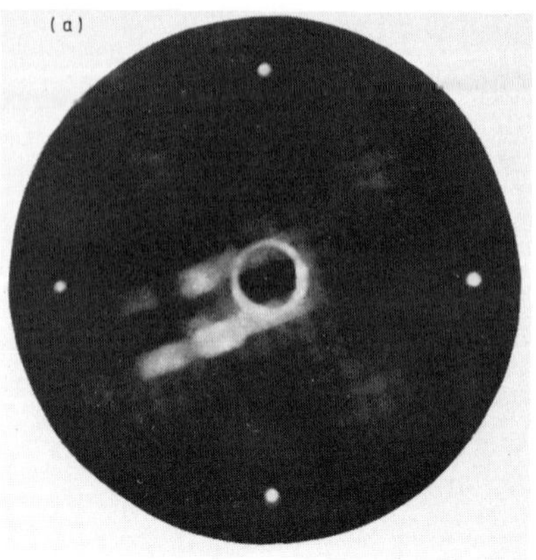

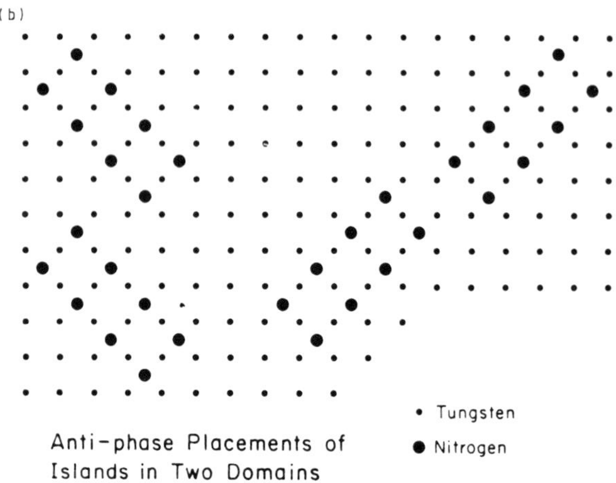

Fig. 37. (a) LEED pattern at 44 eV, showing spots characteristic of clean W{100}, plus orthogonal streaks in $h + \frac{1}{2}$, $k + \frac{1}{2}$ positions. Nitrogen coverage = 0.1. (b) Antiphase placements of islands of nitrogen atoms. (From Adams and Germer [203].)

Finally, as the coverage increases to 0.5, the half-order spots sharpen and become more rounder. Annealling the layer to > 800 K produces a sharp c(2 × 2) LEED pattern.

These ideas of overlayer structure and growth were rationalised by Schmidt and co-workers [198, 208] using a valence-band approach in which specific metal orbitals are involved in bond formation with adsorbed species. Band theory calculations [209] show that the tungsten 6*s* and 5*d* bands overlap in energy and form sd_{xy} hybrids (where x and y lie in the surface plane). The linear combination of the *s* and *d* orbitals yields a pair of two-lobed hybrid orbitals (Fig. 39). Each pair of *sd* hybrids is occupied by one electron only. As a result, band formation via the nitrogen $2p_x$ or $2p_y$ orbital will localise the electron into one of the two *sd* hybrids, leaving the second *sd* hybrid empty. Since the *sd* hybrids are symmetrical about each tungsten

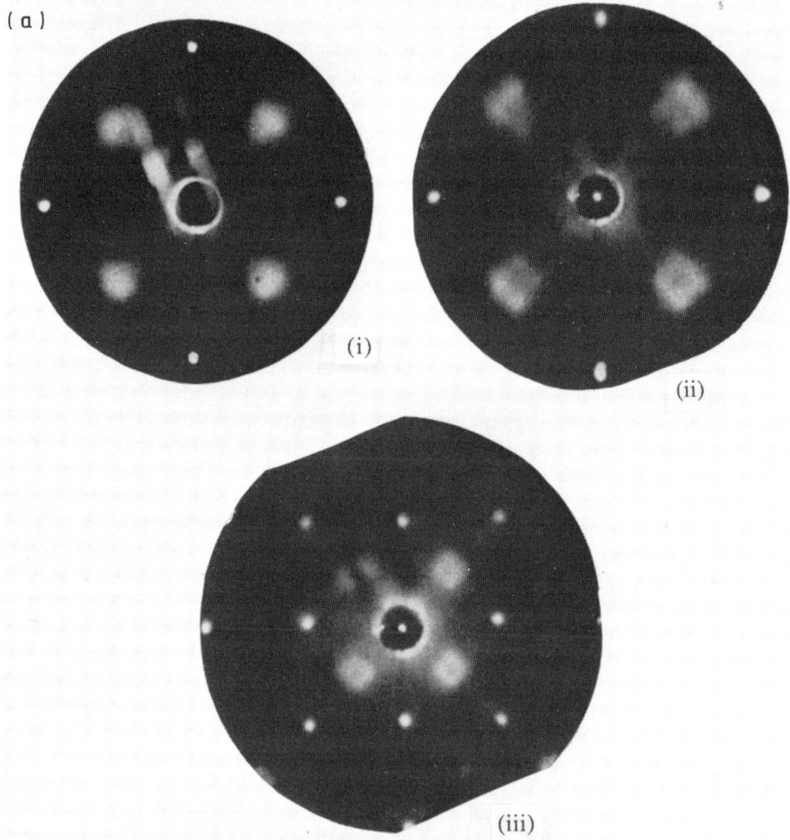

Fig. 38. (a) Typical LEED patterns at θ = 0.3–0.4 for N/W{100}: (i) after room-temperature adsorption, 44 eV; (ii) after (i) and anneal to 900 K, 44 eV; (iii) after (i) and anneal to 900 K, 125 eV. (b) Periodic array of out-of-phase islands. The solid lines indicate the unit cell of the array. The faint lines demarcate out-of-phase boundaries. (The broken lines indicate the unit cell of the alternative placement which is structurally equivalent). Small black circles: W; large black circles: N. (From Adams and Germer [203].)

atom, adsorption sites adjacent to the occupied site in the $\langle 11 \rangle$ direction will be preferred, whereas adjacent sites in the $\langle 10 \rangle$ direction will not be available for further adsorption, hence leading to a c(2 × 2) layer.

Elongated island growth can also be explained by this model. For example, when two N atoms occupy adjacent sites, the vacant sites at the ends of the "chain" are preferred to those either side of it. In fact, Adams and Germer [203] suggest that delocalized π-bands may form in the $\langle 11 \rangle$ direction, leading to long-range interactions causing formation of elongated islands.

Although the discussion above shows the c(2 × 2) structure to be consistent with experimental data, this overlayer structure has come into question since the discovery that the clean W{100} surface undergoes a displacive

Fig. 38 (b)

phase transition on cooling below 370 K [210, 211]. Furthermore, it has been established that a new set of displacive structure are stabilized by hydrogen chemisorption [212]. These observations suggested the need to re-examine the N/W{100} system, which had hitherto been interpreted without regard to possible substrate-surface-atom displacement. Griffiths et al. [213] reinvestigated the system, and below a coverage of 0.3 their observations agreed with those of Adams and Germer [190, 203]. In the coverage range 0.3–0.4, they report that the half-order LEED beams are split into a quartet which shows an asymmetry in intensity [Fig. 40(a)]. The presence of the quartet and its uneven distribution of intensity could only be explained by the model

(a) the overlayer consists of islands which contain, on average, 16 nitrogen adatoms (corresponding to $\theta = 0.39$). This would produce the required splitting of the half-order beams;

(b) each nitrogen adatom occupies a fourfold hollow site on the substrate, but the four surface tungsten atoms in the site are all uniformly displaced towards the nitrogen adatom;

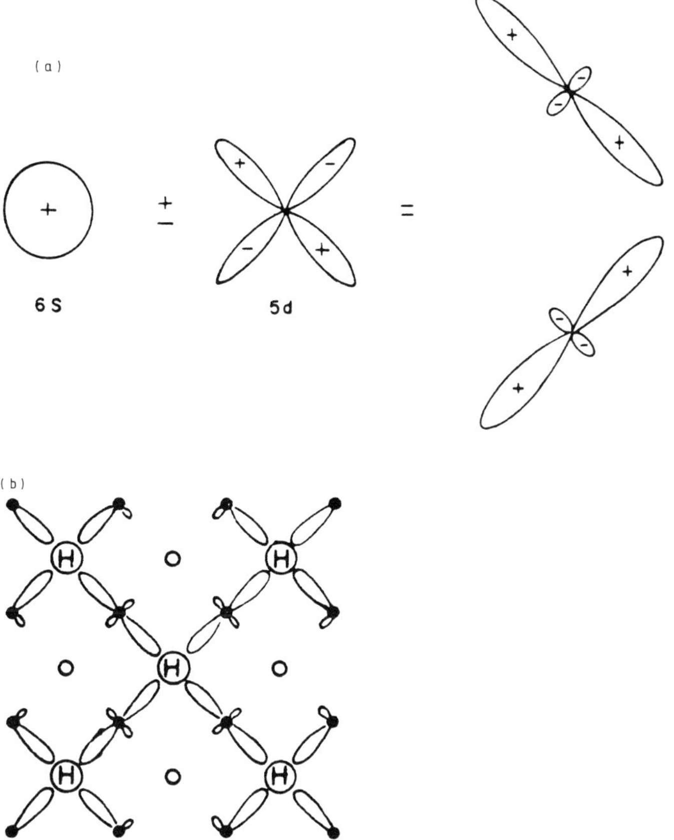

Fig. 39. (a) Schematic diagram of the orbitals formed by hybridisation of a 5*d* orbital with a 6*s* orbital. (b) Structure of the bonding orbitals of W when the β_2 state of H_2 is formed on W{100}. The other orbital shown in (a) is of higher energy and unoccupied. (From Tamm and Schmidt [208].)

(c) within each island the lateral interatomic spacing between surface W atoms is constant.

Essentially, the model, depicted in Fig. 40(b), involves a contracted-domain structure in which the misregistry with the second-layer atoms at the island boundaries is relatively large. Such a structure also explains why the overlayer should favour a (4 × 4) island dimension since the strain induced at the island edges due to the displacement of the W atoms would be minimised for relatively small islands.

Further support for the contracted domain structure has come from a surface core level shift (SCCS) study [214] in which it has also been inferred that N adatoms cause a local displacement of surface W atoms, forming contracted islands.

As the coverage is further increased, Griffiths et al. [213], in agreement

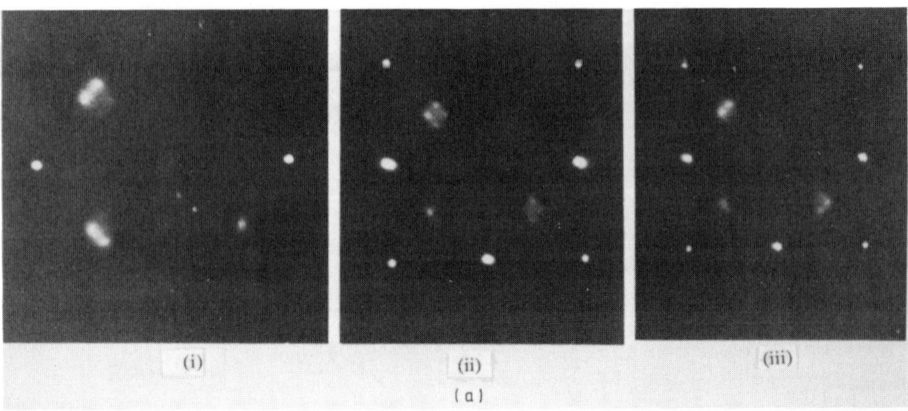

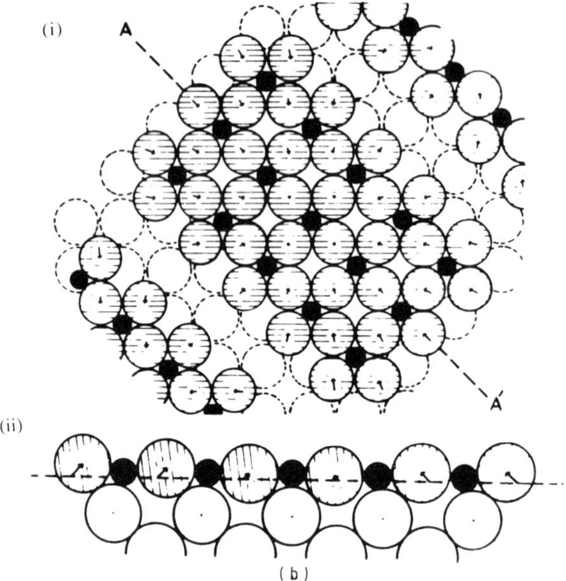

Fig. 40. (a) LEED patterns obtained from W{100} with 4×10^{14} N atoms cm^{-2} at primary beam energies of (i) 32 eV, (ii) 62 eV and (iii) 74 eV. Part of the pattern is obscured by the crystal manipulator. (b) The proposed contracted-domain structure for 0.4 monolayer N on W{100} surface: Large hatched circles, top-layer W atoms; small filled circles, N atoms; arrows, W-atom displacement vectors. The neutral atomic radius for N atoms, 0.59 Å, was used. (i) Plan view, illustrating domain and boundary structure. (ii) Cross-section through the domain along AA′; the broken line would pass through the centres of the undistorted top-layer W atoms. (From Griffiths et al. [213].)

with Adams and Germer [190, 203], find that the quartet at the half-order positions coalesces to form a single beam whose intensity passes through a maximum at exactly $\theta = 0.5$. This is due to boundary sites being filled, leading to a symmetric strain on the W atoms, and thus lifting the displacement and producing a conventional "checkerboard" $c(2 \times 2)$ structure.

Griffiths and King [204] also claim that the adsorbed layer is more stable at $\theta = 0.38$ than at higher coverages. This is demonstrated when a surface with a coverage of 0.5 is annealled to 1000 K. They found that *absorption* of nitrogen occurred into the bulk, until the surface layer coverage is reduced to 0.38. It would seem that the most stable structure corresponds to the contracted domains, where N adatoms are fivefold coordinate to nearest-neighbour W atoms. At $\theta = 0.5$, these W—N bond distances are increased and the chemisorbed layer becomes relatively unstable compared with the bulk nitride. Given that desorption occurs at > 1250 K, it would seem that a simple surface atom recombination process may no longer be relevant as a desorption mechanism. However, it should be noted that Winters [215] was unable to reproduce this result, which may be due to the presence of defects or steps on the crystal used by Griffiths and King [204].

It should be mentioned that the "saturation" coverage at reasonable exposure is now thought to be 0.6 [186, 204, 216], not 0.5 as previously suggested [198, 202, 203]. This increase in coverage is accompanied by the appearance of new one-fifth-order diffraction beams [204, 216].

In a remarkable and unique paper, Winters and co-workers [217] examined the behaviour of W{100} after exposures to N_2 of up to 10^6 L, well in excess of anything previously attempted. On a linear plot they report, in agreement with King and Wells [186], that the sticking probability extrapolates to zero at a coverage of 0.6. However, a process with a low sticking probability (which we estimate from their data as 10^{-5} to 10^{-6}, comparable with N_2 on Fe surfaces) produces a maximum coverage of about 1.15 monolayers, i.e. about 15% higher than that required to fill all fourfold hollow sites at the surface. From a change in slope of their AES N intensity vs. coverage (determined by TPD) plot, they deduce that N adatoms in excess of 0.55 monolayer are situated beneath the top layer of tungsten atoms. At maximum coverage, the LEED pattern becomes a well-defined (6 × 6) structure, and the desorption spectra contain two β peaks. The authors propose a model involving a reconstruction of the surface layer in which the top layer of W atoms is condensed into a close-packed square array with the nearest neighbour W—W distance ~2.713 Å, compared with 2.74 Å in the bulk. The model, shown in Fig. 41, bears a strong resemblance to the 0.38 monolayer structure in Fig. 40(b) except that here N adatoms occupy surface sites in the fourfold hollow sites, and, additionally, underlayer sites in voids between first and second layers.

(b) Kinetic studies

The interaction of nitrogen with W{100} has been the subject of wide-ranging experimental investigation [186, 190, 198, 203]. These data, along with the theoretical models used to rationalise the results, have been critically reviewed by King [155].

The most extensive data pertinent to the kinetics of the system have been obtained by King and Wells [186] using the molecular beam method. Sticking

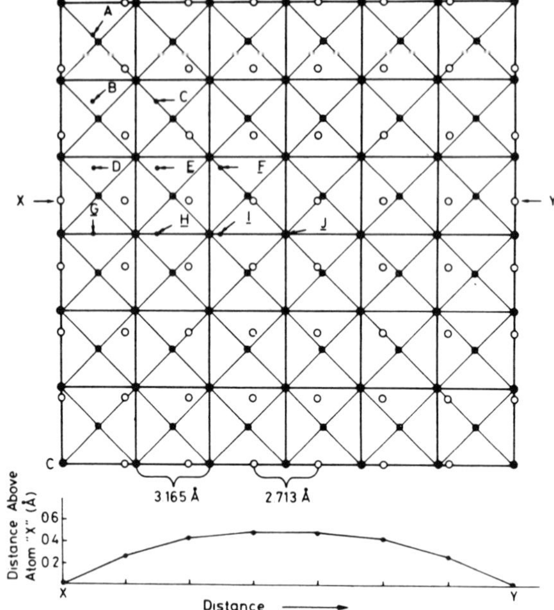

Fig. 41. Unit cell of the reconstructed (6 × 6) W{100} structure. The open circles are the top-layer W atoms compressed into a close-packed square array. The circles with crosses are the second layer of tungsten atoms in their normal b.c.c. positions. The black circles are the third layer of tungsten atoms in their normal positions. Nitrogen atoms are assumed to be adsorbed at the centre of the fourfold sites at a position indicated by the arrows. Sites A–J represent all 49 sites in the unit cell. Nitrogen is assumed to sit one layer beneath the surface for sites E–J and on top of the surface for sites A–D. The tungsten atoms in the top layer are displaced upward in manner suggested by the curve at the bottom of this figure. The normal lattice constant is 3.165 Å. (From Winters et al. [217].)

probability (s) curves, shown in Fig. 42, determined for a range of adsorbent temperatures, followed a general pattern with s varying little with coverage up to half a monolayer. The initial sticking probability, s_0, at room temperature is measured to be 0.59 ± 0.01 and is found to be independent of the angle of incidence of the molecular beam, up to 45° from the normal to the surface. It was also found that s_0 is virtually independent of adsorbent temperature T_s at low temperatures between 195 and 77 K. This gives a condensation coefficient, α, for the interaction of N_2 with W{100} of 0.6. The variation in s_0 with gas temperature, T_g, was also studied and found to fall with increasing T_g. This is consistent with non-activated adsorption and the requirement for trapping to occur into a precursor state prior to chemisorption. In support of the trapping-dominated mechanism, King and Wells also found that at low temperatures the sticking probability is initially independent of surface coverage. The saturation coverage was found to be $6.0 \pm 0.2 \times 10^{14}$ atoms cm^{-2}.

Desorption spectra of the high-temperature states of N_2 on W{100} have

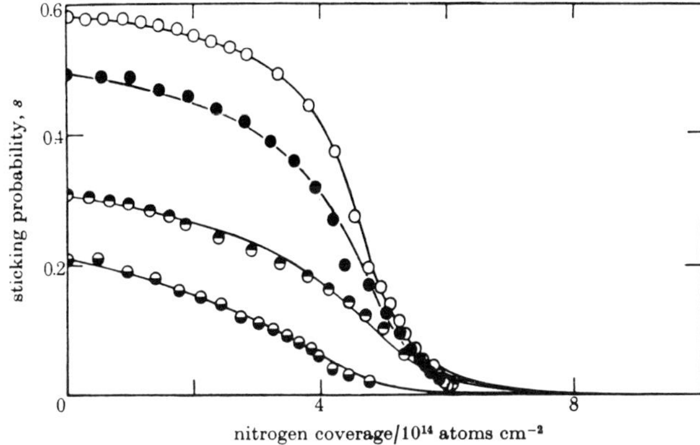

Fig. 42. Sticking probability profiles for nitrogen on W{100}; T_g = 300 K, variable T_s. Points: experimental data; lines: computed best-fit curves from the King and Wells model. ○, 300 K; ●, 433 K; ◓ 663 K; ◒ 773 K. (From King and Wells [186].)

also been recorded [186, 198, 218]. Flash desorption data obtained by Clavenna and Schmidt [198] (Fig. 35) shows a major peak due to the β_2 state. Although, the shift of peak temperatures with coverage suggests that this state obeys second-order desorption kinetics, closer analysis reveals deviation from this behaviour. With low crystal temperatures and high exposures of nitrogen, another state, the β_1, is also observed. This state displays first-order kinetics. King and Wells [186] also report the presence of the β_1 peak in addition to the β_2 peak for exposures corresponding to coverages greater than 0.5, and, as remarked above, the β_1 state becomes heavily populated after N_2 exposures of 10^6 L, occupying underlayer sites [217].

The adsorption kinetics of the N_2 on W{100} are amenable to theoretical modelling, partly because of the wealth of experimental data available and partly because the dissociated chemisorbed β_2 state is formed with a large heat of adsorption compared with the physisorbed and molecularly chemisorbed states. As a result, theoretical models can be considerably simplified for it can be assumed that the lifetimes of all the adsorbed species involved are either very short or very long compared with the timescale of the experiment. For N_2/W{100}, this criterion is valid for temperatures between 200 and 900 K, allowing valid comparison between experiment and theory over this entire temperature range.

As outlined previously, the sticking probability for N_2 on W{100} is independent of N at low coverage, suggesting a trapping-dominant adsorption mechanism. Kinetic models involving short-lived precursor states which transfer to stable chemisorbed states have been constructed by Ehrlich [154] and Kisliuk [219] based on ideas developed by Taylor and Langmuir [220] and Roberts [221]. Clavenna and Schmidt [198], using essentially the Kisliuk formalism, obtained a good correlation between their data for N_2 on W{100} and the precursor state model. However, their formulation included the parameter α (the trapping probability into the precursor state) which had to be varied in an empirical manner in order to obtain a good fit to the data.

Probably the most successful model applied to the $N_2/W\{100\}$ system is that developed by King and Wells [186]. This model, again based on the Kisliuk formalism, is derived for the specific case of the dissociative adsorption of a diatomic molecule on a homotattic, unreconstructed surface with fourfold symmetry and specifically takes into account the ordering in the chemisorbed overlayer resulting from the existence of lateral interactions between chemisorbed species in nearest-neighbour positions.

The assumptions made in this model are

(a) the incident molecule is trapped at the point of impact with a probability α. It is assumed that the molecule is either elastically (or inelastically) scattered or rapidly equilibrated at the incident site, in which case the temperature of the trapped molecule is taken to be that of the crystal, T_s;

(b) the condensation coefficient, α, is independent of the degree of occupation of the surface by the chemisorbed overlayer (this is shown to be true experimentally);

(c) once trapped in the physisorbed state in the immediate vicinity of a pair of empty nearest-neighbour (n.n.) chemisorption sites, the molecule may become dissociatively chemisorbed, with a probability f_c; it may hop to a neighbouring physisorption site, with a probability f_m; or it may be scattered back into the gas phase, with a probability f_d. If empty n.n. pair sites are not available to the physisorbed molecule it may hop (probability f'_m) or be scattered back into the gas phase (probability f'_d);

(d) at all coverages, adatoms in the chemisorbed overlayer occupy specific, identical geometric sites on the substrate. Adatom penetration and substrate reconstruction are precluded;

(e) the chemisorbed atoms are subject to a pairwise n.n. lateral interaction energy, ω, which is independent of the fractional coverage θ; all other lateral interactions are ignored;

(f) the distribution of adatoms on chemisorption sites corresponds to the equilibrium distribution, which is achieved by the hopping of atoms in the chemisorbed state from site to site.

Using the assumptions outlined above, a statistical analysis yields the expression

$$s = \alpha\left(1 - \frac{f_d}{f_a}\right)^{-1}\left[1 + K\left(\frac{1}{\theta_{00}} - 1\right)\right]^{-1} \tag{1}$$

for the sticking probability for dissociative adsorption where $K = f'_d/(f_a + f_d)$, and θ_{00} is the probability that a given n.n. pair of sites is empty at a coverage θ. This term may be obtained from the lateral interaction energy, ω, using order–disorder theory, the quasichemical approximation [222] yielding

$$\theta_{00} = 1 - \theta - \frac{2\theta(1 - \theta)}{[1 - 4\theta B(1 - \theta)]^{1/2} + 1} \tag{2}$$

where $B = 1 - \exp(\omega/kT_s)$ and T_s is the crystal temperature. B is described as the short-range order parameter and possesses the limits $0 \leqslant B \leqslant 1$, where $B = 1$ corresponds to complete disorder (i.e. corresponding to the Kisliuk model).

A degree of success has been achieved in applying eqn. (1) to the N_2/W{100} system for a range of temperatures between 300 and 773 K [186]. The value of α required in eqn. (1) was obtained from the experimentally observed low-temperature limiting value of s_0 since it had been predicted [186] that

$$\lim_{T_s \to 0} s_0 = \alpha \tag{3}$$

Using this value of α, and empirically varying B, sticking probability curves were computed which closely mimicked the experimentally observed curves [186] over the temperature range 300–773 K (Fig. 42). The best fits were obtained for values of B close to unity (the mean value is 0.984 ± 0.005).

However, some inconsistencies between the theoretical model and the experimental data were noted. First, no systematic variation in the best-fit values of B with adsorbent temperature, T_s, was observed [186], whereas the quasichemical approximation, based on the assumption of equilibrium in the overlayer site distribution gives $B = 1 - \exp(\omega/kT_s)$. This inconsistency is believed to arise from the fact that ordering is produced by limited migration of "hot" atoms in the chemisorbed layer rather than by continuous migration of β-nitrogen. This suggestion is supported by the experimental observation that continuous migration of β-nitrogen is not seen below 250 K [223].

Secondly, above 700 K, the parameter K from best-fit analysis to eqn. (1) is found to vary inconsistently from the expected linearity of $\ln[(1/K) - 1]$ vs. $1/T_s$ [186]. This is thought to be due to the onset of continuous migration, leading to the formation of adsorbate islands produced by next-nearest-neighbour attractive interactions.

Despite these inconsistencies, several of the model assumptions were shown to be valid. For example, the independence of α with respect to the surface coverage and T_s was experimentally verified at low substrate temperatures. Also, experiments with variable beam temperatures (T_g) showed K and B to be insensitive to this parameter while α alone varied with T_g. This is consistent with the model assumption that an incident molecule is either scattered at the incident site or rapidly accommodated to the substrate temperature, T_s.

Above all, agreement appears to have been established with the structural studies of the N_2/W{100} system. First, the total number of nitrogen adatom sites, N_s, was taken to be equal to the number of W atoms on the surface. The best-fit value of N_s used to compute the sticking probability curve shows a remarkable consistency and bears a close 1:1 correspondence to the number of W atoms in the surface layer of the {100} plane [186]. The stoichiometry at "saturation" coverage of nitrogen shows that the density of the adlayer is about 60% of the surface density of the tungsten atoms in the {100} surface plane. Since LEED indicates the adlayer is partially ordered in a c(2 × 2) structure, it suggests that the β state is dissociatively adsorbed. (The fact that slightly more than half the sites are filled is probably a consequence of a small degree of disorder in the adlayer.)

The model also provides an explanation for the low saturation coverage of nitrogen adatoms when adsorption takes place via the N_2 molecule, since empty pairs of n.n. sites are required for dissociation. It further explains why higher saturation coverages are achieved when the N_2 molecule is atomized in the gas phase [224] because the requirement for empty n.n. sites is then relaxed.

Theoretical models for desorption have also been proposed and rate equations for desorption from a chemisorbed layer with n.n. interactions have been derived [155, 218, 231]. Using the quasichemical approximation, the desorption rate for associated ("second-order") desorption of a homonuclear diatomic molecule is [155]

$$-\frac{\mathrm{d}N}{\mathrm{d}t} = \nu_2 N_{\mathrm{AA}} \exp(-E/RT) \tag{4}$$

This equation takes into account the requirement that n.n. sites must be filled for the desorption activated complex to be formed.

Thus, E is the activation energy for desorption and N_{AA} is the coverage in n.n. occupied site pairs and is given by

$$N_{\mathrm{AA}} = \frac{zN_{\mathrm{s}}}{4}\left[\theta - \frac{2\theta(1-\theta)}{\{1-4\theta(1-\theta)B^{1/2}\}+1}\right] \tag{5}$$

From eqn. (5), it is clear that the simple concept of reaction order has been lost. It can also be shown [155, 218] that this form of the desorption rate would lead to a θ^2 dependence at low coverages and a linear dependence on θ at high coverages. A two-peak desorption spectrum would be seen; the low coverage, high T_{p} peak approximates to second-order behaviour and the high coverage peak to first-order kinetics [155].

As predicted, the desorption spectra [198] do indeed show two peaks associated with the β state, the high-temperature peak showing "second-order" behaviour while the lower-temperature peak shows "first-order" behaviour. It can also be concluded from this data that adatoms without nearest neighbours are more difficult to desorb than those having nearest neighbours. However, we note that structural studies indicate that the β_1 state derives from an underlayer: presumably due to repulsive interactions between n.n. overlayer N adatoms, the total energy of the system is lowered by distorting the top W layer and incorporating the β_1 state into an underlayer, as in Fig. 41.

Despite the success of the King and Wells model in describing the kinetics of the N_2/W{100} system, its formulations have been brought into question [225, 226]. First, their order–disorder model is based on a "checkerboard" surface model which is clearly not applicable in view of the lateral surface reconstruction which occurs during adsorption. However, a reinvestigation [227] showed no break or discontinuity in the sticking probability vs. coverage curve in the region of 4×10^{14} N atoms cm^{-2} when contracted domains are thought to be present [213]. This suggests there is no significant kinetic effect associated with this reconstruction. The reason for this could be that dissociative chemisorption proceeds in the channels between islands, initially on second layer W atoms and the N atoms formed then "sew" the islands together, finally removing the surface layer reconstruction at $\theta = 0.5$. Therefore, despite the presence of contracted domains, the number of adsorption sites effectively remains constant, explaining the success of the checkerboard model.

Secondly, time-of-flight studies by Janda et al. [225] of N_2 scattered from polycrystalline tungsten showed a significant decrease in the precursor state trapping probability with increasing substrate temperature. This, too, conflicts with the King and Wells model. However, a number of difficulties are associated with the analysis of Janda et al. and their interpretation has been questioned [227]. More recently, these difficulties appear to have been resolved in a supersonic molecular beam study of dissociative chemisorption of N_2 on W{100} by Rettner et al. [228]. By measuring the velocity distributions of scattered molecules in the angle-resolved mode, they were able to demonstrate that scattering can be distinguished between a direct inelastic component, scattered into a quasispecular lobe, and a trapping-desorption component scattered into a cosine law component. The fraction scattered into the cosine component rises from a value close to zero at low substrate temperatures to a value of about 0.44 at high substrate temperatures (> 1000 K). This essentially confirms the King and Wells model: at low temperatures, all trapped molecules are chemisorbed, while at higher temperatures, the rate of trapping-desorption becomes appreciable. The authors conclude that the trapping probability into the precursor state is relatively insensitive to temperature [228]. At the present time there is no disagreement between the two groups concerning the adsorption mechanism.

Wolf et al. [226], too, have questioned the King–Wells model on the basis that it assumes an equilibrium spatial distribution of adatoms. However, since King and Wells also assume that the adatoms are practically immobile when energy-accommodated with the surface, there is no proper mechanism for achieving such an equilibrium distribution. Furthermore, θ_{00} has been shown by King and Wells to be insensitive to crystal temperature, which again disagrees with θ_{00} being an equilibrium quantity, as pointed out in the original paper. Wolf et al. [226] propose a similar model to that of King and Wells, but attribute disorder in the chemisorbed adlayer to the irreversible nature of the adlayer formation process. This model predicts the temperature independence of both the adlayer disorder and the saturation

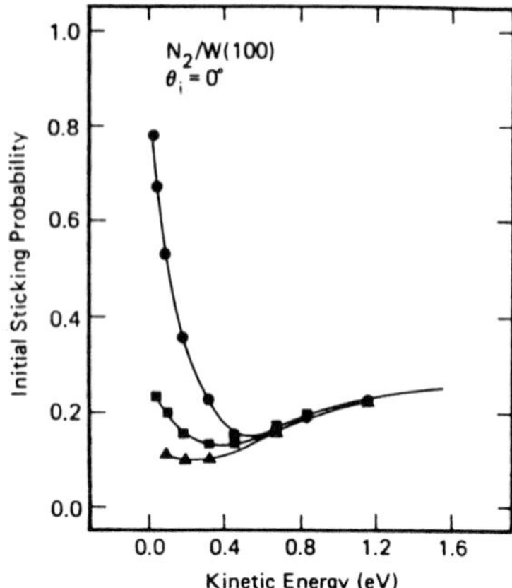

Fig. 43. Effect of incidence energy on the dissociative chemisorption probability of N_2 on W{100} at normal incidence. Results are displayed for surface temperatures of 300 (●), 800 (■) and 1000 K (▲). (From Rettner et al. [229].)

coverage. Both these predictions are consistent with experiment. However, it should be noted that the sticking probability curve theoretically computed by King and Wells far better represents the experimental data for coverages < 0.55 than does the theoretical curve of Wolf et al.

In conclusion, it can be said that, though the King and Wells model does not incorporate all the complicated physical phenomena associated with the N_2/W{100} system, it nonetheless remains a highly successful formulation which is remarkably consistent with the observed kinetics of the system at thermal incident energies. At high incident energies, trapping in the precursor well becomes inefficient, and the chemisorption channel is directly accessed. This transition between precursor-dominated adsorption at low beam energies ($\lesssim 0.4$ eV incident energy) and direct dissociative adsorption at high energies is very clearly demonstrated in Fig. 43, from the work of Rettner et al. [229]. The strong decrease in the measured sticking probability over the range 0–0.4 eV is due to the expected decrease in the trapping probability α with increasing impact energy; the subsequent rise in s is therefore due to direct adsorption, the implication being that this direct adsorption process is inefficient at low impact energies. This may be surprising in view of the zero activation energy barrier, and is not in accord with trajectory calculations based on a ground state, adiabatic potential energy surface.

7.5 NITROGEN ADSORPTION ON HIGH INDEX PLANES OF TUNGSTEN

It is clear from the discussion of N_2 adsorption on W{110}, W{111} and W{100} and from Table 6 and Fig. 44 that the nitrogen–tungsten system exhibits remarkable crystallographic anisotropy in adsorption kinetics at 300 K. Of the low Miller index planes studied, W{100} and W{110} show the

TABLE 6

Summary of sticking probability and coverage data for nitrogen on single-crystal planes of tungsten [187]

Plane	{100} sites (cm^{-2})	Maximum coverage at 300 K (atoms cm^{-2})	s_0 at ~300 K	s_0 at 90 K
{100}	10.0×10^{14}	6.0×10^{14}	0.585	0.60
{310}	6.3×10^{14}	8.0×10^{14}	0.72	0.76
{320}	2.75×10^{14}	6.0×10^{14}	0.735	0.84
{411}	4.2×10^{14}	2.0×10^{14}	0.40	0.60
{111}	0	2.0×10^{14}	0.08	
{110}	0	$\sim 2.0 \times 10^{14}$	$\sim 10^{-2}$	0.23

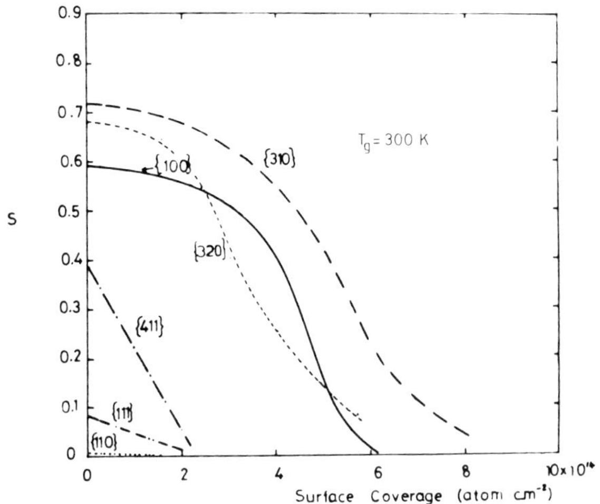

Fig. 44. A comparison of sticking probability profiles for nitrogen on W single crystal planes obtained by the molecular beam technique at normal incidence. In each case, the gas temperature (T_g) is 300 K; the result for the {320} plane refers to a crystal temperature (T_s) of 360 K and the remainder to 300 K. (From Singh-Boparai et al. [187].)

greatest and least reactivity, respectively, to N_2 adsorption. It is, therefore, interesting to study those high index planes which are comprised of both {100} and {110} structures. To this end, the planes of the [001] zone are particularly relevant. Planes located between the (310) and (110) poles are made up of "{100}" steps separated by {110} terraces while planes located between the (310) and (100) poles consist of "{110}" steps and {100} terraces.

Adams and Germer [190] carried out LEED studies of nitrogen adsorption on the {100}, {310} and {210} planes of tungsten. On each plane, the LEED patterns could be interpreted as due to adsorbed nitrogen atoms double-spaced along the [001] direction and single-spaced along the next most densely packed surface direction. Therefore, for the {310} and {210} surfaces, nitrogen atoms were inferred to adsorb on all {100} steps, double spaced in the [001] direction but not at all on the {110} terraces. On the basis of these results, Adams and Germer proposed a simple model to account for the crystallographic anisotropy of the N/tungsten system. They suggested that the β-N adatoms can only be adsorbed into fourfold symmetric, fivefold coordinate sites characteristic of the {100} plane, with occupation of alternate sites. Thus the saturation coverage and reactivity of any W crystal plane for β-N adsorption is expected to be proportional to the number of (100) sites on that plane. All planes on the crystallographic triangle defined by the zones containing {111}, {110} and {211} planes, being devoid of (100) sites, should therefore be inactive.

The Adams and Germer model was not only consistent with LEED data but also with work function data for the planes studied on the [001] zone. The change in work function upon nitrogen adsorption varies markedly with crystal plane. At room temperature, nitrogen adsorption causes the work function of the (100) plane to decrease by ≈ 0.6 eV while on all other planes, the work function is increased, albeit by as little as 0.03 eV on the {110} plane [162, 164, 166, 178, 190, 223]. Delchar and Ehrlich [164] proposed an empirical model to explain the surface potential variability in the β state on different planes. On flat surfaces (e.g. {100}), the centre of positive charge lies above the surface giving a negative work function change, whereas on rough surfaces (such as {111}) it is "inside" the surface giving a positive work function change. Support for this model has come from experiments in which disordered W atoms were deposited on surfaces and the effect on the work function change measured [230].

The Delchar and Ehrlich model for work function change was modified and refined by Adams and Germer [190] on the basis that a fourfold symmetric "(100)" site is required for N adsorption. Given the geometry of the adsorption site, they derived an expression for the maximum work function change due to β-nitrogen adsorption, $\Delta\phi_{\{hk0\}}$, for any plane $(hk0\}$ on the [100] zone

$$\Delta\phi_{\{hk0\}} = \frac{\Delta\phi_{\{100\}}(h - k)[(a_0 - k/2d) - h]}{(h^2 + k^2)} \tag{6}$$

where a_0 is the lattice constant for tungsten and d is the interlayer spacing. With $\Delta\phi_{\{100\}} = -0.6$ eV and $d = 0.038$ nm, this expression gives a close description of experimental work function data for the {100}, {610}, {310}, {210} and {110} planes for which $\Delta\phi$ varies over the range -0.6 eV to $+0.27$ eV. The value assumed for d is remarkably close to the value obtained in a subsequent quantitative LEED study by Griffiths et al. [205] of 0.049 nm. For several reasons, however, this agreement must be considered to be somewhat fortuitous. In the first place, within the framework of the algebraic expression in eqn. (6), only those N adatoms in {100} sites adjacent to or at steps are considered; for planes with {100} terraces, this expression would need to be modified. More importantly, however, the model assumes that the redistribution of charge resulting from N adsorption produces a centre of positive charge at the N adatom itself. This is quite contrary to electronegativity expectations. Surface core level shift spectroscopy is sensitive to charge transfer with adsorbed species, and the data of Jupille et al. [214] on the W $4f_{7/2}$ surface core level shifts demonstrates clearly that charge *is* transferred to the N adatom for N/W{100}. At this stage there is therefore no satisfactory explanation for the observed decrease in work function for N/W{100}, within the framework of the currently accepted structural model [205].

An important test for the general model of Adams and Germer lies in the direct measurement of absolute sticking probabilities and surface coverages. In addition to the data obtained for W{110}, {111} and {100}, as discussed previously, an extensive study on some high index planes has been carried out by Singh-Boparai et al. [187] using the molecular beam technique. The initial sticking probabilities and maximum coverages obtained at 300 K for both the low and high Miller crystal planes are shown in Table 6. Also shown is the surface density of "(100)" sites on each plane. From this data, it is clear that the Adams and Germer model is not fully confirmed, i.e. a direct correlation between "(100)" site density and either the initial sticking probability or the saturation coverage on the various planes is not present. Singh-Boparai et al. [187] extended the Adams and Germer model somewhat and proposed, on the lines of the King and Wells model [186], that dissociation of nitrogen molecules into the β state occurs only at pairs of nearest-neighbour (100) sites, but the chemisorbed atoms thus formed may subsequently migrate out into areas devoid of (100) sites, thus populating other areas of the crystal with β adatoms. In particular, for planes on the [001] zone which are comprised of steps and terraces of (100) and (110) sites, the (110) sites may be populated by this mechanism. Since continuous migration in the β state only occurs at temperatures above 650 K [223], the movement of β adatoms away from "(100)" site pairs must result from the exoergicity of the chemisorption process. This heat of adsorption would produce vibrationally excited adatoms which can make a limited number of diffusion hops, before the excess energy is dissipated to the lattice.

This model does not conflict with the LEED data of Adams and Germer [190] since the presence of adatoms on the terraces would only alter the contents of the overlayer unit mesh but not its size and shape.

Singh-Boparai et al. [187] also extended the kinetic model of King and Wells [186] to describe the situation on stepped planes, by introducing a parameter, ξ, which is the probability that a molecule colliding with the surface is in a physisorption trap in the vicinity of a (100) site pair, i.e. ξ is the ratio of (100) sites to total sites on the surface. This yields a general expression for N_2 adsorption on all W crystal planes [187]

$$S_0 = \alpha \left(1 + \frac{f_d}{f_a}\right)^{-1} \left[1 + K\left(\frac{1}{\xi\theta_{00}} - 1\right)\right]^{-1} \tag{7}$$

where K, the so-called precursor state parameter, is given by

$$K = \frac{f'_d}{(f_a + f_d)} \tag{8}$$

However, both f_d/f_a and K tend to zero as $T_s \rightarrow 0$. Therefore, α, the condensation coefficient, is again experimentally accessible as the low-temperature limit of the initial sticking probability, i.e.

$$\alpha = \lim_{T \rightarrow 0} s_0$$

With the experimentally determined values of α and the best-fit parameters shown in Table 7, for a variety of crystal planes, eqn. (7) provides a quantitative description of all the data over a wide range of substrate temperatures [187].

The experimentally determined parameter α shows significant variation

TABLE 7

Kinetic parameters for nitrogen adsorption on ⟨001⟩ zone planes of tungsten [187]

Plane	$E_d - E_a$ (kJ mol^{-1})	ν_d/ν_a	$E_{d'} - E_{m'}$ (kJ mol^{-1})	$\nu_{d'}/\nu_{m'}$	α	ζ	B
{100}	18	30	7	1	0.60	1	0.985
{310}	18	50			0.76	1	
{320}	18	20	7	1	0.84	1/3	0.985

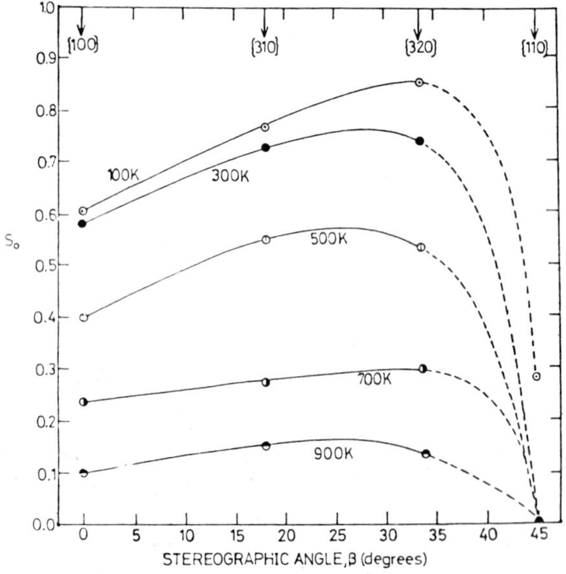

Fig. 45. The variation of the initial sticking probability, s_0, with stereographic angle β across the ⟨001⟩ zone for nitrogen on tungsten single crystal planes. (From King [155].)

across the crystal planes and provides a clue as to why the {100} plane is *not* the most reactive in the [100] zone. Singh-Boparai et al. [187] concluded that the condensation coefficient is highest for planes with the highest surface density of W atoms. Therefore, α increases across the ⟨001⟩ zone, from {100} to {110}, while the {100} n.n. site pair density decreases in the same direction. These two factors together produce a maximum in s_0 near the {320} plane as shown in Fig. 45.

Unfortunately, the above model cannot account for the kinetic data on the {411}, {111} and {110} surfaces. The {411} surface only possesses (100) sites in next-nearest-neighbour positions while the {111} and {110} planes are totally devoid of (100) sites. Despite this, β-N adsorption does occur on these surfaces. It has been suggested that this could be due to the existence of surface defects creating (100) sites. The density of surface defect atoms does not have to be large to produce the results observed. Although the {111}, {411} and {110} planes are not unreactive as predicted, their reactivity is considerably less than that of the {100}, {310} and {320} planes. Insofar as this great difference in reactivity between the two sets of planes is observed, and the relative difference in reactivity of the {100}, {310} and {320} planes is predicted correctly, the kinetic model is consistent with the experimental data.

8. Summary

Since we started writing this review, the number of relevant papers has grown rapidly. It soon became apparent that we could not comprehensively review nitrogen adsorption data on all metals, which led to the strategy of selecting relatively few metals for a thorough review. However, we have chosen our systems on the basis of their popularity as well as the need to cover a representative range of behaviour. The most extensively studied metals for nitrogen adsorption are nickel, iron and tungsten, and, by including these metals, we have also approached comprehensiveness.

Dissociatively adsorbed nitrogen forms strong bonds to all surfaces investigated. This even applies to metals to the right of the periodic table, where the equivalent bulk nitrides are relatively unstable (Table 2). Nitrogen atomised in the gas phase and then adsorbed on Cu has a high desorption temperature (~ 800 K), whereas no uptake has been observed from gaseous dinitrogen. This would imply a large activation energy barrier for dissociative adsorption. For iron, such a barrier has been observed and measured; there is also a barrier for dissociative adsorption on rhenium. To the left of these metals in the periodic table, however, there is generally no activation energy barrier for dissociative adsorption (an exception is the close-packed {110} plane of W) and to the right of these metals only molecular adsorption is observed. The strong bond formed between atomic N and metal atoms leads to reconstruction, surface nitride formation, or, as in the case of N on

References pp. 124–129

W{110}, underlayer formation. Dinitrogen adsorption on metals displays a range of bonding configurations, from on-top and bridged upright to lying down. The heat of adsorption of dinitrogen is always lower than for the carbon monoxide analogue.

The results obtained for nitrogen adsorption on iron, which are of such vital importance in developing an understanding of the ammonia synthesis reaction, presented some difficulties to the reviewer. These difficulties, we believe, arise in part from the complexity of the adsorption phenomena occurring on the various single crystal planes of iron, and in part from the difficulty of carrying out definitive experiments in this field. Although a number of different laboratories have been involved in this work, it is worth noting that many of these laboratories have used the same crystal! (The reason is that producing a clean crystal is quite a mammoth task.) As a result, the usual checks of consistency from one crystal sample to another have not been carried out. There are also difficulties arising from the uptake of nitrogen into the bulk of the crystal, and from the very low sticking probabilities for N_2 adsorption, which leads to high dosing pressure and hence susceptibility to impurities. Nevertheless, there is quite a coherent body of work on this system.

As we go to press, further highly relevant papers continue to appear. Of particular interest are the range of sophisticated studies using supersonic molecular beam sources related to the chemical dynamics of the nitrogen–metal interaction which will undoubtedly continue to stimulate interest in these systems well into the future.

References

1 M. Grunze, in D.A. King and D.P. Woodruff (Eds.), The Chemical Physics of Solid Surfaces and Heterogeneous Catalysis, Vol. 4, Elsevier, Amsterdam, 1982, p. 143.
2 K. Horn, J. DiNardo, W. Eberhardt, H.-J. Freund and E.W. Plummer, Surf. Sci., 118 (1982) 465.
3 G. Blyholder, J. Phys. Chem., 68 (1964) 2772.
4 P.S. Bagus, C.R. Brundle, K. Hermann and D. Menzel, J. Electron. Spectrosc. Relat. Phenom., 20 (1980) 253.
5 P.S. Bagus, K. Hermann and M. Seel, J. Vac. Sci. Technol., 18 (1981) 435.
6 G. Doyen and G. Ertl, Surf. Sci., 43 (1974) 197.
7 A.D. Allen, R.O. Harris, B.R. Loescher, J.R. Stevens and R.N. Whiteley, Chem. Rev., 73 (1973) 11.
8 J.C. Green and M.L.H. Green, in J.C. Bailar, H.J. Emeléus, R. Nyholm and A.F. Trotman-Dickenson (Eds.), Comprehensive Inorganic Chemistry, Pergamon Press, Oxford, 1973, p. 446.
9 H. Huber, E.P. Kündig, M. Moskovits and G.A. Ozin, J. Am. Chem. Soc., 95 (1973) 332.
10 E.P. Kündig, M. Moskovits and G.A. Ozin, J. Mol. Struct., 14 (1972) 137.
11 E.P. Kündig, M. Moskovits and G.A. Ozin, Can. J. Chem., 51 (1973) 2710.
12 C. Agte and K. Moers, Z. Anorg. Allgem. Chem., 198 (1931) 233.
13 R.E. Rundle, Acta Crystallogr., 1 (1948) 180.
14 S.N. L'vov, V.F. Nemchenko and G.V. Samsonov, Dokl. Akad. Nauk SSSR, 135 (1960) 577.
15 A.R. West, Solid State Chemistry and its Applications, Wiley, New York, 1984, p. 661.

16 P. Schwarzkopff and R. Kieffer, Refractory Hard Metals, MacMillan, London, 1953.
17 H. Bilz, Z. Phys., 153 (1958) 338.
18 A. Ozaki and K. Aika, in R.W.F. Hardy, F. Bottomley and R.C. Burns (Eds.), Treatise on Dinitrogen Fixation, Wiley, New York, 1979, p. 169.
19 R. Van Hardeveld and A. Van Montfoort, Surf. Sci., 4 (1966) 396.
20 D.A. King, Surf. Sci., 9 (1968) 375.
21 K. Kunimori, T. Kawai, T. Kondow, T. Onishi and K. Tamaru, Surf. Sci., 59 (1976) 302.
22 E. Miyazaki, I. Kojima and S. Kojima, Langmuir, 1 (1985) 264.
23 Y. Kuwahara, M. Jo, H. Tsuda, M. Onchi and M. Nishijima, Surf. Sci., 180 (1987) 421.
24 Y. Kuwahara, M. Fujisawa, M. Jo, M. Onchi and M. Nishijima, Surf. Sci., 188 (1987) 490.
25 B.M.W. Trapnell, Proc. R. Soc. (London) Ser. A, 218 (1953) 566.
26 R.N. Lee and H.E. Farnsworth, Surf. Sci., 3 (1965) 461.
27 G.G. Tibbetts, J. Chem. Phys., 70 (1979) 3600.
28 D. Schmeisser, K. Jacobi and D.M. Kolb, Vacuum, 31 (1981) 439.
29 G.G. Tibbetts, J.M. Burkstrand and J.C. Tracy, Phys. Rev. B., 15 (1977) 3652.
30 R.E. Kirby, C.S. McKee and L.V. Renny, Surf. Sci., 97 (1980) 457.
31 M.H. Mohamed and L.L. Kesmodel, Surf. Sci., 185 (1987) L467.
32 H.C. Zeng, R.N.S. Sodhi and K.A.R. Mitchell, Surf. Sci., 188 (1987) 599.
33 V. Higgs, P. Hollins, M.E. Pemble and J. Pritchard, J. Electron. Spectrosc. Relat. Phenom., 39 (1986) 137.
34 M. Salmerón, A.A. Baró and J.M. Rojo, Phys. Rev. B., 13 (1976) 4348.
35 J. Perdereau and G.E. Rhead, Surf. Sci., 24 (1971) 555.
36 J.M. Burkstrand, G.G. Kleiman, G.G. Tibbetts and J.C. Tracy, J. Vac. Sci. Technol., 13 (1976) 291.
37 G.G. Kleiman and J.M. Burkstrand, Solid State Commun., 21 (1977) 5.
38 R. Franchy, M. Wuttig and H. Ibach, Z. Phys. B., 64 (1986) 453.
39 M.H. Matloob and M.W. Roberts, J. Chem. Soc. Faraday Trans. 1, 73 (1977) 1393.
40 G.G. Tibbetts, J.M. Burkstrand and J.C. Tracy, J. Vac. Sci. Technol., 13 (1976) 362.
41 A.B. Anderson, Chem. Phys. Lett., 49 (1977) 550.
42 J.R. Smith, F.J. Arlinghaus and J.G. Gay, Solid State Commun., 24 (1977) 279.
43 H.L. Yu and E.E. Whiting, Surf. Sci., 82 (1979) 301.
44 S. Ferrer and J.M. Rojo, Solid State Commun., 24 (1977) 339.
45 D. Heskett, A. Baddorf and E.W. Plummer, Surf. Sci., 195 (1988) 94.
46 P. Feulner and D. Menzel, Phys. Rev. B, 25 (1982) 4295.
47 A.B. Anton, N.R. Avery, B.H. Toby and W.H. Weinberg, J. Electron. Spectrosc. Relat. Phenom., 29 (1983) 181.
48 H. Pfnür, D. Menzel, F.M. Hoffmann, A. Ortega and A.M. Bradshaw, Surf. Sci., 93 (1980) 431.
49 R.A. de Paola, F.M. Hoffmann, D. Heskett and E.W. Plummer, Phys. Rev. B, 35 (1987) 4236.
50 P. Hollins and J. Pritchard, Surf. Sci., 89 (1979) 486.
51 K. Hermann, P.S. Bagus, C.R. Brundle and D. Menzel, Phys. Rev. B, 24 (1981) 7025.
52 P.S. Bagus, C.J. Nelin and C.W. Bauchlicher, Phys. Rev. B, 28 (1983) 5423.
53 D. Saddei, H.-J. Freund and G. Hohlneicher, Surf. Sci., 95 (1980) 527.
54 C.M. Kao and R.P. Messmer, Phys. Rev. B, 31 (1985) 4835.
55 E.S. Hood, B.H. Toby and W.H. Weinberg, Phys. Rev. Lett., 55 (1985) 2437.
56 R.P. Eischens and J. Jacknow, in Proc. 3rd Int. Congr. Catal., North-Holland, Amsterdam, 1965, p. 627.
57 R. Van Hardeveld and A. Van Montfoort, Surf. Sci., 17 (1969) 90.
58 A.M. Bradshaw and J. Pritchard, Surf. Sci., 19 (1970) 198.
59 B.E. Nieuwenhuys, Surf. Sci., 105 (1981) 505.
60 C.R. Brundle and A.F. Carley, Discuss. Faraday Soc., 60 (1975) 51.
61 C.R. Brundle, J. Vac. Sci. Technol., 13 (1976) 301.
62 G. Ranga Rao, K. Prabhakaran and C.N.R. Rao, Surf. Sci., 176 (1986) L835.
63 M. Grunze, P.A. Dowben and R.G. Jones, Surf. Sci., 141 (1984) 455.

64 E. Umbach, Surf. Sci., 117 (1982) 482.
65 J.C. Fuggle, E. Umbach, D. Menzel, K. Wandelt and C.R. Brundle, Solid State Commun., 27 (1978) 65.
66 D. Saddei, H.J. Freund and G. Hohlneicher, Surf. Sci., 102 (1981) 359.
67 P.A. Dowben, Y. Sakisaka and T.N. Rhodin, Surf. Sci., 147 (1984) 89.
68 J. Stohr and R. Jaeger, Phys. Rev. B., 26 (1982) 4111.
69 T. Kihara and A. Koide, Adv. Chem. Phys., 33 (1975) 51.
70 M.J. Grunze, J. Fuhler, M. Neumann, C.R. Brundle, D.J. Auerbach and J. Behm, Surf. Sci., 139 (1984) 109.
71 C. Kemball, Adv. Catal., 2 (1952) 233. D.H. Everett, Proc. Chem. Soc., 37 (1957) 38.
72 T.D. Thomas, J. Chem. Phys., 53 (1970) 1744.
73 K. Siegbahn, C. Nordling, G. Johansson, J. Hedman, P.F. Heden, K. Hamrin, U. Gelius, T. Bergmark, L.O. Werme, R. Manne and Y. Baer, ESCA Applied to Free Molecules, North-Holland, Amsterdam, 1971.
74 C.R. Brundle, P.S. Bagus, D. Menzel and K. Hermann, Phys. Rev. B., 24 (1981) 7041.
75 E. Umbach, Solid State Commun., 51 (1984) 365.
76 K. Schönhammer and O. Gunnarsson, Solid State Commun., 23 (1977) 691; 26 (1978) 399.
77 O. Gunnarsson and K. Schönhammer, Phys. Rev. Lett., 41 (1978) 1608.
78 K. Hermann and P.S. Bagus, Solid State Commun., 38 (1981) 1257.
79 O. Jepson, J. Madsen and O.K. Anderson, Phys. Rev. B., 26 (1982) 1790.
80 W.F. Egelhoff, Jr., Surf. Sci., 141 (1984) L324.
81 W.F. Egelhoff, Jr., Phys. Rev. B., 29 (1984) 3681.
82 Y. Wu and P. Cao, Surf. Sci., 179 (1987) L26.
83 E. Umbach, A. Schichl and D. Menzel, Solid State Commun., 36 (1980) 93.
84 R.L. Dubs and V. McKoy, Chem. Phys. Lett., 142 (1987) 237.
85 H.J. Freund, R.P. Messmer, C.M. Kao and E.W. Plummer, Phys. Rev. B, 31 (1985) 4848.
86 R.P. Messmer, J. Vac. Sci. Technol. A, 2 (1984) 899.
87 M.J. Breitschafter, E. Umbach and D. Menzel, Surf. Sci., 178 (1986) 725.
88 X.H. Feng, M. Yu, G. Meigs and E.L. Garfunkel, to be published.
89 A. Quick, V.M. Browne, S.G. Fox and P. Hollins, Surf. Sci., 221 (1989) 48.
90 M. Golze, M. Grunze and W. Unertl, Prog. Surf. Sci., 22 (1986) 101.
91 M. Grunze, R.K. Driscoll, G.N. Burland, J.C.L. Cornish and J. Pritchard, Surf. Sci., 89 (1979) 381.
92 M.E. Brubaker, I.J. Malik and M. Trenary, J. Vac. Sci. Technol. A, 5 (1987) 427.
93 M.E. Brubaker and M. Trenary, J. Chem. Phys., 85 (1986) 6100.
94 M.E. Brubaker and M. Trenary, J. Chem. Phys., 90 (1989) 4651.
95 B.J. Bandy, N.D.S. Canning, P. Hollins and J. Pritchard, J. Chem. Soc. Chem. Commun., (1982) 58.
96 M. Golze, M. Grunze, R.K. Driscoll and W. Hirsch, Appl. Surf. Sci., 6 (1980) 464.
97 J. Stohr, R. Jaeger and J.J. Rehr, Phys. Rev. Lett., 51 (1983) 821.
98 T.A. Carlson and M.O. Krause, Phys. Rev., 140 (1965) A1057.
99 M. Grunze, P.H. Kleban, W.N. Unertl and F.S. Rys, Phys. Rev. Lett., 51 (1983) 582.
100 M. Golze, M. Grunze and W. Hirschwald, Vacuum, 31 (1981) 697.
101 S.N. Coppersmith, D.S. Fisher, B.I. Halpern, P.A. Lee and W.F. Brinkman, Phys. Rev. B, 25 (1982) 349. J. Villain and P. Bak, J. Phys. (Paris), 42 (1981) 657. S.T. Chui, Phys. Rev. B, 23 (1981) 5982. M. Kardar and A.N. Berker, Phys. Rev. Lett., 48 (1982) 1552.
102 Y. Kuwahara, M. Fujisawa, M. Onchi and M. Nishijima, Surf. Sci., 202 (1988) 17.
103 A. Schenk, M. Hock and J. Küppers, Surf. Sci., 217 (1989) L367.
104 H. Ueba, Surf. Sci., 188 (1987) 421.
105 J.B. Benziger and R.E. Preston, Surf. Sci., 141 (1984) 567.
106 G.L. Price, B.A. Sexton and B.G. Baker, Surf. Sci., 60 (1976) 506.
107 E. Roman and R. Riwan, Surf. Sci., 118 (1982) 682.
108 W. Heiland and E. Taglauer, Nucl. Instrum. and Methods, 194 (1982) 667.

109 N. Shamir, D.A. Baldwin, T. Darko, J.W. Rabalais and P. Hochmann, J. Chem. Phys., 76 (1982) 6417.
110 T. Darko, D.A. Baldwin, N. Shamir, J.W. Rabalais and P. Hochmann, J. Chem. Phys., 76 (1982) 6408.
111 W. Daum, S. Lehwald and H. Ibach, Surf. Sci., 178 (1986) 528.
112 L. Wenzel, D. Arvanitis, W. Daum, H.M. Rotermurd, J. Stöhr, K. Baberschle and H. Ibach, Phys. Rev. B, 36 (1987) 7689.
113 G. Ertl, Sci. Eng. (Catal. Rev.), 21 (1980) 201.
114 R. Brill, E.-L. Richter and E. Ruch, Angew. Chem. Int. Ed. Engl., 6 (1967) 882.
115 J.A. Dumiesic, H. Topsøe, S. Khammouma and M. Boudart, J. Catal., 37 (1975) 503.
116 J.A. Dumiesic, H. Topsøe, S. Khammouma and M. Boudart, J. Catal., 37 (1975) 513.
117 N.D. Spencer, R.C. Schoonmaker and G.A. Somorjai, J. Catal., 74 (1982) 129.
118 F. Bozso, G. Ertl, M. Grunze and M. Weiss, J. Catal., 49 (1977) 18.
119 F. Bozso, G. Ertl and M. Weiss, J. Catal., 50 (1977) 519.
120 G. Wedler, D. Borgmann and K.-P. Geuss, Surf. Sci., 47 (1975) 592. G. Wedler, G. Steidl and D. Borgmann, Surf. Sci., 100 (1980) 507.
121 K. Kishi and M.W. Roberts, Surf. Sci., 62 (1977) 252.
122 D.W. Johnson and M.W. Roberts, Surf. Sci., 87 (1979) L255.
123 M. Grunze, M. Golze, J. Fuhler, M. Neumann and E. Schwarz, Proc. 8th Int. Congr. Catal., W. Berlin, Vol. IV, 1984, p. 133.
124 P. Gundry, J. Catal., 1 (1968) 363.
125 V. Ponec and Z. Knor, J. Catal., 10 (1968) 73.
126 G. Strasser, M. Grunze and M. Golze, J. Vac. Sci. Technol. A, 3 (1985) 1562.
127 G. Ertl, S.B. Lee and M. Weiss, Surf. Sci., 114 (1982) 515.
128 L.J. Whitman, C.E. Bartosch, W. Ho, G. Strasser and M. Grunze, Phys. Rev. Lett., 56 (1986) 1984.
129 M.-C. Tsai, U. Seip, I.C. Bassignana, J. Küppers and G. Ertl, Surf. Sci., 155 (1985) 387.
130 I.D. Gay, M. Textor, R. Mason and Y. Iwasawa, Proc. R. Soc. London Ser. A, 356 (1977) 25.
131 K.S. Love and P.H. Emmett, Z. Phys. Chem. Abt. B, 13 (1931) 401.
132 M. Grunze, M. Golze, W. Hirschwald, H.-J. Freund, H. Pulm, U. Seip, M.C. Tsai, G. Ertl and J. Küppers, Phys. Rev. Lett., 53 (1984) 850.
133 H.J. Freund, B. Bartos, R.P. Messmer, M. Grunze, H. Kuhlenbeck and M. Neumann, Surf. Sci., 185 (1986) 187.
134 M. Grunze, G. Strasser, M. Golze and W. Hirschwald, J. Vac. Sci. Technol. A, 5 (1987) 527.
135 D. Tomanek and K.H. Bennemann, Phys. Rev. B, 31 (1985) 2488.
136 R.P. Thorman and S.L. Bernasek, J. Chem. Phys., 74 (1981) 6498.
137 J. Böheim, W. Brenig, T. Engel and U. Leuthäusser, Surf. Sci., 131 (1983) 258.
138 C.T. Rettner and H. Stein, Phys. Rev. Lett., 59 (1987) 2768.
139 M. Bowker, I.B. Parker and K.C. Waugh, Appl. Catal., 14 (1985) 101.
140 M. Bowker, I.B. Parker and K.C. Waugh, Surf. Sci., 197 (1988) L223.
141 P. Stoltze and J. Norskøv, Phys. Rev. Lett., 55 (1985) 2502.
142 P. Alnot, A. Cassuto and D.A. King, Surf. Sci., 215 (1989) 29.
143 R. Brill, J. Catal., 16 (1970) 16.
144 G. Ertl, M. Grunze and M. Weiss, J. Vac. Sci. Technol., 13 (1976) 314.
145 R. Imbihl, R.J. Behm, G. Ertl and W. Moritz, Surf. Sci., 123 (1982) 129.
146 G. Gafner and R. Feder, Surf. Sci., 57 (1976) 37. R. Feder and G. Gafner, Surf. Sci., 57 (1976) 45.
147 G. Broden, G. Gafner and H.P. Bonzel, Appl. Phys., 13 (1977) 333.
148 G. Ehrlich, in W.M.H. Sachtler, G.C.A. Schuit and P. Zwietering (Eds.), Proc. 3rd Int. Congr. Catal., Amsterdam, 1964, North-Holland, Amsterdam, 1965, p. 113.
149 P.A. Dowben, M. Grunze and R.G. Jones, Surf. Sci., 109 (1981) L519.
150 D.O. Hayward, in J.R. Anderson (Ed.), Chemisorption and Reactions on Metallic Films, Vol. 1, Academic Press, London, New York, 1971.

151 M.W. Roberts and C.S. McKee, Chemistry of the Metal-Gas Interface, Chap. 10, Clarendon Press, Oxford, 1978, Chap. 10.
152 G. Ehrlich, Annu. Rev. Phys. Chem., 17 (1966) 295.
153 T.E. Madey and J.T. Yates, Jr., J. Vac. Sci. Technol., 8 (1971) 525.
154 G. Ehrlich, Adv. Catal., 14 (1963) 225.
155 D.A. King, CRC Crit. Rev. Solid State Mater. Sci., 7 (1978) 167.
156 G. Ehrlich, J. Phys. Chem., 60 (1956) 1388.
157 G. Ehrlich, J. Chem. Phys., 36 (1962) 1171.
158 T.W. Hickmott and G. Ehrlich, J. Phys. Chem. Solids, 5 (1958) 47.
159 P. Kisliuk, J. Chem. Phys., 30 (1959) 174.
160 P. Kisliuk, J. Chem. Phys., 31 (1959) 1605.
161 P.A. Redhead, Vacuum, 12 (1962) 203.
162 T. Oguri, J. Phys. Soc. Jpn., 18 (1963) 1280.
163 L.J. Rigby, Can. J. Phys., 43 (1965) 532.
164 T.A. Delchar and G. Ehrlich, J. Chem. Phys., 42 (1965) 2686.
165 T.E. Madey and J.T. Yates, Jr., J. Chem. Phys., 44 (1966) 1675.
166 T. Oguri, J. Phys. Soc. Jpn., 19 (1964) 83.
167 J.T. Yates, Jr. and T.E. Madey, J. Chem. Phys., 43 (1965) 1055.
168 R.P.H. Gasser, C.P. Lawrence and D.G. Newman, Trans. Faraday Soc., 61 (1965) 1771.
169 J.L. Robins, W.K. Warburton and T.N. Rhodin, J. Chem. Phys., 46 (1967) 665.
170 G. Ehrlich, J. Chem. Phys., 34 (1961) 29.
171 D.O. Hayward, D.A. King and F.C. Tompkins, Proc. R. Soc. London Ser. A, 297 (1967) 305, 321.
172 T. Tamura and T. Hamamura, Surf. Sci., 95 (1980) L293.
173 T.E. Madey, J.T. Yates, Jr. and N.E. Erickson, Surf. Sci., 43 (1974) 526.
174 G. Ehrlich and F.G. Hudda, J. Chem. Phys., 36 (1962) 3233.
175 A. Van Oostrom, J. Chem. Phys., 47 (1967) 761.
176 D.A. King and M.G. Wells, Surf. Sci., 29 (1972) 454.
177 G.J. Dooley, III and T.W. Haas, J. Vac. Sci. Technol., 7 (1970) 590.
178 T.E. Madey and J.T. Yates, Jr., Nuovo Cimento (Suppl. 2), 5 (1967) 483.
179 B.J. Hopkins and S. Usami, in G.A. Somarjai (Ed.), The Structure and Chemistry of Solid Surfaces, Wiley, New York, 1968.
180 P.W. Tamm and L.D. Schmidt, Surf. Sci., 26 (1971) 286.
181 C. Somerton and D.A. King, Surf. Sci., 89 (1979) 391.
182 M. Bowker and D.A. King, J. Chem. Soc. Faraday Trans. 1, 75 (1979) 2100.
183 J.C. Fuggle and D. Menzel, Surf. Sci., 79 (1979) 1.
184 R. Liu and G. Ehrlich, Surf. Sci., 119 (1982) 207.
185 J.T. Yates, Jr., R. Klein and T.E. Madey, Surf. Sci., 58 (1976) 469.
186 D.A. King and M.G. Wells, Proc. R. Soc. London Ser. A., 339 (1974) 245.
187 S.P. Singh-Boparai, M. Bowker and D.A. King, Surf. Sci., 53 (1975) 55.
188 K. Besocke and H. Wagner, Surf. Sci., 87 (1979) 457.
189 D.L. Adams and L.H. Germer, Surf. Sci., 32 (1972) 205.
190 D.L. Adams and L.H. Germer, Surf. Sci., 27 (1971) 21.
191 R.C. Cosser, S.R. Bare, S.M. Francis and D.A. King, Vacuum, 31 (1981) 503.
192 J.E. Lennard-Jones, Trans. Faraday Soc., 28 (1932) 333.
193 J. Lee, R.J. Madix, J.E. Schlaegel and D.J. Auerbach, Surf. Sci., 143 (1984) 626.
194 D.J. Auerbach, H.G. Pfnür, C.T. Rettner, J.E. Schlaegel, J. Lee and R.J. Madix, J. Chem. Phys., 81 (1984) 2515.
195 H.E. Pfnür, C.T. Rettner, J. Lee, R.J. Madix and D.J. Auerbach, J. Chem. Phys., 85 (1986) 7452.
196 A. Kara and A.E. De Pristo, Surf. Sci., 193 (1988) 437.
197 G.P. Derby and D.A. King, Faraday Discuss. Chem. Soc., 89 (1990) in press.
198 L.R. Clavenna and L.D. Schmidt, Surf. Sci., 22 (1970) 365.
199 M.A. Chesters, B.J. Hopkins and R.I. Winton, Surf. Sci., 59 (1976) 46.

200 W. Ho, R.F. Willis and E.W. Plummer, Surf. Sci., 95 (1980) 171.
201 D.A. King and C. Somerton, unpublished results.
202 P.J. Estrup and J. Anderson, J. Chem. Phys., 46 (1967) 567.
203 D.L. Adams and L.H. Germer, Surf. Sci., 26 (1971) 109.
204 K. Griffiths and D.A. King, Proc. 4th Int. Conf. Solid Surf., Suppl. Rev. Vide, 201 (1980) 237.
205 K. Griffiths, D.A. King, G.C. Aers and J.B. Pendry, J. Phys. C, 15 (1982) 4921.
206 N.R. Palmer and D.A. King, Phys. Scr., T4 (1983) 122.
207 J.E. Houston and R.L. Park, Surf. Sci., 21 (1970) 209.
208 P.W. Tamm and L.D. Schmidt, J. Chem. Phys., 51 (1969) 5332.
209 L.F. Mattheis, Phys. Rev. A., 139 (1965) 1893.
210 M.K. Debe and D.A. King, J. Phys. C., 10 (1977) L303; Phys. Rev. Lett., 39 (1977) 709; Surf. Sci., 81 (1977) 193.
211 T.G. Felter, R.A. Barker and P.J. Estrup, Phys. Rev. Lett., 38 (1977) 1138.
212 D.A. King and G. Thomas, Surf. Sci., 92 (1980) 201.
213 K. Griffiths, C. Kendon, D.A. King and J.B. Pendry, Phys. Rev. Lett., 46 (1981) 1584.
214 J. Jupille, K.G. Purcell and D.A. King, Solid State Commun., 58 (1986) 529.
215 H. Winters, private communication, 1980.
216 S. Usami, N. Tominaga and T. Nakajima, Vacuum, 27 (1976) 11.
217 H.F. Winters, P. Morgan, S. Tongaard and J. Onsgaard, Proc. 4th Int. Conf. Solid Surf., Suppl. Rev. Vide, 201 (1980) 303.
218 D.L. Adams, Surf. Sci., 42 (1974) 12.
219 P. Kisliuk, J. Phys. Chem. Solids, 3 (1957) 95.
220 J.B. Taylor and I. Langmuir, Phys. Rev., 44 (1933) 423.
221 J.K. Roberts, Proc. R. Soc. London Ser. A, 152 (1935) 445.
222 R.H. Fowler and E. A. Guggenheim, Statistical Thermodynamics, Cambridge University Press, London, 1960.
223 G. Ehrlich and F.G. Hudda, J. Chem. Phys., 35 (1961) 1421.
224 K. Besocke and H. Wagner, Phys. Rev. B, 8 (1973) 4597; Surf. Sci., 53 (1975) 351.
225 K.C. Janda, J.G. Hurst, C.A. Becker, J.P. Cowin, L. Wharton and D.J. Auerbach, Surf. Sci., 93 (1980) 270.
226 N.O. Wolf, D.R. Burgess and D.K. Hoffmann, Surf. Sci., 100 (1980) 453.
227 P. Alnot and D.A. King, Surf. Sci., 126 (1983) 359.
228 C.T. Rettner, E.K. Schweizer, H. Stein and D. Auerbach, Phys. Rev. Lett., 61 (1988) 986.
229 C.T. Rettner, H. Stein and E.K. Schweizer, J. Chem. Phys., 89 (1988) 3337.
230 S.P. Singh-Boparai and D.A. King, Surf. Sci., 61 (1976) 275.
231 C.G. Goymour and D.A. King, J. Chem. Soc. Faraday Trans. 1, 69 (1973) 736.

Chapter 3

The Initial Interaction of Oxygen with Well-defined Transition Metal Surfaces

C.R. BRUNDLE and J.Q. BROUGHTON*

IBM Almaden Research Center, San Jose, CA 95120 (U.S.A.)

1. Introduction

1.1 GENERAL

The original intention of this chapter was to review critically all the work relating to the interaction of oxygen with single-crystal metal surfaces. Whereas this would have been feasible four or five years ago the tremendous growth in activity in this area plus the propensity of the authors for detail now precludes a thorough review in the space available. We therefore decided, by default, to restrict ourselves to the transition metals and even within this group we have been very selective in our coverage. Comparison is made between the elements of the Group VIB metals, Cr, Mo and W, which is the most well-studied group, and Ni, the next most studied metal after W. In addition literature coverage is complete only up to the end of 1984, the date when this review was completed and submitted.

What types of properties are we trying to compare and contrast? Traditionally we can attempt to split the initial interactions of oxygen with metal surfaces into several different sequential steps. These would be sub-monolayer overlayer adsorption (where overlayer adsorption is defined as involving the formation of an oxygen overlayer without disruption of the crystal lattice); penetration of the lattice to give a reconstructed surface (defined here as the top two-dimensional layer in which both O atoms and metal atoms exist in a coplanar or nearly coplanar arrangement); penetration deeper into the lattice; and the growth of oxide nuclei or layers in forms ranging from amorphous to well-defined epitaxial relationships with the substrate. It would greatly simplify this review if the interaction of oxygen with metals *did* always follow this (traditional) sequence. In practice the sequence is often not realized. There are examples where there is no stable chemisorbed overlayer, penetration occurring from the start of the interaction. This appears to be true for Be [1], Zn [2] and Pb [3]. Other possibilities are: an overlayer structure stabilized at high coverage, θ, whereas a lower coverage converts more easily to a penetrated structure; and overlayer

* Present Address. Complex Systems Theory, Naval Research Laboratories, Washington, DC 20375, U.S.A.

References pp. 381–388

adsorption only *after* reconstruction or penetration; i.e. adsorption on a metal–oxygen substrate lattice. The definition used above for reconstruction is itself somewhat arbitrary. Figure 1(a) indicates schematically a situation where the reconstruction definition is clear; the top metallic layer has only half the number of metal atoms as the next layer down, the missing half being occupied by O atoms. In this simplest representation the metal atoms remaining in the top layer are in the same positions as in the bulk. This may not be true, of course, leading to considerable registration possibilities for the reconstructed layer with the bulk substrate. Even the distinction between the different regimes are somewhat arbitrary; for example, overlayer adsorption, reconstruction, and oxide formation are not always straightforwardly separable. Figure 1(b) illustrates a situation (FeO{100}/O) [4], where the distinction between overlayer adsorption and reconstruction and oxide formation is becoming blurred. The "overlayer adsorption" illustrated dif-

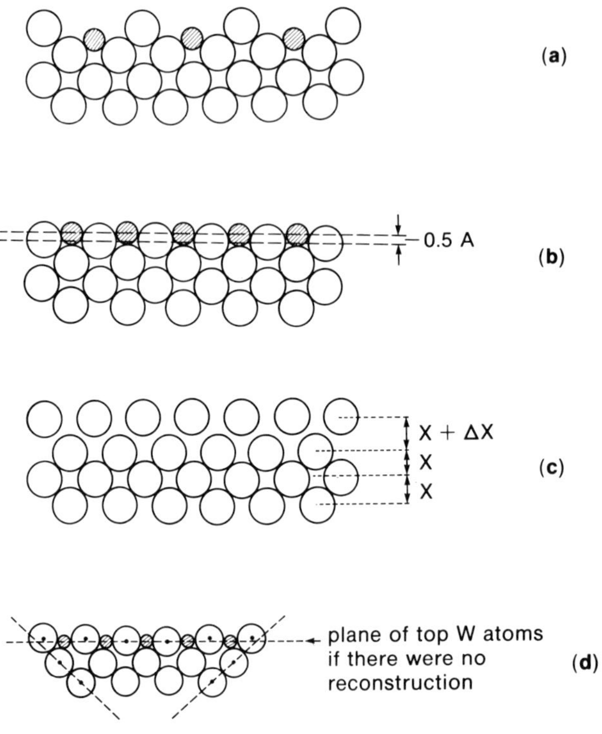

Fig. 1. Some possible distortion arrangements of the top layer of metal substrate atoms under the influence of an adsorbate. Open circles = substrate atoms; hatched circles = adsorbate atoms. (a) Replacement of one half of the top substrate atoms by adsorbate atoms. Depending on relative sizes, the adsorbate atoms will tend to lie above or below (as shown) the plane of the substrate top atomic plane. (b) "Overlayer" oxygen atom adsorption on Fe{100} [4]. (c) Outward relaxation of top layer of substrate atoms under influence of adsorbate overlayer (not shown), such as occurs for c(2 × 2) O on Ni{100} [5]. (d) Proposed "displacive reconstruction" model for nitrogen on W{100} [6, 7].

fers from a regular FeO{100} single plane by only a 0.5 Å displacement of the O atoms normal to the surface, since the Fe–Fe spacing in an FeO(100) plane is practically identical to the Fe–Fe spacing in the Fe(100) plane. In addition if the O atoms were coplanar with the Fe, but the Fe–Fe spacings did not change, would the surface be termed reconstructed? Technically no, since the term is usually reserved for situations where the substrate top atom layer undergoes a change in periodicity. Clearly, when displacements are very small, distinctions between one formal description and another are rather meaningless. The phenomena of relaxation of a clean topmost metal layer [either outwards or inwards, e.g. Fig. 1(c)] is well known [5]. In the presence of an adsorbate the magnitude or even the sign of this relaxation may change. There also exists the possibility of small lateral movements of the top layer substrate atoms, sometimes accompanied by small vertical movements also, under the influence of adsorbate atoms. Such a proposed example [6, 7] is shown in Fig. 1(d). It is generally termed displacive reconstruction.

In addition to all the possible variations of geometry there are always questions of the degree to which any of the individual arrangements coexist with others. For example, when an intensity versus voltage analysis of a LEED pattern [4] determines that oxygen atoms reside in a particular overlayer ordered pattern, to what extent does this preclude there being oxygen atoms also present in a disordered arrangement on the surface or beneath it or beneath the surface in an ordered structure? This aspect, the possible presence of an additional structure, is one of the hardest subjects to deal with, even using modern surface spectroscopic tools. This can be either because interpretation becomes difficult when two types of oxygen are present, such as in angular-resolved photoemission where features representing the electronic structure of one type of metal–oxygen bonding severely overlap those of another type, or because the particular technique being used is only sensitive to one type, even if there are significant amounts of another type present. The latter case fits LEED, of course, where only ordered structures are observable. It is possible that it can also apply more severely to several of the other surface-sensitive techniques, since there is evidence that, at least in some cases, the signal observed originates only or predominantly from a very minor concentration species on the surface. Conversely, of course, such properties might make a technique ideal for studying minority species. There is one technique, thermal desorption, which has been routinely used to distinguish different states of the same species on surfaces [8]. The distinction is in terms of binding energy of the species to the surface, which may or may not be relatable to structure, site of adsorption, etc. The strength of the technique is its ability to distinguish states only a fraction of a kcal apart. The problem with this technique is that there is no guarantee that a species desorbed at temperature T_d existed in the same form at the temperature T_{ads} at which it was adsorbed and in which we may be interested. It may have undergone several changes during heating to the desorption temperature. In addition, for the specific case of oxygen, heating often drives oxygen into the bulk rather than desorbing it. Thermal desorption studies on the W/O_2 system illustrate some of these problems in that oxide species are observed to desorb ($\sim$ 1700 K) from room temperature adsorbed states which apparently show no evidence, by other techniques, of having undergone penetration/oxidation at the adsorption temperature (see Sect. 2.9). The kinetics and mechanism of these interconversions from one type of oxygen species to another, particularly conversion from overlayer structures to oxide structures, are, of course, an important area of study in their own right.

Despite the complications raised above concerning categorizing the oxygen interaction in terms of overlayer, reconstructed, and penetrated structures we shall try to follow this general scheme. Our discussions will center on such things as the kinetics and mechanism of the dissociative adsorption process (nearly all the adsorption studies of oxygen on metals refer to dissociative interaction, see Sect. 1.3); the geometric and electronic structures of the adsorption products; the oxidation states or stoichiometries of the initial oxidation products; and the mechanism of the interconversion from adsorbed oxygen to oxide structures. Overall we are looking for trends within Group VIB which might be explained in terms of differences in crystal structure (b.c.c., f.c.c.); metal–metal bond strengths; metal–metal bond lengths and packing arrangements of the low index faces (only the low index faces and stepped surfaces involving these faces are considered in this review); and electronic structure variations of the metal atoms. For Ni there are well-documented differences from face to face which we would like to be able to explain.

In the remainder of the review we will refer heavily to data obtained using the following techniques: Auger spectroscopy (AES); X-ray and ultraviolet photoelectron spectroscopy (XPS and UPS); Low-energy electron diffraction (LEED); reflectance high-energy electron diffraction (RHEED); low-energy ion scattering (IS); Rutherford backscattering (RBS); electron and photon stimulated desorption (ESD and PSD) and their associated ion angular and ion energy distribution techniques (ESDIAD, ESDIED, PSDIAD, PSDIED); secondary ion mass spectroscopy (SIMS); thermal programmed desorption spectroscopy (TPD); energy loss spectroscopy (ELS); surface vibrational spectroscopy either by high-resolution energy loss (HRELS) or infrared reflectance (IR); and work function change ($\Delta\phi$) measurements. It is assumed that the reader is familiar with the basics of these techniques. Reviews concerning some of these techniques can be found in this series of volumes.

Several further points, general to M/O_2 interactions, need to be discussed before going on to the individual metal case histories.

1.2 THE NATURE OF THE CLEAN SURFACE

We have implicitly been assuming so far that the surface layer of metal atoms simply represents the termination of the bulk structure. This is certainly not always the case. There can be contractions or expansions normal to the surface in the interlayer spacing for the top layer (surface relaxation) [5] and, more dramatically, there are now several known examples where the clean surface has a different symmetry from the underlying substrate atoms (i.e. a displacive reconstructed surface) [9, 10]. The different symmetry shows up in LEED. In some cases it was originally wrongly thought to be representative not of the clean surface but of some impurity-stabilized structure. The real-space geometries giving rise to the LEED patterns are often a subject of controversy, though it seems that whatever metal atom movement is involved it is probably rather small. The significant point for oxygen interaction is that it seems that small amounts of oxygen adsorption may be sufficient to restabilize the regular bulk termination structure or to cause disruptions of this structure.

1.3 MOLECULAR OXYGEN ADSORPTION

As mentioned above there are very few confirmed examples of molecular oxygen having been observed on well-characterized metal surfaces. The only one known with confidence at room temperature is the Pt/oxygen system where Fisher et al. have identified a strongly bonded species that bears a close resemblance to epoxy radicals which exist in bulk structures [11]. At lower temperatures (77 K) chemisorbed molecular oxygen species have been observed on Ag surfaces [12] and they possibly also exist on Pd surfaces. On all other metal surfaces it appears that no stable molecular structure exists for temperatures above 77 K at sufficient concentration to be detectable by the spectroscopic techniques available. If adsorption takes place at a sufficiently low temperature, clearly one can obtain significant concentrations of molecular oxygen. In this limit these would be weakly bound physisorbed and condensed species, though it is still common that the first monolayer dissociates and the molecular species is adsorbed on top of this passivated surface. Opila and Gomer have shown that several molecular oxygen species exist on W between 20 and 40 K but that they *all* form on top of a dissociated O first layer [13]. (There is much ESR data in the literature referring to molecular oxygen surface species. It is difficult to put any of the ESR work in context with this review because much of it concerns high-area powder oxide surfaces and because it is not generally known whether the ESR signals observed represent anything more than minority species at very specific surface sites.) Does this lack of observation of measurable quantities of chemisorbed molecular oxygen at higher temperatures imply that the role of the molecular species can be ignored at these temperatures? We do not believe so. In specific cases there is evidence from kinetics to show that dissociative adsorption takes place via adsorption into a weakly bound mobile "precursor" state (which may or may not be physisorbed oxygen, but is presumably molecular) even though the steady-state concentration of this species is too low to detect at the adsorption temperature [14]. The important question is the role any molecular state plays in the mechanism and kinetics of the dissociation processes. These questions are addressed in some detail in individual cases in this review.

A strong effect of the precursor state may be implied from a sticking probability which is high at zero coverage, θ, and does not fall as θ increases. The sticking probability can be maintained because the precursor state oxygen migrates over the surface to an available dissociation site. Note, however, that a constant sticking probability alone is not conclusive evidence of precursor mobility. If penetration of the metal lattice by O_{ad} leads to regeneration of metal sites at the surface, a high sticking probability may be maintained for low adsorption pressure. If there is no precursor state, or if its surface diffusion versus desorption probability is low, then the sticking probability will be proportional to the number of dissociation sites available on the surface (Langmuir kinetics) [14].

1.4 CONVERSION TO "OXIDE"

Having discussed in brief general terms the initial nature of the metal surface and the nature of the dissociation process, it remains to indicate the mechanisms by which lattice penetration and oxide formation might occur. In general bulk oxide structures are more thermodynamically stable than overlayer structures, so it is a question of overcoming some activation energy barrier to produce them. The height of the barrier will depend critically on the nature of the metal surface. For the clean surface the major factor is probably the metal–metal bond energy which must be overcome. The lower this energy the more likely is the activation energy barrier to be low or zero. In such a case penetration of the lattice at room temperature could be rapid and no stable overlayer structure on the clean metal surface formed. Thermal treatment for non-zero barrier height systems can also help achieve oxidation, of course, but there are many examples where oxidation occurs at room temperature following formation of stable overlayer structures by simply increasing oxygen exposure. In these cases the activation barrier height may be changing as a function of the changing nature of the surface. This leads us to the field-assisted description of initial oxygen penetration and fast thin oxide growth contained in the theories of Cabrera and Mott [15] and Fehlner and Mott [16]. Examples of stable overlayers which convert to oxide at some coverage will be given later in the case of Ni. They illustrate the importance and interplay of substrate order and adsorbate overlayer order, mobility, and coverage in determining the route followed during the oxygen interaction. In many cases the conversion from stable overlayer to oxide can be ascribed to the presence of *defects* at the surface, which act as nucleation sites for oxide growth.

A rather large body of data exists for some of the transition metals on the facetting of single-crystal surfaces under the influence of oxygen and on the epitaxial growth of oxides on both the smooth and facetted surfaces. Historically much of this work preceded the submonolayer adsorption-type studies. Since these areas are, perhaps, less appropriate in a volume entitled "Chemisorption" we only discuss those aspects where they obviously relate and have importance to the adsorption and initial penetration processes.

1.5 UNWANTED CO-ADSORPTION REACTIONS

We have implicitly assumed so far that one is always sure that an M/O_2 reaction is being studied. Unfortunately there are many examples in the literature where subsequent careful examination reveals that there were problems with co-adsorption of minority impurity species. This is a problem which is particularly difficult with oxygen since oxygen is particularly prone to generate impurity species from wall reactions. Co-adsorption and reactions with H_2O, CO and CO_2 have all caused problems and misidentifications of "new" oxygen states. The most common problem is perhaps the co-adsorption and reaction of H_2O to generate OH instead of O species on the surface [17].

2. Chromium, molybdenum and tungsten

2.1 GENERAL

These three metals belong to the same group (VIB) of the periodic table and have nearly half-filled d shells (Cr, Mo d^5s^1; W d^4s^2). They are all b.c.c. metals. They have very high melting points, particularly Mo and W, and their oxides have large negative free energies of formation. There are, however, some significant differences in their properties. The Cr–Cr bond length is much shorter than Mo–Mo or W–W, resulting in higher surface atom densities for Cr for any given face. The three low index faces of Cr and Mo are represented in Fig. 2 together with oxygen atoms of covalent and ionic radii for comparison purposes. Clearly if one defines a surface coverage, θ, in the conventional manner as the ratio of surface species to substrate surface atoms, the O atom density in atoms cm^{-2} achieved for $\theta = 1$ on Cr surfaces must be much greater than on Mo or W surfaces. This should be kept in mind when comparing O coverage, θ, between the different metals. The other important variation is the heat of atomization of the metal. That of Cr is considerably less than those of Mo or W. A summary of these properties is given in Table 1. The oxide data in this table pertains to the nature of the oxides grown in typical UHV conditions. These oxides are also those which form a protective coating in more severe corrosive conditions [21]. Cr always appears to oxidize to Cr_2O_3, Mo to MoO_2 and W to WO_3.

Despite the strong free-energy driving force to oxidation for all three metals, the behavior in the early stages of oxygen reaction differs down the group. Cr oxidizes easily, even at room temperature, with a minimum of

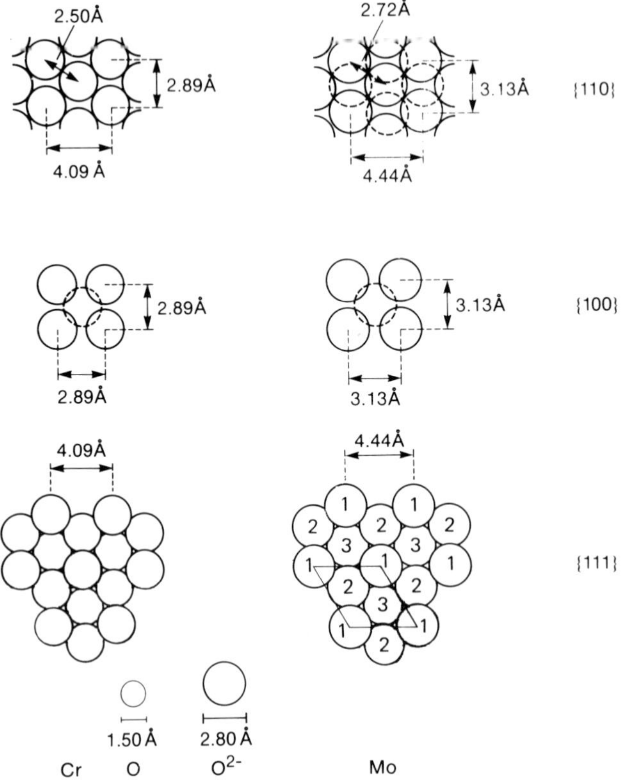

Fig. 2. Hard-sphere representations of the low index faces of Cr and Mo. For the {100} and {110} faces the second layer atom positions are represented by dotted circles. For the open {111} face, large portions of third layer atoms are also "visible" to adsorbates. The atomic and ionic diameters of oxygen atoms are also shown.

intervening chemisorption states/structures. W, on the other hand, forms stable ordered overlayer structures at room temperature, at least for the {110} and {100} surfaces, and rather drastic conditions are required to produce bulk oxide. The most obvious reason for the difference would appear to be that, although the heat of adsorption of oxygen is similar on W and Cr (see Table 1), the Cr–Cr bond energy is much lower than that of W–W. Therefore the large amount of energy released upon adsorption of oxygen is capable of easily disrupting Cr–Cr bonds but not W–W bonds. Mo should occupy an intermediate position between Cr and W, though from consideration of the parameters in Table 1 one might expect it to be much closer to W in behavior. The detailed comparisons referred to later in this section suggest that this is a false expectation; Mo is intermediate between Cr and W, but from several points of view seems to be closer to Cr than to W, even though its behavior is more often compared to W.

The reason for the high heat of adsorption of oxygen on all three metals

TABLE 1

Metal–metal and metal–oxygen bond data [18]

Property	Metal		
	Cr	Mo	W
Crystal bond length (Å)	2.4980	2.7251	2.7409
Enthalpy of atomization ($kcal\,mol^{-1}$)	95.0	157.3	203.1
Free energy of formation of oxide at 298 K	− 250.2 (Cr_2O_3)	− 127.4 [20] (MoO_2)	− 182.5 (WO_3)
Approximate heat of adsorption of O_2 on metal film [19] ($kcal\,mol^{-1}$)	170	170	190
Surface atom density ($\times 10^{-15}$)			
{110}	1.70	1.43	1.41
{100}	1.20	1.01	1.00
{111}	0.70	0.58	0.58

is thought to be the large number of unpaired electrons in these metals available to form M–O bonds. This, of course, holds true for other adsorbates. It is well known, for example, that low index faces of W will crack most simple molecular adsorbates into their constituent atoms at room temperature [22].

The three low index faces chosen for review here are not strictly in the order of closest packing for the b.c.c. system. The order, in fact, goes {110}, {100}, {112}, {310} followed by {111}. As Table 1 indicates, the surface density of the {111} is approximately 40% of that of the {110}; it has a very open structure. Thus, we might expect the adsorption behavior of oxygen on the {111} face to be rather different from that on the {110} and {100} faces.

The clean {110} and {111} surfaces of these metals terminate in the bulk structures [23–29] although Cr{110} very easily facets to {100} planes [1, 30, 31]. The {100} surfaces of Mo and W, on the other hand, have been shown by LEED [29, 32, 33] to have a 6–12% contraction normal to the surface of the unreconstructed (1 × 1) surfaces, and also both metals reconstruct to c(2 × 2) structures near or below room temperature [34–36]. One therefore has the possibility of starting oxygen adsorption on either the reconstructed or unreconstructed {100} surfaces. The Cr{100} surface has also been reported as reconstructed to a c(2 × 2) structure at room temperature [37], but subsequent studies indicate that the "reconstruction" was due to a combination of C and O impurities [38]. The reconstruction phase transition on W and Mo is thought to be related to the presence of surface electronic states on the {100} surfaces [39, 40] which have been the subject of numerous studies [41–48].

Much more work has been published on W than on Mo and Cr though much of it is early work and should be viewed with caution because surface purity and structure checks were not available. Two pre-1970 reviews are available [49, 50]. Though Mo and Cr are less well studied than W the information available for them is, in fact, more complete than for many other metals and enough data exist to make meaningful comparison between Cr, Mo and W.

2.2 CHROMIUM

LEED, AES and RHEED studies have been performed on the {110}/O_2 system [21, 27, 30, 31, 51], and the {100}/O_2 system [21, 30, 31, 42, 51, 52] but not on the {111} surface, so there is no information available as to the geometry and orientation in the {111} case until very heavy oxidation (1070 K, damp hydrogen atmosphere) where some X-ray diffraction data is available [53]. At room temperature the {110} surface shows no ordered overlayer structures [27, 31], and the {100} surface shows, at best, very weak ordering [27, 31, 47]. This is consistent with the interpretation, given later, that lattice disruption occurs from the start without the formation of a stable adsorbate overlayer. Detailed photoemission (XPS and UPS) results combined with LEED/AES and $\Delta\phi$ measurements have been reported by Peruchetti et al. [47] on the Cr{100}/O_2 system at room temperature. From these the authors derive a phenomenological model for reaction from initial adsorption through to passivation under UHV conditions. A similar study was reported earlier by the same group [54] for Cr{111}/O_2, though without the LEED/AES, allowing a comparison to be made between the two faces.

2.2.1 Cr{110}

Most of the information available on the {110} surface relates to the high coverage, initial oxidation stages. At room temperature and for oxygen exposures up to 10 L (saturation coverage according to AES [31]), no oxygen-induced superstructures are observed in LEED, the {110} Cr spots merely decrease in intensity as the background intensity increases. No sticking probability or oxygen coverage estimates are available, in contrast to the {100} and {111} surfaces, but in RHEED diffuse streaks are observed [27] over this exposure range and have been interpreted as resulting from a sheet-like pseudomorphic Cr_2O_3 overlayer. Thus it seems that saturation coverage corresponds to a thickness of several atomic layers. This is consistent with results on the {100} and {111} surfaces (see later) where saturation coverages between 2.8×10^{15} and 5.5×10^{15} atoms cm^{-2} (i.e. several monolayers) at 15–20 L exposure were estimated. Since there is no evidence for a chemisorbed *molecular* state on the {100} or {111} surfaces it is likely that *all* exposure conditions on {110} at room temperature also produce only dissociated O species. There has been no specific investigation of this question, however.

Heating a room-temperature exposed surface to 900 K, or giving low

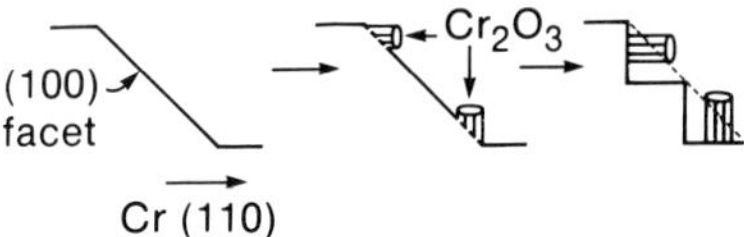

Fig. 3. Schematic representation of the growth of Cr_2O_3 on a facetted Cr{110} surface [27].

exposures to a clean surface at 900 K, produces O-covered Cr{100} facets (not oxidized facets) as observed both by LEED [30, 51] and RHEED [27]. Once formed, these facets are hard to eliminate. Exposure at temperatures somewhat lower than 900 K leads to the growth of epitaxial Cr_2O_3 on the flat (unfacetted) surface, as indicated by the RHEED and LEED data [21, 27, 31, 51]. The RHEED data of Kennett and Lee [27] indicate that the Cr_2O_3 nuclei orientation is {114} ∥ Cr{110}. Heating this epitaxial oxide above the facetting temperature of 900 K produces a degradation into a new structure consisting of unoxidized, but presumably O-covered, {100} facets and a new oxide epitaxy. This new epitaxy is actually a mixture of two, corresponding to Cr_2O_3 nuclei growing both vertically and horizontally on the {100} facet "saw teeth," which exist at 45° to the original {110} flat surface (Fig. 3). The same structures form if sufficiently high exposures are used at ~900 K, i.e. the mechanism of this epitaxial oxide growth requires first the growth of the {100} facets (which requires the presence of oxygen and temperatures near 900 K), followed by oxide nucleation which eats into the facet planes to give the resultant (110) and (1$\bar{1}$0) oxide structures (see Fig. 3).

For Cr{100} and {111} (Sect. 2.2.2 and 2.2.3), the available XPS and UPS data allow discussion of the oxidation state changes in the Cr as a function of oxygen coverage. These data are not available for the {110} surface, but similar information is *potentially* available from the behavior of the low-energy Cr $M_{23}M_{45}M_{45}$ Auger peak at 36 eV. Eklund and Leygraf [31] report that the metallic 36 eV Cr peak disappears after 1 L exposure at room temperature and is replaced by peaks at 31 and 46 eV, which reach maximum intensity at around 10 L exposure. The authors' interpretation of the new features are in terms of Cr^{3+} of bulk Cr_2O_3 (46 eV peak) and Cr atoms which are still metallic in nature, but strongly influenced by neighboring O atoms (31 eV peak). There are a number of anomalies in the details of these spectroscopic assignments, which cannot be addressed here, but the escape depth is very short (~3 Å) for these Auger kinetic energies and so the strong changes observed do at least indicate that oxidation state changes in the Cr top layer start to occur within the first 1 L of exposure.

No other surface spectroscopy studies of the Cr{110}/O system in the early adsorption stages exist.

2.2.2 Cr{100}

At room temperature it is clear, from the observed O(1*s*) and O(2*p*) photoemission binding energies (BEs) [47], that oxygen interaction with Cr{100} always results in dissociation. It was reported in the LEED study of Peruchetti et al. [47] that the clean Cr{100} surface was reconstructed from p(1 × 1) to c(2 × 2) and that the latter structure was stable to ~800 K. Subsequent work by Foord et al. [38] showed that this c(2 × 2) structure was in fact due to a mixture of C and O impurities of between 0.1 and 0.4 monolayer coverage. The ordering is established by the sputtering and annealing procedures during cleaning attempts. At impurity coverages of 0.01 monolayer or less (apparently very difficult to achieve) the expected unreconstructed p(1 × 1) Cr{100} structure is observed above room temperature [38]. The other LEED [31, 51] and RHEED oxygen adsorption studies [27] do not mention an initial c(2 × 2) structure, so presumably the starting point of these adsorption studies was always the unreconstructed clean p(1 × 1) surface. The Peruchetti studies, however, start with the contaminated c(2 × 2) structure, which must complicate the interpretation of their results. The two LEED studies [31, 51] disagree on whether oxygen adsorption at room temperature ever produces any ordered structures. The early study of Eklund and Legraf [31] did not report any ordered structures, but the more recent Michel and Jardin study [51] reported a supposedly c(2 × 2)O weak structure. The quality of their LEED data is very poor, however, and no other surface techniques were used to check for contamination, so the question of whether ordered O structures can be formed by room-temperature adsorption on a clean p(1 × 1) Cr{100} surface is really still open.

Despite the problem of having started with an apparently contaminated surface, the combined XPS, UPS, LEED, AES and $\Delta\phi$ studies of Peruchetti et al. [47] form the most complete body of data on the Cr{100}/O system and discussions of this data and the authors' interpretations take up most of the remainder of this section. Their first major conclusion was that O atom penetration of the surface occurred from $\theta = 0$, i.e. no classical chemisorption overlayer stage exists. This conclusion is based on the analyses of the kinetics of the reaction and on the $\Delta\phi$ data. These two aspects are summarized in Fig. 4(a) where the XPS O(1*s*) growth and the $\Delta\phi$ data are plotted against exposure. The conversion of O(1*s*) intensities into absolute coverages were based on the O(1*s*)/Cr(2*p*) ratios and certain assumptions concerning l_e, the mean free path length for inelastic scattering of the photoelectrons, and the relative O(1*s*)/Cr(2*p*) photoionizations cross-sections. The procedure could easily be in error by a factor of up to two. This would affect the estimated initial sticking probability, S_0, (0.34 according to Peruchetti et al.) and the coverage, θ, at which S is observed to decrease, but does not alter the most important point derivable from the O(1*s*) versus exposure curve, namely that S remains constant initially and even at high coverage only decreases by a factor of ~2. As this review progresses it will become clear that we regard the behavior of S as a function of θ as one of the most informative of measurements. Two explanations for a constant S behavior are possible, as discussed briefly in Sect. 1.3. The first is adsorption into a weakly bound mobile molecular precursor state which subsequently

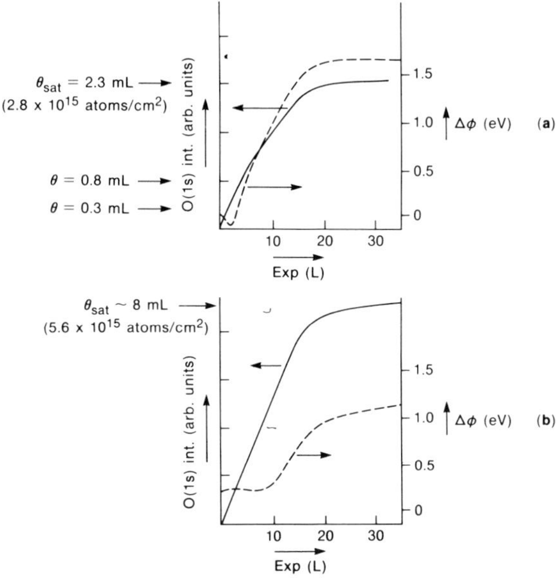

Fig. 4. O(1*s*) intensity and $\Delta\phi$ versus oxygen exposure at 300 K for (a) Cr{100} and (b) Cr{111}. The O(1*s*) scale is arbitrary [47, 54].

dissociates into the observed strongly bound oxygen. The value of S can then remain constant even though available dissociation sites are being used up as θ increases through population of the precursor which can diffuse to an unoccupied site for dissociation. The whole philosophy of the involvement of precursor states [14] and the effect they can have on the functional form of coverage versus exposure curves (which can be converted to sticking probability versus coverage) is discussed in some detail later, particularly in the sections on W and Ni where extensive S versus θ data are available. We do not discuss it further here because the second explanation is favored by Peruchetti et al. and we agree with this interpretation. This alternative is qualitative and simple: penetration of the Cr{100} lattice is rapid enough from $\theta = 0$ such that the number of clean dissociative Cr metal sites at the surface does not change. This suggestion is supported by the $\Delta\phi$ data. Looking at Fig. 4(a), we see that $\Delta\phi$ is negative in the 0–2 L exposure range. This is usually interpreted as implying, for the adsorption of an electronegative species, that the adsorbate atom is below the plane of the metal surface. If the $O^{\delta-}$ were present as an overlayer, $\Delta\phi$ would be expected to be positive. Simple dipole layer arguments for smooth, structureless surfaces must certainly predict such behavior. In practice surfaces are not structureless and are not always smooth (steps, defects) and there are a few cases where there is evidence for an overlayer electronegative adsorbate and yet $\Delta\phi$ is negative [55, 56]. Nevertheless we are going to rely on $\Delta\phi$ behavior rather heavily in this review as a good guide to overlayer versus penetration-type adsorption. Many authors go much further and attempt to calculate an in-

References pp. 381–388

ternuclear distance, d, from a point charge model, assuming some charge separation, Δq. Alternatively Δq can be calculated assuming d. These procedures have little, if any, validity since a point-charge model is completely unrealistic for oxygen atoms where the O(2p) orbitals can be very diffuse for an O^{2-} ion. Thus one cannot use localization of charge on the O nucleus in any sensible way to determine surface metal–oxygen bond ionicity/covalency and readers should disregard such published efforts. Peruchetti et al. try to refine the general positive or negative $\Delta\phi$ arguments by suggesting that a low concentration of O beneath the Cr surface might be expected to give a $\Delta\phi$ of zero, whereas a large amount of oxide (Cr_2O_3) should give ϕ characteristic of the bulk oxide (negative $\Delta\phi$ compared with clean Cr). We consider it over-ambitious to imply anything more from the $\Delta\phi$ measurements than that they argue against significant build-up of O overlayer coverage in the 0–2 L (θ = 0–0.3 mL, according to the Peruchetti coverage calibration) range.

Above 2 L, $\Delta\phi$ swings sharply positive and increases linearly to about 5 L. The authors attribute this to the build up of overlayer oxygen in significant quantities between 2 and 5 L, in addition to continuing penetration and incorporation below the surface. S does not start decreasing till around 5 L, however. If O atoms are taking up overlayer sites above 2 L, S should drop unless a precursor mechanism is in effect. When S does drop (above 5 L exposure) it does not go quickly to zero but maintains a constant, fairly high (0.18) value up to $\sim$ 15 L exposure. Over this exposure range (5–15 L) there is a reduction in the rate of increase of $\Delta\phi$ with θ [see Fig. 4(a)]. The explanation offered by the authors, which we consider highly plausible, is that the overlayer coverage has reached a saturation value at $\sim$ 5 L and the rate-limiting step to further uptake is the transport of overlayer oxygen through the metal–oxygen surface layer. The authors claim the 6–15 L stage kinetics fits the Mott–Fehlner constant field oxidation model [16]. The electrostatic field across the metal–oxygen subsurface region may certainly be the driving force for the incorporation of further oxygen atoms by cation/anion diffusion, but we think it rather fanciful to try and fit the kinetics between $\sim$ 1–2.5 monolayer coverage to the Mott–Fehlner model. It seems likely to us that some islanding and lateral growth of oxide nuclei is possible, as has been found for many other metal/oxygen systems [1, 57–60]. This, of course, is not included in the Mott–Fehlner treatment of the kinetics [16]. If lateral spread of oxide patches does occur this would account for the sharp fall in S between 15 and 20 L (θ = 2–2.3 mL). In this region the oxide patches (2 mL thick) completely cover the surface and to incorporate further O atoms requires a thickening normal to the surface which must involve diffusion through the highly stable thin Cr oxide layer now present. For more eleborate discussions of oxide island growth the reader is referred to the second half of this review on Ni.

What is the electronic nature of the Cr during the adsorption/oxidation sequence? The nucleated oxide formed at higher temperatures is certainly

Cr_2O_3 (i.e. Cr^{3+}) as indicated by the LEED [31, 51] and RHEED [27] studies. The orientation is less clear with fibrous growth [27], oxide powder (i.e. random nuclei) [51] and epitaxial relationships [31] being claimed. Nor is it clear what the thickness is at the higher temperatures, though it is certainly greater than at room temperature. It is clear that the {100} Cr face does not facet in the presence of oxygen [21, 27, 30, 31, 51]. At room temperature no ordered Cr_2O_3 is formed, at least in sufficient quantity or in sufficiently large domains to be observed by LEED or RHEED. At the room-temperature saturation coverage, the XPS Cr(2*p*) lines and UPS Cr(3*d*) band do show the presence of Cr^{3+}, however. At what stage between $\theta = 0$ and 2.3 mL Cr^{3+} features in the XPS/UPS first appear is unknown. Our own opinion, based on experience with other metal/oxygen systems, is that for the chromium to exhibit Cr^{3+} character in XPS or UPS, it is necessary for it to be heavily coordinated by O^{2-} ions, so that metallic screening of the charge on Cr^{3+} is eliminated. Put another way, "oxide" nuclei of some minimum size are required. This can probably not occur until a significant concentration of O^{2-} exists within the ~3 Å affected layer. To establish the onset of Cr^{3+} formation requires careful grazing emission angle XPS work to enhance the surface sensitivity. Whether any lower oxidation state exists prior to Cr^{3+} formation or whether the subsequent overlayer adsorption stage can be detected by a change in the nature of the XPS and UPS Cr signals will require ever greater effort. In the lower-energy Auger spectra it is possible that the two oxidation-related Cr peaks at 31 and 46 eV, observed by Eklund and Leygraf [31], hold the key to the question of the electronic nature of the surface Cr atoms in the early coverage stages, but this awaits satisfactory interpretation of the origin of these peaks.

The suggested sequences of the reaction on Cr{100} at room temperature are summarized in Fig. 5. It is clear that all available data support the idea that surface reconstruction or penetration occurs immediately on dissociative adsorption and that at saturation a thin amorphous layer of Cr_2O_3 has formed. The intermediate region suggestions are more speculative but the data are consistent with the idea of overlayer oxygen building up significant concentration between $\theta = \sim 0.3$ and ~ 0.8 mL, beyond which the overlayer concentration remains constant but the underlying "oxide" either thickens or grows laterally at constant thickness by cation or anion migration under the influence of an electric field until the entire top layer of Cr atoms has been consumed by the reaction. Further thickening then becomes very slow because it now requires diffusion through the strongly cohesive Cr_2O_3 layer. The electronic nature of the Cr is unclear in the stages prior to saturation.

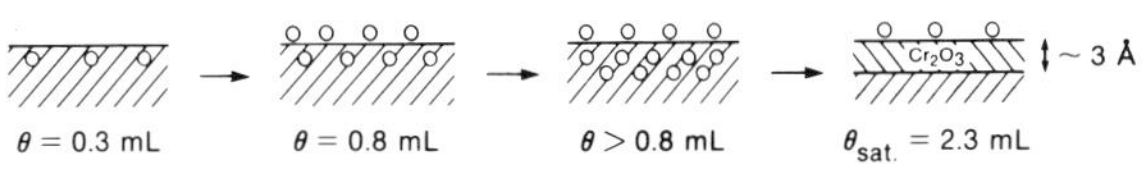

$\theta = 0.3$ mL → $\theta = 0.8$ mL → $\theta > 0.8$ mL → $\theta_{sat.} = 2.3$ mL (Cr_2O_3, ~3 Å)

Fig. 5. Suggested schematic sequence for the reaction of oxygen with Cr{100} at 300 K. Open circles represent oxygen atoms [47].

At higher temperatures, bulk Cr_2O_3 oxide nuclei are detectable by RHEED but the orientation is unclear. The {100} Cr surface is stable to facetting.

2.2.3 Cr{111}

No LEED or RHEED studies have been performed and the only structural information we have is the X-ray diffraction data of Sailors et al. [53] who showed that at 1070 K damp hydrogen at atmospheric pressure caused the growth of bulk {111} Cr_2O_3.

The photoemission studies [54] on Cr{111} may be directly compared with those on Cr{100} [47]. The results are qualitatively similar but with some significant differences. The most important is that the saturation thickness at 300 K is estimated at ~9 Å compared with the ~3 Å on the {100} face. An increase in thickness might well be expected for such an open face (see Table 1). The saturation product has the Cr^{3+} oxidation state and a stoichiometry of Cr/O ≈ 0.56. Adsorption at higher temperatures produces a much thicker film (20–40 Å) with the expected Cr_2O_3 stoichiometry. The $\Delta\phi$ and S curves for the {111} surface [Fig. 4(b)] may be compared with those on {100} in Fig. 4(a). The main differences are that $\Delta\phi$ does not become positive until ~10 L ($\theta \approx$ 1.5 mL) compared with ~2 L ($\theta \approx$ 0.3 mL) for Cr{100} and S does not fall until near saturation (~20 L compared with ~6 L for Cr{100}). The qualitative explanation for these differences would seem to be that, because of the very open structure of the {111} face, several layers of Cr can easily be attacked and place-exchanged with O, compared with one for Cr{100}. Oxygen atom adsorption on top of the surface only starts to occur after this facile process is complete.

In the photoemission study a set of annealing experiments were performed during one adsorption run. After each incremental adsorption the surface was annealed to ~800 K. XPS showed that this pushed the O distribution further into the bulk and the $\Delta\phi$ measurements [Fig. 4(b)] demonstrated that it forced all overlayer O into the subsurface since a large positive $\Delta\phi$ was converted to a large negative $\Delta\phi$ (characteristic of the oxide surface without adsorbed overlayer O). Once recooled to room temperature, the overlayer O could be replaced by further adsorption, thereby producing a large positive $\Delta\phi$ again. This behavior on annealing has its parallel in the much better studied W{100} and W{110} studies. There (see Sect. 2.6.2), annealing above a critical T causes a sharp change from complete overlayer adsorption to reconstruction.

2.2.4 Summary of Cr results

Exposure of the Cr{110}, {100} and {111} surfaces to oxygen at room temperature leads directly to reconstructed and penetrated surfaces and film growth of 3–9 Å thickness (probably amorphous) without any evidence for ordered overlayers, with the possible exception of a weak c(2 × 2) structure on Cr{100}. The sequence seems to be immediate O penetration with a building up of the subsurface O concentration by place exchange until the place exchange becomes so slow that overlayer O concentration begins to increase. The latter occurs when the first Cr layer has been consumed for Cr{100} and when the first two or three layers have been consumed on the much more open Cr{111} surface. In all cases the bulk oxide is Cr_2O_3. At elevated temperatures Cr{110} facets to {100} planes in the presence of oxygen and the epitaxial Cr_2O_3 produced with further oxygen exposure is "derived" from the {100} planes. The Cr{100} surface is stable to facetting and at elevated temperature epitaxial Cr_2O_3 layers can be grown. These results are summarized in Tables 2–5. Tables 2 and 3 summarize room-temperature and high-temperature overlayer structures. Table 4 summarizes

TABLE 2

Room-temperature adsorption overlayer structures

Metal	Property		
	LEED/RHEED patterns (298 K)	Ref.	Comments
Cr{110}	No superstructures. Background increases with coverage.	27, 31	Kennett and Lee [27] observe broad streaks in background. Interpreted as sheet of Cr_2O_3.
Cr{100}	No superstructures. Background increases with coverage.	27, 51	Michel and Chardin [51] observe c(2 × 2) at low exposure.
Cr{111}	N/A		
Mo{110}	p(2 × 2) (1/4); p(2 × 1) (1/2); p(2 × 2) (3/4); (1 × 1) or 1/6th order patterns.	25, 61–63	Hayek et al. [62] observe (1 × 1) pattern; others observe 1/6th-order patterns at high exposures. The Zukov et al. [63] c(2 × 2) pattern is the p(2 × 2) (1/4) structure.
Mo{100}	c(2 × 2), (6 × 2) (1/3), (6 × 1) (2/3), (3 × 1) (1), (1 × 1) (> 1).	63, 64–66	Riwan et al. [66] did not observe (6 × 1).
Mo{111}	No superstructures. Background increases with coverage. Facets (?)	61, 67	Ferante and Barton [67] observe diffuse reflections from (112) facets at high exposures.
W{110}	p(2 × 1) (1/2); p(2 × 2) (3/4); (1 × 1) (1).	23, 68–72	Bauer and Engel [69] obtained patterns above $\theta = 0.5$ by dosing with WO_2 and annealing.
W{100}	p(2 × 1), p(4 × 1) (1/2), c(8 × 2) (5/8), p(8 × 1) (7/8), streaked (4 × 1) (1–1.4).	73–82	Desplat [74] observed p(2 × 1). Bauer et al. [77] observed c(8 × 2) and p(8 × 1), and report p(4 × 1) (1/2) is composite of p(2 × 1), p(4 × 1) and c(8 × 2). Early papers report no superstructures.
W{111}	No superstructures. Background increases with coverage.	83–85	Three-lobed ESDIAD patterns exist whereas LEED superstructures do not.

TABLE 3

High-temperature adsorption overlayer structures

Metal	Property		
	LEED/RHEED patterns (298 K)	Ref.	Comments
Cr{110}	{100} facets. (3 × 1), c(3 × 1). Two epitaxial forms of Cr_2O_3.	21, 27, 31, 51	Hexagonal precursors to oxide formation seen by LEED. See also Table 4.
Cr{100}	p(2 × 2), p(4 × 1). No facets, Cr_2O_3 (epitaxial?).	21, 27, 31, 51	RHEED [27] suggests epitaxial fibrous oxide growth. LEED [51] gives oxide powder pattern.
Cr{111}	N/A	53	Atmospheric oxidation of Cr{111} gives epitaxial Cr_2O_3 (X-ray diffraction data).
Mo{110}	p(2 × 2) (1/4), p(2 × 1) (1/2) (?) complex, {100} facets, epitaxial MoO_2.	62 21, 27, 61, 63, 25, 86	Heating RT structures gives their inverse sequence. Authors disagree on pattern sequence, thermal stability, facetting.
Mo{100}	c(2 × 2), (1 × 1), ($\sqrt{5} \times \sqrt{5}$), p(2 × 2), (1 × 1) or c(4 × 4), p(2 × 1), (5 × 5), p(2 × 1), c(2 × 2). {110}, (112) facets, epitaxial MoO_2.	64, 61, 87, 65, 87, 66, 61, 65, 21	Heating gives reverse order of structures. Bauer and Poppa [65] report c(4 × 4) is c(4 × 2) + p(2 × 1). $\sqrt{5} \times \sqrt{5} \equiv 5 \times 5$. Kennett and Lee [61] and Bauer and Poppa [65] report p(2 × 2) is two domains of p(2 × 1).
Mo{111}	Complex, (112) facets, epitaxial MoO_2.	61, 67	(4 × 2) patterns reported by Kennett and Lee [61]. 1/4th-order patterns reported by Ferante and Barton [67].
W{110}	Complex, epitaxial WO_3. No facets.	23, 83, 48, 89	RT structures stable up to ~1000 K; then with increasing temperature c(14 × 7), c(21 × 7), c(48 × 16), c(2 × 2), p(2 × 1).
W{100}	p(2 × 1) (1/2), p(4 × 1) (1), p(2 × 1) (1), p(2 × 2) (1–1.4), p(3 × 1) (1.1–1.4), p(3 × 3) (1.3). {110}, (112) facets, epitaxial WO_3.	70, 74–80, 88, 90, 91	Bauer et al. [77] report p(2 × 2) is combination of c(2 × 2) and p(2 × 1) structures.
W{111}	(4 × 4). (112), {110} facets. Epitaxial oxide (?).	84, 70, 83	(112) facets obtained at lower exposures and temperatures than {110}. Oxide data nonexistent.

the facet and oxide structure data at high temperature. Table 5 compares the oxygen reactivity trends of the three metals towards facetting and oxidation.

In comparison of Cr with Mo and W the most striking differences are the lack of overlayer structures for Cr and the corresponding progression to bulk oxidation under much milder conditions.

TABLE 4

Facet and oxide formation conditions at elevated temperatures

Metal	Property						
	Facet				Oxide		
	Temp. (K)	Exposure (L)	Face	Ref.	Temp. (K)	Exposure (L)	Epitaxy
Cr{110}	870	12	{100}	27, 31, 51	570	100	Cr_2O_3(114) ‖ (110)
						100	Cr_2O_3(110) ‖ (110)
							Cr_2O_3(001) ‖ (110)
Cr{100}			Stable	27, 31, 51	850	60	Cr_2O_3($\bar{1}$11) ‖ (001)
Cr{111}			N/A		1070	Atm.	Cr_2O_3(1$\bar{1}$0) ‖ (111)
Mo{110}	900–1300	10 000	{100}	61	900	100–10 000	MoO_2-(1$\bar{1}$00) ‖ (110) MoO_2-(0001) ‖ (110)
Mo{100}	1000–1300	1–5	{112} + {110}	61, 88	850–1070	100–10 000	MoO_2-(1$\bar{1}$01) ‖ (100)
	> 1300	1–5	{110} only	61, 88			
Mo{111}	1100	~ 100	{112}	26, 61	900–1100	1000	No principal MoO_2 planes ‖ (111). But MoO_2-(1$\bar{1}$00) ‖ (112) MoO_2-(0001) ‖ (112)
W{110}			Stable	92, 21, 88, 89, 69	650–850	10^6	WO_3(100) ‖ (110)
					800–1100	> 20 000	WO_3(111) ‖ (110)
W{100}	1200	~ 5	{110} + {112}	21, 88, 92, 77, 80	990	~ 4000	WO_3(100) ‖ (100)
W{111}	1020	0.5	{112}	70, 84, 83			N/A
	> 1020	> 0.5	then {110} on {112}				

TABLE 5

Oxygen reactivity trends
Line (2) refers to room temperature reaction.

Metal		Face: {110}	{100}	{111}
Cr	(1)	Facet {100} A,B	Stable to facetting	N/A
	(2)	Oxide film	Oxide film	Oxide film
	(3)	Epitaxial oxide MT, LE Derived {110}	Epitaxial oxide MT, LE	N/A
Mo	(1)	Facet {100} B	Facet (undefined) A Facet {112} B Facet {110} B	Facet {112} A Facet (undefined) B
	(2)	Non-reconstructed	Reconstructed	Penetration?
	(3)	Epitaxial oxide HT, ME	Epitaxial oxide HT, ME Derived {110}	Epitaxial oxide HT, ME Derived {112}
W	(1)	Stable to facetting	Facet {112} B Facet {110} A, B	Facet {112} A, B Facet {110} B
	(2)	Non-reconstructed	Non-reconstructed	Non-reconstructed
	(3)	Epitaxial oxide HT, HE	Epitaxial oxide HT, HME Derived {110}	N/A

A = room-temperature exposure followed by heating.
B = adsorption at high temperatures.
MT = moderate temperature (~800 K).
HT = high temperatures (~1000 K).
LE = low exposure (~100 L).
ME = moderate exposure (~1000 L).
HE = high exposure (~10 000 L).

2.3 MOLYBDENUM

In the low coverage regime ($\theta < 2\,\text{mL}$) there have been detailed LEED/AES/$\Delta\phi$ studies [65, 66] on Mo{100} at room temperature which can be compared in a fairly direct manner with the XPS/UPS/LEED/$\Delta\phi$ study on Cr{100} [47] described in the previous section and work on W{100} discussed later. These studies also discuss initial adsorption at elevated temperatures. There are also various other LEED [62, 64, 86, 87], ELS [93], UPS [94], ESD [95] and SIMS [96] data which provide little additional information to the two major recent studies [65, 66] and so will not be discussed much here. The initial interaction stages for Mo{110} are restricted to LEED studies [25, 61–63, 86], some $\Delta\phi$ studies [62, 63, 97] and a brief UPS report [98]. The LEED and $\Delta\phi$ provide some limited basis for comparison with the Mo{100} system. Data on Mo{111} are sparse, consisting of an early LEED study [26] and a LEED–AES study on a badly contaminated surface [99].

At higher temperatures, exposures and coverages, Kennett and Lee have studied all three faces by RHEED and have given extensive discussions of the epitaxial oxide growth and facetting behavior involved in this more extensive reaction regime [61].

2.3.1 Mo{110}

At room temperature, ordered LEED structures have been reported in all studies [25, 63], in contrast to Cr{110}. As a function of oxygen exposure different authors are in agreement that a p(2 × 2) followed by a p(2 × 1) structure are observed. No coverage data are available but these structures have been arbitrarily given their simplest coverage interpretations, i.e. $\theta = 1/4$ and 1/2 mL. Using the p(2 × 2) as a calibration point, Haas and Jackson [25] estimated the initial sticking probability as greater than 0.5. There is considerable disagreement on the sequence of higher-coverage structures observed (Table 2) and the reported exposures required to reach saturation vary from 6 to 60 L. Hayek et al. [62] report a final (1 × 1) structure (10 L exposure), to which they assign monolayer coverage, followed by a gradual background increase and spot decrease (up to 60 L) which they attribute to continued amorphous growth. Other authors find a final complex 6th-order structure [25, 61] instead of the (1 × 1), but it is not clear in these studies whether exposure was continued further and, if it was, whether any evidence for an amorphous thickening was observed.

The only indication of the location of the oxygen (superstructures or reconstructed/penetrated) during the room-temperature adsorption comes from $\Delta\phi$ data. Zukov et al. [63] report a small negative $\Delta\phi$ (~ -0.04 eV) on initial adsorption. Hayek et al. [62] and Hopkins and Ibrahim [97] report no change up to ~0.3 L exposure (maximum intensity of the p(2 × 2), θ = 1/4 mL, structure) followed by an approximately linear increase with coverage up to ~8 L exposure. To be consistent with the interpretation of the Cr work function changes in the previous section and with the more detailed Mo{100} studies discussed later, we argue that initially (up to 0.3 L exposure or the $\theta = 1/4$ mL p(2 × 2) structure), penetration at least into the first Mo layer is occurring and after this a mixture of penetration and overlayer adsorption gives rise to the positive $\Delta\phi$. At the saturation value of $\Delta\phi$ (+1.4 eV at ~8 L; $\theta \approx 1$ mL) the overlayer coverage is presumably saturated. Further uptake has to occur by place exchange, increasing the subsurface O content and giving rise to the observed LEED pattern weakening (the thin amorphous layer thickening suggested by Hayek et al.) [62]. From this interpretation, based on the $\Delta\phi$ behavior, the p(2 × 2) O, $\theta = 1/4$ mL LEED structure must represent a reconstructed surface, as originally suggested by Hayek et al. These authors, however, also assign the higher-coverage p(2 × 1) O structure to a reconstructed surface, even though $\Delta\phi$ is going positive in this coverage region. They do so because the same LEED patterns are observed with greater intensity with adsorption at high temperature and under such conditions there is *no* increase in ϕ. At room temperature, therefore, the suggestion is that, beyond the p(2 × 2) O structure, the p(2 × 1) O formation represents that fraction of O atoms which have penetrated the surface, rather than the fraction in the overlayer (presumably disordered). At higher temperatures, all the O atoms go into and below the surface (hence no increase in ϕ).

If the room temperature p(2 × 1)1/2 or p(2 × 2)1/4 structures are heated to greater than 750 K they disappear but return on cooling [61, 63]. If one starts with high coverage and maintains a temperature of > 750 K it is found [61–63] that the inverse sequence of LEED patterns through to clean Mo{110} is observed. For exposure actually at high temperature (900 K) there is disagreement. Hayek et al. [62] report intense 1/4- and 1/2-order LEED patterns with $\Delta\phi$ remaining at zero, interpreted as reconstructed surfaces. Kennett and Lee [61] do not report the 1/4- or 1/2-order structures but at (presumably) higher exposures complex LEED/RHEED structures are observed and assigned to a thin oxide layer. At temperatures greater than 900 K they eventually observe oriented MoO_2 growth [61]. (Although MoO_3 has been reported [100], Kennett and Lee point out that MoO_3 is unstable at these temperatures.) The epitaxial relationships are reported in Table 4. Both the work of Kennett and Lee [61] and of Zukov et al. [63] indicate that facetting to {100} surfaces accompanies the oxide formation.

Summarizing the data for Mo{110} then, it seems clear that, at high temperature and high exposure, epitaxial MoO_2 is formed together with Mo{100} facet plane growth. It seems likely that ordered reconstructed surface structures precede this. Conclusions from the room-temperature chemisorptive regime remain more speculative until more extensive data exist, but on the basis of the $\Delta\phi$ data it is likely that O atoms initially penetrate the top layer, giving rise to the ordered fractional monolayer LEED and that subsequently (between 1/4 and 1 mL) the overlayer coverage becomes significant until it eventually saturates around monolayer coverage. After this, place exchange incorporation occurs thickening the altered surface region into an amorphous oxide layer. Qualitatively this sequence is the same as that already discussed for Cr, though in the Cr case the surface disruption is easier and more extensive. An alternative way of stating this is that the temperature regime at which facile reconstruction and penetration occurs for Mo{110} is higher than for Cr. More quantitative comparisons between Mo{110} and Cr cannot yet be made. Many authors have made comparisons between Mo{110} and the very well-studied W{110} surface, noting strong similarities. We think this is rather misleading, since our conclusions on W{110} are that overlayer adsorption dominates until O desorption temperatures are approached at which point reconstruction can occur (Sect. 2.5).

2.3.2 Mo{100}

Riwan et al. [66] monitored adsorption at room temperature through to saturation using LEED, AES and $\Delta\phi$ measurements. Relative oxygen coverages were determined from O Auger intensities and these were converted to an absolute scale by calibrating against the O Auger intensity for dissociated CO on Mo{100}, (β state) which gives a sharp c(2 × 2) LEED pattern assumed to represent 1/2 monolayer coverage (i.e. 1/4 monolayer C atoms, 1/4 monolayer O atoms). The c(2 × 2) C layer C Auger intensities were consistent with this interpretation, being twice as large as for the c(2 × 2) CO case. Bauer and Poppa [65] essentially repeated these measurements in greater detail while also performing thermal desorption and ESD measurements as a function of coverage. They also used O Auger intensities for relative coverages but converted them to an absolute coverage scale by a different procedure from Riwan et al. They simply assigned compatible coverages to the respective LEED patterns observed. Though similar LEED and $\Delta\phi$ sequences are observed by both sets of authors, there are significant discrepancies in the absolute exposures and derived coverage scales. The method of absolute calibration cannot account entirely for the discrepancies since they are non-linear and cannot be eliminated by simply normalizing one scale against the other. Neither author addresses directly the assumption that O Auger intensity is proportional to O coverage, though Riwan et al. [66] indicate in one of their figures that they have included an (unknown) O Auger attenuation factor for coverages above θ = 1 mL, and Bauer and Poppa [65] advise caution in interpreting their Auger intensities above 1 mL because of an observed line shape change. In Fig. 6 we have converted the coverage versus exposure data of Riwan et al. into an S versus θ plot. It can then be directly compared with the same plot from the data of Bauer and Poppa, also shown in Fig. 6. Neither set of authors mentions any observation of c(2 × 2) structures on the clean surface. We assume therefore that the starting point was the clean non-reconstructed p(1 × 1) Mo{100} surface, rather than the c(2 × 2) reconstructed surface which is known to be stable at lower temperatures. Bauer and Poppa determine S_0 to be unity so we have

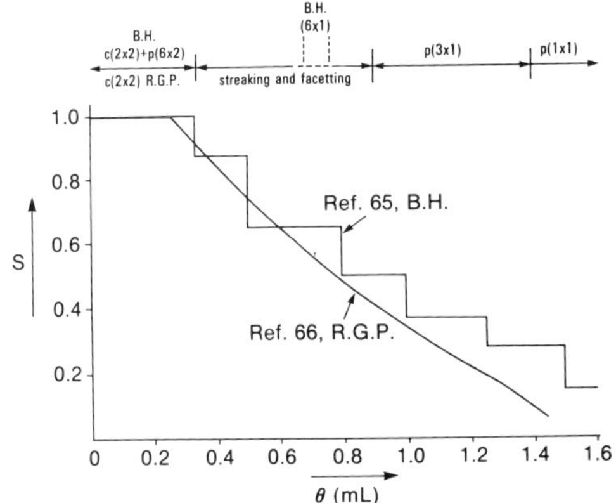

Fig. 6. S versus θ for Mo{100} at 300 K. Coverage regimes of LEED patterns are also shown [65, 66].

assumed a value of unity for S_0 in both plots in Fig. 6. The stepped nature of the curve in the Bauer and Poppa data arises through fitting straight line segments to the O Auger versus exposure data. They apparently place more significance in this than we do. In our opinion the data warrant only a smooth curve and step discontinuities in S make little sense theoretically. The general features of the two curves are an initially constant $S \approx 1$ region, extending no further than $\theta = 0.35$ and possibly less, followed by a monotonic decrease towards zero. Riwan et al. place $S = 0$ at $\theta = 1.5\,\mathrm{mL}$ (12 L exposure) whereas Bauer and Poppa have $S \approx 0.2$ at $\theta = 1.5\,\mathrm{mL}$ (5 L exposure) with $S = 0$ and saturation occurring at $\theta = 2\,\mathrm{mL}$ (25 L exposure).

The LEED sequences, also marked in Fig. 6, are rather complex. They have been discussed in considerable detail by Bauer and Poppa who offer suggestions in terms of complicated superstructure/reconstruction geometries for each pattern, based largely on the correlations with their $\Delta\phi$ and ESD measurements. As suggested by the authors, these suggestions are rather speculative considering the lack of (a) any intensity versus voltage (IV) data and theory, (b) any definite indication of the O coverage at which Mo atoms start to move (such as could be provided by medium-energy ion scattering [7]) or (c) any definitive indication of when oxide nucleation starts (see the Ni/O_2 section for detailed discussions on determining this point). In addition it seems likely from both studies that facetting occurs which complicates any LEED interpretation further. We will therefore consider the $\Delta\phi$ behavior first, since it appears to offer some guide as to where the oxygen is going, and then give a minimal interpretation of the LEED consistent with the $\Delta\phi$ changes. More detailed interpretation of the LEED does not seem justified at this point. $\Delta\phi$ is negative up to $\theta = 0.35$ or $0.5\,\mathrm{mL}$, depending on the author (Fig. 7). It seems, therefore, that oxygen penetrates at least the top layer in the early stages. The LEED patterns observed up to this coverage therefore represent reconstructed surfaces. Riwan et al. observe only a diffuse c(2 × 2) pattern with its maximum intensity at the $\Delta\phi$ minimum, whereas Bauer and Poppa observe c(2 × 2) and p(6 × 2) structures with the latter sharpest at the $\Delta\phi$ minimum. According to Bauer and Poppa the p(6 × 2) represents a c(2 × 2) geometry, but with large numbers of ordered vacancies. $\Delta\phi$ then goes sharply positive over the next ~0.4 monolayer uptake and then exhibits a local maximum or plateau. Bauer and Poppa observed streaking owing to facetting over the region plus a streaked p(6 × 1) structure. Riwan et al. observe only streaking which they interpret as {110} microfacetting with reconstructed oxygen structures on the facets. The facetting, in their view, accounts for the sharp increase in $\Delta\phi$. Bauer and Poppa prefer an interpretation where the O atoms are now all on the surface [the p(6 × 1) structure] to account for the rising $\Delta\phi$ and some aspects of their ESD data. This re-reconstruction seems unnecessary to us and we prefer the less specific interpretation, in line with the previous Cr and Mo{110} that a rise in $\Delta\phi$ represents some fraction of the additional O uptake now going into on-top sites. The sharpness of the rise and the "hump"

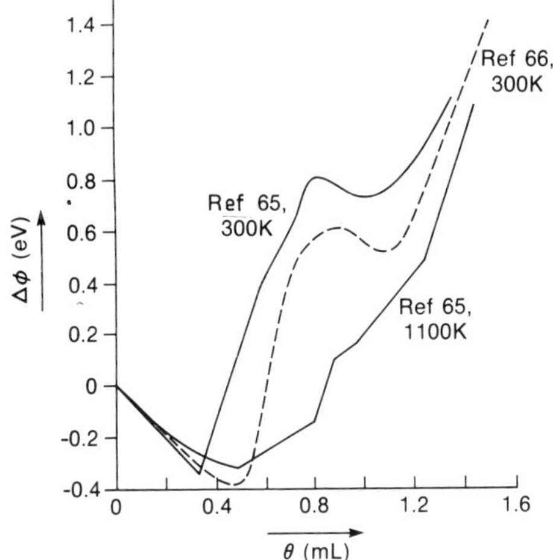

Fig. 7. $\Delta\phi$ versus θ for Mo{100} [65, 66].

at around $\theta \approx 0.8$–0.9 mL (Fig. 7) are then attributable to the additional perturbing effects of the {110} facetting.

At around $\theta = 1$ mL in both studies, a p(3 × 1) LEED pattern is observed which weakens with increasing coverage, giving way to a high background and a weak p(1 × 1) structure at saturation. (This stage supposedly goes from $\theta = 1$ to 1.5 mL in the Riwan et al. study and from $\theta = 1$ to 1.4 mL in the Bauer and Poppa study.) Both LEED structures could be interpreted in terms of epitaxial oxides and on the basis of a strong similarity of ELS features in this region to those of heavily oxidized Mo surfaces, Riwan et al. suggest an oxide interpretation. Bauer and Poppa point out that pseudo-tetragonal MoO_2 with its (001) plane parallel to the Mo(100) plane fits the p(3 × 1) LEED pattern. They then dismiss this interpretation in favor of one involving mixed layers of Mo and O atoms (see also the W section) with the reasoning that the oxygen coverage at the point where the p(3 × 1) structure is sharpest ($\theta \approx 1$ mL) is too low for an MoO_2 structure. We believe that above ~1 mL the structures are almost certainly three-dimensional oxide in nature, whether or not the O/Mo stoichiometry conforms to MoO_2. Excess adsorbed oxygen could account for a ratio that is greater than 2. The weakening of the p(3 × 1) as the oxide thickness increases probably represents the loss of epitaxy, until at saturation only the substrate Mo{100} is weakly observable together with the high background due to a thin amorphous or badly disordered oxide overlayer. The final high positive $\Delta\phi$ value at saturation would then represent the work function of the oxide. The ESD data [65] can be taken to support the suggestion of oxide above 1 mL. In the ESD a very-high-yield O^+ state is observed and oxide nuclei are known to give rise to large ESD signals. (See Sect. 2.8 for a detailed discussion of ESD for O on W surfaces.) It is quite likely, however, that the mixed Mo/O layer configuration of Bauer and Poppa would also give a high ESD signal. One might be tempted to use the thermal desorption data to support the oxide layer conclusion. Below ~1 mL, only O atom desorption is observed, the intensity being proportional to coverage. Above ~1 mL, large amounts of MoO, MoO_2 and MoO_3 desorb. However, to allow such an argument would be inconsistent with our later discussions of thermal desorption for W (Sect. 2.9). There, it is very clear that the onset of WO_x species in desorption *does not* signify the formation of stoichiometric oxides for the adsorption process, but is merely an indicator of a high oxygen coverage. Confirmation of the nature of the oxide or mixed Mo/O overlayer structures for room-temperature UHV saturation coverage (stoichiometric layer?, oxidation state of Mo?, geometric structure?) must therefore await more appropriate measurements such as low-angle XPS and ion scattering.

Bauer and Poppa investigated the low-coverage regime at elevated temperatures, using both LEED and $\Delta\phi$. Riwan et al. restricted their observations under these conditions to the LEED structures and observed that between 350 and 500 K the c(2 × 2) structure converted to a c(4 × 4) pattern. Above 1 L exposure a ($\sqrt{5} \times \sqrt{5}$) structure was observed. The perfection of both structures increased with annealing in vacuum. Bauer and Poppa examined the LEED sequences and $\Delta\phi$ for both adsorption at 300 K followed by annealing to 1100 K and recooling to 300 K, and adsorption at 1100 K followed by recooling to 300 K. There were relatively small differences between these two procedures. A temperature of 1100 K was as high as could be used owing to the onset of desorption. The $\Delta\phi$ plot is shown in Fig. 7. The main difference from the room-temperature $\Delta\phi$ plot is the movement of the minimum in $\Delta\phi$ to higher coverage and the lack of the sharp rise and "hump" following the minimum. Our interpretation of this, consistent with the previous Cr and Mo{110} discussion, is that at 1100 K initial O penetration into the subsurface region is more extensive than at 300 K so that any on top adsorption or formation of a high O density reconstructed layer right at the surface comes at a higher coverage. The "hump" is completely missing because the {110} facetting does not occur at 1100 K. The Bauer and Poppa interpretation of the 1100 K heated LEED sequences is more complex but basically similar. They show that the c(4 × 4) pattern below 1 L is actually a mixture of c(4 × 2) and p(2 × 1) and the c(4 × 2) is again basically a c(2 × 2) with ordered vacancies (cf. room temperature). This "c(2 × 2)" and p(2 × 1) co-exist in patches and are considered to be reconstructed surfaces. The (5 × 5) pattern emerging at higher coverage is considered by the authors to represent either a distorted hexagonal O overlayer or a reconstructed layer with O atoms on top of it. Bauer and Poppa favor the former interpretation. We see no valid reason for suggesting that reconstructed O moves back above the surface and so we favor the latter interpretation. Above 1 mL, the LEED patterns observed, p(2 × 1), c(2 × 2) are consistent with reconstructed and/or oxide layers two or more layers thick. At θ = 1.25 mL, facetting sets in. The observation of facetting and of oxide nucleation growth at these higher coverages was investigated in some detail by Kennett and Lee [61] and Kennett et al. [21] by LEED and RHEED. The behavior is critically dependent on temperature and pressure. Both {110} and (112) facets are formed above ~ 1000 K and below 1300 K, but above 1300 K only {110} facets are observed. Heating in vacuo above desorption temperatures flattens the surface again and leads to the reverse sequence of high-temperature LEED structures as oxygen/oxide is lost from the surface. Because of the strong temperature and pressure dependence of facetting, it is possible to grow epitaxial oxide at high exposures on both the flat and facetted surfaces. Below 1000 K, MoO_2 nuclei grow on the flat surface, whereas surfaces which have been facetted by treatment above 1000 K will grow MoO_2 nuclei on the facets. The epitaxial relationship for growth on both facetted and unfacetted surfaces is a derived one, that is the orientation of the oxide with respect to the substrate is the same as that found for Mo{110}. The kinetics of the oxidation process on the {100} surface and the exposures and temperatures required are very similar to the {110} surface, in sharp contrast to the situation for W where the W{110} surface requires much higher exposures than the {100} to achieve oxidation (see later).

Summarizing for Mo{100} it seems that, despite the much larger volume of data than for Mo{110}, including detailed coverage and exposure information, the detailed interpretation, particularly for the ordered low-coverage structures observed in LEED, remains rather speculative. The general scheme of events is similar to that proposed for Mo{110}; initial penetration of oxygen with reconstructed ordered surfaces resulting in a decreasing $\Delta\phi$. This is followed by a rising $\Delta\phi$, indicating significant on-top adsorption, and on the basis of the LEED, microfacetting (leading to the "hump" in the $\Delta\phi$ curve). Finally, a LEED structure interpretable as either an oriented thin MoO_2 overlayer or a mixed Mo/O bilayer is observed which fades on thickening. Annealing to, or adsorbing at, elevated temperatures has the effect of pushing the $\Delta\phi$ minimum to higher coverage values (cf. Mo{110}), interpretable as increasing the concentration of subsurface O allowable before on-top O concentration sets in. At high exposures at elevated temperatures, further facetting may or may not occur, depending on temperature and pressure, and epitaxial bulk MoO_2 nuclei are formed, in contrast to room temperature where any epitaxial relationship is apparently lost within a few layers.

2.3.3 Mo{111}

The only experimental data available for oxygen interaction with Mo{111} are the early LEED experiments of Ferante and Barton [26], the RHEED experiments of Kennett and Lee [61], and the LEED experiments on a heavily contaminated surface (up to 2/3 mL coverage) of Lambert et al. [99]. No sound information on coverage versus exposure of $\Delta\phi$ behavior as a function of coverage (or exposure) exists. In the Lambert et al. work an initial sticking probability of ~1 was estimated for the clean portions of the surface. No low-exposure ordered structures were observed at room temperature, merely an increase in background intensity. For high exposures (5–500 L) Ferante and Barton observed diffuse spots from {112} facet planes.

Heating the room-temperature low-exposure (~1 L) surface to 1000 K does give ordered structures which are rather complex [26, 61] and which have not been interpreted in terms of the O location and geometry. High exposure at room temperature followed by heating produces the {112} facets with ordered oxygen structures on/in them. Further exposure at 1000 K gives rise to a sharp RHEED pattern of oriented MoO_2 nuclei. The epitaxial relationship is derived from the {112} facet planes. As for Mo{110} and {100}, the oxide could be removed by heating in vacuo (~1300 K) whereupon a reverse sequence of complex lower-coverage patterns was observed [61]. Unlike the other two low index faces, oriented MoO_2 nuclei can be produced on Mo{111} by very high exposure (10^4 L) at room temperature followed by annealing at ~1200 K. This suggests that the saturation coverage and therefore the thickness of the affected surface is greater for Mo{111} than for the other faces.

2.3.4 Summary of Mo results

Insufficient data are available on Mo{110} and particularly Mo{111} for a detailed comparison with Mo{100} for room-temperature low-coverage adsorption to be made. The general comparison suggests that Mo{110} and Mo{100} are very similar in behavior with Mo{111} somewhat different. In the former cases initial penetration occurs, resulting in ordered reconstructed surfaces up to a coverage of ~1/4 to 1/2 mL. Subsequent to this, on-top adsorption becomes significant with facetting complications. Beyond ~1 mL the affected layer thickens, presumably by a place-exchange mechanism with a low sticking probability, resulting in ordered structures which may represent very thin layer epitaxial MoO_2 or two or more mixed Mo/O reconstructed layers. The final stage is the loss of epitaxial relationship with the substrate during a further thickening. The maximum O coverage at this point for Mo{100} is apparently $\theta \leqslant 2$ mL.

Since Mo{110} is more close-packed than Mo{100} we might expect that it has less propensity for oxygen penetration. To be consistent with this suggestion would require the $\Delta\phi$ behavior to start to rise at an earlier coverage than for Mo{110}. The available $\Delta\phi$ data are, however, inadequate to provide any indication whether this is so or not.

Since Mo{111} is the most open structure of the three, we assume that initial O penetration must also occur here, though corroborative $\Delta\phi$ data are lacking. Apparently no ordered fractional monolayer structures are formed on Mo{111} which is consistent with a greater ease of penetration into the subsurface region, and also with the case of W{111} which also lacks ordered structures. The hypothesis that oxidation is easier for Mo{111} is supported by the other major reported difference in behavior from Mo{110} and {100}, namely that high exposure at room temperature followed by annealing will produce oriented MoO_2 nuclei observable by RHEED.

High exposures at high temperatures on all three surfaces produce oriented MoO_2 structures with facetting complications and "derived" epitaxies. Heating in vacuo can re-establish flat surfaces by loss of volatile oxides and take the surfaces back through the low-coverage stages.

In an intercomparison of Mo/O with Cr/O and W/O, it seems that Mo occupies a genuine intermediate position, but is actually phenomenologically rather closer to Cr. Thus, like Cr initial O penetration occurs at room temperature followed later by overlayer build-up. For W, overlayer adsorption dominates at room temperature. The area where Mo is similar to W is in the occurrence of room-temperature ordered fractional coverage O structures for the {110} and {100} faces. These are, however, considered to be reconstructed or reconstructed/overlayer mixtures for Mo, whereas they are apparently all true overlayers for W.

2.4 TUNGSTEN. GENERAL

A tremendous range of data exists on the behavior of oxygen on the low index faces of W. This has certainly led to a more detailed knowledge of the interaction. It has led, in some cases, to increased controversy either because of conflicting data or because of the lack of a sound basis for interpretation. The largest body of data is for the W{110} surface for which we have the most complete understanding of any of the faces in the Cr, Mo, W series. There is almost as much on the W{100} surface, but our understanding is less detailed. Much fewer data have been accumulated for the W{111} surface because of the lack of ordered LEED structures on this face with which to correlate other measurements.

Several groups have been active over many years in W/O studies and it is these groups which have provided the most reliable and important information. Bauer and co-workers [68, 69, 77, 90, 101–106] have LEED, S versus θ, ESD, TDS, $\Delta\phi$ and ISS studies on the {110} and {100} surfaces (including stepped surfaces for {110}) for adsorption at 300 K and above. Gomer and co-workers [107–112] have S versus θ and TDS work on W{110} and {100} for adsorption temperatures as low as 20 K up to 1000 K, in addition to some photoemission [113] and isotopic labelling data on W{110} [113]. Madey et al. [85, 114–118] have published a variety of ESD, PSD and ESDIAD papers and reviews involving all three faces. Finally, Lagally and co-workers [28, 119–127] have used LEED spot intensity and angular profiles as a function of T to derive information on the phase transitions occurring between oxygen overlayer structures on the W{110} surfaces. In addition to the series of papers by the above authors, there exists other important work on LEED dynamical calculations [128, 129], ESDIAD [83, 91, 130–132], surface vibrations [133], photoemission [76, 135–141], surface diffusion and kinetics [70, 142–147]. In attempting to establish what has been learned about the W/O systems from the various spectroscopic techniques, it becomes clear that, in some cases, the understanding of the information provided by the techniques is poorer than the existing knowledge concerning the W/O surface. It then becomes a case of using this existing knowledge to establish the true nature of the information provided by the technique. We consider that the uses of TPD, ESD and SIMS tend to fall into this category. We have therefore made the minimum reference to these techniques during the main body of discussion on the W/O interaction and included separate sections on their interpretation at the end of the section on W.

2.5 W{110}

For the {110} face a small, but important, body of work on adsorption down to liquid helium temperatures exists. We will discuss some aspects of this first since there are, as yet, no published counterparts on any other transition-metal single-crystal surfaces and because the work clarifies some important concepts. Following this we review all the work in the 0–0.5 monolayer adsorption range. There is far more of this than for $\theta > 0.5\,\text{mL}$, mainly because an apparently simple LEED pattern [p(2 × 1)] has fully formed by $\theta \sim 0.5\,\text{mL}$ and the sticking probability of oxygen drops precipitously at this coverage, making a natural calibrated "end-point" for adsorption. The discussion is split into several sections. The S, θ and kinetics-related measurements are taken first. Since S measurements are available over a wide temperature range, including down to liquid helium temperatures, this section follows rather naturally the preceding one on very low temperature adsorption. Next we discuss the evidence for the overlayer versus reconstruction location of the oxygen under various conditions, and after that the structural interpretation of the LEED patterns observed between 0 and 0.5 monolayers. LEED has also been used extensively by Lagally and co-workers [28, 119–127] to follow order–disorder transitions of O on W{110} involving island growth of the ordered adsorbate. The information obtained on the W{110}/O phase diagram between $\theta = 0$ and $0.5\,\text{mL}$ is discussed following the structural interpretations of the LEED data. Finally, the information available from photoemission, AES, and ELS concerning electronic structure aspects of W{110}/O at $\theta \leqslant 0.5$ is reported.

After completing the discussion on the 0–0.5 monolayer range, we deal with the smaller amount of data for $\theta > 0.5\,\text{mL}$, as far as possible in a parallel manner. Following this there is a short section on the diffusion of O adatoms on W{110} surfaces at all coverages. A fairly extensive set of data on adsorption of oxygen at stepped W{110} surfaces exists and we include a section contrasting the stepped W{110} with the flat surface. The stepped data raise interesting questions concerning the kinetics and dynamics of the oxygen dissociation process.

2.5.1 Some aspects of very low temperature adsorption

Leung and Gomer [109] studied adsorption below 40 K and on the basis of TPD, $\Delta\phi$ and ESD data suggested that adsorption on the clean surface was molecular. For a series of increasing oxygen exposures at 20 K, the following sequence was observed in desorption. For low exposure, oxygen was desorbed as atoms between 1700 and 2200 K in exactly the same fashion as if the initial adsorption had been done at 300 K. Increasing the exposure at 20 K produced, sequentially, O_2 desorption at 45, 27 and 25 K. The 25 K peak did not saturate with further exposure and clearly represented oxygen condensed on the other species. The 27 K peak is then oxygen physisorbed on top of whatever gives rise to desorption at 45 and 1700–2200 K. It was then postulated on the basis of $\Delta\phi$ changes and ESD data that initial adsorption on the clean W{110} was into a molecular chemisorbed state which, for coverages up to half its saturation value, converted at 45 K to the atomic species and subsequently desorbed as atoms at 1700–2200 K (i.e. it was the long-sought molecular precursor to dissociative adsorption). Excess coverages above half saturation desorbed at 45 K as molecular oxygen. An absolute value of the total coverage in this chemisorbed state at 20 K was determined by Wang and Gomer [111], using a calibrated effusion source, to be 8.8×10^{14} O_2 molecules cm^{-2} and it was confirmed that half this amount, 4.4×10^{14} molecules cm^{-2} desorbed at 45 K. The atomic coverage which desorbs between 1700 and 2200 K is therefore 8.8×10^{14} atoms cm^{-2} or $\theta_0 = 0.62$ mL. Later XPS/UPS work by Opila and Gomer [13] and isotopic labelling experiments by Michel et al. [113] have conclusively shown, however, that the Leung and Gomer postulate of initial *molecular* adsorption with partial conversion at 45 K is incorrect, and the simpler following interpretation is correct (Fig. 8). Initial absorption at 20 K results in a *dissociative* species, saturating at $\theta_0 = 0.62$ mL. A weakly chemisorbed molecular species (termed α-O_2 by Opila and Gomer) adsorbs on top of the atomic O saturated surface to a saturation value of $\theta(O_2) = 0.31$ mL. It desorbs at 45 K and can be re-adsorbed on recooling. Physisorbed oxygen adsorbs on top of the α-O_2 layer, again to a saturation coverage of around

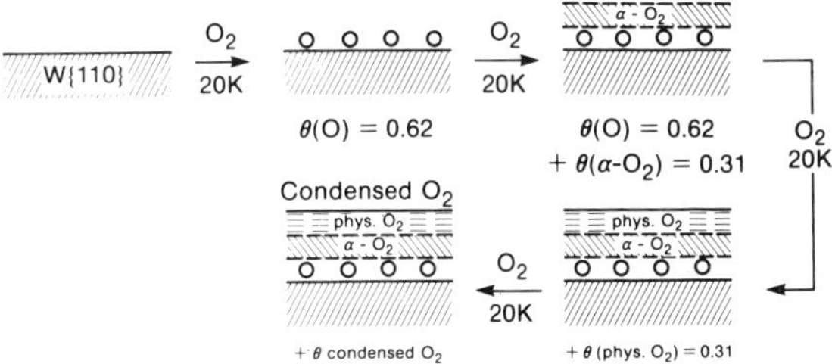

Fig. 8. Schematic adsorption scheme for O_2 on W{110} at 20 K. The condensed O_2 desorbs at 25 K, physisorbed O_2 at 27 K, α-O_2 at 45 K, and the atomic species between 1700 and 2200 K [13, 109, 113].

$\theta(O_2) \sim 0.3$ mL. It desorbs at 27 K. Finally, oxygen condenses on the physisorbed layer. It desorbs at 25 K. The main experimental evidence for the Leung and Gomer original suggestion of conversion at 45 K of a molecular to an atomic state seems to have been small $\Delta\phi$ changes on heating a surface with less than $\theta_0 = 0.62$ mL from 20 to 45 K. Now that it is known that adsorption at 20 K to $\theta_0 < 0.62$ mL results only in chemisorbed *atomic* oxygen, these small changes must be ascribed to some slight redistribution of the O atoms during annealing from 20 to 45 K. The apparent inability to isolate a mobile molecular precursor to the dissociated chemisorbed O, even at 20 K, deserves some comment since the kinetics of adsorption at low and high temperatures (see later) conclusively demonstrate its influence on the dissociative adsorption process. The point would seem to be that, unless the dissociative process is activated any molecular precursor to it cannot be isolated as long as any significant areas of bare metal surface remain, even at low T, because it will always diffuse to an appropriate site and dissociate. Once all the dissociation sites are used up the precursor molecule ought to be isolatable on top of the covered surface at sufficiently low temperatures. We therefore suggest that the α-O_2 state is, in fact, the molecular precursor to the final dissociated product, even though it can only be observed in any concentration on top of a saturated O layer. The dissociated species is thus reached by two possible routes. Molecules striking bare W{110} pass through any weakly bound molecular state without being accommodated and so dissociate. Molecules striking patches of surface already covered with atomic O are trapped into the α state. (This state *may* have the same well-depth of the O_2 molecule on the bare surface, but this is not a requirement of the model.) They diffuse across the patches and dissociate when they reach bare W{110} surface. The concentration in the α state builds up when the size of the O-covered patch they are on is large enough and T low enough for them to remain trapped there. This type of precursor, i.e. one which only forms on top of already chemisorbed patches, is usually called an "extrinsic" precursor. The concepts involved are discussed by Grunze et al. [148]. Several other interesting points emerge from the low-temperature work of Gomer and co-workers. In the XPS/UPS work [13], though the molecular species were easily distinguished from the dissociated species by their core and valence level B.E.s, any distinction between the three proposed types of molecular species, α-O_2, physisorbed and condensed, was beyond the instrumental resolution. Thus α-O_2 is sufficiently weakly bound that its valence and core orbital energies are essentially unaffected by the bonding. From an analysis of the desorption data, Michel et al. obtained a desorption activation energy of 1.9 kcal mol^{-1} for α-O_2. Since this is no larger than the heat of sublimation of O_2 and yet the desorption temperature is about twice that for O_2 sublimation, the authors concluded that the desorption process was two-step, a vibrational excitation to some critical level (1.9 kcal) at which point coupling to lattice phonons can occur causing excitation to the desorption limit.

Other interesting features are the saturation coverage values of the atomic, physisorbed and α-O_2 layer at 20 K. Wang and Gomer [111] established the α-O_2 and atomic coverages both at $\theta_0 \approx 0.6$ mL and from the XPS data it can be inferred that the physisorbed layer desorbing at 27 K also has this saturation coverage. For adsorption at 300 K (discussed later), S drops by several orders of magnitude at $\theta \approx 0.5$ mL, since the completion of the ordered p(2×1) O structure at that coverage leaves only rather unfavorable W sites for further dissociative adsorption (see later sections). Full coverage, $\theta = 1$ mL, can be reached, however, with sufficiently long exposures. The fact that the easily attained atomic coverage (0.6 mL) is the same or only slightly greater at 20 K than at 300 K indicates that there must still be considerable short-range adsorbate order with each O_{ad} effectively blocking two W sites from further facile dissociative adsorption.

The saturation coverage of α-O_2, 4.4×10^{14} molecules cm^{-2}, coincides exactly with that expected if α-O_2 is assigned the oxygen molecule van der Waals' diameter, 3.6 Å, and forms a hexagonal close-packed structure. This suggests that α-O_2 is randomly oriented on the surface. The same statement holds for the physisorbed state underscoring the similarity of these two states deduced from their identical XPS/UPS spectra and the inability to distinguish either from condensed O_2.

2.5.2 0–0.5 Monolayer O range. S, θ measurements

A very extensive set of data exists on S as a function of θ between 0 and 0.5 mL over a wide range of substrate and gas temperatures, T_s and T_g. The most complete set of data is that of Wang and Gomer [107] and Kohrt and Gomer [108] and we will discuss these first. If one is to fully interpret the S, θ behavior it is of course necessary to know something about the location and ordering of the adspecies, matters which we deal with in later sections. This may therefore seem a peculiar sequence of presentation but it is necessary since information about location and ordering are derived from a combination of many types of data which are often related to each other through the S, θ information. The Wang and Gomer and Kohrt and Gomer data were obtained using a reflection technique. Oxygen from a calibrated effusion source strikes with W{110} surface and the amount reflected is detected by adsorption on a field emitter and the consequent effect on its IV characteristics. Some of the representative S versus θ data are shown in Fig. 9.

Let us consider the $T_s = 20$ K data first since it relates directly to the previous section. Here we consider a higher coverage than 0.5 mL because of the build-up of the *molecular* α state on top of the O layer. The atomic coverage does not go above $\theta \approx 0.6$, however (see previous pages). We have the advantage of now knowing that formation of α-O_2 only builds up on the surface above ~0.5 mL (for $T_g \geqslant 300$ K), which simplifies the interpretation. S_0 is less than unity for all but the lowest T_g and decreases with increasing T_g. This could mean that the probability of reflection on first impact, r, is significant unless the O_2 molecule impinges with very little energy. In

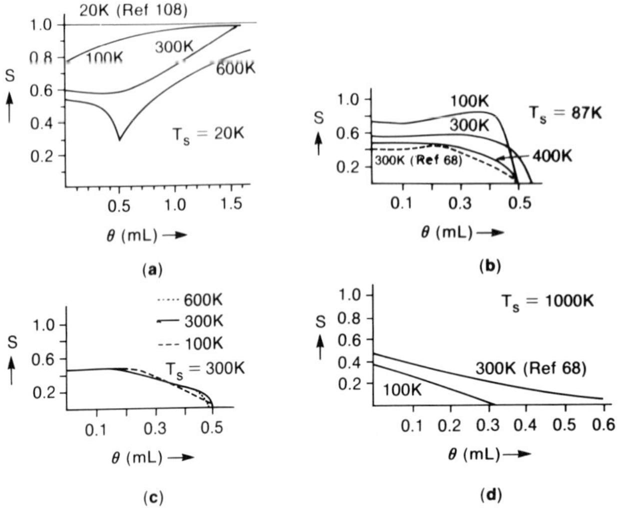

Fig. 9. S versus θ for oxygen adsorption on W{110} over a wide range of substrate temperatures, T_s, and gas temperature, T_g. All data are from ref. 107 except where marked. All temperatures not marked T_s are T_g.

addition, S_0 at $T_s = 20$ K approaches unity at very low T_g and for T_s up to at least 300 K, S initially remains constant (for any given T_g) with increasing θ. These facts are usually considered to imply that the dissociative adsorption takes place from a mobile precursor molecular state and that the surface lifetime of that precursor is not a limiting factor until θ_{sat} ($\sim$0.62 mL at $T_s = 20$ K) is approached. Extensive discussion of precursor state models and their implication concerning adsorption mechanisms and kinetics can be found elsewhere in the section on Ni, so we will restrict our discussion here to an explanation of why the facts that S_0 approaches unity at low T_s and T_g and S remain constant with increasing θ imply a precursor state. If the fate of a molecule striking bare metal could only be reflection, or dissociation, then for a constant flux of molecules the rate of production of dissociated species would be proportional to the concentration of active sites on the surface. Provided there is no movement of the dissociated species below the surface (which does not occur for W{110} until $\sim$1600 K, see later) the concentration of active sites on the surface must decrease as they become covered by O atoms. The functional form of the decrease of concentration of active sites with θ_0 depends strongly on the type of site required for dissociation to occur. For instance, if every W surface atom could dissociate an O_2 molecule on impact the concentration of active sites, and therefore S, should decrease as $(1 - \theta)$. It is commonly assumed in modelling dissociative kinetics that two nearest-neighbor bare sites are required [14]; in this case S should decrease as $(1 - \theta)^2$ if the distribution of sites occupied by O atoms is random. Many other behaviors can be obtained using different assumptions about the site requirements for dissociation. They all require that S

decreases from its S_0 value as θ increases. If the number of active sites were initially small, and if the atomic oxygen could diffuse away from them after dissociation, then it is possible that S would not decrease much with θ initially (see, for example, the section on stepped W surfaces). However, for a molecule whose only choices are to reflect or dissociate, S_0 could not approach unity under such an assumption. Under these conditions the only mechanism whereby S can be maintained at its (high) S_0 value is one which allows diffusion of the oxygen *molecule* prior to dissociation to reach remaining dissociation sites. This implies that the gaseous molecule is trapped into a weakly bound, and therefore mobile, molecule state, in addition to the possibility of dissociating on directly striking an active site. In this manner S remains at S_0 until quite high θ values, and then falls as the dissociation sites are exhausted. Now from the preceding section on adsorption at very low temperature, we have established that, in fact, the trapping of molecules into the precursor state occurs only on O-covered parts of the surface, even at $T_s = 20\,K$, at least for $T_g \geqslant 300\,K$. Also, for $T_s > 45\,K$, no significant build up of this α-O_2 will occur even though its low steady-state concentration controls the kinetics of the atomic oxygen growth. The S versus θ curves for $T_s = 20\,K$, $T_g > 300\,K$ should therefore reflect the build up of α-O_2 on the O-covered patches for $\theta > 0.5\,mL$. For $T_s = 20\,K$, S behaves in a typical precursor-driven manner over the $\theta = 0$–0.5 range. It remains constant and then decreases when $\theta = 0.5\,mL$ is approached (Fig. 9). The fact that it decreases rapidly near 0.5 mL rather than 1 mL implies that one dissociated O atom renders two W atom sites unreactive to further molecular dissociation. It does not imply that a surface with $\theta > 0.5\,mL$ is thermodynamically unstable. It is quite possible to prepare stable higher coverage surfaces by using high exposures, O atomic beams, or evaporating W oxide on the surface and annealing [69]. The large increase in S_0 at $T_s = 20\,K$ with decrease in T_g is also typical of a precursor-driven mechanism. The increase is explained as a decrease in reflectivity, i.e. at $T_g = 600\,K$ the reflectivity from bare W is ~ 0.5 whereas at 100 K it is only ~ 0.25. Qualitatively this is what one would expect for a small potential well depth of the precursor. At 600 K T_g is much larger than the well-depth and the molecule cannot as effectively accommodate as at 100 K T_g. At $\theta = 0.5\,mL$ S has only decreased to ~ 0.3, however, instead of approaching zero [Fig. 9(a)]. It then starts to rise again rapidly as molecular α-O_2 is accumulated on the O-covered surface. We interpret this data as meaning that the initial sticking coefficient for adsorption into the α-state on top of W/O patches (local coverage, 0.5 mL) is ~ 0.3. The probability of molecular reflection from W/O patches must then be ~ 0.7. Up to 0.5 mL, essentially all the α-O_2 adsorbed on W/O patches eventually finds empty sites by diffusion to bare W regions and dissociates to O_{ad}. Above 0.5 mL no easy dissociation sites are left and α-O_2 starts to accumulate on the surface. The rise in S as α-O_2 accumulates implies that O_2 is physisorbed on top of α-O_2 patches with higher probability than α-O_2 was adsorbed on O_{ads} covered patches; i.e. the probability of reflection on impact is smaller when an incoming O_2 molecule strikes an already α-O_2 covered region. It may even be zero. As the total coverage increases, further condensed O_2 starts to form on the physisorbed O_2 layer and S continues to increase until it reaches unity. This means that the reflection probability must be zero on top of physisorbed O_2.

Once T_g is decreased from a high value, the shape of the S versus θ curve at $T_s = 20$ K changes strongly. At intermediate T_g (300 K), S no longer decreases as 0.5 mL is approached. At still lower T_g it actually rises as θ increases from 0 [Fig. 9(a)]. For this to occur Wang and Gomer suggested that the effect of reducing T_g must have been to make the probability of molecule reflection from an O_{ads}-covered surface become less than from the bare surface whereas at high T_g they are either equal or reflection from the O-covered surface is higher. This proposed change in reflectivity with T_g seems to be a possible behavior but perhaps insufficient to explain the very strong changes observed on changing T_g at $T_s = 20$ K. Note that by changing from $T_g = 600$ K to $T_g = 100$ K [Fig. 9(a)] the sticking probability for a surface of $\theta \sim 0.5$ mL changes from ~ 0.3 to ~ 0.9. If a change in reflectivity of impinging molecules on the O_{ads}-covered regions of the surface were to be entirely responsible, the reflection probability would have to be decreased by a factor of 7 on going from $T_g = 600$ K to $T_g = 100$ K (from 0.7 to 0.1). This seems unlikely. A more plausible explanation would seem to be that at $T_s = 20$ K and low T_g there is a build up of trapped α-oxygen on the surface prior to 0.5 mL, even though the XPS/UPS and isotopic labelling work of Gomer and co-workers showed that this was not so for $T_g \geqslant 300$ K. The value of S would then rise smoothly with θ because of the reduced reflectivity (possibly zero) at α-O_2 covered sites. The α-O_2 build up prior to 0.5 mL could occur by accumulation on top of O_{ads}-covered regions because the low T_g does not supply sufficient energy for α-O_2 to diffuse to bare regions. Alternatively, it could actually be trapped on bare W sites, which would imply that, rather than the dissociation process being unactivated, there is a small activation energy which is overcome by all but the lowest T_g. Either suggestion signifies a breakdown of the usual classical assumption in models of adsorption kinetics; namely, that there is an "instantaneous" equilibration of the gas-phase molecule with the surface and therefore T_g plays no role in the kinetics other than determining S_0 through establishing the probability of reflection. To get further with these questions would first require photoemission or isotope labelling experiments at $T_s = 20$ K and $T_g < 100$ K to definitively establish whether α-O_2 is formed prior to $\theta = 0.5$ mL. In the section dealing with stepped W{110} surfaces we will see, however, that similar doubts arise concerning assumptions of instantaneous accommodation.

Turning now to higher values of T_s, for T_s increasing between 20 and 400 K, T_g variation has a decreasingly small effect. By $T_s = 300$ K the different T_g curves are practically indistinguishable [Fig. 9(c)]. S_0 has also converged on a lower limit of ~ 0.45, about the same value as for high T_g and low T_s. This is exactly as expected; as T_s is raised it becomes more difficult for the impacting molecule to lose its energy and become accommodated (so r increases and S_0 decreases), independent of the value of T_g if $T_g < T_s$, the effect being approximately the same as for high T_g on a low T_s substrate. Finally, at very high T_s ($\geqslant 1000$K) the residence time of the molecular precursor becomes so small that the precursor ceases to be effective and the S, θ curve becomes Langmuir-like [Fig. 9(d)]. θ_{lim} for $T_s = 1000$ K appears to be ~ 0.35 mL rather than 0.5 mL, a fact which is unexplained, especially as oxygen adsorbed to $\theta = 0.5$ mL at lower T_s is stable up to $T_s = 1000$ K. For all other T_s in these data, θ_{lim} is ~ 0.5 mL.

Other recent measurements of S, θ have been obtained from AES [68, 142, 149], by molecular beam reflection, and by XPS. Taking the AES data first, those of Engel et al. [68] are the most complete with measurements at $T_g = 300$ K and $T_s = 300$, 1000 and 1350 K. The O Auger intensity was determined as a function of exposure time and the raw data turned into absolute S, θ curves by calibrating the O_2 impingement rate and assigning a coverage of 0.5 mL at the exposure where the sharpest p(2 × 1) structure formed. The $T_s = 300$ K data are very similar to that of Wang and Gomer, but the 1000 K data (identical to the 1350 K data) are different in that S tends to zero at $\theta \sim 0.7$ mL compared with ~ 0.35 mL for Wang and Gomer [Fig. 9(d)]. The differences remain unexplained and the true high-T_s S, θ behavior therefore remains in doubt.

Butz and Wagner [142] also took Auger intensity versus exposure data ($T_s = T_g = 300$ K). They calibrated their data by comparison with similar data on a W{110} surface pre-dosed with W atoms for which they assumed $\theta_{max} = 1$ mL and $S_0 = 1$ (see later). This calibration puts the 300 K S, θ data into perfect agreement with Engel et al. [68], therefore apparently confirming the assumption that $S_0 = 1$ and $\theta_{max} = 1$ mL for the W atom-covered W{110} surface.

Besocke and Berger [149] compared S, θ curves for flat and stepped W{110} surfaces. They calibrated their coverage scale against the p(2 × 1) O structure at $\theta = 0.5$ mL for the flat surface and assumed $S_0 = 1$ for the stepped surface showing the largest relative S_0. The resulting S, θ curve for the flat surface agreed in shape with Wang and Gomer, but S_0 was ~ 0.28 not 0.45. In this work all the stepped surfaces and the flat surface were cut on the same crystal, forming a polyhedron. This makes the intercomparisons rather reliable and gives credence to the low S_0 value for the flat surface. Madey [115] obtained a similar polyhedron to Besocke and Berger and confirmed the general trends of S, θ among the stepped and flat surfaces. He obtained a value of $S_0 \approx 0.38$ for the flat surface, however. Bowker and King [143], in a study of oxygen diffusion on W surfaces, measured S, θ curves for W{110} and the {320} surface. The shape of their curve ($T_s = 300$ K, $T_g = 300$ K) for W{110} agrees well with all the other data and their value of S_0 is ~ 0.35. As for all the other measurements on non-flat surfaces, the S_0 for the {320} surface, which consists of {110} terraces and {100} steps, is much higher at

~1. We will return to all these data in the section on stepped W{110} surfaces but we want to make the point here that the results of Butz and Wagner, Besocke and Berger, Madey, and Bowker and King make it obvious that non-flat W{110} surfaces can have much higher sticking probabilities than flat surfaces. Therefore it is likely that moderate discrepancies in absolute sticking probabilities on "flat" W{110} surfaces may be caused by differences in the number of defect sites. This in turn means that when apparently reliable, but contradicting, data exist it is likely that the lower reported S_0 values are more representative of a truly flat surface. It may also mean that the arguments given above concerning the interpretation of the S, θ data on "flat" {110} surfaces may need substantial modification to include the effect of the defects. This is considered in the section on stepped surfaces.

Finally in this section we mention the XPS O(1*s*) intensity versus exposure results of Fuggle and Menzel [136]. They provide data up to ~500 L exposure at T_s = 300 and 100 K. Taking the 300 K "θ_{lim}" of 0.5 mL at an exposure of ~20 L, their curves indicate that θ_{lim} 100 K is ~0.6 mL, in agreement with the data of Wang and Gomer. However, the wide exposure range of their data really makes it clear that, though at 100 K S genuinely becomes very small over a short exposure range, the rate of decrease at 300 K is much less. Without the availability of another parameter, such as "the sharpest p(2 × 1) structure", θ_{lim} is therefore somewhat arbitrary. These data also indicate that by ~500 L the θ–exposure curves at 100 and 300 K have merged; the coverage is ~0.8 mL and is still increasing very slowly.

2.5.3 0–0.5 Monolayer O range. Location of O normal to surface

Having established that reliable data for S, θ exists over a wide T_s range we can review the $\Delta\phi$ data in terms of coverage. As was the case for Cr and Mo we will see that the very simple argument of the sign and magnitude of $\Delta\phi$ are used as the primary evidence for the vertical location of the oxygen with respect to the W surface. Low-energy ion scattering data seem to confirm the interpretation of much of the $\Delta\phi$ data. Figure 10 shows $\Delta\phi$ versus θ at T_s = 300 K, as obtained by Engel et al. [68]. Adsorption at 1450 K is qualitatively similar. $\Delta\phi$ is very weakly negative between $\theta = 0$ and ~0.2 mL (*not* CO impurity-produced [68]) with the minimum (−0.05 eV) at ~0.1 mL. It then rises monotonically to the value of +0.20–0.25 eV. Taking any 300 K adsorption coverage between 0 and ~0.4 mL and heating produces no change in $\Delta\phi$ until just below the desorption temperature (1700–1900 K) at which point $\Delta\phi$ at 300 K measured after recooling is abruptly reduced to the slightly negative value (see Fig. 10, broken curve), together with a slight loss of oxygen from the surface. Most of the decrease in ϕ can be regained by further adsorption at 300 K. Based on the arguments advanced for the $\Delta\phi$ work on Cr and Mo, the $\Delta\phi$ data is consistent with the interpretation that, for temperatures up to ~1700 K, adsorption is always overlayer for coverages above θ = 0.1 mL. Above 1700 K, O penetrates/reconstructs the W

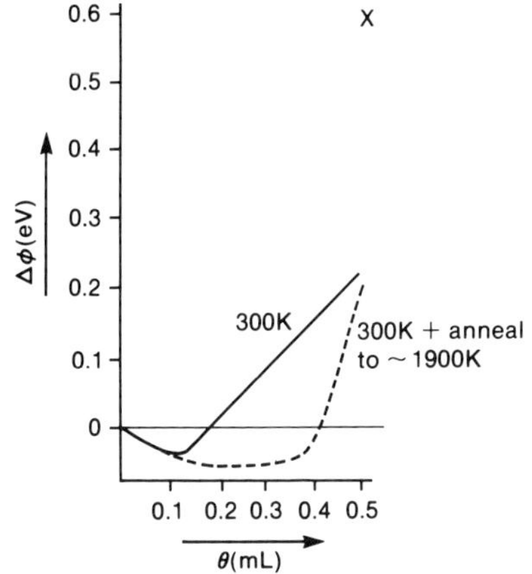

Fig. 10. $\Delta\phi$ versus θ for oxygen adsorption on W{110} [68]. The cross at $\theta = 0.5$ represents the work function obtained by adsorption at $T_s = 20$ K followed by annealing to 300 K [107, 113].

surface, freeing new W sites at the surface. Subsequent 300 K adsorption on these sites is again overlayer.

To be consistent with the Cr and Mo interpretation, the slight negative $\Delta\phi$ between $\theta = 0$ and 0.1 mL should be interpreted as involving reconstruction or penetration of the surface. The small coverage range of the negative $\Delta\phi$ suggests, however, that the initial adsorption may be strongly affected by the presence of defects. This would be in keeping with our remarks concerning the anomalies in the S_0 data. It should be noted, though, that merely suggesting that O atoms are initially located at steps would produce a positive rather than a negative $\Delta\phi$ because ϕ for a clean, stepped surface is lower than for the flat surface and adsorption at the step would tend to smooth out the effect of the step. An alternative explanation for the effects below $\theta = 0.1$ mL could involve the details of the phase diagram for W{110}/O. As will be discussed later, ordered p(2 × 1) islands of O form on the surface for coverages between $\theta = 0.1$ and 0.5 mL. These islands are not detectable below 0.1 mL. There may be a work function change associated with the order–disorder transition. Remember also that a $\Delta\phi$ of +0.4 eV was observed when a $\theta = 0.5$ mL surface was annealed from the adsorption temperature of 20 K to 45 K, again suggesting that an ordering or rearranging of a fixed amount of O_{ad} atoms can cause $\Delta\phi$ effects of this magnitude.

In a much older study of W{110}/O_2 Tracy and Blakely [70] reported $\Delta\phi$ versus exposure at $T_s = 300$ K. They found an essentially flat initial region and then a monotonic increase in ϕ. Their value of $\Delta\phi$ for a well-defined p(2 × 1) structure is +0.7 eV, however, three times larger than that of Engel et al. The reason for the discrepancy in magnitude is unknown, but does not affect the general argument made for the location of the oxygen normal to the surface.

Wang and Gomer [107] and Michel et al. [113] report $\Delta\phi$ measurements at low T. To avoid confusion in deciding when and how much α-O_2 might be present on an O_{ads}-covered surface (see earlier discussions) we consider only their data for adsorption at 50 K < T < 100 K and for adsorption at 20 K subsequently heated to 90 K. Under these conditions only atomic O can be present. In both cases a monotonic increase in ϕ is observed with $\Delta\phi$ at θ = 0.5 mL being ~0.6. The value barely changes on heating to 300 K, i.e. it is again considerably higher than the Engel et al. θ = 0.5 mL value at 300 K. $\theta_{lim} \approx 0.62$ can easily be obtained by adsorption at low T, however, as discussed in Sect. 2.5.1 and the $\Delta\phi$, θ data over the θ = 0.5–0.62 mL region shows a sharp further increase of ~ +0.4 eV. It seems, therefore, that the limiting value of $\Delta\phi$ may be strongly dependent on small changes in θ near 0.5 mL and that certainly the dipole per atom changes significantly in this region. This may be connected with the perfection of the order in the p(2 × 1) structure, which, as discussed later, grows as islands, plus the problem of accommodating the excess oxygen above θ = 0.5 mL.

ISS data are also available from Bauer's group [103]. Using AES-calibrated coverages they plotted the W and O ISS signal intensities as a function of coverage at He^+ energies between 250 and 1400 eV. In the lower energy range where the He^+ penetration beyond the outermost atomic layer is small the W signal falls rapidly with increasing θ. At 500 eV primary energy it had fallen to 3% of its clean value by θ = 0.5. This at first seems very convincing evidence that the O atoms are in overlayer positions and that they shadow the underlying W from the He^+ beam. It should be noted, however, that the decrease in the W ISS signal and increase in the O signal are non-linear with coverage; the W signal falls very rapidly over the first ~0.25 monolayer and much more slowly after that. Very probably, as suggested by the authors, there are unknown and strongly coverage-dependent ion neutralization effects occurring which have to be folded in with the simple shadowing phenomena. The key experiments to properly establish the reliability of ISS for determining the adatom vertical locations would seem to be to look at the ISS of surfaces of ~0.5 mL coverage annealed above 1700 K, where it is believed that the O atoms have penetrated the lattice (see earlier $\Delta\phi$ data), and to examine the ISS of W oxide surfaces in comparison with O covered W surfaces.

2.5.4 0–0.5 Monolayer O range. Geometric structure

Having discussed the vertical location of the oxygen atoms in the previous section we now turn to the surface crystallography. Several phase transitions have been observed as a function of θ and T and considerable effort has gone into establishing a phase diagram for the system. These efforts are discussed in Sect. 2.5.5. Here we consider the periodic structures involved in these various phases ($\theta \leqslant 0.5$ mL).

At θ = 0.5 mL a sharp intense p(2 × 1) O structure is observed between ~200 and ~660 K, provided exposure is done at less than 1600 K. This situation is the most easily obtainable experimentally: all that is required is to expose at less than 1600 K until the oxygen uptake rate drops rapidly and record the LEED between 200 and ~660 K. At coverages below 0.5 mL additional factors such as island formation, and a coverage-dependent low-temperature phase transition must be considered, so we will deal with the "perfect" p(2 × 1) structure at $\theta = 0.5$ first.

The "saturation" p(2 × 1) structure was first reported by Bauer [150] and by Germer and May [23]. The latter authors assigned a coverage of 0.5 mL and, on the mistaken assumption that the high intensity of the half-order spots implied that they must be due to W scattering rather than O, they proposed a reconstructed surface, as in Fig. 11(a), with the outer atomic plane consisting of alternate rows of W and O atoms. Bauer alternatively proposed an overlayer structure [Fig. 11(b)], with O atoms in the site of maximum coordination (referred to hereafter as "center" site), the same site that would be occupied by a W atom in an additional plane of the crystal. Tracy and Blakely [70] supported the overlayer model but for reasons now known to be suspect (the lack of oxide desorption in TDS and the apparent necessity for considerable mass transport of W atoms during reconstruction. TDS is discussed in Sect. 2.9; reconstruction dynamics in the W{100} part of this review).

Matysik [151] measured the RHEED p(2 × 1) intensities and on the basis of a kinematic analysis concluded that the reconstruction model was incompatible with the experimental intensities. Lagally and co-workers [28, 152]

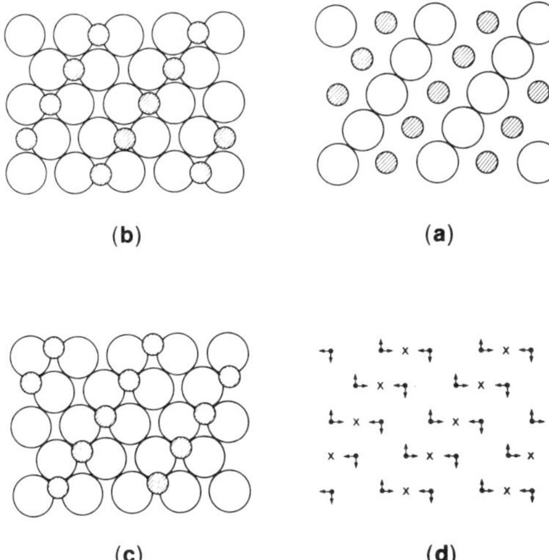

Fig. 11. Proposed structures for p(2 × 1) O structure on W{110}. O atoms are represented by hatched circles. W and O atoms are drawn with a size appropriate to their atomic radii. (a) Reconstructed surface; every other close-packed W row replaced by oxygen [23]. (b) Overlayer adsorption, center site [150]. (c) Overlayer adsorption, threefold site (actually a pseudo-threefold site) [128, 129]. (d) Displacive reconstruction model. W atoms (filled circles) undergo small lateral displacements under the influence of the overlayer O atoms (crosses) in center sites [153].

performed the first LEED IV calculations on the p(2 × 1) structure. They concluded that the W surface was not reconstructed and the displacement of the W atoms from their clean positions was ⩽0.05 Å vertically and ⩽0.5 Å horizontally. The calculations, which extracted the single-scattering intensities, were insensitive to the O atom positions in the electron energy range treatable. The same conclusion concerning reconstruction was later reached from full dynamical calculations, with the additional results that the O atoms were determined to be in the threefold sites [Fig. 11(c)] at 1.25 ± 0.03 Å above the surface [128, 129]. It is not clear, however, whether the quality of fit excludes the Bauer center site of Fig. 11(b).

At this point, then, we can say that all the LEED and RHEED data exclude reconstruction as a possibility, in agreement with the $\Delta\phi$, ISS and photoemission interpretations discussed earlier, and that the adsorption site predicted is the threefold site, with the center site as a reasonable alternative. Some authors have had conceptual difficulties with both of these sites. Examining Fig. 11 it can be seen that, whereas threefold or center sites running along the $(\bar{1}11)$ and $(1\bar{1}1)$ directions of the W{110} substrate are crystallographically equivalent, a p(2 × 1) O structure in the threefold or center sites apparently distinguishes between an O–O interaction in the $(\bar{1}11)$ direction and an O–O interaction in the $(1\bar{1}1)$ direction. We will return to this point in Sect. 2.5.5 but for now simply make the comment that the conceptual difficulties are really with the Monte Carlo calculations that consider (artificially) effective lateral interactions of an O lattice in the absence of a substrate. There is no a priori reason for excluding any of the sites of Fig. 11 as possibilities for the p(2 × 1) structure.

Theodoreau [153] started with the (intuitive) assumption that O atoms occupy center sites and used bond-stretching and bond-bending arguments to predict lateral displacements of W atom of ~0.076 Å [see Fig. 11(d)]. This would be an example of *displacive reconstruction*. The force constants for the model were taken from bulk W and from gas-phase O-containing molecules. While the argument is very qualitative it is very appealing on chemical grounds and is quite likely since it is now known that displacive reconstruction can occur on clean and H-covered W{100} surfaces [5, 10]. No lateral displacements of W were allowed in the LEED dynamical calculations and it is not known whether the quality of fit for center sites might improve if they were. The exact location of the O atoms in the p(2 × 1) overlayer is, therefore, still in question. On intuitive chemical grounds we favor the center sites, as originally proposed by Bauer [150], with the very real possibility of small horizontal W displacements as suggested by Theodoreau [153].

Below ~200 K, no LEED pattern is observed. This of course does not rule out short-range order and indeed the low T_s S measurements (Sect. 2.5.1) indicated that such order must still occur since 40% of the surface W atoms were still excluded as sites at which dissociative adsorption could easily occur ($\theta_{lim} \approx 0.6$ mL). It is reasonable to conclude then that, even below

200 K, most O atoms go into a local p(2 × 1) arrangement with very few occupying the "wrong" repulsive positions between O rows.

At ~ 660 K an order–disorder phase transition occurs for the p(2 × 1)O 0.5 mL situation. The LEED manifestation of this, a rapid decrease in the half-order spot intensities with increase in T_s, has been observed by several groups. Above the transition temperature the new phase is a lattice fluid with the repulsive O–O interactions, which produce the empty rows in the p(2 × 1) arrangement, being completely overcome. We reserve discussion on this to the next section on the phase diagram for W[110]/O.

At temperatures approaching desorption (~ 1600 K), the $\Delta\phi$ data discussed earlier indicated that O in the 0.5 mL structure penetrates the lattice and that this process is not reversible by recooling [68]. The new high-temperature-formed structure is again p(2 × 1) but is clearly different from the overlayer structure ($\Delta\phi$ lower than clean surface; no phase transition on heating to ~ 660 K). It seems quite likely that this is the missing W row reconstructed surface originally proposed by Germer and May [23] for the 300 K adsorption situation. Very little work has been done on the details of this structure, however.

Let us now consider coverages significantly below 0.5 mL. Germer and May [23] noticed that p(2 × 1) spots at 300 K were observable well below 0.5 mL and suggested that islanding was occurring. Tracy and Blakely [70] claimed the pattern was observable as low as 0.1 mL and concluded that the symmetrical fuzziness of the spots indicated small, round anti-phase islands. Engel et al. [68] showed there was a sharp decrease in spot width between $\theta = 0.05$ and 0.2 mL at $T_s = 300$ K, but no further change up to 0.5 mL. On annealing to ~ 450 K and recooling a slight further sharpening occurs. They also monitored the intensity of the scattered background as a function of coverage and found that it increased up to $\theta \approx 0.15$ mL and then rapidly decreased to zero above this coverage. From this they were able to conclude that below $\theta \approx 0.25$ mL growth occurred by a mixture of statistical adsorption [p(2 × 1) lattice sites statistically occupied] and p(2 × 1) island growth. Both processes contribute to spot broadening, but only the former causes an increase in background. Above $\theta \approx 0.25$ mL, island growth dominates as the background goes to zero. Having shown that, above $\theta \approx 0.25$ mL, islands dominate it is immediately obvious that there must be many islands rather than a single one, which would be the expected thermodynamic equilibrium situation (minimization of island boundary length). Clearly if all of even 0.1 monolayer coverage were in one island the size would be many orders of magnitude larger than the coherence width of the LEED beam and therefore the spots would be as sharp as the instrumental response.

Why are many islands present, even after annealing to 450 K for extended times? It is possible that equilibrium is not reached, either because diffusion rates are low or because the change in free energy on coalescing small islands is small. A more likely explanation is that substrate defects, such as steps, divide the "single crystal" surface into many small essentially non-interacting regions. Information on the average island size can, in principle, be obtained by calculating the LEED spot angular profile as a function of average island size and comparing with experiment as described by Lagally and co-workers [119–127]. Unfortunately, the calculated average island size for a given LEED spot angular width is quite sensitive to the choice of island size distribution (Poisson, linear, single) and to correlations between islands which cannot be ignored for significant total coverages. The average heterogeneity of the substrate prior to adsorption can also, in principle, be determined from the variation of the angular beam width of the substrate spots as a function of electron energy. For the W{110} crystal used by Lagally and co-workers it seems as though the derived average size and therefore number of p(2 × 1) O islands is compatible with the derived substrate surface defect density, such that each island might be located on a terrace separated from the next terrace by a step, i.e. the surface heterogeneity argument for the existence of many, as opposed to one, island is supported. As a point of reference the number of defects was determined as $\sim 10^{11}\,\mathrm{cm}^{-2}$, about the number of steps expected if the crystal were cut to 0.2° accuracy. We will return to the significance of defects/steps in the section on stepped surfaces (Sect. 2.5.14).

Summarizing this section on the geometric structure of O for $\theta \leqslant 0.5$ mL we can say that at 0.5 mL O atoms are in an overlayer position up to ~1600 K. Below 200 K they exhibit only local order, presumably p(2 × 1)-like. Between ~200 and 660 K they are in an ordered p(2 × 1) phase which covers the surface, are ~1.25 Å above the surface and are either in threefold or center sites. Between $\theta \approx 0.1$ and 0.5 mL the O atoms exist as islands of p(2 × 1) structure for temperatures below ~450 K. The islands appear to be thermodynamically isolated and each island may be nucleated by a substrate defect. Below $\theta = 0.1$ mL the p(2 × 1) ordered structure is not observable.

2.5.5 0–0.5 Monolayer range. Phase diagram

In Fig. 12 we have reproduced the phase diagram proposed by Lagally and co-workers. The experimental points were established by locating the temperature at which the LEED half-order spot intensities decayed rapidly. It is likely that there could be errors of at least 0.1 monolayer in the coverage calibration since they were all derived by assuming $\theta = 0.5$ mL for the maximum intensity p(2 × 1) spots for 300 K exposure and relying on the Wang and Gomer absolute S values to establish the other coverages. A qualitative understanding of the phase diagram, as drawn, is straightforward. To form p(2 × 1) islands requires at least three significant pairwise interactions. This is illustrated in Fig. 13(a). The nn attractive interaction, ω_1, is necessary to form a close packed row, the repulsive interaction, ω_2, is necessary for the rows to propagate alternatively, and the attractive interactions, ω_4, is required for the overall p(2 × 1) structure to grow as islands rather than statistically. Since the islands are approximately round the attractive ω_1 interaction must be similar to the attractive ω_4 interaction.

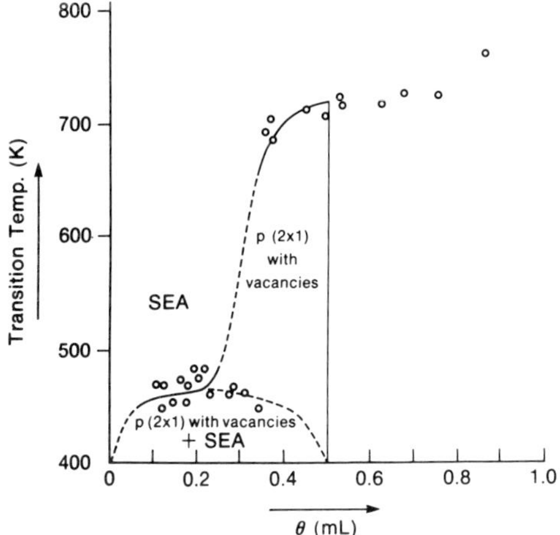

Fig. 12. Proposed phase diagram for W{110}/O based on the observation of a low T (~ 450 K) and a high T (700–750 K) phase transition in LEED. The open circles are the data points. See ref. 122 and other references quoted therein.

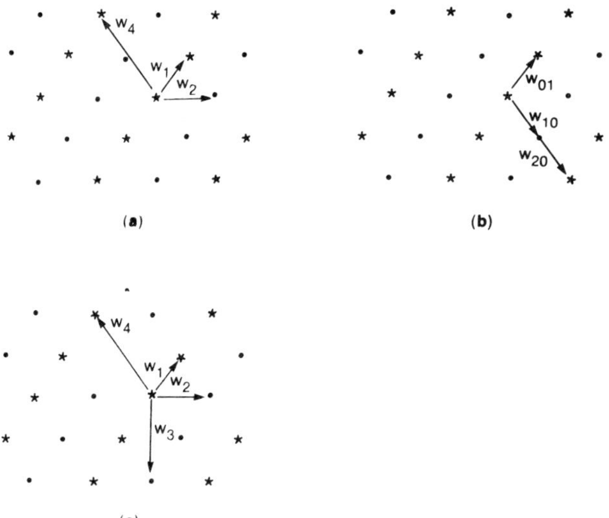

Fig. 13. Pairwise interactions between O adatoms used to simulate a p(2 × 1) lattice on a {110} substrate. Stars represent filled sites; dots empty sites. (a) Minimum requirement for a p(2 × 1) lattice to form; w_1 attractive, w_2 repulsive, w_4 attractive. (b) Interaction scheme used by Ertl and Schillinger [154] and Lu et al. [120, 125]. (c) Interaction scheme used by Williams, et al. [155]. In all these schemes, no assumptions are necessary concerning the registry of the p(2 × 1) O lattice with the W{110} lattice, i.e. they do not directly distinguish between possibilities (a)–(d) in Fig. 11.

With these interactions at $T = 0\,K$ the equilibrium situation would be with all the O atoms in a perfect p(2 × 1) island (no vacancies) with the rest of the surface an empty sea. At finite low T for a given coverage the surface would have a p(2 × 1) island with some vacancies and a sea with some O atoms randomly distributed. In principle the concentration of vacancies in the islands, and oxygen atoms in the sea, and relative amount of oxygen in the islands and in the sea can be established at any θ and T by constructing tie lines in the phase diagram. There is insufficient data to know the true shape of the phase boundary, however, so such estimates are very approximate. Lagally et al. [122] have estimated that, at ~450 K for $\theta = 0.25$ mL, 60% of the surface is sea with one out of 10 sites in the sea containing oxygen, while the other 40% is p(2 × 1) islands with vacancies. If T is raised above the phase-transition temperature for the island plus sea two-phase region, one of two things can happen. Below $\theta = 0.25$ mL one moves into a one-phase region consisting of lattice gas with random occupation of p(2 × 1) sites (the "sea"). To achieve this the attractive interactions holding the adatoms in rows have to be overcome. For $\theta > 0.25$ mL the single-phase region consists of the p(2 × 1) structure with vacancies. The density of vacancies in the p(2 × 1) phase or of O atoms in the lattice gas depend on coverage θ. For coverages close to $\theta = 0.5$ mL the phase transition to the "sea" is at much higher T because at this coverage the strong repulsive interaction must be overcome to force adatoms into "wrong" sites. Unfortunately the experimental data do not fully support the simple diagram as drawn. The data between $\theta = 0.25$ and 0.35 mL indicate that the low T transition is still observable over this region, yet this transition should now be from the p(2 × 1) island plus sea phase to the p(2 × 1) plus vacancies single-phase structure, rather than to the lattice gas (see Fig. 12). Contrary to experiment no sharp decrease in LEED intensities should be observable for this transition since the p(2 × 1) structure still exists. In fact since there will be no possibility for antiphase boundaries in the single-phase p(2 × 1) region the LEED intensity should even increase. We conclude that either the phase diagram is more complex than that proposed or the coverage scale needs adjusting. If experimental coverages were decreased by 20% the fit to the simple phase diagram would be much better.

Early attempts at extracting the O–O interaction energies were made before the low T phase-transition part of Fig. 12 had been discovered. Working at $\theta = 0.5$ mL and assuming only a repulsive nearest-neighbor interaction Buchholz and Lagally [119] matched the experimentally determined angular widths and intensities of the half-order LEED spots at a given T in the order–disorder transition region to a short-range order model which extracted the order parameter, p, the probability of having unlike nearest neighbors along a chain. The parameter p is related to the enthalpy difference, ΔH, between ordered and random configurations by

$$\frac{(1-p)}{p} \propto \frac{e^{-\Delta H}}{kT} \tag{1}$$

allowing ΔH to be extracted from a plot of ln $(1 - p)/p$ versus $1/T$. They obtained values of 0.43 or 0.68 eV atom^{-1}, depending on whether they used the angular widths or the intensities. Ertl and Schillinger [154] performed a Monte Carlo simulation of the equilibrium configuration as a function of T at $\theta = 0.5$ mL using the pairwise interactions shown in Fig. 13(b). After obtaining the Monte Carlo equilibrium for each T the LEED intensity distribution was evaluated at each T by summing up all the scattered amplitudes from each site. The calculated relative LEED intensity as a function of T was then compared with the experimentally available data. In order to reduce the number of variable parameters in the Monte Carlo simulation ω_{01} was always kept equal to ω_{20} (they should be similar because of the observation of circular islands) and from the result of trial runs ω_{01} was set at $-0.9\omega_{10}$. A best fit to the experimental I/I_0 versus T curve through the order–disorder transition at ~ 700 K then yielded a value of $\omega_{20} = \omega_{01} = -1.1$ kcal mol^{-1} and $\omega_{10} = 1.2$ kcal mol^{-1}. The value for ω_{10} is about an order of magnitude less than the nn repulsive interaction obtained by Buchholz and Lagally. Ertl and Schillinger suggest that this is due to the short-range order model of Buchholz and Lagally being inappropriate [154].

Once the low T phase transition had been discovered for coverages much less than 0.5 mL, Lu et al. [120, 125] were able to calculate the attractive interaction $-\omega_{01}$ independently of the repulsive interaction ω_{10} since the low T phase transition involves dissolving O atoms from p(2×1) islands into a lattice gas, which requires overcoming only the attractive terms causing the p(2×1) structure to island. They again set $\omega_{01} = \omega_{20}$ and used a two-dimensional version of the magnetic Ising model to calculate the pair correlation functions using the Onsager solutions for an infinite two-dimensional lattice for different values of ω_{01}. The LEED intensities were then obtained from the summation of the Fourier transforms of the pair correlation functions and compared with the experimental data as a function of T. The best fit was obtained for $\omega_{01} = -0.065$ eV atom^{-1}. This value was then used to analyze the p(2×1) $\theta = 0.5$ mL high-temperature order–disorder transition in which the ω_{10} repulsive interaction is important. In the Ising model for the transition both ω_{10} and ω_{01} are required. Using the value of 0.065 eV atom^{-1} for ω_{01} a best fit to the I versus T data for the high-temperature transition was obtained for $\omega_{10} = 0.15$ eV atom^{-1}. The results are much closer to, but still higher than the Monte Carlo-derived results of Ertl and Schillinger [154]. Williams et al. [155] pointed out that the interaction pairs used in the calculations of Ertl and Schillinger and Lagally and co-workers violated fundamental symmetry restrictions in making the ($\bar{1}11$) and ($1\bar{1}1$) axes inequivalent if only two-body adatom interactions are involved, as is assumed in the simulation models used. [Physically, of course, the interactions which force the p(2×1) O structure are not necessarily two-body and occur predominantly through electronic interaction with the substrate W atoms anyway. The distances between O atoms are too large for anything else to be dominant. An inequivalence of the ($\bar{1}11$) and ($1\bar{1}1$) direction may then be unimportant.] Williams et al. used a different set of interaction pairs, shown in Fig. 13(c), and derived values of $\omega_1 = -2.1\,\mathrm{kcal\,mol^{-1}}$, ω_2 (set equal to ω_3) $= +1.7\,\mathrm{kcal\,mol^{-1}}$ and $\omega_4 = -0.7\,\mathrm{kcal\,mol^{-1}}$ from fits of Monte Carlo calculations to both the high and low T phase transitions. They also showed that these parameters were not a unique set but could be used to generate total interaction energy, heat capacity, and entropy curves as a function of T for the $\theta = 0.5\,\mathrm{mL}$ situation.

Ching et al. [123] also performed Monte Carlo calculations using the same pair interactions as Williams et al. [155]. They obtained similar values except for ω_3, which they found to be zero. This emphasizes the non-uniqueness of the derived values. They also addressed the question of the importance of three-particle adatom interactions. A p(2 × 2) phase is formed above $\theta = 0.5\,\mathrm{mL}$ (see later) which has no counterpart below $\theta = 0.5\,\mathrm{mL}$. For two-particle interactions the phase diagram should be symmetric about $\theta = 0.5\,\mathrm{mL}$. The fact that it is not is a clear indication that three-particle interactions are important. Ching et al. concluded from their Monte Carlo calculations including three-particle interactions that the latter were only of moderate importance for $\theta < 0.5\,\mathrm{mL}$ but exert increasing influence on LEED intensities above $\theta = 0.5\,\mathrm{mL}$. Einstein [156] demonstrated that there are many tri-interactions that might be expected to have comparable magnitudes, the tris chosen by Ching et al. being arbitrary, and warned against taking any Monte Carlo derived pairwise interaction energies too seriously.

Summarizing this section we conclude that the W{110}/O phase diagram below $\theta = 0.5\,\mathrm{mL}$ is qualitatively understood, even though it has not been fully mapped. The significance of attractive and repulsive interactions in determining the phase diagram is clear and if one wishes to think in terms of pairwise adatom interaction, then values are around $1\,\mathrm{kcal\,mol^{-1}}$. In reality these are really effective interactions since the substrate transmits the adatom interactions. From the asymmetry of the phase diagram above $\theta = 0.5\,\mathrm{mL}$ it seems likely that three (or n, where n is odd) particle interactions are not, in general, ignorable though a reasonable description below $\theta = 0.5\,\mathrm{mL}$ can be given without considering them.

2.5.6 0–0.5 Monolayer range. Electronic structure effects

Having dealt with all the location and structural aspects of W{110}/O up to $\theta = 0.5\,\mathrm{mL}$, it remains to discuss what is known about the nature of the W–O bonding over this range. The techniques used have been XPS and UPS (including synchrotron studies), AES and ELS. Reference to some of this work has already been made in the context of clarifying distinctions between molecular and atomic oxygen, the presence or not of more than one adsorption state or as a means of following the kinetics, etc. Here we concentrate on the information provided on electronic structure.

In XPS the O(1*s*) spectrum at $\theta = 0.5\,\mathrm{mL}$ for 300 K adsorption is a single sharp peak at 529.5 eV [110, 136]. In a study specifically looking for any variation in shape or position of this peak with coverage, none was observed

[157]. This O(1s) BE is the same, within a few tenths of an electron volt, to that found for atomic oxygen on many metal surfaces. It therefore serves as a confirmation that the species is indeed dissociated oxygen, and also suggests that the charge transfer is large and similar on all these metal surfaces (i.e. the O atom, at submonolayer coverage always end up as $O^{\delta-}$ where the $\delta-$ is similar from metal to metal). Owing to the difficulty of interpreting O(1s) BE values in detail, because of three major contributing factors (initial state chemical shift, final state relaxation and Madelung potential effects), it is not possible to be any more specific (a full discussion of theoretical considerations of O(1s) BEs for metal–oxygen bonding is given elsewhere [158]).

If a W surface is heavily oxidized ($\sim$20 Å thickness) a WO_3 layer is formed in which the W(4f) levels undergo a $\sim$4.3 eV chemical shift to higher BE from its metallic value of $\sim$31 eV [159]. For adsorption on W{110} to $\geqslant$0.5 mL, no shifts or peak shape changes are observed using laboratory X-ray AlKα (1486 eV) or MgKα (1251 eV) anodes [157]. Since the KE of the emitted W(4f) electrons is large (1455 eV), only some $\sim$15% of the W(4f) signal comes from the top layer. In addition the resolution is very limited because of the $\sim$1 eV linewidth of the laboratory X-ray source. These factors mean that, in looking for a shifted W(4f) level, one is looking for a small unresolved shoulder on the side of the bulk peak. The W{110}/O system has also been examined using synchrotron radiation, with $h\nu$ between 50 and 150 eV. In this range the escape depth is short ($\sim$2 atomic layers) and the resolution of the monochromator is good ($\sim$0.1 eV), and the surface chemical shifts are now observed [135, 137, 140, 160]. In fact it was found that, for the clean W{110} surface, a $\sim$0.3 eV negative shift existed for the top W atomic layer [135]. On low exposures of oxygen the intensity of this peak decreased [137]. At higher exposures a new peak, with a $\sim$0.5 eV positive chemical shift appears [137, 160]. This latter shift is therefore associated with the bonding effect of the overlayer O to the top W layer. Its value, about one-ninth that for WO_3, is also reasonable since at $\theta = 0.5$ mL every surface W is essentially coordinated to one-half an O atom only, compared with six in WO_3. One should note, however, that there is not a one-to-one correspondence between the loss of intensity in the clean surface chemical shifted peak and the appearance of intensity in the O-overlayer derived peak. The negative shift of the clean surface W atoms probably represents delocalized surface states, whereas the positive shift for W/O probably is a more localized effect. This would explain why the clean surface shift can apparently be quenched with a lower oxygen exposure than is required to build up the W/O peak.

A recent ELS study of the W{110}/O system at 300 K and above has been reported by Rawlings [144]. He used a 40 eV energy primary beam. Under these conditions the sampling depth should be even shorter than the synchrotron photoemission study. Some of the results are hard to understand in the light of the photoemission results. The W(4$f_{7/2}$) loss peak is reported as shifting continuously with coverage to 1 eV higher BE by $\theta \approx 0.5$ mL. Above this coverage the shift ceases. The only way we can rationalize these results with the photoemission is to suggest that the poorly resolved differentiated ELS data represent the superposition of peaks. In Fig. 14, it can be seen that, for an unresolved spectral situation, but with high surface sensitivities, the quenching of a surface peak ~0.3 eV to lower BE and simultaneous growth of a W/O peak ~0.5 eV to higher BE would look like a gradual shift with coverage. The total shift by photoemission, 0.35 + 0.5 = 0.85 eV is close to the 1 eV ELS value. The fact that the ELS W($4f_{7/2}$) peak ceases to move for $\theta > 0.6$ mL is perhaps consistent with the photoemission data showing that the W/O chemically shifted peak remains in a constant position (+0.5 eV) once it starts to grow. This would suggest that the charge transfer from each surface W does not increase on increasing θ above 0.5 mL. This in turn seems compatible with the proposed geometric interpretation for the LEED data for $\theta > 0.5$ mL (see later) that at saturation coverages the O atoms are trying to achieve a close-packed arrangement and that the effects of registry with the substrate atoms become less dominant. The interpretations offered here for the W($4f_{7/2}$) ELS shifts are not identical to those suggested by the author. Rawlings suggests that the ~1 eV shift up to $\theta \sim 0.5$ mL is continuous and due to a mixture of incorporation of O into the surface (up to $\theta \approx 0.2$ mL) and the O overlayer, and that the lack of further shift above $\theta = 0.5$ mL is due

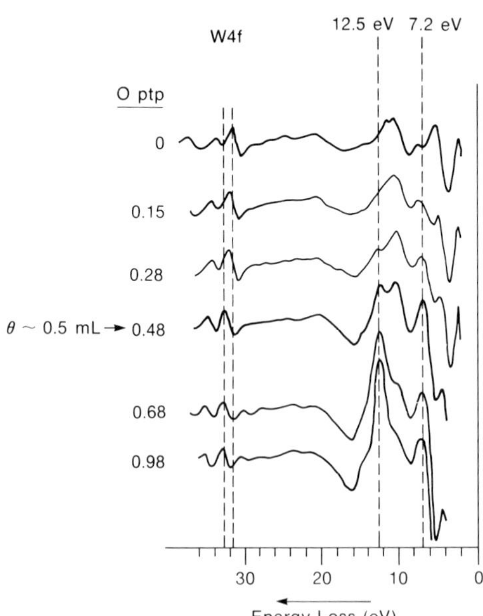

Fig. 14. ELS spectra for oxygen interaction with W{110} at 300 K [144]. Oxygen coverages are given in terms of O AES peak-to-peak height as a fraction of saturation value. In a parallel kinetic study [144], the p(2 × 1) plateau was achieved at Optp ≈ 0.5Optp(sat.), from which we may construe a value of $\theta \approx 0.5$ mL.

to a compensation between further charge transfer and an increased final state relaxation.

He I and He II UPS data for oxygen coverages up to very high oxygen exposures show a single O(2*p*) feature in the 0–0.5 monolayer range which moves from 6.7 to 6.4 eV BE with increasing coverage [76]. For $\theta > 0.5$ mL a second feature appears which is discussed in Sect. 2.5.12. Avery [162] deconvoluted the fine structure in the NVV O Auger spectrum to yield an adsorbate level moving from 7.4 to 6.9 eV over the $\theta = 0$–0.5 mL range, in reasonable agreement with the UPS data. UPS data for adsorption at 20 K and annealed to 100 K ($\theta \approx 0.6$ mL) also shows a single adsorbate feature at ~6.6 eV [13]. The UPS and Auger data therefore confirm the O(1*s*) results: a single, dissociated species is observed. The ~0.5 eV decrease in BE observed for O(2*p*) on increasing coverage from 0 to 0.5 mL may be a band-structure dispersion effect.

The ELS data shows several loss features in the valence region which can be associated directly with UPS adsorption features with the assumption of a single dominant final state for the ELS transitions [144]. A feature at 7.2 eV is associated with the O(2*p*) UPS level (7.4–6.9 eV). A feature at 12.5 eV has no clear counterpart in UPS below $\theta = 0.5$ mL. It was suggested by Rawlings that the peak is missing in UPS due to a low transition probability. For $\theta < 0.5$ mL this 12.5 eV feature disappears reversibly on heating to temperatures well below the reconstruction temperature. On this basis Rawlings concludes that its removal represents disruption of the O–O attractive interactions as the p(2 × 1) islands dissolve into the lattice gas at the order–disorder transition. Examination of Fig. 14 makes it clear that the 12.5 eV feature grows mainly for $\theta \geqslant 0.5$ mL, however. We therefore believe that if it is related to O–O interactions it represents the repulsive O–O interactions that must occur for $\theta \geqslant 0.5$ mL. In accord with this suggestion, the peak does not disappear on heating a $\theta \leqslant 0.5$ mL surface.

Summarizing this section, all the electronic structure data support the idea of a single dissociated O species below $\theta = 0.5$ mL. The high-resolution synchrotron photoemission is able to identify the W(4*f*) chemical shift induced by this species under the $\theta \leqslant 0.5$ mL condition as +0.5 eV with respect to a bulk W atom [135, 137, 160]. This is to be compared with the ~4.3 eV shift of bulk WO_3 [159]. The valence band effects suggest that band-structure dispersion of about 0.5 eV occurs in the O(2*p*) level, representing O–O interactions. A peak in the ELS at 12.5 eV may be specifically associated with O–O interactions, but it is not entirely clear whether they are attractive or repulsive.

2.5.7 Greater than 0.5 monolayer O range. General

The greater than 0.5 monolayer coverage range is not nearly so well studied as the 0–0.5 mL range. After completion of the p(2 × 1) 0.5 monolayer structure at 300 K, which takes only a few tens of L exposure, it is necessary to expose to hundreds of L before the next LEED structure, a diffuse p(2 × 2) pattern is observed (optimum coverage ~0.75 mL) and anywhere between 300 and 10 000 L (depending upon the author) to achieve saturation at around 1 monolayer, where a (1 × 1) pattern has been reported. Owing to the experimental difficulty of achieving high coverages with O_2 exposure the situation has been simulated by other means such as evaporating WO_2 on to the surface and annealing [69], or pre-adsorbing W atoms, exposing to O_2, and annealing [142]. These techniques do not, of course, provide information on the kinetics and mechanism of O_2 interactions above $\theta = 0.5$ mL, but they do seem to be able to produce identical LEED structures and work function behavior as for the high-coverage O_2-exposed surfaces, and are therefore assumed to represent the equilibrium adsorption structures at these coverages.

2.5.8 Greater than 0.5 monolayer O range. S, θ measurements

In the original LEED study, Germer and May [23] reported an exposure of 160 L to reach the best p(2 × 2) condition and 300 L to reach the final p(1 × 1). From this they give an average S for adsorption between 0.5 and 0.75 monolayers [the coverage they arbitrarily assigned to the p(2 × 2) structure] of 0.003. This is actually likely to be at least a factor of 2 too large and their exposures 2 times too small since they quote an exposure of 1.6 L to form the fully developed p(2 × 1) structure, whereas most later authors would say that at least 13 L are required. No other S value has been quoted for the $\theta > 0.5$ mL situations, although estimates can be made from the various uptake versus exposure curves given in the literature. AES data from Madey [115] and from Rawlings [144] and XPS data from Fuggle and Menzel [136] are reproduced in Fig. 15 for intercomparison. The data of Fuggle and Menzel is the most informative in that it covers the widest exposure range at both 300 and 100 K. Using the S versus θ data of Wang and Gomer [107] at $T_s = 87$ K the sharp break in the Fuggle and Menzel 100 K

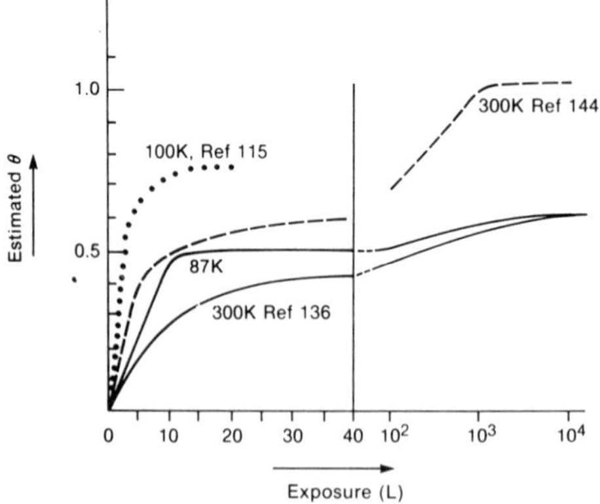

Fig. 15. Oxygen exposure versus estimated coverage on W{110} for exposures into the high coverage regime. We have estimated the coverage scale for the data of ref. 136 by assuming that the sharp break in the 87 K curve comes at $\theta \approx 0.5$ mL (see text).

curve should come at a coverage of 0.5 mL. This would make the "θ_{lim}" of the 300 K p(2 × 1) structure 0.4 rather than 0.5 mL, and the saturation value at both 300 and 100 K after ~5000 L exposure only 0.6 mL, not unity. Madey assumes a coverage of 0.5 mL after 20 L exposure at 300 K [sharp p(2 × 1) observed] and obtains a saturation coverage of 0.75 mL for 80 K exposure [p(2 × 2) observed after warming to 300 K]. All this points up the difficulty of making relative and absolute coverage calibrations over these huge exposure ranges. It seems likely (a) that the coverage achieved in all UHV O_2 adsorption situations is less than unity, (b) that the "best" p(2 × 1) value at 300 K may well be nearer 0.4 rather than 0.5 mL, and (c) that S at 300 K in the $\theta = 0.5$ mL vicinity does not decrease rapidly enough to allow an accurate coverage determination on the basis of a "θ_{lim}" value. In fact "θ_{lim}" is a rather arbitrary value which is the reason for possibly large coverage error bars in the phase diagrams of Lagally and co-workers. From a comparison of the initial slope of the 100 K curve and the slope in the p(2 × 2) to full coverage region at 300 K we have made a rough estimate of 3×10^{-5} for the average value of S over the p(2 × 2) coverage range. Germer and May [23] reported that S in the p(2 × 2) formation region could be increased by ~10 by exposing at 550 K although no detailed study of the kinetics at elevated T_s has ever been reported.

2.5.9 Greater than 0.5 monolayer O range. Location of O normal to the surface

Although Bauer had originally proposed that the 300 K $\theta = 0.5$ mL p(2 × 1) structure was an overlayer structure, and was subsequently vindicated in this claim, he also originally proposed that the LEED structures formed at higher coverage were from reconstructed surfaces. This belief was based largely upon the change in thermal desorption products above $\theta = 0.5$ mL from O atoms only to W oxides, and on similar effects in SIMS. It is now realized that although changes in TDS and SIMS are certainly indicative that structural changes are occurring on the surface, one cannot make the simple correlation that ejected oxide species imply reconstruction or oxide formation (see Sects. 2.9 and 2.10). Bauer himself realised this in later papers [69, 77]. Our best indications of the vertical location of the oxygen are, as for the < 0.5 monolayer situation, $\Delta\phi$ measurements supplemented by ISS data and any evidence on the electronic structure of the outermost W atoms that might be available from photoemission, Auger, or energy loss measurements.

Work function changes as a function of coverage above $\theta = 0.5$ mL have been measured in two ways, either directly as a function of O_2 exposure or by establishing a coverage by the WO_2 or W pre-dosing techniques referred to earlier. Since a relatively high-temperature anneal is required in the latter cases to order up the surface and produce the same LEED structures as observed by direct O_2 exposure, the measured $\Delta\phi$ values obviously refer only to an annealed situation.

The 300 K data by direct O_2 adsorption are rather sparse and old, the most recent measurements being by Hopkins et al. [161]. They give data only as a function of exposure, not coverage, and find that, following the ~ 0.3 eV increase up to the p(2×1) $\theta = 0.5$ mL exposure condition, a further monotonic increase of 0.5 eV occurs up to saturation coverage. This compares with the WO_2 deposited and 1200 K annealed data of Bauer and Engel [69] where a 0.2 eV increase is obtained over the first 0.5 monolayers (reproducing the 300 K direct O_2 exposure results, see earlier) plus a further linear increase with coverage of 0.8 eV up to 1 monolayer [AES calibrated against the p(2×1) $\theta = 0.5$ mL situation]. Similar results were obtained using the W pre-deposition method [142]. It is likely that the difference in the $\Delta\phi$ values at saturation represent a difference in saturation coverage (less than unity in the O_2-exposed case) rather than a difference owing to annealing procedures in the WO_2 deposition and W pre-deposition methods. Bauer and Engel conclude from the WO_2 deposition $\Delta\phi$ results that, for all coverages (and LEED structures) between $\theta = 0.5$ mL and unity, the oxygen is present in an overlayer form with no reconstruction and that the same is therefore true for the O_2-exposed coverages between 0.5 and saturation. Hopkins et al. [161] also give 80 K exposure $\Delta\phi$ data. The curve is identical to their 300 K data up to $\theta = 0.5$ mL and then a sharp decrease with further exposure was observed. They attribute this to adsorption of weakly bound molecular oxygen, which from the results of Gomer and co-workers is certainly incorrect (α-O_2 desorbs at 45 K) [13, 113]. They noted that the new species could be desorbed by the LEED beam or by warming to 300 K. We suggest it is contaminant physisorbed H_2O. Michel et al. give $\Delta\phi$ data up to $\theta \approx 0.6$ mL for 20 K exposure followed by annealing to 90 K (i.e. all α-O_2 desorbed) [113]. Between 0.5 and 0.6 mL, $\Delta\phi$ increases by ~ 0.2 eV in contradiction to the data of Hopkins et al.

Bauer and Engel [69] quote the He^+ scattering ISS data [103] in support of their claim of non-reconstruction between 0.5 and 1 monolayer. The W signal is already $\sim 1/70$ its clean value by 0.5 monolayer coverage. This decreases almost linearly by a further factor of ~ 5 up to 1 monolayer while the O signal continues to rise smoothly (WO_2 deposition technique). If reconstruction occurred above $\theta = 0.5$ mL one might expect the W signal to increase over its $\theta = 0.5$ mL value, or at least that there should be some discontinuity in the W and O curves. Since neither happens, the ISS data do seem to support the idea that the extra O is present largely in overlayer positions. Of course if some O penetrated below the surface at the same time that overlayer O was building up, ISS would not be able to detect this.

Studies for $\theta > 0.5$ mL at elevated temperatures up to desorption are sparse. The suggestion is that reconstruction of the surface can occur near the desorption temperature, as it does for $\theta < 0.5$ mL [69]. Since O is also partially lost by desorption and the desorption temperature is lower for $\theta > 0.5$ mL, the coverage can be reduced below 0.5 mL again before reconstruction occurs.

In summary, all the available data suggest overlayer adsorption up to the final saturated coverage at temperatures $\leqslant 300$ K, at least for the ordered phases observed in LEED. Some caution should be maintained, however. Clearly exposure to atmosphere does not stop at overlayer adsorption; penetration does occur very slowly. The suggestion is that it occurs predominantly at defects. One therefore expects that minority "pre-oxidation" of this sort will occur under UHV conditions, but the concentration of oxygen involved may be too small to be easily observed by those techniques which respond equally to all O atoms (XPS, AES) and will also not be observable by the techniques favoring ordered structures (e.g. LEED). These possible minority "pre-oxide" species may well dominate other techniques, however, such as SIMS and ESD (or PSD), where the mechanism of the processes giving rise to the observed signal are still not clearly understood. Bauer and Engel [69] point out that if adsorption is continued beyond $\theta = 1$ mL under UHV conditions and at ~ 1000 K, microcrystallites of WO_3 can clearly be observed by LEED. Small amounts are probably present well before they are observable by LEED. Recent synthrotron work by Brundle et al. [160] tends to support this since chemical shifts in the W(4f) levels intermediate between those appropriate for overlayer O and those for WO_3 layers are seen at quite low exposures. They are, however, of weak intensity, even for photon energies giving the best surface sensitivities.

2.5.10 Greater than 0.5 monolayer O range. Geometric structure

Having established that the available evidence supports the suggestion that at all coverages between $\theta = 0.5$ and 1 mL only overlayers are formed up to at least 1300 K exposure temperatures, or subsequent annealing temperatures, we turn to the interpretation of the geometry of the several LEED patterns observed. The interpretations in terms of adsorption sites are speculative since no dynamical LEED calculations have been performed and no experimental IV data are available. The kinematic analysis taken together with the progressive manner in which one structure converts to another with change in coverage seem sufficient to establish the geometries of the structure, even though the registry of this structure with the W substrate cannot be established [69]. The key to the analysis is knowing the coverages for each LEED structure sufficiently accurately to test the coverage expected for the kinematic model of this structure. For several years these were not known accurately enough, leading to wrong interpretations, but the work of Bauer and Engel [69] finally established the coverages using the WO_2 dosing technique. They demonstrated, using AES, that the O coverage was directly proportional to their WO_2 dosing times, thereby easily establishing the relative coverages, and then they used the optimum p(2 $\times$ 1) LEED structure as an absolute calibration point. Remember that all their structures were formed by annealing to 800–1250 K. The LEED pattern sequence they observed could be interpreted kinematically as in Fig. 16. The compatible coverages for these interpretations are shown in the figure. The Auger p(2 $\times$ 1) calibrated experimental values were in good agreement at 0.72, 0.97

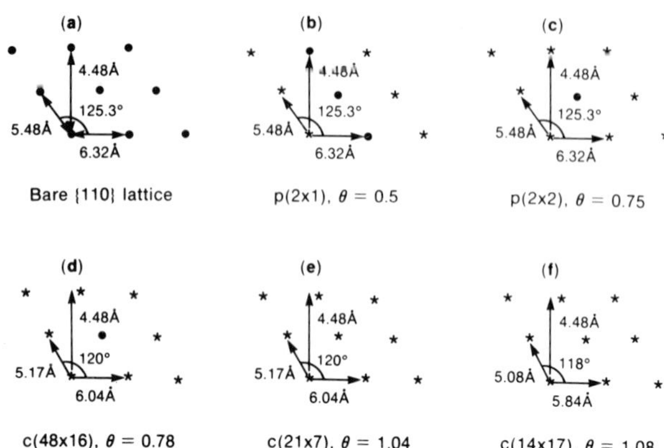

Fig. 16. Structures proposed by Bauer and Engel [69] for the $\theta > 0.5$ mL coverage situations for W{110}/O. The structures are based on a kinematic analysis of the LEED data and compatible oxygen coverages as determined from AES. Dots represent bare sites and stars adsorbate atoms.

and 1.05 for structures (d), (e) and (f), respectively, related to the p(2 × 1) of (b). The LEED patterns themselves, which are complex, were first reported by Germer and May for high O_2 exposure at high temperature and given the descriptions c(48 × 16), c(21 × 7) and c(14 × 7), respectively [23]. Readers are referred to the original paper for the description of these patterns. No analysis was attempted though it was suggested then that the coverages were 0.5–0.7 mL, less than 1 mL, and greater than 1 mL, respectively. Note that in Fig. 16 we have not indicated the registry of the p(2 × 2) structure with the substrate, since there is no dynamical analysis available, but the strong implication is that the same sites as the p(2 × 1) are involved and that simply every other site in the missing row has been filled. The "c(48 × 16)" LEED pattern can be generated from the p(2 × 2) structure by mild annealing, without change in O coverage or change in work function. This suggests only a minor rearrangement between p(2 × 2) and "c(48 × 16)", as is proposed in the kinematic analysis [Fig. 16(d)]. The main change is the imposition of a more hexagonal O structure on the substrate lattice. The c(21 × 7) and c(14 × 7) structures evolved smoothly from the "c(48 × 16)" when the coverage was increased in the WO_2 doping procedure and the kinematic derived structures of Fig. 16(e) and (f) suggest that the main effect in both structures is to fill the empty site at the centre of the O hexagon. The reverse sequence of structures back to the p(2 × 2) can be obtained by annealing at temperatures sufficiently high to desorb some oxygen. It should be emphasized that the registry of these proposed structures with the substrate is unknown and that it is also possible that there are local displacements within the superstructure unit mesh to allow optimum coordination of the O atoms to the W substrate.

Since the p(2 × 1) $\theta = 0.5$ mL structure reconstructs on heating to just below the desorption temperature, one might expect that these structures do the same. As mentioned in the previous section, the desorption temperature is lower in the > 0.5 mL range, however, and since the same complex series is regenerated in reverse on heating the c(14 × 7) structure it seems that desorption occurs prior to any possible reconstruction until a coverage of $\leqslant 0.5$ mL is reached.

2.5.11 Greater than 0.5 monolayer O range. Phase diagram

The few experimental data points for the high-T order–disorder transition for $\theta > 0.5$ mL were shown in Fig. 12 together with the suggestion of Lagally and coworkers of what the features in the phase diagram might look like. The coverage scale is unlikely to be at all accurate above 0.5 mL and in fact it is not clear how it was established. Since the c(48 × 16), c(21 × 7) and c(14 × 7) structures all exist for $\theta > 0.5$ mL in addition to the p(2 × 2) and the p(1 × 1) the authors indicate that the true phase diagram is likely to be extremely complex and no real attempt has been made so far to study it either experimentally or theoretically. About the only sure point that can be made is that the asymmetry about the $\theta = 0.5$ mL position is a strong indication of the importance of three-body interactions above 0.5 mL. It is also likely that the origins of the interactions giving rise to the observed phases above 0.5 mL are more complex than below 0.5 mL. In the latter case we suggested that electronic through-substrate interactions were dominant. Above $\theta = 0.75$ mL, the suggested structures of Fig. 16 would say that this is no longer true and that there is a competition between through-substrate interactions to establish the overlayer geometry and direct O–O interactions which are tending to force a hexagonal O registry on top of a non-hexagonal surface.

2.5.12 Greater than 0.5 monolayer O range. Electronic structure effects

Several brief UPS and XPS studies above a θ value of 0.5 mL exist. In the He II UPS a clear shoulder at $\sim$7.5 eV is observed in addition to the main O(2p) structure at $\sim$6.0 eV for $\theta = 0.5$–0.9 mL. Bradshaw et al. [76] suggests that the two features represent a splitting due to O–O lateral interactions above 0.5 mL. Avery [162] deconvoluted levels at $\sim$6.0 and 7.4 eV in the monolayer coverage regime from the NVV W Auger spectrum. A small feature in the UPS at $\sim$11 eV is observable for $\theta > 0.5$ mL, in addition to the splitting of the main O(2p) peak [76]. It is not clear whether this is related to the 12.5 eV transition observed prominently in ELS for $\theta > 0.5$ mL. As mentioned in Sect. 2.5.6, the latter was attributed by the author [144] to attractive interactions within the p(2 × 1) $\theta = 0.5$ mL islands. Our opinion is that if the 7.4 and 11 eV features in UPS and the 12.5 eV features in ELS are related to lateral O interactions, they represent the repulsive interactions for $\theta > 0.5$ mL.

There is no evidence either in UPS or the NVV Auger for anything dramatic happening to the W d-bands above $\theta = 0.5$, which is a strong support of the suggestion of no reconstruction. In the XPS data the lack of any observation of a $W(4f)$ shifted peak using a laboratory X-ray source [157], and the lack of movement of the 0.5 eV shifted peak observed using synchrotron radiation for $\theta \leqslant 0.5$ on increasing θ above 0.5 [160] also suggest that no reconstruction occurs. In similar studies on W{100}, where reconstruction can be more easily achieved by annealing, the $W(4f)$ shift increased from 0.5 to 1.4 eV on going from overlayer to reconstruction [160]. The ELS $W(4f)$ loss data also support the no-reconstruction hypothesis: no further shift is observed above ~60% saturation coverage [144].

2.5.13 Diffusion of O on W{110}

In trying to establish the elementary processes involved in oxygen dissociative adsorption from kinetic data it is obviously necessary to have information on the mobility of the chemisorbed O atoms as well as information on the mobility and residence time of any molecular precursor. It was assumed in early studies that the O atoms were essentially immobile at temperatures below ~600 K. If this were true the p(2 × 1) LEED pattern observed at temperatures below 600 K would always have to occur by O_2 molecules dissociating at the final p(2 × 1) sites for $T < 600$ K. This is clearly not true, as evidenced by the phase diagram of Fig. 12. Also, adsorption at low T (~100 K) produces no LEED pattern up to saturation coverage (0.6–0.7 mL) but on warming, a pattern becomes evident above 200 K [115], demonstrating that long-range order is developing by O atom diffusion at that temperature, though if a local p(2 × 1) arrangement already exists the O atoms do not have to move far. Of course O adatom mobility might be expected to depend strongly on θ, i.e. the availability of suitable hopping sites. Several investigations of O adatom diffusion on W{110} flat surfaces have been carried out and are summarized below.

Butz and Wagner [142] measured the coverage-dependent diffusion coefficient, D, in the 1060–1180 K range for O on W{110}. They started with a sharp boundary between sections of the surface having different initial oxygen concentrations and determined the concentration profiles across the boundary as a function of annealing time at the diffusion temperature. This was done by measuring work function changes with a movable Kelvin probe with a lateral resolution of ~50 μm. The work function was related to coverage through AES measurements. At 1060°C they found D was $\sim 3 \times 10^{-7}\,\mathrm{cm^2\,s^{-1}}$ at $\theta = 0$ to ~0.3 mL, peaked to a maximum of $\sim 2 \times 10^{-6}\,\mathrm{cm^2\,s^{-1}}$ at $\theta \sim 0.4$ mL, then decreased sharply above $\theta \sim 0.4$ mL to $\sim 2 \times 10^{-8}\,\mathrm{cm^2\,s^{-1}}$ at $\theta \sim 0.9$ mL. At 1180°C the values were about a factor of 5 higher. A constant E_{diff} of $27 \pm 2\,\mathrm{kcal\,mol^{-1}}$ was derived from Arrhenius plots for $\theta > 0.4$ but < 0.9 mL. Below $\theta = 0.4$ mL the data were too scattered to obtain E_{diff} values. At the temperatures used in this study, the O is in the lattice gas–lattice fluid phase (Fig. 12) at all coverages. The sharp decrease in diffusion constant above $\theta \sim 0.4$ mL can then be rationalized as being due to the necessity of O atoms moving into sites with repulsive adatom interactions during the diffusion process. The peaking at $\theta = 0.4$ mL is less understandable. Chen and Gomer [112] determined diffusion constants and activation energies using a field emission microscope technique. Here, microscopic surface areas are used and diffusion over Å rather than μm are considered. This allows measurements to be made at lower temperatures and removes the possible effects of defects and competing desorption processes. The measurements also refer to diffusion on homogeneous surfaces rather than at sharp concentration boundaries. Temperatures between 500 and 770 K were used. For 0.56 mL (the highest θ reached) an E_{diff} of ~ 22 kcal mol^{-1} is found, in agreement with the Butz and Wagner [142] values for θ between 0.4 and 0.9 mL. Extrapolating the linear Arrhenius plots to a temperature of 1050 K, the minimum temperature used by Butz and Wagner, the diffusion coefficients are found to be 2–3 orders of magnitude smaller than those found by Butz and Wagner, i.e. the pre-exponential diffusion constant, D_0, is 2–3 orders of magnitude smaller. The reason for this discrepancy is not clear. Chen and Gomer also extracted E_{diff} values for θ values between 0 and 0.5 mL. Below $\theta \approx 0.3$ mL a much lower constant value of ~ 14 kcal mol^{-1} was found. Above $\theta \approx 0.3$ mL the value rose towards the $\theta = 0.56$ mL value. Below $\theta \approx 0.3$ mL, in the temperature range considered, E_{diff} represents essentially isolated adatom diffusion, since the dilute lattice gas situation obtains (see Fig. 12). Around $\theta \approx 0.3$ mL, a phase change occurs, the surface now being covered entirely by the p(2×1) plus statistical vacancies phase. E_{diff} therefore rises because of the necessity to overcome the lateral adatom interaction involved.

Bowker and King [143] investigated diffusion in the W{110}/O system between 950 and 1350 K using a molecular beam dosing arrangement and secondary electron emission changes with a scanning electron beam as a means of following O concentration changes at the dosed O patch boundary. The principle of the measurements are therefore very similar to those of Butz and Wagner. They obtained an E_{diff} value at $\theta = 0.25$ mL of 23 kcal mol^{-1}, rising to 25 kcal mol^{-1} at $\theta = 0.5$ mL, i.e. in good agreement with the higher coverage values of Chen and Gomer and Butz and Wagner. The low value of 14 kcal mol^{-1} observed by Chen and Gomer was not found in this work, but it is possible, given some (unspecified) error bars in the relative coverage determination in the different experiments, that the Bowker and King measurements do not really extend to sufficiently low coverage to observe the isolated atom diffusion behavior. The pre-exponential D_0 values were similar to those of Butz and Wagner, i.e. considerably higher than those of Chen and Gomer.

Summarizing the diffusion results, then, there is good agreement in the determined E_{diff} values for θ between ~ 0.3 and ~ 0.9 mL, values ranging from 22 to 27 kcal mol^{-1}. This constant value is considered to represent diffusion in the situation where the adatom lateral interactions must be overcome. The determined diffusion constants, D_0, and hence the actual diffusion coefficients at any temperature over this coverage range do not agree so well and are, anyway, coverage-dependent with values ranging from 10^{-4} to

$0.4\,cm^2\,s^{-1}$. The reasons for these discrepancies are not understood. For low coverages above ~ 450 K, the one determined value of 14 kcal mol^{-1} is considered to represent diffusion of isolated adatoms in the absence of lateral interaction effects. Actual diffusion rates are not higher at these lower coverages, however, since D_0 apparently also decreases dramatically. Again, the reasons are not understood.

2.5.14 Stepped W{110} surfaces

In the preceding sections several references were made to the strong effects of defects on the W{110} surface behavior; for instance in the kinetics and "θ_{lim}" values and in the formation of many small p(2 × 1) islands instead of one large one. In this section we consider the work that has been done on {110} surfaces containing characterized defects. Three types of such surfaces have been studied; flat {110} surfaces on which W adatoms have been deposited [142], W{110} surfaces which have been deliberately misoriented by a few degrees to produce particular types of steps and {110} terraces [104, 115, 149], and higher index surfaces which essentially consist of two surfaces, one of which is the {110} plane.

The most detailed study is that of Engel et al. [104]. They compared kinetics, work function behavior, desorption, and p(2 × 1) O domain growths for W{110} surfaces miscut by $\sim 2°$ and $\sim 5°$ such that steps run parallel to a close packed direction, as shown in Fig. 17. From the spot splitting observed in LEED, the average terrace widths were found to be 27 and 11 atoms, respectively, with an average single atom step height.

The comparison of the S, θ characteristics of the stepped and flat surfaces was rather crude, with only a $\pm 10\%$ relative accuracy claimed, but no difference between the stepped and flat surfaces to within $\pm 15\%$ was found. Engel et al. [104] concluded from this equality that the adsorption rate was determined by the lifetime of the precursor molecule which was independent of the step density. They did observe strong effects on the p(2 × 1) LEED formation, however (see below). Taken together with the fact that even their flat surface S, θ characteristics are not in very good agreement with those

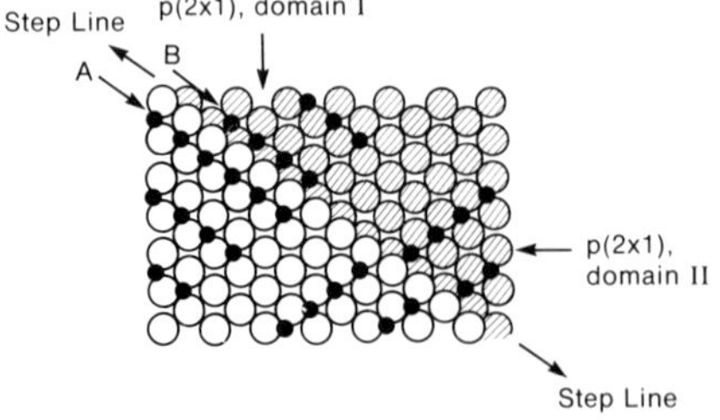

Fig. 17. Stepped W{110} surface with close-packed step edge. Large open circles = upper W{110} plane; large hatched circles = lower W{110} plane; small filled circles = oxygen atoms. Engel et al. [104] assumed the atomic oxygen to be located in the threefold site, but for interpretation of the effect of the step this is not a required assumption.

of Wang and Gomer [107] or their own earlier data [68], and that subsequent studies on (different) stepped {110} surfaces show very strong effects in the S, θ characteristics, these results must be considered somewhat ambiguous. The type of terrace edge exposed in this study is close-packed and it may be that the effect of the steps on the kinetics was therefore comparatively small and therefore undetected (see later). The $\Delta\phi$ measurements on the 11-atom terrace crystal indicated a sharp increase at very low coverages (< 0.005 monolayer) of ~ 0.1 eV followed by a plateau to ~ 0.15 monolayers and then a linear increase with coverage to $\Delta\phi \sim +0.5$ mL at $\theta = 0.5$ mL. Since steps of adatoms decrease the work function of a flat {110} surface, the initial sharp rise was attributed to adsorption at step sites, resulting in an electronic smoothing which returns ϕ to its flat clean surface value. The effect on $\Delta\phi$ from there on is approximately the same as for a flat surface. These results also have significance for "flat" {110} studies. They suggest that, for a crystal which is slightly misoriented, the minimum in the $\Delta\phi$ curve could be missed, and the $\Delta\phi$ value at $\theta = 0.5$ mL could be several tenths of an electron volt higher than it should be. This may explain the discrepancies referred to in Sect. 2.5.3 between the different $\Delta\phi$ studies for "flat" {110} surfaces.

The growth of the p(2 × 1) islands as a function of θ is the most interesting aspect of the Engel et al [104] study. From the presence and orientation of LEED spot splitting in the (1/2, 1/2) spots and the absence of such a splitting in the ($\bar{1}$/2, 1/2) spots, Engel et al. showed that domains of types I and II (Fig. 17) were formed and concluded that for type I there was a fixed phase relationship from one terrace to the next, whereas the relation was random for type II. They also showed that the I_{max} LEED intensities for I were greater than for II, the effect increasing with T and with decreasing terrace width (a factor of 30 for the 11-atom terrace crystal at 460 K), and that I_{max} came significantly below $\theta = 0.5$ mL for both domains with II reaching a maximum first at $\theta = 0.23$ mL. A distorted p(2 × 2) structure was observed to start for this domain after the p(2 × 1) maximum. The authors offered the following explanation, which is consistent with the LEED observations and also the $\Delta\phi$ data. Initial dissociative adsorption for domain II occurs randomly from the diffusing precursor, resulting in no fixed-phase relationship from terrace to terrace. Initial adsorption for domain I takes place at step sites (positions along lines A and B, Fig. 17); domain I islands then grow outermost from the steps. This would explain why the initial LEED intensity from domain I is lower than from II (the "islands" are initially one-dimensional) and why there is a fixed phase relationship from one terrace to the next for domain I. The values of I_{max} for domains I and II are considered to pass through a maximum below $\theta = 0.5$ mL because the p(2 × 1) structure becomes unstable with respect to the p(2 × 2) earlier than for the flat surface. While the interpretations offered seem consistent with the data they are not unique and leave a number of puzzling questions. Given what we know about the phase diagram and the lateral interactions that have to be overcome to form the p(2 × 2) structure, why does the presence of steps apparently destabilize p(2 × 1) islands against the p(2 × 2) structure? Also, why do domain I islands apparently only grow out from the terrace edges? This seems to imply that precursor molecules dissociating in the center of the terraces always go into type II domains. An alternative adsorption mechanism, which removes some of the anomalies, will be presented after consideration of the other stepped surface work.

When the stepped surfaces in the study of Engel et al. [104] are heated to > 1000 K for θ > 0.5 mL, the terraces agglomerate into much wider terraces (> 100 Å from the LEED angular profile) and the p(2 × 2) structure of domain type II dominates. Heating to 1560 K causes desorption and the p(2 × 1) structure is regenerated, but type II are now as numerous as type I since the steps are now very far apart and exert little relative influence. This desorption temperature is about 200 K lower than for the flat surface. As a consequence of the reduced desorption temperature, reconstruction which competes with the desorption on the flat surface is not achieved easily on the stepped surface.

Besocke and Berger [149] studied the influence of stepped surfaces on the adsorption kinetics by using a polyhedral W crystal which had a flat {110} surface and four stepped surfaces. Two exposed stepped surfaces were parallel to the (100) direction, and two parallel to the (110) direction as in Fig. 18(a). The former had {110} average terrace lengths of 6 and 12 atoms (Sections L and R), the latter 8 and 16 atoms (Sections T and B). O Auger intensities were measured as a function of exposure and the data converted into S/S_0 versus θ curves (Fig. 19) by assuming θ = 0.5 mL at 6 L exposure for the flat surface. This calibration is almost certainly wrong, their point at 40 L exposure being the correct θ = 0.5 mL calibration. A corrected θ scale is also shown in Fig. 19. The fact that the relative S_0 of the {110} flat surface is only 0.28 of the most reactive stepped surface is compatible with an absolute value of S_0 of 0.45 as found by Wang and Gomer [107], as mentioned in Sect. 2.5.2. Madey [115] repeated these measurements on a nearly identical polyhedron, finding qualitatively similar O Auger versus θ curves except that the coverage scale, calibrated by actual observation of the optimum p(2 × 1) LEED pattern on the flat surface, more nearly coincided with the corrected scale of Fig. 19. Quantitatively there are significant differences in S_0, however. The S_0 values obtained by Madey and by Besocke and Berger are compared in Table 6. Madey's value (S_0 = 0.38) for the flat surface is somewhat nearer the Wang and Gomer value of 0.45. The important factors are illustrated by both sets of data, however; S_0 is much higher for stepped surfaces and θ_{lim} is nearer unity than 0.5 mL. The kinks that occur around 0.5 mL in the stepped curves also suggest that either two independent processes are occurring or a phase transition occurs near θ = 0.5 mL. Madey also took O AES versus exposure data at 100 K. For the stepped surfaces, S/S_0 decreased less rapidly than at 300 K, but θ_{lim} was hardly affected. For the flat surface, S_0 was higher than at 300 K, remained high, and θ_{lim} was significantly higher than 300 K (in agreement with Wang and Gomer). In the light of

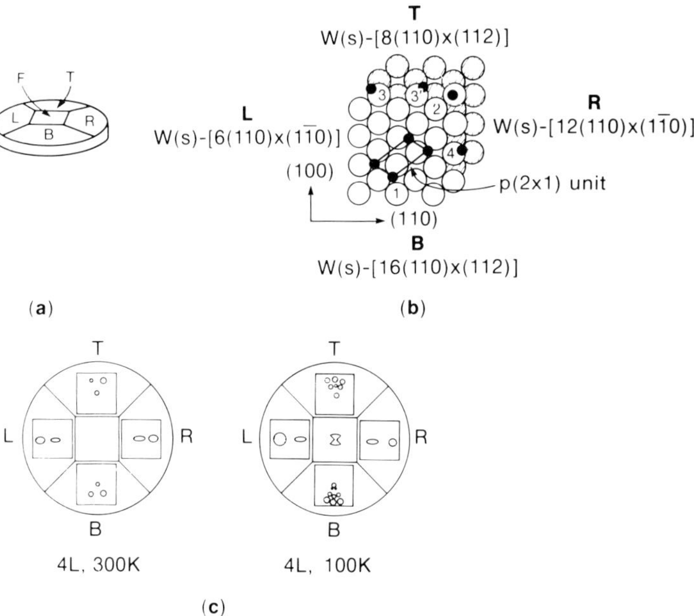

Fig. 18. (a) Sketch of polyhedral W crystal used by Besocke and Berger [149] and by Madey [115] to study the relative kinetics of oxygen adsorption on flat (F) and stepped (L, R, T, B) surfaces. (b) Atomic structure of polyhedral W surface used by Besocke and Berger [149] and Madey [115]. The nomenclature indicates the widths and types of the terraces and steps on the L, R, T and B facets used by Besocke and Berger [149]. The Madey values are slightly different (see text). The small filled circles represent oxygen atoms on the flat terrace [p(2 × 1) structure, assuming threefold adsorption site] and sites at the steps which are possible candidates for the ESDIAD patterns observed by Madey [115] (see text Sect. 2.8). (c) ESDIAD patterns observed from the five surfaces of the polyhedral crystal in (a) [115], following oxygen exposure at 300 and 100 K.

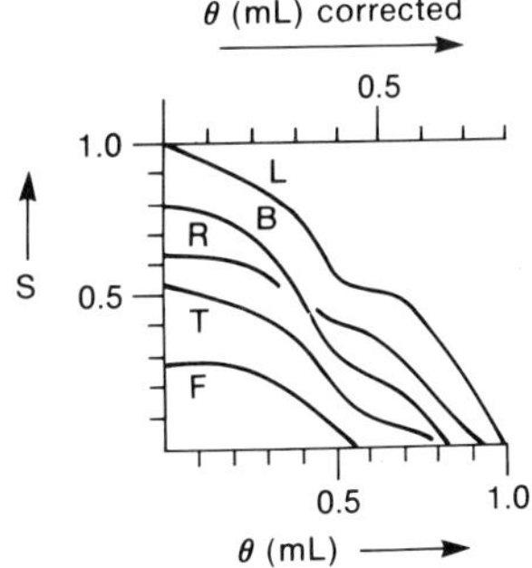

Fig. 19. S versus θ for the five facets of the polyhedral W surface of Besocke and Berger [149]. The notation corresponds to that in Fig. 18.

TABLE 6

Relative initial sticking probabilities[a], S_0, for oxygen on the flat and stepped surfaces of a W polyhedral crystal at 300 K

Notation in Figs. 18 and 19	Orientation		S_0	
	Ref. 149	Ref. 115	Ref. 149	Ref. 115
F	W{110}	W{110}	0.28	0.38
R	W(S)-[12{110} × {1$\bar{1}$0}]	W(S)-[10{110} × {1$\bar{1}$0}]	0.65	0.79
L	W(S)-[6{110} × {1$\bar{1}$0}]	W(S)-[6{110} × {1$\bar{1}$0}]	1.0	1.0
T	W(S)-[16{110} × {112}]	W(S)-[15{110} × {112}]	0.53	1.0
B	W(S)-[8{110} × {112}]	W(S)-[8{110} × {112}]	0.8	0.97

[a] The maximum value observed is assumed to be unity.

this kinetic data on stepped surfaces, Madey suggested the following model for the adsorption process: The precursor-adsorbed molecule can dissociate either at a pair of bare W sites on the {110} terrace or at step sites. Limited diffusion at 300 K results in the p(2 × 1) islands on the terrace (both from dissociation on the terrace sites and at steps followed by diffusion away from the steps). Madey points out that the suggested model, in which steps act as "dissociative" sites which "feed" the terraces with O_{ads}, but are not themselves consumed, is similar to that suggested by King and co-workers [163] for N_2 adsorption on W{110} except that, for nitrogen, the entire dissociation occurs at steps. It is implicit in Madey's model (but not mentioned by Madey) that the increase in S_0 for stepped, compared with flat, W{110} surfaces is due to the higher probability for dissociation of the precursor at the stepped surfaces. The value of S_0 for the {110} surface is $\leqslant 0.45$ at 300 K*, but for the stepped surfaces is unity or close to unity. If S_0 is unity, it implies that at zero coverage every molecule striking the surface at a terrace site either dissociates (maximum probability $\leqslant 0.45$ at 300 K) or somehow reaches a step where it dissociates with unit probability. Thus the reflectivity on the terrace region must be zero. This flatly contradicts the explanation for $S_0 = 0.45$ on flat surfaces given by Wang and Gomer and discussed in Sect. 2.5.2, namely that the reason for S_0 being less than unity at 300 K is due to a large reflection probability. With such a suggestion, S_0 on a stepped surface could never be unity.

A radical suggestion that overcomes the dilemma is to abandon the usual starting assumption of "classical" precursor models of instantaneous accommodation of the molecule into the precursor state, i.e. the molecule either reflects or transfers all its energy to the surface at the site it strikes and is trapped into the precursor. Thus, in this "classical" precursor model T_g can only affect the value of S_0 (by altering the reflectivity probability) and can have no effect on the subsequent mechanism. If this concept of instantaneous energy accommodation is abandoned we can suggest than an incoming O_2 molecule striking a {110} terrace never undergoes a first bounce reflection, but is partially accommodated (normal component of momentum) with the surface. It can then "skate" across the surface for long distances before either becoming fully accommodated into the precursor state or returning to the gas phase. For stepped surfaces of 16 atoms terrace width (the largest in the W{110} stepped surface studies) the partially accommodated molecule may reach a step and dissociate there if it does not dissociate on the terrace on the way. For a flat surface, S_0 is then much less than unity, not because the reflectivity is significant, but because there is a small activation barrier to dissociation between the precursor state and the O_{ads} state. In this situation, with an activation barrier on the flat surface but the possibility of the O_2 partly accommodated molecule traveling a long way

*Actually, it is probably $\leqslant 0.28$, the lowest of the values reported here. Umbach [164] also gives this valuo.

across the surface, it is likely that a significant contribution to the S_0 experimental value of $\leqslant 0.45$ at $T_s = 300$ K is dissociation taking place at defects. Thus the defects on the nominally flat surface act as "feeders" of O_{ads} to the terraces, just as the steps do on the deliberately stepped surface.

Madey's results [115] and his suggested model for the "open" type of steps differ from those of Engel and Bauer [69] on the close-packed steps, where no significant effect on the kinetics were found and the suggestion was made that O atoms stuck at steps rather than diffused from them. Kleban and Flagg [165] performed Monte Carlo calculations for O on the Engel et al.-type stepped surface (10 atom terrace) using the flat {110} adatom interaction parameters from Ching et al. [123] as input and varying the signs of the changes in adsorption energy at terrace edge sites. They then calculated the intensities of the LEED domains I and II and the spot-splitting characteristics as a function of these variables and showed that it is not necessary to have domain I formed with the terrace edge sites always occupied to explain the data. In fact a much better fit was obtained if those sites were not occupied, i.e. the adsorption energy at the terrace edges is lower than on the terrace. The studies by Engel et al. and Madey then come more into agreement: terraces are preferable equilibrium positions for O atoms for *both* types of stepped surfaces. The difference in the kinetics and the θ_{lim} obtainable between the different types of stepped surface would then imply that, whereas the open terrace edge sites act as sites with greatly enhanced probability for dissociation, the close-packed edge sites offer no significant increase of dissociation probability, though they are effective in causing type I domains to dominate.

Butz and Wagner [142] in their study of oxygen atom diffusion on W{110} surfaces evaporated W onto the flat surface, adsorbed oxygen and then annealed to 1000 K briefly as a way of obtaining a high O coverage easily. Such surfaces appear to be identical to flat {110} surfaces which have achieved the same O coverage by prolonged 300 K exposure. During the work, however, Butz and Wagner also determined the S, θ characteristics for both the flat surface and the surface with W atoms deposited on it at 300 K. The latter apparently has small W islands of $\sim$ 10 Å diameter present. The flat-surface S, θ curve shape is in good agreement with the Engel et al. and Gomer et al. results and the W island exhibits an S, θ curve very similar to that of the reactive surfaces shown in Fig. 19. The value of S_0 was determined to be just over twice that for the flat surface, so if the latter is set at 0.45 the value for the W island covered surface is 1. Again we see that defects increase S_0 dramatically and allow $\theta = 1$ mL to be reached easily, presumably with the defects acting as feeders of O atoms to the terraces.

Bowker and King [143] studied the diffusion characteristics of oxygen on flat {110} surfaces; {320} surfaces which consist of {110} terraces (two unit meshes wide) and {100} steps; and on the {411} surface. The method and the {110} flat surface results were discussed earlier. For the {411} surface, diffusion rates were found to be much lower than on the flat surface, as might be expected because of its open rough structure. On the stepped {320} surface, it was found that at the temperature required to monitor the diffusion by the method used ($> \sim 1100$ K) reconstruction or O penetration of the surface occurred in a similar manner to the flat {100} results (see Sect. 2.6.5). Bowker and King also measured the S, θ characteristics of the {320} surface using a molecular beam technique. Their curve shape was again similar to that of the most reactive stepped {110} surface in the Besocke and Berger study (Fig. 19), with S_0 determined as 0.92 ± 0.03.

Summarizing this section on stepped W{110} surfaces, we can say that, in addition to providing information on the role of the step-atoms in determining the adsorption characteristics, the results also impose some severe restrictions for the adsorption mechanism on the flat surfaces. The overall model we have for the elementary processes at 300 K is as follows. For zero coverage surfaces, O_2 molecules striking the {110} terraces are partially accommodated into the precursor molecular state with immediate access to the steps, even if they impact at the terrace centers, owing to the energy they still possess. The probability for dissociation at {110} terrace sites is presumed to be low, the process being activated, and a significant contribution to the dissociation process occurs by the "hot" precursor reaching either an unintentional (random) defect site on the terraces or by reaching the steps. Both these types of site have to be non-close-packed since we know that close-packed steps have little or no effect on the kinetics and that the "θ_{sat}" achievable before S drops dramatically is still close to 0.5.

These suggestions imply that the variability in experimental S_0 on flat surfaces (0.28–0.45) may be genuine and in fact represent different defect densities. It also implies that these active defect sites and active deliberate steps do not get used up in a one-to-one manner by the dissociation process since $S \sim S_0$ up to $\theta \approx 0.3$ on both flat and stepped surfaces, i.e. the O atoms move away from the steps/defects as proposed by Madey [115] and by King and co-workers [163] for N_2 adsorption on W. There can be at least two reasons why S starts to decrease above $\theta \approx 0.3$. (i) With increasing coverage the p(2 × 1) island structures on the terrace begins to "crowd" the active sites and affect their ability to dissociate, possibly by trapping freshly dissociated O atoms actually at the active sites. (ii) The region of "immediate access" to the incoming O_2 molecule may be less on p(2 × 1) covered regions than on bare W, i.e. as the number and size of p(2 × 1) islands grow, O_2 molecules striking the islands fail to access the active defect sites before returning to the gas phase. One still has to explain why on highly (open) stepped surfaces θ goes to 1 mL whereas on flat surfaces and close-packed step surfaces S becomes very small at $\theta = 0.5$ mL. A plausible explanation is as follows. First we have not claimed that dissociation on {110} terraces is zero, just that it is a much lower probability activated process. We know that, after completion of the p(2 × 1) structure at $\theta = 0.5$ mL, only adsorption sites with repulsive adatom interaction are available for the location of further O atoms. The activation energy for dissociation at such sites is therefore probably much higher than for bare {110} terraces and so any contributions to S from terrace dissociation drops out at $\theta = 0.5$ mL. Only dissociation via active defects followed by diffusion away remains. The diffusion studies on flat {110} surfaces have shown that, above $\theta = 0.4$ mL, O diffusion coefficients drop by two orders of magnitude, presumably owing to the necessity to move O atoms into unfavorable sites once the p(2 × 1) structure is completed. So even if some active dissociation sites remain at $\theta = 0.5$ mL on flat surfaces, only very small areas around them can achieve a higher coverage than 0.5 mL because of the now very limited diffusion capability of O atoms. On the highly stepped surfaces, O_{ads} diffusion is required over only a few lattice spaces from an active step to access the entire terrace surface.

Clearly a lot of further study varying T_s, T_g, and the concentration and nature of the steps needs to be done before any further detailed understanding is achieved.

2.6 W{100}

The {100} surface behaves quite differently from the {110}. Sticking probabilities are higher ($S_0 \approx 1$) [77, 107] and the θ_{lim} reached easily is greater than 1 mL, rather than 0.5 mL. The LEED behavior is very complex, in contrast to the simple p(2 × 1) structure observed for the {110} surface below $\theta = 0.5$ mL. There is no body of data existing for the W{100}/O phase diagram, though a high-temperature order–disorder transition has been claimed [90]. This transition is for a surface which has already reconstructed at a lower T, however. Because of the complexity of the LEED data and the complete lack of any IV data and dynamical calculations, it is not yet possible to give complete interpretations for all the structures between zero and saturation coverage for adsorption at room temperature or for annealing to temperatures below reconstruction temperatures. Despite this we can say that ordered adsorption below a critical temperature (coverage-dependent) is overlayer only, even up to $\theta = 1.4$ mL [77]. At the higher coverages, O_{ads} tries to impress a close-packed hexagonal structure on the W lattice [77]. The proof of overlayer adsorption rests largely, as for W{110} on the consistent interpretation of the $\Delta\phi$ [77] and ISS [105] data. In the {110} case, however, this data was strongly supported by the kinematic analysis of all the LEED structures and their consistency in terms of logical development of one from another with the expected coverage values. Because this is partly lacking in the {100} case the proof of overlayer adsorption only is weaker than for W{110}, though we believe it to be correct. One should bear in mind, however, that as with W{110} the possibility of minority species in reconstructed, penetrated or even oxide crystallite environments always exists, and some techniques may be particularly sensitive to these species. Following completion of most of the work reviewed here, studies on the clean W{100} surface have shown that it reconstructs to a c(2 × 2) structure below 400 K [10]. Most likely the reconstruction is of the displacive type, involving small lateral distortions of the W{100} lattice together with possible buckling to optimize the bonding arrangement. It has been demonstrated that H adsorption stabilizes this displacive reconstruction to temperatures well above 400 K [166, 167]. It is very possible, therefore, that small displacive reconstructions also occur during oxygen adsorption, although no direct evidence of this has been claimed. In fact with two exceptions [77, 168] all the W{100}/O_2 studies reported here fail to mention any reconstruction of the clean surface prior to adsorption.

No stepped {100} surfaces have been studied and since S_0 is already ~1 for flat surfaces it is not likely that much additional information on the adsorption mechanism could be extracted from such studies. A lot of ESD, ESDIAD, and SIMS work has been carried out on the {100} surface. We do not review this in this section but defer it to the separate section on these topics since, as for the {110} surface, we consider that the value of these studies is more in helping to understand the nature of the techniques on already defined surfaces rather than in providing information on unknown adsorption situations.

2.6.1 S, θ measurements

Wang and Gomer [107] have determined S, θ curves for T_s = 90–1050 K and T_g = 110–440 K by the reflection detection method described in Sect. 2.5.1. Representative results are shown in Fig. 20. In comparison with the W{110} results of Fig. 9, the most striking difference, in addition to the higher S and θ_{lim} values, is the obvious division of the S, θ characteristics into two sections, below and above θ = 0.5 mL. Only for a combination of low T_s and low T_g does this not occur. Above a T_s value of 200 K, variation of T_g has virtually no effect on the S, θ curves. Since the concentration of O_2 on the surface does not become a factor for W{110} until a T_s of $\leqslant$25 K we suggest it will not be a factor at T_s = 90 K for W{100}, the lowest value

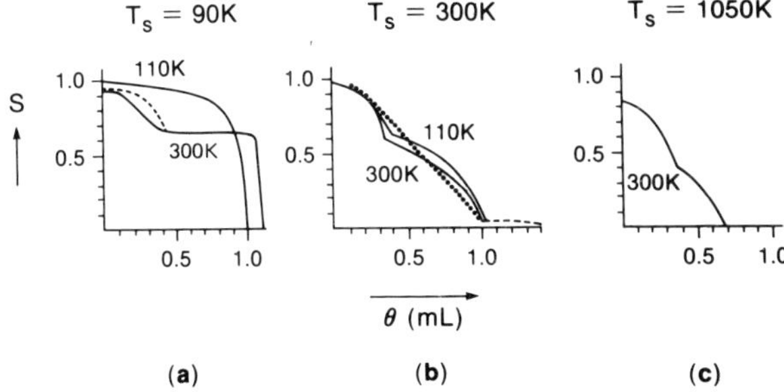

Fig. 20. S versus θ for oxygen adsorption on W{100}. T_g values are marked on the individual curves. (a) T_s = 90 K [107]. The broken curve is what we would have anticipated at T_g = 300 K (see text). (b) T_s = 300 K [107]. The broken curve approximately represents data from ref. 77, showing that the saturation coverage is close to 1.4 monolayers. The dotted curve is data at T_g = 300 K from ref. 170. (c) T_s = 1050 K [107].

determined. As was the case for W{110}, Wang and Gomer interpret their data in terms of large reflectivity changes as a function of O coverage on the surface. For example, the $T_s = 90\,K$, $T_g = 300\,K$ curve [Fig. 20(a)] is interpreted as r increasing strongly from an initial value of zero coverage up to a value of ~0.4 at 0.5 mL, rather than a constant r of zero and a precursor lifetime-limited fall-off in S with increasing θ. They do this because, within their model, which only allows variation in r and precursor lifetime, it is obvious that once above $\theta = 0.5\,mL$ at $T_s = 90\,K$ and $T_g = 300\,K$, the precursor lifetime *is* sufficient for the available dissociation sites always to be found, hence the constancy of S above $\theta = 0.5\,mL$. For a θ value between 0.5 mL and saturation, the effect of increasing T_s is then as expected: the precursor lifetime is gradually reduced and the shape changes towards Langmuir-like kinetics [Fig. 20(b) and (c)]. The effect of decreasing T_g at low T_s [Fig. 20(a)] is not really explainable with this model, however. It requires the very unsatisfactory set of conditions that r increases rapidly with θ between $\theta = 0$ and 0.5 mL for $T_g = 300\,K$ at all values of T_s, whereas it does not increase at all for $T_g = 100\,K$ and low T_s (90 K), and yet for $T_g = 100\,K$ and high T_s (300 K) it does increase strongly. We consider that, since the explanations based on reflectivity changes and an instantaneous precursor accommodation model were shown to be inadequate for W{110}, similar explanations for W{100} are also in error. The missing ingredients from the argument of Wang and Gomer [107], in addition to the necessity of abandoning the instantaneous precursor accommodation concept, are a proper consideration of the effect of order in the chemisorbed layer and the admittance that over some coverage ranges adsorption can become an activated process.

Let us consider the $T_s = 300\,K$ adsorption first. At this T_s it is known from LEED that an ordered structure forms above $\theta \approx 0.3\,mL$ and is sharp at $\theta = 0.5\,mL$ [77]. Unfortunately, the real-space interpretation is not straightforward since it has been shown that the LEED pattern consists of a composite of several independent structures (see Sect. 2.6.3). Since it is sharp at $\theta = 0.5\,mL$, however, we can assume that the dominant lateral interactions are such that each adsorbed O uses up two available W sites [cf. the p(2 × 1) structure on W{110}]. The shape of the S, θ curve between $\theta = 0$ and 0.5 mL is then consistent with $r = 0$ and dissociation via the lifetime-limited diffusing precursor state until all such dissociation sites are used up at $\theta = 0.5\,mL$. If these sites were the only possibility for dissociation to occur S would reach zero at $\theta = 0.5\,mL$. Note also that for this 0–0.5 mL stage the dissociation process must be unactivated since S_0 is essentially unchanging at unity for all T_g and T_s values up to at least 400 K. Thus this process corresponds to the dissociation on the W{110} surface which takes place only at open steps or defects. At $\theta = 0.5\,mL$ no unactivated sites remain. The dominant structure at this coverage has been suggested to be a p(2 × 1) [77]. Since S does not become zero at $\theta = 0.5\,mL$, the remaining empty sites must be available for dissociation, but with a significant activation energy such that S_0 for these sites is only ~0.6. Consistent with the interpretation of the

W{110} surface and the unactivated 0–0.5 mL region for W{100} we suggest that r is zero on the $\theta = 0.5$ mL covered surface and the precursor has immediate access to a wide area (i.e. it is not instantaneously accommodated). S_0 is not unity because of the activation energy to dissociation. Assuming equal pre-exponential factors, an S_0 value of ~ 0.6 suggests that the activation energy, E_a, for dissociation is approximately equal to the desorption energy, E_d, from the precursor state. The $\theta \geqslant 0.5$ mL regime is therefore also a precursor search-for-sites process, just as the $\theta = 0$–0.5 mL is, the only difference being the type of site involved. Since we have suggested that "flat" {110} surfaces have only an activated dissociation process but {100} surfaces have both unactivated ($\theta \leqslant 0.5$ mL) and activated dissociation processes ($\theta \geqslant 0.5$ mL), we might expect that the S, θ curves for the {100} surface for $\theta > 0.5$ mL should behave very similarly to those for {110} from $\theta \geqslant 0$ mL. Inspection of Figs. 19 and 20 shows that this expectation is born out. Further, we might expect that stepped {110} surfaces, which have both unactivated and activated sites possible might show similar (S, θ) behavior to W{100}. Inspection of Figs. 19 and 20 show that this is also true.

If T_s is lowered to 90 K, keeping T_g constant at 300 K, the only change should be that S does not become precursor lifetime-limited until somewhat higher values of θ because the precursor surface lifetime is greater at $T_s = 90$ K than at 300 K. This occurs in the $\theta \geqslant 0.5$ mL range but there is no obvious change in the $\theta = 0$–0.5 mL range. The "anticipated" curve at 90 K [see Fig. 20(a)] and the measured curve do not differ by very much, however. Note that at $T_s = 90$ K and for $T_g \geqslant 300$ K the effect of one O_{ads} excluding two W is still apparent, indicating that short-range order still exists. For $T_g = 110$ K this effect disappears; S shows no sign of dropping as 0.5 mL is approached. Within the model we are proposing this implies that the short-range order is considerably reduced (local θ_0 can become much larger than 0.5 mL) such that the number of unactivated dissociation sites does not become zero at $\theta = 0.5$ mL. This is quite different from W{110} where no such T_g effect on short-range order is apparent. This strong T_g effect can surely only be explained by suggesting that, for low T_g on W{100}, there is insufficient energy available to O_{ads} atoms from the dissociation process for them to make the few hops necessary to order up on a local scale, hence there is no discontinuity at $\theta = 0.5$ mL. This is rather startling because again it suggests that the usual assumption of instant energy accommodation is incorrect. With the usual assumption, T_g cannot affect such things as the ordering process in the resulting O_{ads} layer. The difference between W{100} and W{110} in their response to T_g for ordering of O_{ads} is consistent, however, with the expectation that O_{ads} in the open W{100} fourfold site is expected to have a deeper potential well than O in the center site on the close-packed W{110} surface. It should, therefore, take more energy for O_{ads} to diffuse on W{100} than W{110}, which is known to be the case [145, 169], and so the effects of a limited diffusion of O_{ads} should be felt more easily on W{100}.

Bauer et al. measured sticking probabilities for $W\{100\}/O_2$ by AES as part of their detailed LEED studies [77]. Their coverage calibration point was the high temperature reconstructed p(2 × 1) O pattern which they assumed corresponded to $\theta = 0.5$ mL. The 300 K data are basically in good agreement with those of Wang and Gomer, except that Bauer et al. chose an interpretation in terms of discrete steps in S versus θ rather than smooth curves. Bauer et al., however, found almost identical S, θ curves at 1050 K to 300 K, in sharp contrast to Wang and Gomer who observed lower S values and a lower limiting θ. The reason for this discrepancy is unknown and the true high T_s S, θ must remain in doubt. Bauer et al. carried out adsorption to higher exposure values than did Wang and Gomer, and showed that saturation coverage was $\sim\theta = 1.4$ mL at about 45 L exposure, i.e. considerably higher than one might think from examining Fig. 20. This indicates that S does not go smoothly to zero in Fig. 20 but that there is another break in the curves around $\theta = 1$ mL as schematically indicated by the broken line in Fig. 20. Thus, even above $\theta = 1$ mL there are still sites available for dissociation. We will refer to the $\theta > 1$ mL region again in the LEED section but it is interesting to note here that at $\theta = 1.4$ mL on the W{100} surface the O atom density is identical to that for $\theta = 1.05$ mL on the W{110} surface (the saturation value) at 1.43×10^{15} atoms cm^{-2}. This fact seems to have gone unnoticed before.

Several other S measurements [145, 170, 171] at $T_s = T_g = 300$ K had been made prior to the two studies discussed above. They tend to show a smoother S, θ curve than that of Wang and Gomer in Fig. 20 with only a hint of the two separate sections, though an S_0 value of ~ 1 and a saturation coverage of $\theta = 1.2$–1.4 monolayers is agreed upon. A curve by Clavenna and Schmidt [170] is shown for comparison in Fig. 20. We assume the discrepancies are due to lack of precision in determining absolute S values in the earlier work. On the basis of this earlier work it is often reported that the functional form of S goes approximately as $(1 - \theta)$ i.e. that expected of simple Langmuir kinetics. The complexity of the more detailed and accurate later work shows how misleading a suggestion this can be.

2.6.2 Location of O normal to the surface

The $\Delta\phi$ versus θ measurements, as reported by Bauer et al. [77] are shown in Fig. 21(a). Though many $\Delta\phi$ measurements had been made prior to this study, an accurate coverage (as opposed to just exposure) scale was lacking. The small breaks in the curve for 300 K exposure are genuine but not quantitatively reproducible. The slightly smaller slope between 0.5 and 0.75 monolayers is associated with structural changes observed in LEED over this range, but the nearly linear increase in ϕ with θ over the whole coverage range is indicative of overlayer oxygen at all coverages at 300 K. The exposure at 1050 K and recooled to 300 K curve is drastically different and clearly indicates reconstruction or penetration at $\theta = 0.5$ mL, the coverage where a sharp p(2 × 1) LEED pattern is observed for this temperature

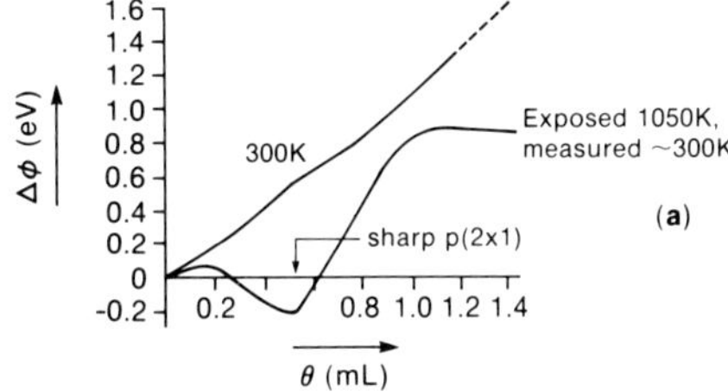

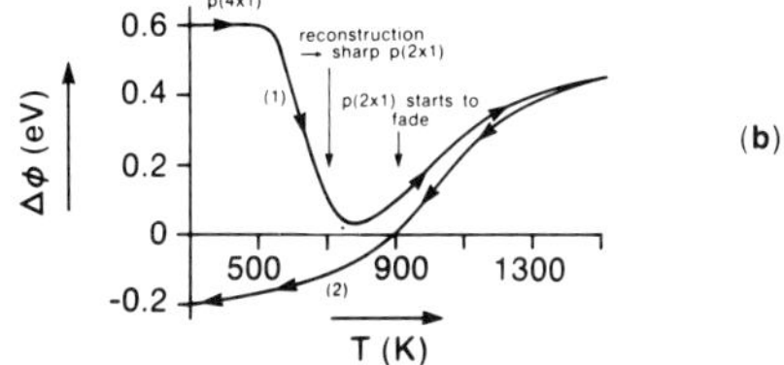

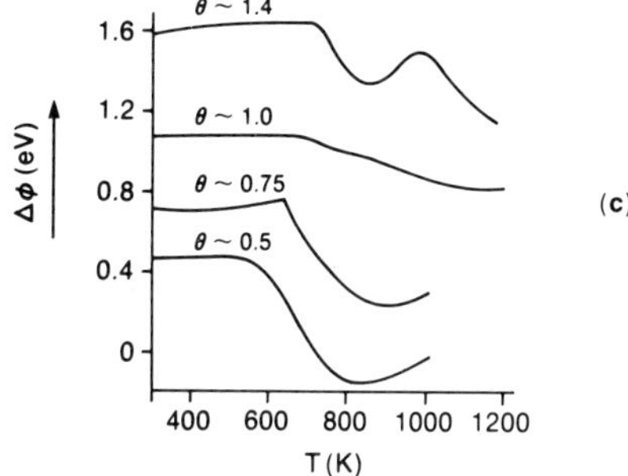

Fig. 21. (a) $\Delta\phi$ versus θ for W {100}/O_2 [77]. (b) $\Delta\phi$ versus T for W{100}/O at $\theta = 0.5$. Measurements made at temperature (see text) [90]. Note that a split-spot p(2 × 1) LEED is obtained for insufficient heating time. See text later. (c) $\Delta\phi$ versus T for W {100}/O for a variety of θ values as a function of annealing temperature. Measurements made at room temperature [77].

exposure. Beyond 0.5 mL, $\Delta\phi$ is sharply positive, indicating that further adsorption is overlayer. Bauer et al. suggest that, for $\theta \geqslant 1$ mL, the increase in ϕ is so large that all the O atoms must again be in an overlayer position. This is possible but we do not consider the $\Delta\phi$ measurement capable of distinguishing the reconstruction/penetration plus overlayer situation from the entirely overlayer situation. If the $\Delta\phi$ measurements are performed at 1050 K without recooling to 300 K, the initial decrease in $\Delta\phi$ does not occur. The difference is greatest at $\theta = 0.5$ mL where the minimum value is obtained for measurement at 300 K. This effect is delineated more clearly in Fig. 21(b) where the $\Delta\phi$ behavior is shown for a $\theta = 0.5$ mL surface prepared at 300 K as a function of subsequent annealing temperature [90]. Here the measurements are made at temperature. On the initial heating cycle, curve 1, the $\Delta\phi$ value is unchanged till ~600 K, when it decreases sharply, reaching a minimum at ~ 800 K. It then rises again. On recooling to 300 K it decreases monotonically (curve 2) ending up ~ 0.8 eV lower than the initial 300 K value. Further cycles retrace curve 2. Clearly two things happen. On the initial heating an irreversible reconstruction/penetration occurs at ~ 600 K. Once this has occurred the subsequent position of the O atoms normal to the surface is apparently strongly dependent on T. At high T they are further out from the surface than at 300 K. The transition temperature for this process is ~ 900 K which correlates well with the temperature at which the sharp p(2×1) pattern observed after the initial reconstruction fades. On recooling, the pattern reappears. There is therefore, apparently a reversible phase transition for the already reconstructed surface where the O atoms disorder at ~ 900 K and apparently move out from the surface while doing so. The irreversible reconstruction and subsequent reversible phase transition have considerable variation in their characteristics with coverage. Only the 0.5 mL case has been discussed in detail but at $\theta = 1$ mL and above the decrease on initial heating [Fig. 21(c)] is small and spread over almost 500 K (700–1200 K). This was interpreted initially [77] as structural changes in the overlayer rather than reconstruction, but later the same group states the reconstruction occurs at all exposures [90]. At $\theta = 1.25$ mL and above, high temperatures can lead to facetting to {110} surfaces (LEED), the degree depending on the pressure conditions as well as on T. In the limit, the saturation $\Delta\phi$ then agrees with that for the saturated W{110} surface, and the saturation coverage agrees with that expected for the geometric area of a surface fully facetted to W{110} [77].

Discussions on what exactly are the structural overlayer, reconstructed, and facetted arrangements on the {100} surface are reserved for the LEED section. Here we just summarize the $\Delta\phi$ measurements, as they relate to the question of vertical location of the O atoms, to say that they imply that at 300 K adsorption results in overlayers only at all coverages, whereas adsorption above ~ 600 K, or annealing above 600–800 K (depending on coverage), causes reconstruction. Above a coverage of $\theta = 0.75$ mL some, if not all, of the O is again in an overlayer arrangement and for $\theta = 1.25$ mL and above facetting can occur. Once the surface is reconstructed a reversible phase transition involving O atom disordering and movement away from the surface apparently occurs at ~ 900 K.

Work function changes as a function of coverage during annealing have been studied in a quite different manner by Wells and King [145]. They were attempting to study diffusion of O on W{100} by molecular beam dosing a patch (to various coverages) on a W{100} surface and then monitoring the behavior of the patch as a function of annealing by measuring the total secondary electron yield (SEY) across the patch with a scanning electron beam. The SEY can be related directly to $\Delta\phi$. Provided no reconstruction occurs, $\Delta\phi$ is directly proportional to θ [see Fig. 21(a)] and the aim was to use this fact to monitor O diffusion away from the patch perimeter. No macroscopic diffusion was in fact observed, but typical SEY profiles across an O

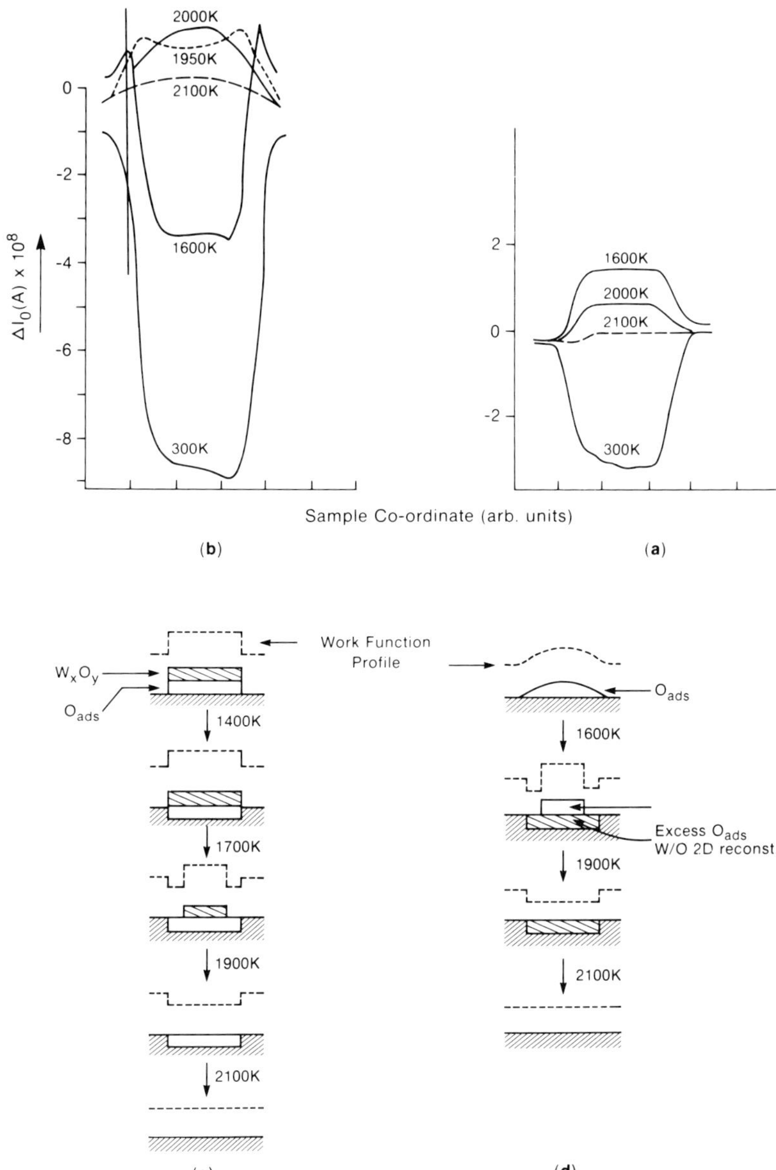

Fig. 22. (a) Secondary electron yield (SEY) of a spot on a W{100} surface beam dosed with O_2 to low coverage (< 0.5 mL), at 300 K and subsequently annealed [145]. (b) As (a) but for saturation coverage [145]. (c) The Wells and King [145] schematic interpretation of the SEY effects observed in (b). (d) Our interpretation of the effects observed in (b).

patch as a function of annealing are shown in Fig. 22(a). The low coverage ($\theta \leqslant 0.5$ mL) effects seem straightforward. At the lowest annealing temperature reconstruction has already occurred, which leads to $\Delta\phi$ becoming negative, and hence the inversion for the ΔSEY [Fig. 22(a)]. At temperatures above 1900 K, desorption occurs, returning $\Delta\phi$ and ΔSEY to zero. This was essentially the interpretation of Wells and King also. For high coverage ($\theta > 0.5$ mL), the ΔSEY (and $\Delta\phi$) reversal at the perimeter of the patch occurred earlier than at the center, giving rise to the "wings" in the profile [Fig. 22(b)]. Wells and King interpreted this in terms of a double layer adsorption model at 300 K above $\theta = 0.5$ mL as schematically shown in Fig. 22(c). They attributed the earlier reversal in ΔSEY in the "wings" on heating to preferential "second layer" oxide desorption at the perimeter. Based on the current interpretation that 300 K adsorption produces overlayer O only, even above $\theta = 0.5$ mL, our interpretation is rather different, as summarized in Fig. 22(d): Heating causes reconstruction of the layer and the wings are caused simply by the fact that the original 300 K adsorption boundary is not sharp, leading to a gradual fall-off in θ at the perimeter. This is supported by the fact that, in the wings of the high-exposure case [e.g. the vertical line marked in Fig. 22(b)], the SEY behavior at 300 K, and on heating to 1600 K, is identical to the behavior in the patch center of the low-exposure case [Fig. 22(a)]. Looking back at Fig. 21(c), we can see that, whereas at $\theta < 0.5$ mL reconstruction changes the sign of $\Delta\phi$, for θ between 0.75 mL and saturation only a decrease in the positive value of $\Delta\phi$ occurs. This is because at high coverage excess oxygen stays on the surface after reconstruction. The SEY behavior observed by Wells and King simply monitors this coverage difference between center and perimeter. Once T is raised high enough, desorption starts, but from the excess O in the center first. Eventually the difference in coverage between center and perimeter is thus eliminated. It should also be pointed out that, even if a high θ did lead to reconstruction at 300 K, the inverted double layer arrangement suggested by Wells and King [Fig. 22(c)] is not the correct one. Their choice of this arrangement is really based on the observation that, for an initial $\theta > 0.5$ mL, W_xO_y species are observed in thermal desorption, whereas for θ below 0.5 mL only O atoms desorb. The erroneous conclusion is thus drawn that W_xO_y species imply oxide formation (see Sect. 2.9), leading to their proposed double layer model of Fig. 22(c). As we shall see later the reconstructed W{100}/O surface actually has a two-dimensional mixed W/O top layer only.

ISS data are also available from Bauer's laboratory [105]. It is much more complicated than for the W{110}/O system where the almost complete attenuation of the W signal was taken to indicate overlayer adsorption (see the W{110} section). Effects which can only be interpreted as strong variation in neutralization probabilities and in shadowing efficiency as a function of chemical nature and coverage become apparent. In addition, double scattering processes, whereby He^+ ions once scattered by O atoms are scattered again by W atoms are observed. These are also dependent on the relative location of O and W atoms. All these effects mean that there are strong variations in the nature of the scattering spectra with polar and azimuthal angle and with impact energy. It is fair to say that though the results do confirm the interpretations placed on the $\Delta\phi$ data concerning vertical location of the O atoms, this conclusion might not have been reached with a less detailed ISS study or if the object of the work had not primarily been to help corroborate the previous $\Delta\phi$ and LEED work.

The piece of ISS data that provides the best evidence of the overlayer-reconstruction transition at ~600 K and the reversible phase transition at ~950 K is shown in Fig. 23. The data are for $\theta = 0.5$ mL and are taken using a low incident He^+ energy to minimize the sampling depth, low polar angle to increase the shadowing effects of O on W, and along the (011) azimuth. The O signal decreases and the W signal increases at the reconstruction temperature. At temperatures above the reconstruction temperature the W signal taken at the annealing T is smaller than on recooling to 300 K and vice versa for the O signal, in agreement with the $\Delta\phi$ data, suggesting that the reversible phase transition at 900 K involves movement of O atoms out of the plane of the surface. Two additional points should be made though. First, data taken under other conditions of polar angle, azimuth and energy do not show such a simple behavior. This was attributed to the fact that the internal relative position of O and W with respect to each other and themselves are important in determining shadowing and self-shadowing effects in addition to the simple idea we are pursuing here that overlayer O atoms attenuate the W substrate signal. In fact variations with azimuthal angle were actually used to suggest a real space structure for the p(2 × 1) reconstruction LEED pattern (see Sect. 2.6.3). Secondly, some of the ISS data at 1 mL coverage as a function of annealing temperature are very similar to those at 0.5 mL. This supports the idea that reconstruction on annealing takes place at all coverages.

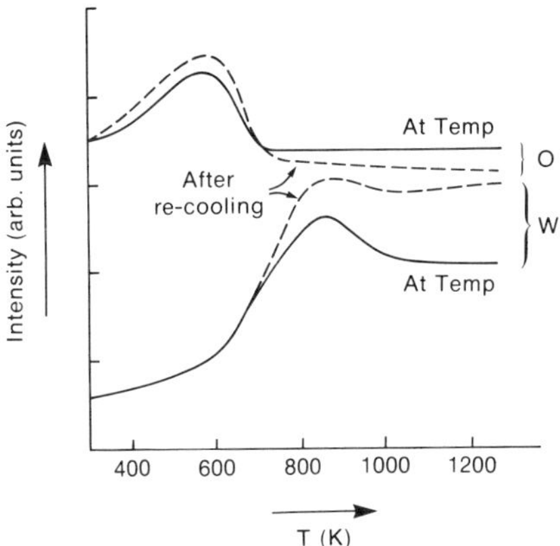

Fig. 23. Temperature dependence of the integrated O and W ISS signals from W{100}/O at $\theta = 0.5$ mL. He^+ energy = 200 eV; polar angle = 18°; (011) azimuth [105].

Recently, Hölzl and Schäfer [172] measured the angular-resolved secondary electron emission spectra for the $\theta = 1$ mL W{100}/O system as a function of annealing temperature. They extracted a parameter from these spectra which they relate to the number of induced surface states within the gap region. This parameter undergoes a very sharp change at ~ 800 K coincident with the simultaneously measured $\Delta\phi$ decrease. The authors ascribe this drastic change in surface states to reconstruction of the surface. There thus seems little doubt that reconstruction takes place on heating at high coverages [90, 172] not just for $\theta \leqslant 0.5$ mL as originally proposed [77].

Spin-polarized electron scattering measurements carried out on the W{100}/O system have shown that the spin polarization is very susceptible to the reconstruction processes [173]. Changes in the spin polarization as a function of annealing temperature are a good indicator of when reconstruction occurs. In fact for θ as low as 0.1 mL, reconstruction could be detected this way, even though no significant change in LEED IV profiles could be observed at such a coverage.

2.6.3 Geometric structure

The ordered structures observed by LEED for the W{100}/O system are a good deal more complex than the W{110}/O system and many of the interpretations are less sure, despite the fact that ISS [105] and surface vibrational spectroscopy (HRELS) [133] have been brought to bear on the subject in addition to LEED [71]. The coverages for which the various LEED structures exist is known from the work of Bauer et al. [77], which limits the possibility for real space interpretation, but unfortunately it appears that very often two or more structures can co-exist at a given coverage under some conditions. The geometry which seems to be best established is that of the $\theta = 0.5$ mL reconstructed surface (heated above $T_c \approx 600$ K) where the p(2×1) O represents a mixed W/O surface where every other W row has been replaced by an oxygen row [Fig. 24(a)]. In addition, the details of how this is achieved during the heating also appear to be well established [90]. As in the case of W{110} it has been suggested [77] that, below reconstruction temperatures, there is a tendency to a hexagonal close-packed overlayer of O atoms (or ions) with increasing θ. Even assuming this interpretation is correct, which is certainly not beyond question, the tendency is never as well-developed as for the W{110} surface, implying that W{100} substrate enforces its geometry constraints on the overlayer more strongly than does W{110}. In addition, as we shall see later, there is electronic structure evidence [76, 174] that above $\theta = 0.5$ mL there are two distinct O species present. The hexagonal-like kinematic analysis of the LEED data below the reconstruction temperature does not account for this distinction in O species. Alternative analyses of this coverage range suggest that the overlayer oxygen may pass through double row, triple row and quadrupole row oxygen atom arrangements as the coverage increases [133].

We will consider the LEED sequence observed at several temperatures

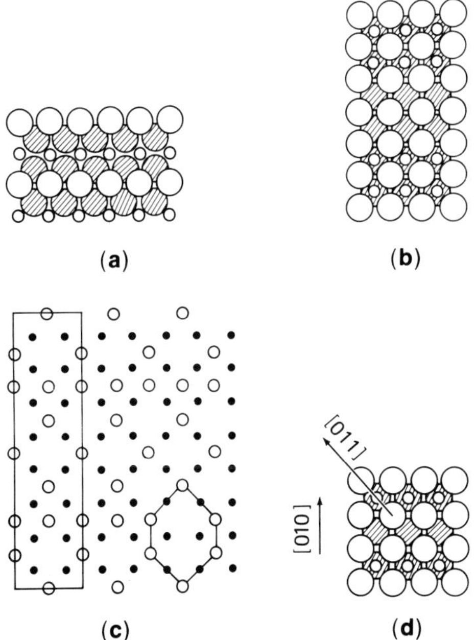

Fig. 24. Suggested real space structures corresponding to LEED patterns observed for the W{100}/O system. Large circles = W atoms; small circles = O atoms (drawn with approximate relative covalent radii). (a) θ = 0.5 mL reconstructed surface (heated above ~ 700 K) p(2 × 1) O structure in which every other top row of W atoms has been replaced by O atoms [77, 90]. Open circles represent the top layer of atoms and hatched circles the second layer. (b) Suggested structure for the "p(4 × 1)" O LEED pattern observed for adsorption at 300 K to θ = 0.5 mL [73]. Symbols as for (a). (c) Suggested overlayer structure for the c(8 × 2) O LEED pattern which co-exists as part of the "p 4 × 1" structure [77]. Closed circles = sites of W{100} substrate atoms; open circles = site of overlayer O atoms. (d) Alternative structure for the "p(4 × 1)" O LEED pattern suggested from ISS work [105]. Symbols as for (a). Note that the pattern is actually p(2 × 1) (see text).

and the conversions observed for given coverages when changing temperature.

(a) Adsorption at 130 K

Luscher [168] has reported on oxygen adsorption at 130 K. He observed the c(2 × 2) reconstruction of the clean surface at that temperature and noted that oxygen adsorption to less than 0.5 mL removed it and returned the original p(1 × 1) pattern. Beyond this coverage the background increases, but no new spots were observed. Annealing a 0.5 mL covered surface to ~ 370 K produced the well-known p(4 × 1) pattern. Compared with the low-temperature adsorption on W{110} the development of long-range order, as evidenced by ordered LEED structures, is clearly less facile. On W{110} annealing to 270 K after-low temperature exposure was sufficient to produce a LEED pattern. This is in accord with the expected deeper potential well of the more open {100} surface.

(b) Adsorption at 300 K

All authors except Bauer et al. [77] have apparently started with the unreconstructed (1 × 1) W{100} clean surface at 300 K. Bauer et al. mention the observation of a weak and variable c(2 × 2) structure. Oxygen adsorption on either surface produces no order till $\theta \approx 0.25$ mL, above which the "p(4 × 1)" pattern mentioned above develops, reaching a maximum intensity at $\theta \approx 0.5$ mL and persisting up to $\theta \approx 0.75$ mL. Papageorgopolous and Chen [73] suggested a structure for this pattern as in Fig. 24(b), with a residual attractive interaction between O atoms pairs. Bauer et al. showed [77], however, that the pattern resulted from a co-existence of p(2 × 1), p(4 × 1) and c(8 × 2) structures. They have not offered a real-space analysis for the p(2 × 1) and p(4 × 1) components. The co-existing c(8 × 2) structure exhibits a maximum contribution at around $\theta = 5/8$ mL and then fades. The kinematic analysis of this structure based on spot intensities, shapes and optimum coverage was suggested by Bauer et al. as that of Fig. 24(c). Note that it involves attempted hexagonal close-packing by the oxygen atoms. At higher coverages a streaked p(4 × 1) develops which persists till saturation. The interpretation offered by Bauer et al. is that, with changing θ, it represents a changing mixture of the many structures observed for $\theta \geqslant 0.75$ mL following mild annealing (450 K, see later).

The HRELS data of Froitzheim et al. [133] should be considered in the light of the LEED data. The surface vibrational spectra observed for adsorption at 300 K are shown in Fig. 25(a). For $\theta \leqslant 0.25$ mL a single W–O stretch is observed, compatible with statistical adsorption and in agreement with the LEED data that no ordered structure exists over this coverage range. The assumption is that the O atoms sit in fourfold hollow sites as in Fig. 25(b), I, and a crude ball and spring estimate of the W–O vibrational frequency for this site agrees with experimental value. Above $\theta = 0.25$ mL a second vibration appears and grows with coverage. Froitzheim et al. [133] associate this two-peaked spectrum with the LEED "p(4 × 1)" pattern. They consider the paired row model of Papageorgopolous and Chen to be the correct interpretation of this pattern, except that they place the O atoms in the local threefold, instead of center, site [Fig. 25(b), II]. They argue that in the paired row arrangement the W/O arrangement is similar to that in WO_2 and that this fact is essentially the driving force for the formation of a double row arrangement. Remembering that the "p(4 × 1)" is actually a composite [77], the assignment of the $\theta = 0.5$ mL vibrational spectrum to the structure of Fig. 25(b), II, cannot be entirely correct and the WO_2-like reasoning is therefore suspect. In fact at $\theta = 0.5$ mL, and even perhaps 0.3 mL, a third surface vibration with frequency around 90 meV is already present [see Fig. 25(a)]. At $\theta = 0.7$ mL it dominates. This could represent O in the c(8 × 2) O structure, though Froitzheim et al., being unaware of the composite nature of the p(4 × 1), favored a three-row structure [Fig. 25(b), III] at this coverage. At higher coverages, corresponding to the streaky p(4 × 1) region (0.75 mL to saturation) the vibrational spectrum becomes very complex.

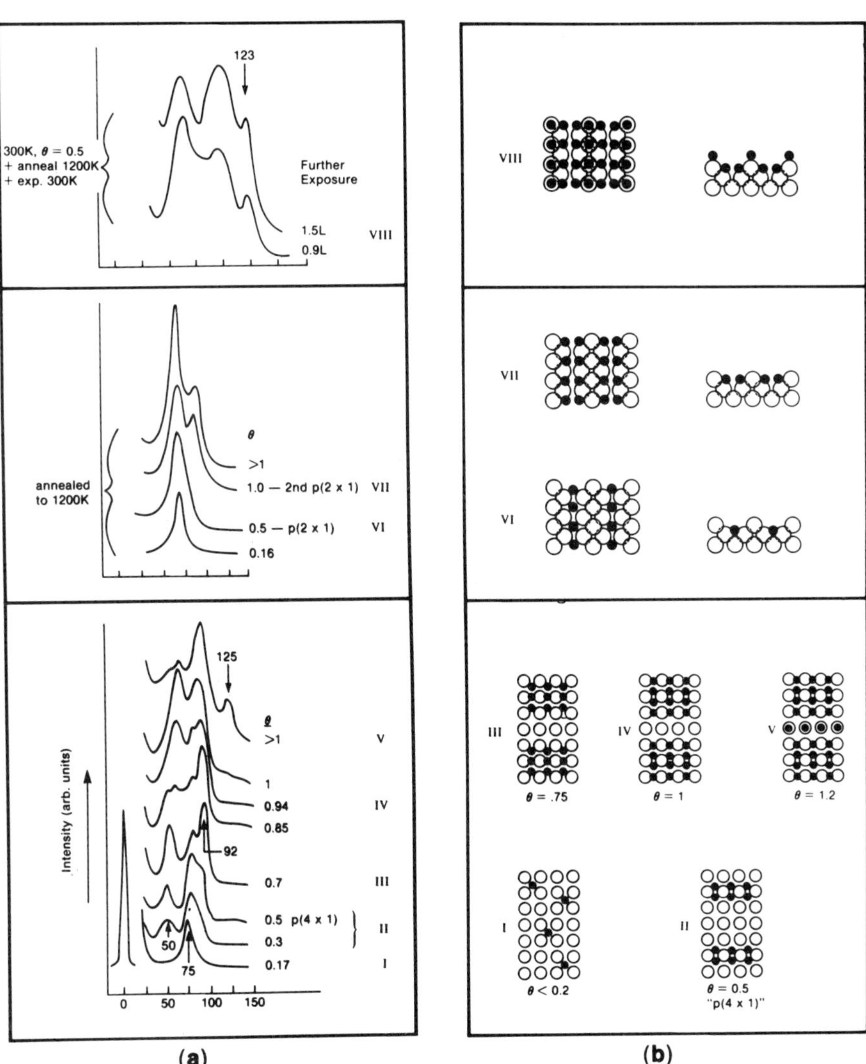

Fig. 25. (a) HRELS data for W{100}/O_2. (b) Authors' suggestion for structures giving rise to HRELS data [133]. Open circles = top row W atoms; filled circles = O atoms. Cross-sectional views also shown for upper two panels.

Froitzheim et al. suggest a four-row structure [Fig. 25(b), IV], but most likely the Bauer et al. mixture of structures could also be made to fit the complex data. At saturation a vibration at 125 meV appears. This frequency seems too high to be anything other than an "on-top"-type O, as suggested by Froitzheim et al. [Fig. 25(b), V]. They also correlate it with a high-yield ESD, β_1 state. Since this ESD state is thought to be a minority species (see Sect. 2.8.2) this correlation is not likely to be correct unless the 125 meV peak has an anomalously high cross-section and therefore does not represent a significant amount of surface oxygen. Prigge et al. [106], however, *also* associate the 125 meV frequency with the β_1, ESD state and suggest that the frequency represents O in on-top positions on WO_3 crystallites, since it is these crystallites which are considered responsible for the β_1 ESD state [106]. This assignment does not remove the problem that the WO_3 crystallites are a minority species whereas the 125 meV ELS peak is quite pronounced.

(c) Adsorption at 300 K plus annealing to 450–600 K

For adsorption at 300 K up to 0.75 mL coverage, followed by a mild anneal at 450–600 K, a well-formed p(4 × 1) is achieved [72]. On increasing the coverage at 300 K from 0.75 mL to saturation in steps, each time with the 450–600 K anneal, a series of more complex, but related, LEED patterns is observed. In earlier studies many of these structures after annealing have been attributed to oxide formation or reconstruction. However, in the interpretation of Bauer et al. [77] they are all overlayer structures attempting hexagonal symmetry. The readers are referred to the original paper for the authors' suggestions as to the real space lattices concerned. The HRELS study of Froitzheim et al. did not consider this intermediate annealing regime. It would be interesting to do so since the Bauer et al. interpretation suggests that the main difference from the 300 K streaky p(4 × 1) is in the proportion of the different structures present, and perhaps their perfection. Presumably, therefore, only the relative intensities of peaks in the HRELS for the annealed situation would be different from the 300 K situation. An interpretation involving reconstruction or double layer oxide formation, however, should produce drastic changes (see below).

(d) Adsorption at 300 K plus annealing above T_c

Once a W{100} surface which has been exposed to oxygen is heated above a critical temperature, T_c, an irreversible reconstruction occurs, as discussed in Sect. 2.6.2. T_c is somewhat coverage-dependent but is $\sim$600 K for $\theta = 0.5$ mL [77, 90]. The reconstruction is accompanied by a drastic drop in work function [Fig. 21(b)] and concomittant changes in the LEED patterns. The reconstruction process also occurs if the exposure is performed directly at high temperature and the sample recooled. One has to be careful to distinguish between measurements made at the annealing temperature, and after recooling to 300 K because, though the reconstruction is not reversible there is apparently significant reversible movement of O and W atoms as a function of T above the reconstruction temperature as discussed in Sect. 2.6.2.

The reconstruction process for θ at and close to 0.5 mL has been studied in detail by Kramer and Bauer [90]. The 300 K adsorption p(4 × 1) pattern at first reverts to a (1 × 1) with high background and then a split p(2 × 1) structure is formed. The temperature at which the split p(2 × 1) emerges coincides with the drastic decrease in $\Delta\phi$ of Fig. 21(b). The width of the splitting decreases with increasing annealing temperature, or time of annealing at a fixed temperature above T_c, and eventually vanishes. In the accompanying AES measurements no change in the absolute or relative W and O Auger signal was detected during the annealing sequence, indicating that the O content and distribution within the escape depth does not change. Kramer and Bauer were able to show that the p(2 × 1) splitting was caused by the production of a stepped surface (Fig. 26). They proposed an initiating place-exchange mechanism between O and W atoms (Fig. 26) which would produce a minimum width terrace. Longer annealing times or higher temperatures cause the terrace widths to grow and hence the p(2 × 1) spot splitting to decrease. They studied the growth rate of the terraces as a function of annealing temperature, time, and exact oxygen coverage (θ = 0.4–0.6 mL). They found that longer annealing times or higher temperatures produced equivalent behavior, but that the growth rate was sensitive to θ, with a maximum in rate at exactly θ = 0.5 mL. The real space geometry of the reconstructed p(2 × 1) (Fig. 26) differs from the ideal structure proposed in Fig. 24(a) only by the presence of the single atom high steps.

If the θ = 0.5 mL reconstructed surface which has been heated once to ~1500 K and recooled to 300 K is recycled to 1500 K and back, $\Delta\phi$ changes reversibly as in Fig. 21(b). From the increase in $\Delta\phi$ at ~950 K Kramer and Bauer concluded that the O atoms are moving out from the surface at that temperature and the concomittant destruction of the p(2 × 1) O superstructure is caused by this movement. The LEED characteristics of the order–

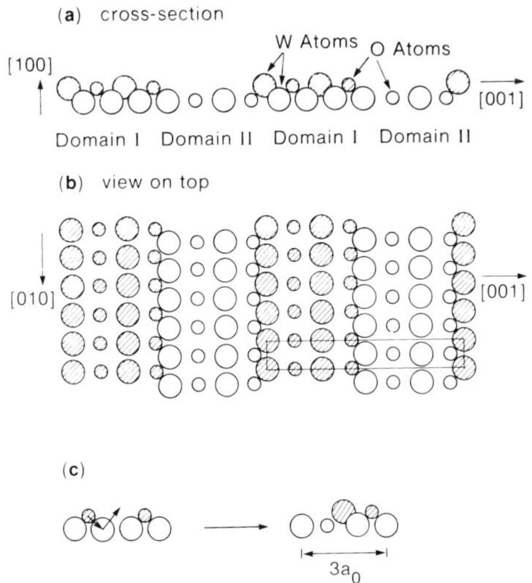

Fig. 26. Step and terrace structure corresponding to the split-spot p(2 × 1) reconstructed LEED pattern observed for a θ of 0.4–0.6 mL at 300 K annealed above T_c (~650 K) for W{100}/O [90]. (a) Cross-sectional view. (b) Top view. Terrace width depends on annealing time (see text). (c) Proposed initiating place exchange process for producing minimum periodicity step and terrace structure.

disorder transition are similar to the order–disorder transition observed for $\theta = 0.5$ mL W{110}/O discussed in Sect. 2.5.5, the important difference being that in the latter case the order–disorder transition referred only to the overlayer O atoms, whereas here it refers to loss of order for both the O and W atoms in the 2D-reconstructed top layer.

The Froitzheim et al. HRELS study [133] also considered the $\theta = 0.5$ mL reconstruction process, and independently arrived [Fig. 25(b), VI] at the same real-space structure for the p(2 × 1) as Bauer et al. Heating the $\theta = 0.5$ mL situation through T_c eliminated the two (three)-peaked structure, replacing it with a single vibration at ~75 meV [Fig. 25(a), middle section]. Starting with a θ of 0.16 mL gave the same final spectrum with 16/50 of the intensity, indicating that the O environments were identical. This is entirely consistent with the proposed reconstruction model since at $\theta = 0.5$ mL one is going from the 300 K "p(4 × 1)", which is a superimposition of three structures, to the single reconstructed p(2 × 1) structure where each O atom is identical and in a site [Fig. 25(b), VI or Fig. 24(a)] somewhat similar in local bonding (and therefore vibrational frequency) to the proposed $\theta < 0.25$ mL 300 K situation where it sits in a fourfold hollow [Fig. 25(b), I].

The effects on the LEED patterns of heating $0.5\text{ mL} < \theta < 1\text{ mL}$ 300 K adsorption coverages to high temperatures (and remeasuring at 300 K) are too complicated to review in full here. The reader is referred to the original articles for a full discussion. Here we restrict discussion to the most important and clearest points for $\theta \geqslant 1$ mL.

For a $\theta \approx 1$ mL, the streaked p(4 × 1) at 300 K becomes sharp at around 1000 K and converts to a p(2 × 1) pattern at higher temperatures. This structure is clearly not the same p(2 × 1) as that formed by heating the 0.5 mL 300 K adsorbed surface, since it evolves without any oxygen loss from the $\theta = 1$ mL streaky p(4 × 1) surface and the p(2 × 1) structure of Fig. 26 cannot accommodate 1 mL of O. Bauer et al. propose two real-space alternatives for this structure (Fig. 27). In their original publication [77] they favored the oxygen overlayer structure [Fig. 27(a)] because the authors originally believed that for $\theta \geqslant 1$ mL the $\Delta\phi$ measurement did not support reconstruction on heating. In the later papers of Kramer and Bauer [90], however, the authors explicitly suggest that reconstruction occurs at all coverages, which agrees with our inclination based on the $\Delta\phi$ and ISS data (see earlier). We therefore favor the "double reconstruction layer" model of Fig. 27(b) as the more likely candidate. It is interesting to compare this structure with the step and terrace arrangement proposed for the $\theta = 0.5$ mL p(2 × 1) reconstructed arrangement (Fig. 26). An extension of the W/O arrangements in the "upper" and "lower" terraces over the whole surface yields the proposed $\theta = 1$ mL p(2 × 1) structure. Froitzheim et al. offer a different real space interpretation for the $\theta = 1$ mL p(2 × 1) structure based on their HRELS data. In contrast to the 0.5 mL p(2 × 1) single-peaked spectrum discussed earlier, the 1 mL p(2 × 1) structure shows two W–O

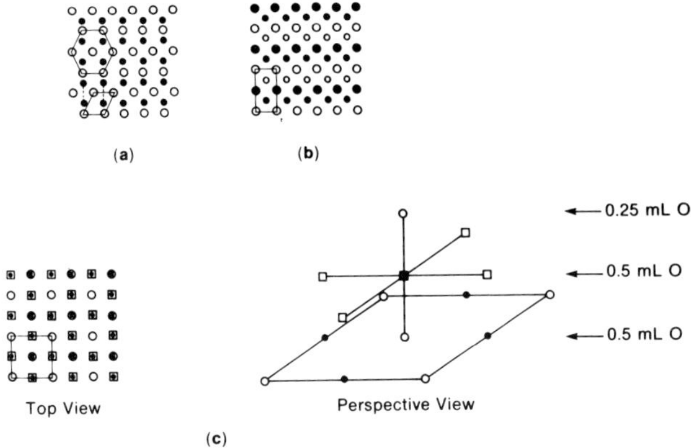

Fig. 27. Proposed real space structure for $\theta > 0.5$ after annealing to high T (> ~ 110 K, depending on coverage) for W{100}/O [77]. (a), (b) Alternative proposals for $\theta = 1$ p(2 × 1) structure. Open circles are oxygen, filled circles W. Top layer = large circles; second layer = small circles. Note (a) is essentially an overlayer hexagonal O structure and (b) is a double layer reconstructed structure. (c) Proposed structure for $\theta = 1.25$ "p(2 × 2)" structure. Large circles represent atoms in the first and third layers; squares represent atoms in the second layer.

vibrations [Fig. 25(a)]. Froitzheim et al. interpret this in terms of a paired row reconstruction [Fig. 25(b), VII], i.e. all the O is in one missing W row. We have no idea at this point whether or not the Bauer et al. double layer reconstruction model of Fig. 27(b) is also compatible with the HRELS data. For 1.25 mL > θ > 1 mL, a well-known "p(2 × 2)" pattern develops from the complex 300 K structures on heating above 1000 K. Bauer et al. showed it to be a co-existing mixture of a p(2 × 1) [presumably the same p(2 × 1) as at $\theta = 1$ mL] and a weaker c(2 × 2) structure. The c(2 × 2) component is most pronounced at $\theta = 1.25$ mL, just below the coverage where facetting occurs on heating. Though the "p(2 × 2)" LEED pattern is well-known, the coverage it corresponded to, $\theta = 1.25$ mL, was not known prior to the work of Bauer et al. [77]. The coverage rules out some previously suggested structures [78, 80], but any interpretation is still speculative. The three-layer unit cell of Fig. 27(c) was tentatively suggested by Bauer et al. [77]. It corresponds to a θ_{max} value of 1.25 mL and resembles the WO_3 structure, implying that it is the precursor to true oxidation. Froitzheim et al. [133] found no new vibrations for the 300 K θ > 1 mL plus annealing situation compared with $\theta = 1$ mL, however, whereas a high-frequency vibration would be expected for the on-top O atom of Fig. 27. It is not clear whether Froitzheim et al. ever recorded the HRELS of the true $\theta = 1.25$ mL p(2 × 2), however. Such an "on-top" peak case was found on further exposure at 300 K after a high-temperature anneal of a lower coverage [Fig. 25(a), top section]. Froitzheim et al. propose the structure in Fig. 25(b), VIII, i.e. further 300 K adsorption occupies on-top sites. The qualitative difference between this structure and that of Fig. 27(c) is that the Fig. 25(b), VIII structure has a full monolayer of O atoms in the top reconstructed plane, whereas the Fig. 27(c) structure has half a monolayer in each of the two reconstructed planes, to account for the c(2 × 2) component of the "p(2 × 2)" LEED pattern [77]. Each structure has 0.25 mL O in the overlayer plane to account for the p(2 × 2) part of the LEED pattern.

(e) Facetting at high oxygen coverage

The saturation coverage at 300 K is $\theta \sim 1.3$–1.4 mL, as determined by AES, with a complex LEED pattern consisting of coexisting structures [77]. If a surface with $\theta \geqslant 1.25$ mL is heated to 1050 K, or sufficiently long exposure is made at 1050 K to reach $\theta > 1.25$ mL, the surface breaks up into {110} facet planes exhibiting a (1 × 1) pattern. This is preceded by an increasing roughening of the surface, indicated by spot splitting and streaking in LEED. The work function behavior is erratic in this region because the coarseness of the roughening, which involves massive W transport, depends on the details of oxygen pressures, heating rates, etc. The surface finally becomes so rough that the {110} facetted planes become energetically more stable. The maximum O coverage on the {110} plane is 1.05 monolayers, or 1.41×10^{15} atoms cm^{-2} (see Sect. 2.5.10), but since the facetted area is geometrically larger than the flat {100} surface the coverage, θ, per unit area of original {100} surface can go higher. Therefore continued exposure at 1050 K leads to a final maximum coverage of $\sim 2 \times 10^{15}$ atoms cm^{-2}, or equivalent to 2 mL for a flat {100} surface [77].

If the facetted surface with 2 mL $> \theta \geqslant 1.25$ mL is heated to higher temperatures, desorption occurs between 1450 and 1800 K. With loss of oxygen the surface passes back through the "p(2 × 2)" structure ($\theta \approx 1.25$ mL) and the p(2 × 1) structure ($\theta \approx 0.5$ mL), the LEED always being recorded after cooling to near room temperature.

The remaining set of data that should be considered is the ISS data [105], referred to earlier in Sect. 2.6.2. There it was concluded that, though the W and O ISS peak behavior as a function of coverage in general supported the contention of overlayer adsorption for 300 K adsorption and reconstruction above the critical temperature, T_c, the interpretation of the data was complicated by a significant degree of double scattering and by shadowing and self-shadowing effects which were a strong function of azimuthal scattering angle and ion energy. These effects are more important on the W{100}/O surface compared with the W{110}/O because of the more open and rougher structure of the former. Whereas the interpretation for the W{110}/O surface was apparently unambiguous in terms of the decrease of W and the increase of O signals with increasing θ, this is not the case for W{100}/O. Prigge et al. [105] have attempted to extract structural information from the strong azimuthal and energy-dependent behavior. In particular, they consider dependencies for a θ of 0.5 mL below the reconstruction temperature and above it. Though they claim that the data supports a description of the p(2 × 1) reconstructed geometry as claimed by Bauer et al. and Froitzheim et al. [Fig. 24(b) or 25(b), VI], they also suggest that the dominant arrangement before reconstruction is p(2 × 1) [Fig. 24(d)], even though the observed LEED pattern is the "p(4 × 1)" consisting of a mixture of p(2 × 1), p(4 × 1) and a weak c(2 × 8). We summarize some of their arguments here for the sake of completeness, but we find them mostly rather speculative and conclude that, until the details of shadowing, self-shadowing, double scattering and ion-neutralization effects are better understood, the ISS data, though consistent with the geometries proposed by the authors, cannot be considered as conclusive evidence for them.

For low ion energy (E_0 = 200 eV) scattering in the (011) azimuth, it was found that heating above T_c caused the O/W ISS signal ratio to decrease considerably (Fig. 23). The explanation put forward was that, before heating, the top W atoms were all effectively shadowed by the O atoms lying in the (011) azimuth [Fig. 24(d)] and therefore the observed W signal came largely from double scattering (O first, then W in the top W plane, then back out); on reconstruction the O atoms are coplanar with the W atoms, leading to some O ISS decrease by W shadowing. The W atoms in the top layer are now the major contributor in a single-scattering manner to the W signal. Hence the W signal increases considerably and also moves to the single scattering energy (higher by ~6 eV). In the (010) direction heating above T_c causes both the O signal and the W signal to decrease by about 35%. The W peak scattering energy is unchanged on passing through T_c, corresponding to the single scattering position in both cases. In addition the absolute magnitude of the W signal is much smaller in the (010) direction than in the (011), leading to a smaller value for the W/O ratio in the former case. The explanation given by Prigge et al. is that, before reconstruction, the W signal in the (010) direction is small because of self-shadowing caused by the short W–W distance in this azimuth, rather than because of shadowing by O atoms. There is no shadowing by O atoms and no O–W double scattering contribution in this plane because the O rows are laterally displaced with respect to the W rows [Fig. 24(d)]. After reconstruction the W signal is decreased because of the decrease by a factor of 2 in the number of surface W atoms contributing to the scattering.

One final interesting point is that the O signal growth as a function of θ in the (010) azimuth levels off as θ = 0.5 mL is approached, and then rises again immediately $\theta \geqslant$ 0.5 mL, both for the unreconstructed and, particularly, the reconstructed situation. This is exactly what would be expected if rows of O atoms are completed at θ = 0.5 mL and new ones started above θ = 0.5 mL, since the O atoms completing rows would suffer a lot of shadowing from other O atoms in the row, while the O atoms starting new rows would be very "visible" to the ion beams.

(f) Summary

Summing up this section on ordered structures it seems that for only one structure is there substantial agreement on the real-space geometry. That is the reconstructed, half-monolayer coverage, p(2 × 1) O structure [Fig.

References pp. 381–388

24(a)]. It also seems well established that it forms from the $\theta = 0.5$ mL unreconstructed surface by a place-exchange mechanism, and that as a result it is a step and terrace surface with terrace width growing with annealing time or temperature. For the higher coverage reconstructed surfaces [p(2 × 1), $\theta = 1$ mL; "p(2 × 2)", $\theta = 1.25$ mL] there is no agreement on the real-space geometries, or even on whether there are one or more reconstructed layers (the limit being an incipient oxide structure). From the high-frequency vibration observed at saturation coverage on the reconstructed surface, it does seem very likely that the final fraction of adsorbed oxygen occupies positions "on-top" of W atoms in the outermost mixed W/O reconstructed layer.

For adsorption below T_c the reconstruction temperature (which varies by several hundred degrees, depending on coverage), very little is firmly established about the real space geometries. Below $\theta \approx 0.25$ mL no ordered structure is formed and the single-peaked vibrational spectrum and the frequency of that vibration offer fairly convincing evidence that all O atoms are equivalent and sit in the fourfold sites. Above $\theta \approx 0.25$ mL and above ~300 K the well-known "p(4 × 1)" O structure develops but real-space interpretations are ambiguous because of the composite nature of the pattern (three independent patterns). The original double O row interpretation of Papageorgopolous and Chen [73] [Fig. 24(b)] cannot therefore be correct (or at least can only represent a fraction of the oxygen). Prigge et al. [105] believe, on the basis of the ISS data, that an overlayer p(2 × 1), with real space geometry as in Fig. 24(d), is the dominant structure, but this also seems somewhat speculative. At coverages above ~0.5 mL when the streaked p(4 × 1) develops, Froitzheim et al. [133] offer interpretations in terms of triple O rows and then quadrupole rows (Fig. 25) which they claim are compatible with the observed vibrational spectra. Since this interpretation is based on the incorrect double row model for the p(4 × 1) $\theta = 0.5$ mL structure of Papageorgopolous and Chen, the interpretation is highly suspect. Bauer et al. [77] favor interpretation in terms of a sequence of mixed complex structures with increasing θ where the driving force is the attempt by the O overlayer to establish a hexagonal close-packed structure on the surface. This interpretation seems equally speculative, though it seems well-established that this is the correct sequence of events at high θ for the close-packed W{110} surface. Interestingly the saturation coverage $\theta = 1.4$ mL on the W{100} surface is equivalent to 1.4×10^{15} atom cm^{-2} which is identical to the saturation coverage on the W{110} surface and is also equal to the density of a hexagonal close-packed layer of oxygen ions (ionic radius 1.4 Å). Just to confuse matters further, there seems rather good electronic-structure evidence that above $\theta = 0.5$ mL there are two distinct types of O present (see later). Neither the multiple-row model nor the pseudo-hexagonal model effectively account for this, so our final conclusion is that, despite all the effort expended, the real space geometries for adsorption below T_c are unknown. We are not even sure that at high coverage

($\theta > 0.5$ mL) some amount of reconstruction does not already occur at 300 K, even though it may be disordered and therefore not observed in LEED. In this regard the recent synchrotron photoemission study of Brundle et al. [160] of the W(4f) level indicates that substantial O penetration *can* occur for high coverages. This work is referred to in the next section. Two other points worth remembering are that low-temperature (130 K) [168] low-coverage adsorption is disordered but destroys the c(2 × 2) W{100} clean surface reconstruction, and that high-coverage ($\theta \geqslant 1.25$ mL) adsorption plus heating leads to a {110} facetted surface with O present on the facets up to the normal saturation coverage for a {110} surface [77].

2.6.4 Electronic structure effects

UPS spectra have been reported by Bradshaw and coworkers [76, 174], Waclowski et al. [175], and Feuerbacher and coworkers [138, 139]. XPS data has been reported by Bradshaw et al. [76, 174] and Yates et al. [176]. The W(4f) core levels have also been studied by synchrotron radiation ($h\nu$ between 70 and 150 eV) by Van der Veen et al. [140] and Brundle et al. [160]. In the XPS work an O(1s) BE of 530 eV is reported at low exposures at 300 K, which develops a pronounced shoulder at ~531.5 eV for exposure above 2 L [76, 174] [in the streaky p(4 × 1) LEED pattern region, see later]. At saturation exposure the 531.5 eV peak is about two thirds the intensity of the 530 eV peak, and the same result is found for adsorption at 100 K. An O(1s) BE of 531.5 eV is very high for atomic oxygen on a transition metal and it is rather unusual to find two resolved O(1s) features from genuine O_{ads} species. But for the intensity and apparent reproducibility one would be tempted to ascribe the 531.5 eV BE to OH impurity (see, for example, the Ni/O_2 section of this review). Since it appears genuine, the only possible explanation seems to be that there are distinct O species present. This might suggest that reconstruction/penetration occurs at high exposures at 300 K, giving rise to a surface oxygen species and a subsurface species. Alternatively it could represent two distinct surface oxygen species. Bradshaw et al. heated the W{100}/O 300 K saturated surface to 1500 K, resulting in a sharp p(2 × 2) O structure [76, 174]. The two resolved O(1s) features coalesced into a broad feature at ~530.7 eV. Reconstruction should certainly have occurred at this temperature so we assume that the 530.7 eV value represents O in the reconstructed surface. The situation is complicated by the fact that saturation coverage was used, however, since the evidence (see Sect. 2.6.2) suggests that the annealing may produce a mixed reconstruction and overlayer situation. This may explain the broadness of the 530.7 eV peak. If the heating had been done at $\theta = 0.5$ mL, where only one O(1s) existed at 300 K, to give the sharp p(2 × 1) O reconstructed surface with only one type of oxygen atom, the distinction in terms of O(1s) values between reconstruction and overlayer situations would have been clearer.

The W(4f) core levels have been studied using both laboratory and synchrotron photon sources. Using AlKα radiation Bradshaw et al. [76, 174]

were able to detect a slight (0.25 eV) chemical shift for the surface component in the W(4f)'s by taking difference spectra between the clean and 300 K saturated spectra. Van der Veen et al. [140], using 70 eV synchrotron photons and a monochromator resolution of $\sim$0.1 eV, reported that the surface W(4f) levels for the clean W{100} surface were shifted $\sim$0.3 eV lower in BE than the bulk value. At $\theta \approx 0.5$ mL this surface peak was eliminated and a broad weak structure built up at $\sim$0.5 eV higher BE than the bulk 4f position. This clearly represents first-layer W atoms bonded to overlayer O atoms. The value of the chemical shift in the W(4f)'s is practically the same as that induced on W{110} by overlayer O atoms. The sample was then annealed, above the reconstruction temperature, and a slight sharpening and increased intensity in the 0.5 eV shifted peak was observed which was ascribed to the reconstruction process. Since O atoms replace W neighbors in the reconstructed surface a much larger chemical shift might have been expected. More detailed work by Brundle et al. [160] using $h\nu = 130$ eV indicated that Van der Veen et al. [140] had not, in fact, achieved reconstruction of a significant fraction of the surface (remember it is a transport-limited process and the rate depends critically on θ). When reconstruction is achieved a shift of 1.4 eV is found for the W(4f) levels. This effect was found to be quite reproducible and was accompanied by the formation of the reconstructed p(2 $\times$ 1) O LEED pattern [160]. This same study also showed that, for high exposures ($>$ 5 L), small amounts of O start to penetrate the surface at 300 K, giving rise to a $\sim$ 1.9 eV shifted W(4f) peak in addition to the main 0.5 eV overlayer shifted peak. Finally measurement at high temperature after reconstruction failed to produce any movement in the $\sim$1.4 eV reconstruction-shifted W(4f) peak. This last point casts some doubt on the simple interpretation of the reversible order–disorder transition observed by Kramer and Bauer [90] that O moves back above the surface at high T. If it were this simple one would expect the $\sim$1.4 eV W(4f) shifted peak to move back toward the $\sim$0.5 eV position.

UPS studies of the valence region using both HeI and HeII have been reported, but they are fragmentary. Waclawski et al. [175] reported HeI spectra for 5 L exposure at 300 K and after annealing to 1500 K. $\Delta\phi$ values were taken which allow a direct comparison with the data of Bauer et al. [77, 90]. At 300 K $\Delta\phi$ was $+1.4$ eV and after heating it dropped to $+0.7$ eV. Comparison with the $\Delta\phi$ versus θ data of Fig. 21 indicates that saturation coverage ($\theta = 1.4$ mL) had been achieved at 300 K. The large decrease of $\Delta\phi$ on heating indicates that substantial changes occurred, but we cannot be sure what mixture of facetting, desorption, or reconstruction was involved. Again a study at exactly $\theta = 0.5$ mL would have been more informative. Two oxygen-induced features at 5.2 and 6.3 eV of approximately equal intensity were observed at both temperatures. As a function of increasing the detection angle away from normal emission the O(5.2 and 6.3 eV)/W(0–3.5 eV) intensity ratio decreased sharply for the heated surface but not for the 300 K surface. This could be indicative of a deeper distribution of O atom for the heated surface, or it could be caused by the different symmetries of the O and W orbitals because of bonding changes. Variations in the relative intensities of the 6.3 and 5.2 eV O(2*p*) features were also observed with angle for which similar, but ambiguous interpretations could be put forward.

Bradshaw et al. [76, 174] reported angle-integrated HeI UPS data as a function of exposure. At low exposures (< 1 L) one O 2*p* feature is seen at ~ 5.5 eV and at high exposures the two ~ 5.5 and ~ 6.5 eV features reported by Waclawski et al. are seen. The O(2*p*) behavior as a function of exposure therefore parallels the observed O(1*s*) behavior referred to earlier. On heating the saturated surface to 1500 K the 5.5 eV feature was depressed, the 6.5 eV feature increased and chemically shifted W *d* band features became apparent, suggesting reconstruction/penetration or facetting which leaves the outermost W atoms in a more oxide-like environment. This is in agreement with the results of the synchrotron W(4*f*) study by Brundle et al. [160]. (Note that none of these effects was observed on heating in the Waclawski et al. results. They may be dependent rather critically on small differences in coverage and in the length of heating.) Bradshaw and Menzel [174] also report HeII ($h\nu = 40.8$ eV) UPS data for saturation exposure at 300 K. Here the higher BE O(2*p*) feature dominates, which is not the case for the HeI ($h\nu = 21.2$ eV) results. Again the interpretation is ambiguous. HeII data have a shorter escape depth than HeI, which could suggest that the higher BE O(2*p*) feature was from O located only at the surface while the lower BE feature was for O with a deeper distribution. Alternatively it could imply differences in valence electron distribution of two types of O which are reflected in cross-section variations as a function of $h\nu$.

Avery [162] deconvoluted NVV W Auger spectra to obtain the oxygen-induced valence features for 300 K exposure. He found two features in approximately the same positions as for UPS. He also reported that the adsorbate levels shifted ~ 2 eV on heating, in disagreement with the UPS data. The discrepancy remains unexplained.

Luscher [168] has made a very extensive ELS study of the W{100}/O_2 surface. He performed adsorption experiments at 130 and 300 K from zero to saturation coverages as well as subsequent annealing experiments up to desorption temperatures. He correlated his experiment directly with LEED and $\Delta\phi$ measurements and so was able to make extensive comparison with other studies in terms of coverage and adsorbate order. For W the final states for the ELS excitation process are flat, narrow (~ 0.2 eV), *d* bands lying about 3 eV above the Fermi level. The presence of these unfilled, localized orbitals allows the filled orbital energies of the W and the adsorbate to be mapped out by the ELS process without final state complications. A primary beam energy of 115 eV was used, providing a high degree of surface sensitivity. A typical spectrum of the clean W{100} surface is shown in Fig. 28(a). Below 5 eV W *d*-band features, interband transitions, surface and bulk plasmon losses are observed, and W(4*f*) and 5*p* core levels occur between 30 and 40 eV. On adsorption of oxygen additional features due to

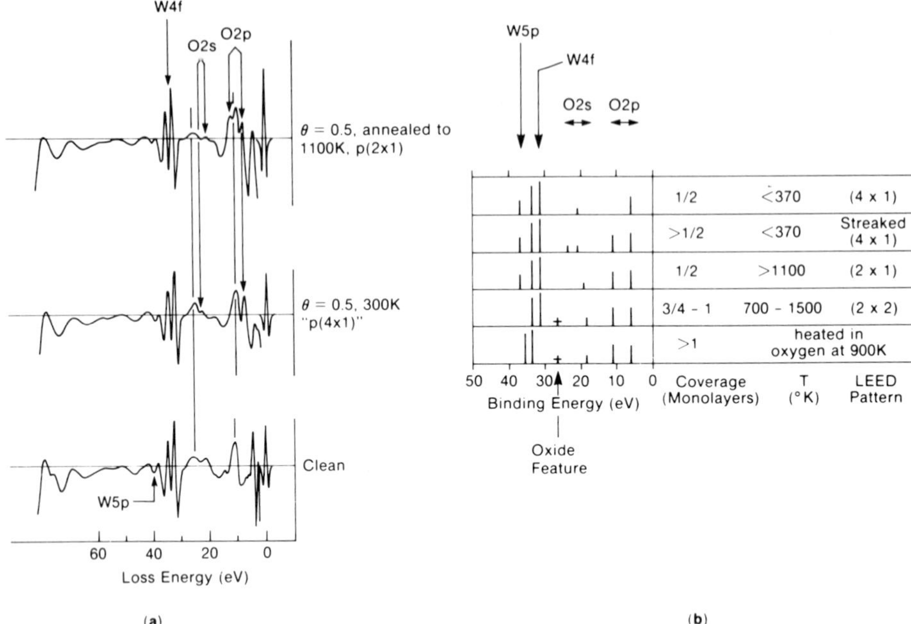

Fig. 28. ELS spectra for the W{100}/O system [168]. (a) Representative spectra using a 115 eV primary energy showing W(4*f*), O(2*p*) and O(2*s*) derived structure. (b) Line diagram of the significant changes observed under different oxygen treatment conditions. Note that no shifts in W(4*f*) levels are reported except for conditions of heavy oxidation.

O(2*p*) and O(2*s*) levels are observed. The actual spectra are too complex to be reproduced here, so we show only two representative examples in Fig. 28(a). A summary of all the important changes observed under a variety of oxygen treatments is shown in a summary line diagram in Fig. 28(b). Below $\theta = 0.5$ mL and for temperatures below ~ 700 K (the reconstruction temperature) a single O(2*s*) feature and a single O(2*p*) feature (at ~ 5.5 eV) are observed. At $\theta = 0.5$ mL and for T above 300 K but less than 700 K a sharp p(4×1) LEED pattern is observed in conjunction with an unchanged ELS spectrum. For $\theta > 0.5$ mL under these temperature conditions a second, higher BE, O(2*s*) feature is observed accompanied by a new O(2*p*) band which is apparently split into two peaks separated by ~ 3.5 eV (at 7.5 and ~ 11 eV). These results confirm the general trend of the XPS and UPS data reported above in that there appears to be an additional distinguishable electronic state of O formed above $\theta = 0.5$ mL. In ELS, however, the O(2*p*) region of the new state shows two peaks at 7.5 and 11 eV, whereas UPS does not show the 11 eV peak. This situation is similar to the ELS W{110}/O results of Rawlings [144] where a 12.5 eV feature was observed to grow prominently for $\theta > 0.5$ mL. The interpretation of that feature, based largely on its reversible behavior with heating below reconstruction temperatures, was that it represented O–O lateral interactions.

Luscher and Propst performed several annealing experiments. For the $\theta = 0.5$ mL case, where only one O(2s) feature and the single ~5.5 eV O(2p) feature exist, annealing to 1100 K produced a sharp p(2 × 1) structure characteristic of the reconstructed surface. The single O(2p) feature of the "p(4 × 1)" structures split into two features, presumably representing the inequivalent $2p_{xy}$ and $2p_z$ O orbitals of the reconstructed O in the p(2 × 1) structure. No noticeable change occurred in the W(4f) or 5p levels, whereas the synchrotron W(4f) photoemission results [160] suggest that a ~1.4 eV shifted component should have appeared. Heating the 300 K saturated surface, exhibiting the ELS features of the two types of O, results in a sharp p(2 × 2) O pattern. The two O(2s) features revert to one [cf. the XPS O(1s) results], but a split O(2p) band remains. This is then consistent with the $\theta = 0.5$ mL behavior: heating results in a single O species with a split O(2p) band. For this higher coverage situation, changes are now observed in the ELS W core-level spectra, however. The W5$p_{3/2}$ peak is eliminated, or strongly broadened and a small feature some 10 eV lower in BE appears. It has some resemblance to a feature observed for a thick W oxide layer produced by heating in oxygen at ~900 K. The W(4f) peaks do not show any clear evidence for a shifted component, but looking at the original data it is not clear that a small shifted component would be observable at the spectral resolution available.

Summarizing this section, the available photoemission (XPS, UPS and synchrotron), AES, and particularly ELS data all suggest that above $\theta = 0.5$ mL, for 300 K adsorption, a distinct second O species forms. Since this species has single O(2s) and O(1s) peaks, but apparently a split O(2p) band (ELS data), it seems likely that it represents oxygen that has penetrated the surface. It would then correlate with the ~1.9 eV W(4f) shift observed in the synchrotron study for exposures of greater than 5 L. These conclusions based on core-level chemical shifts and electronic effects in the valence band contradict the Bauer et al. [77, 90] conclusions that only overlayer adsorption occurs for θ up to 1.4 mL at 300 K. Of course they were primarily concerned with the ordered LEED structures. It is likely that the penetrated O showing up in the electronic effects is not ordered and does not contribute to the LEED. The two sets of data are therefore not necessarily contradictory. For $\theta \leqslant 0.5$ mL the electronic structure results support the interpretations of other data; a single overlayer O species below T_c and reconstruction above T_c.

2.6.5 Diffusion of O on W{100} surfaces

In contrast to W{110} no detailed studies of O diffusion over the W{100} surface exist. The reason is simple: reconstruction occurs below the temperatures required to observe macroscopic movement of O over the W{100} surface. Any macroscopic movement observed at temperatures higher than the reconstruction temperature therefore relates either to the O within the reconstructed mixed W/O surface, or to O adatoms moving over this recon-structured surface. Remembering that W atoms themselves become mobile above 900 K, the interpretation of what exactly is being measured is complex.

Diffusion over small distances (as opposed to macroscopic distances) at temperatures below reconstruction can be observed using a field emission microscope. Gomer and Hulm [169] examined W{100}/O at ~ 500 K and found an activation energy of 95 kcal at low coverages ($\theta \lesssim 0.3$ mL) for a boundary-free diffusion process, whereas at higher coverages boundary diffusion occurred with $E_{diff} \approx 130$ kcal. Wells and King [145] originally interpreted these results in terms of O atom overlayer-only diffusion for the lower coverage case and diffusion over a W_xO_y "layer" at high coverages (see Sect. 2.6.2 for a discussion of this adsorption model). From all our previous discussions we would now expect that both the low and high coverage data relate to overlayer-only situations with the $\theta \lesssim 0.3$ mL situation representing an W{100} surface with O in random fourfold sites and the high coverage situation representing ordered overlayer structures. The results are consistent with those found for W{110} where the low coverage phase also showed a lower activation energy for diffusion. The explanation offered is also similar: in the low coverage situation there are no lateral interactions to overcome. At higher coverage there are both lateral interactions to overcome and there is a lack of available empty "hopping sites" on the surface.

Wells and King attempted to study the macroscopic motion of O on a W{100} surface by a molecular beam patch dosing technique. They studied the behavior of prepared patches of different O concentrations as a function of temperature by scanning across the patches with an electron beam and measuring the total secondary electron yield (SEY) which can then be related directly to the work function (see Sect. 2.5.13 for similar work on W{110}). Assuming a random walk process the average distance travelled per unit time for the low-temperature diffusion processes observed by Gomer and Hulm by field emission can be calculated from their measured diffusion coefficients. Wells and King calculated that the Gomer and Hulm low-coverage diffusion coefficient would give a movement of 1 mm in 60 s at 1200 K, whereas the high-coverage value required 1600 K for the same motion. Wells and King were, however, unable to detect *any* macroscopic broadening of their beam-dosed patches at 1400–1900 K over hundreds of seconds. Clearly, therefore, the diffusion process in operation at ~ 500 K is not present at 1400–1900 K, as expected since reconstruction occurs above ~ 700 K. The changes in the secondary electron emission profiles for the beam-dosed patches clearly showed this reconstruction and indicated that desorption from the reconstructed patch occurred in preference to macroscopic diffusion. The details of the SEY behavior were discussed in Sect. 2.6.2.

Kramer and Bauer [90] followed, by angular LEED, the growth of the reconstructed terraces from the overlayer structure at $\theta = 0.5$ mL as a function of annealing temperature (730–820 K) and extracted an activation energy of 1.9 eV for the whole process. On the basis of the assumption that this value was too high to represent diffusion of O over a reconstructed surface they concluded that the rate-determining step in the domain growth was the simultaneous place exchange motion of Fig. 26 and that the 1.9 eV activation referred to this process. The assumption that 1.9 eV is too high for O_{ads} diffusion over the reconstructed W/O surface was based on an erroneous belief concerning the data of King and Wells. Kramer and Bauer mistakenly believed that King and Wells had observed O_{ads} diffusion over this surface and therefore determined an activation energy much lower than 1.9 eV. In fact, King and Wells observed no diffusion and it is quite possible that the diffusion step is rate-determining.

2.7 W{111}

Unlike W{110} and {100}, little detailed work exists for W{111}/O_2, except in the areas of ESD and ESDIAD. The reason is clear: no ordered O structures are formed except under facetting conditions and the reason for many studies is often the interpretation of LEED patterns. The lack of LEED patterns also seems to be the reason behind much of the effort in ESDIAD. Since LEED fails to reveal any geometric information, because of the lack of long-range order, perhaps ESDIAD, which does not require long-range order, will be particularly useful. Unfortunately, despite the considerable effort expended, it is still not clear whether ESD and ESDIAD on the W{111}/O system ever represents the majority O species in the adsorption regime, as opposed to minority species such as oxide nucleii. The authors providing most information on this subject, Madey et al. [117, 118] and Niehus [83] hold flatly contradictory opinions. In the W{110} and {100} sections we deliberately omitted ESD/ESDIAD data and deferred it to a separate section, largely because of the majority/minority species problem plus other problems concerning ESD/ESDIAD interpretations. For W{111} we will make some reference to the ESD/ESDIAD work here, in addition to a more detailed discussion in Sect. 2.8, because of the lack of other evidence on surface geometry. Our opinion is, however, that minority species probably do dominate the data and therefore any structural evidence based on ESD/ESDIAD should be treated with caution (see Sect. 2.8). A similar situation exists for TPD, also deferred in the previous W{100} and W{100} sections because of the belief that it often provides little information on the state of the adsorption system prior to the desorption temperature. For W{111} we will refer briefly to TPD spectra in this section, again because of the lack of other data. We use it, however, only as a means of coverage determination.

2.7.1 S, θ measurements

The saturation coverage of oxygen on W{111} is not well established. In the most recent determination Niehus [83] followed the O Auger signal as a function of exposure up to 60 L, which he believed corresponded to saturation. He calibrated the coverage at this point by comparing the O(sat)/W(clean) ratio with that found for the W{110}/O system in a different apparatus. The value of 6.5×10^{14} atoms cm^{-2} which he obtained for θ_{sat} is suspiciously low, though it seems to have gone unchallenged. Niehus defines a monolayer as the number of W atoms in the top layer, which is equal to 5.8×10^{14} atoms cm^{-2}. One might have expected that the open W{111} structure could have accommodated up to three times this density of O atoms since the total density of first, second and third row atoms, all essentially exposed to the vacuum, is 1.74×10^{15} atoms cm^{-2} (see Fig. 2). Remember also that the saturation coverage of the other two surfaces is $\sim 1.4 \times 10^{15}$ atoms cm^{-2}. In fact older determinations, based on absolute desorption measurements and on XPS data, do yield much higher values than 5.8×10^{14} atoms cm^{-2}, as discussed below.

Vasko et al. [177] determined the amounts of O, WO, WO_2 and WO_3 flash desorbed from a W{111} oriented ribbon as a function of oxygen exposure by integrating the desorption curves and calibrating the O signal against previous work on W{110}. At high exposure at 300 K (36–360 L: 10^{-8} to 10^{-7} Torr for 60 min) the total amount of oxygen desorbed in the form of oxygen atoms and oxides was 1.2×10^{15} atoms cm^{-2}. It is not clear that true saturation has been reached, however, since the WO_2 and WO_3 desorption intensities appear to be still increasing at the highest exposure.

The XPS coverage determination was done for saturation adsorption at 120 K, which takes $\sim$5 L exposure [141]. It was based on measuring the ratio of the total photoelectron yield from the adsorbate to that from the clean W substrate and using X-ray mass absorption coefficients and an electron mean free path length through W of 13 Å [178]. Though there is no reason to expect this absolute method to be accurate to better than $\sim$50% the number obtained was 1.2×10^{15} atoms cm^{-2}, the same as the flash desorption value. Given the known S, θ characteristics for W{110} and {100} as a function of T, one might expect that for W{111} θ_{sat} really was reached in 5 L exposure at 120 K, but that the exposures were not long enough to reach θ_{sat} in the 300 K AES study of Niehus [83].

We are therefore left with a maximum discrepancy of a factor of $\sim$2 between the AES measurement (6.5×10^{14} atoms cm^{-2}) and the XPS or thermal desorption measurements (1.2×10^{15} atoms cm^{-2}), which might be partly due to saturation not being reached in the AES measurements. Our inclination is to view the 1.2×10^{15} atoms cm^{-2} as being the more likely, since two independent determinations are in agreement and the coverage is in the expected range. The error in the Auger measurement would then have to be ascribed to some combination of incorrect calibration against the W{110} result on a different apparatus using the W 169 eV Auger transition; electron-stimulated desorption effects at high oxygen coverage; or the fact that θ_{sat} at 300 K requires much larger exposures than 60 L.

From the XPS data of Yates and Erickson [141] it appears that S_0 is large (> 0.5) at 120 K and that $S \approx S_0$ up to $\theta \approx 0.9\theta_{sat}$ ($\sim$2 L exposure) at which point it drops rapidly towards zero (Fig. 29). There is insufficient data to establish whether there are any small, but real, decreases in S prior to the $\sim 0.9\theta_{sat}$ point that might correspond to specific fractional coverages (cf. W{100}), and without a proper calibration it is not possible to say whether S_0 is nearer 0.5 (like W{110}) or 1 (like W{100}).

The behavior of S versus θ at 300 K has not been determined directly. The work function increases rapidly as a function of exposure up to 1–2 L exposure [116, 146] (the initial 0.2 L range has not really been monitored), in a similar manner to that at 125 K [116], from which we can assume that S_0 is high and does not fall off rapidly over this range (see Fig. 29). At $\Delta\phi \cong 1.1$ eV (1–2 L) the rate of increase in ϕ at 300 K has dropped rapidly and a gradual rise to $\Delta\phi \approx +1.3$ eV is reached after ~ 50 L exposure. There is no indication that the gradual rise has stopped at 50 L. At 125 K, on the other hand, $\Delta\phi$ continues to increase approximately linearly with exposure to a maximum value of + 1.7 eV at ~ 3 L, at which point it remains constant with further exposure. Since the $\Delta\phi$ versus exposure data at 125 K tracks the XPS coverage versus exposure data it seems that $\Delta\phi$ goes approximately linearly with coverage. From the relative $\Delta\phi$ behavior at 125 and 300 K it therefore appears that S_0 at 300 K is similar to that at 125 K, but S has fallen off considerably by $\theta \approx 0.65\theta_{sat}$ (125 K). Above this coverage θ increases gradually with exposure and the increase is not complete at 50 L where $\theta \approx 0.76$ θ_{sat} (125 K). This behavior is reminiscent of the W{110} situation where very high exposures (1000s of L) were required at 300 K to reach the θ_{sat} value achieved at a much lower exposure at 77 K (see Fig. 15). It also implies that the Niehus "θ_{max}" at 60 L is probably only $\sim 0.76\theta_{sat}$ (125 K). Adjusting the Niehus value at 300 K by 1/0.76 gives $\sim 9 \times 10^{14}$ atoms cm^{-2} for θ_{sat} (125 K),

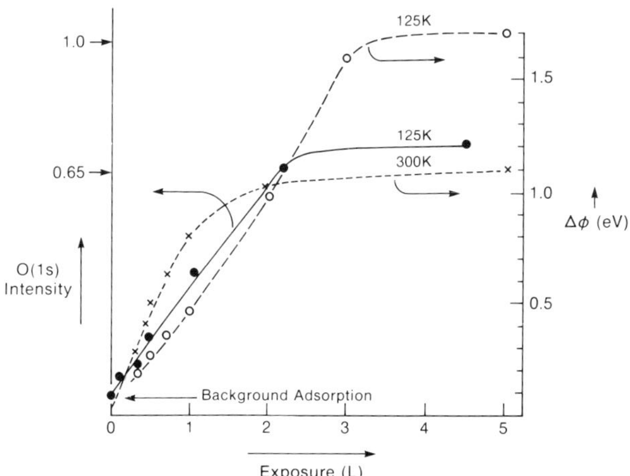

Fig. 29. O(1s) intensity versus exposure at 125 K for W{111}/O_2 [177] and $\Delta\phi$ versus exposure data at 125 and 300 K [116].

much closer, but still smaller than the 1.25×10^{15} atoms cm^{-2} determined from the 125 K XPS measurement. Attempting to obtain relative coverages at 125 and 300 K directly from $\Delta\phi$ measurements in the above manner is not, of course, guaranteed to be correct. It assumes that the oxygen atom vertical location is the same in the two cases, in addition to the assumption that, for a given vertical location, $\Delta\phi$ is proportional to θ.

At high temperatures (> 800 K) facetting occurs under the influence of oxygen [24, 92]. It is clear that adsorbed O stabilizes the facetted surfaces and that the flat W{111} surface is regenerated by annealing at sufficiently high temperature (2600 K) through loss of oxygen and oxides. It is not well established what oxygen coverages are involved in the various facetting stages. However, it has been speculated from LEED that the initial facetting stages involve $\sim 4 \times 10^{14}$ atoms cm^{-2} and the final stages involve saturation coverage on {110} facets ($\sim 1.4 \times 10^{15}$ atoms cm^{-2}, see Sect. 2.5.8).

2.7.2 Location of O normal to the W{111} surface

The information on the vertical location of the O atoms for adsorption below ~850 K (facetting temperature) is sparse, being confined to an interpretation of the $\Delta\phi$ exposure data [116, 146], and controversial interpretations of the ESDIAD data [83, 117]. Clearly with the rough structure of the W{111} surface (Fig. 2) "overlayer" adsorption might be expected to have a more complicated effect on $\Delta\phi$ than for the smoother W{110} and {100} surfaces. One might anticipate that O adatoms located above first-row W atoms would increase ϕ whereas O adatoms above third-row W atoms might have no effect or might even decrease ϕ. In fact ϕ initially increases smoothly with exposure at both 300 and 125 K, as discussed above (see Fig. 29), which at least seems to indicate that during this coverage regime there is no reconstruction or penetration of O into the W lattice. For 125 K this is the situation up to saturation with $\Delta\phi_{\text{lim}}$ being + 1.7 eV and one therefore tentatively concludes that the entire adsorption at 120 K up to θ_{sat} is overlayer. If the XPS and flash desorption coverage calibrations are roughly correct the smoothness of the increase in $\Delta\phi$ at 120 K to saturation implies that either it makes little difference to $\Delta\phi$ where the adatoms are in the overlayer (above first-, second-, or third-row atoms) since 1.2×10^{15} atoms cm^{-2} cannot be accommodated entirely at only one of these types of site, or the distribution of O atoms in these different possible sites does not change with increasing θ.

The only other information available relevant to the vertical locations of O on the unfacetted W{111} surface is the ESDIAD data. On the basis of the angular patterns and ion yields, it has been suggested [83] (see Sect. 2.8.3) that initial adsorption is at bridge positions between first-row W atoms; that at higher coverages all O atoms move to "on-top" of second- or third-row W atom positions; and that at saturation additional oxygen is adsorbed on on-top first-row W atom positions. In addition to the majority/minority species ambiguity, we consider such a specific location assignment based on ion angular distributions and yields as highly speculative, despite apparent agreement with computer simulation models.

The facetting of W{111} under the influence of O has been studied in some detail using LEED by Taylor [24] and by Tracy and Blakely [92]. Above ~850 K at low oxygen exposures (0.5 L) oxygen-covered {112} facet planes develop. Higher exposures lead to oxygen-covered {110} microfacets on top of the {112} facets. Since we believe that oxygen on flat {110} surfaces is present as an overlayer up to temperatures much greater than 850 K it seems likely that O on the facetted W{111} surface is present as an overlayer. Of course the W{100}/O system undergoes reconstruction at only 600 K and it is therefore possible that, prior to facetting, the W{111}/O system also reconstructs. If so, no new LEED pattern has been reported, but it is not clear that LEED in this regime (600–850 K) has ever been studied in any detail.

The changes in $\Delta\phi$ on annealing 300 K adsorbed situations to 1000 K or on adsorbing at 1000 K have been studied. At sufficiently high O coverage (presumably greater than $\sim 4 \times 10^{14}$ atoms cm^{-2}) and for long enough annealing time, large {110} microfacet areas develop and the final ϕ for an O-saturated surface is identical to that for W{110}/O [92]. The low-coverage region preceding the facet-induced $\Delta\phi$ changes is more complex [146]. The effect of annealing to 1000 K after exposure at 300 K as a function of exposure is shown in Fig. 30(a). At low coverages, annealing increases $\Delta\phi$ strongly. On the other hand for coverages at 300 K corresponding to $\Delta\phi$ 1.1–1.3 eV (which was estimated as $\theta \approx 0.76\ \theta_{sat}$ in the previous section) annealing decreases $\Delta\phi$ [Fig. 30(a)]. At higher coverages, *no* change in ϕ is observed on annealing provided the facetting referred to above does not occur (longer annealing times). The interpretation offered by the authors [146] involves the redistribution of the adsorbate atoms on first-, second- and

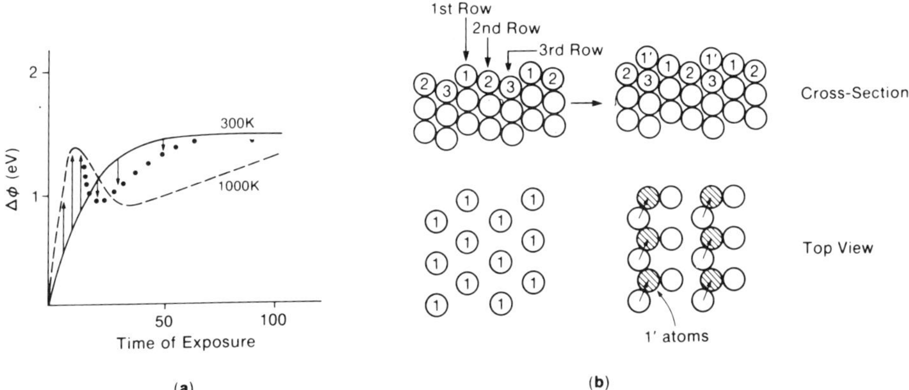

Fig. 30. $\Delta\phi$ versus exposure for W{111}/O_2 at 300 and 1000 K. Arrows represent the effect of annealing a 300 K exposed situation to 1000 K [146]. [b] Suggested reconstruction process occurring for annealing 300 K exposed W{111} to 1000K. W atoms only are shown.

third-row W atoms as a function of θ and T. The suggestion is that at 300 K one O atom of the dissociating molecule adsorbs in "on-top" positions on first-row W atoms, causing the observed $\Delta\phi$, whereas the other cannot diffuse to a further first-row W atom and so resides somewhere in the "well" of second- and third-row atoms close by [Fig. 30(b)]. This latter O atom does not contribute to $\Delta\phi$. Annealing to 1000 K allows diffusion and the second O atom moves to an "on-top" first-row atom position, thereby doubling $\Delta\phi$. The reversal of the effect on $\Delta\phi$ by annealing at 1000 K for higher initial $\Delta\phi$ (i.e. θ) and the null effect at even higher θ, was ascribed to a "pre-facetting" effect. If facetting is initiated by the movement of first-row W atoms as indicated in Fig. 30(b), and at the same time O adatoms which were on these atoms move into the "well", $\Delta\phi$ should decrease. If all "well" sites are already filled (i.e. at higher θ), "on-top" first-row O adatoms cannot move into the well and $\Delta\phi$ will not change. These explanations are both ingenious and plausible. We note in particular that the stage termed "pre-facetting" here is very similar to the place-exchange step and terrace reconstruction process described for the W{100}/O surface at $\theta = 0.5$ mL in Sect. 2.6.3. The increase of $\Delta\phi$ on annealing the low O coverage W{111} surface has no counterpart for W{100}, however. The suggestion that initial adsorption at 300 K produces a distribution of O adatoms half in on-top first-row positions and half in "well" positions is also in direct contradiction to the interpretation of the low-coverage ESDIAD [83] data discussed in Sect. 2.8.3, where the speculation is that all O atoms in the initial stage occupy bridge positions between first-row W atoms. Clearly from the $\Delta\phi$ data something drastic *does* occur on annealing the low-coverage situation, and this is emphasized by the fact that the ESD desorption rates are quite different after annealing [83]. The assumption made by the authors in their interpretation summarized above is that this stage is not correlated with the later facetting or pre-facetting stages in which W atoms move. Actually it seems rather unlikely that O_{ads} cannot diffuse a few lattice spaces at 300 K, though this is a necessary assumption in the argument that only one of the O atoms from dissociating O_2 occupies "on-top" positions on first-row W atoms.

Summarizing this section, it is clear that, while there has been considerable speculation on the vertical location of the O adatoms under various condition of T and θ based on $\Delta\phi$ and ESDIAD data, there are few hard conclusions even though the very open W{111} surface would seem to offer the most chance for variation in vertical location. It is not known whether O adatoms sit "on-top" first-row atoms, in the well, or in a mixture of sites, and it is not known whether reconstruction precedes the facetting process that is known to occur at high O coverage at $T > 850$ K.

2.7.3 Electronic structure effects

There are few relevant studies for W{111}/O. The XPS study of Yates and Erickson [141] at 120 K showed a major O(1*s*) peak which did not change its BE with coverage, and a minor feature ~1.5 eV higher in BE. No indication for any chemically shifted W(4*f*) component was mentioned, supporting the idea that adsorption is overlayer rather than reconstruction or thin oxide formation. The O(1*s*) spectrum is stable on heating to 480 K, indicating that the high BE O(1*s*) feature is unlikely to represent OH contamination. It could still represent CO, however, even though the authors suggest it represents O adsorbed in a different binding site. In general the O(1*s*) and W(4*f*) behavior is more similar to that for W{110} than to W{100}, where the second O(1*s*) feature reached comparable intensity to the first at high coverage and was definitely ascribed to an electronically distinct O.

The second study of relevance is a recent synchrotron radiation study of the W(4$f_{7/2}$) level as a function of oxygen exposure at 300 K [140]. The results are shown in Fig. 31. The use of the low photon energy, $h\nu = 70$ eV, maximizes the surface sensitivity. The interpretation offered for the clean {111} surface is that peak S_1 represents the first-row W atoms (coordination number 4), S_2 represents the second- and third-row atoms (coordination number 7), and B represents the bulk (coordination number 8). The chemical shifts, compared with bulk W atoms, are therefore -0.43 and -0.10 eV for first-row and second/third-row atoms, respectively. Exposure to 2 L O_2 suppresses and shifts S_1 and probably S_2 while producing a new broad peak at $\sim$0.4 eV higher BE than the bulk peak. This peak represents W atoms bound to O_{ad}. Our comment would be that 2 L exposure is nowhere near saturation, as the

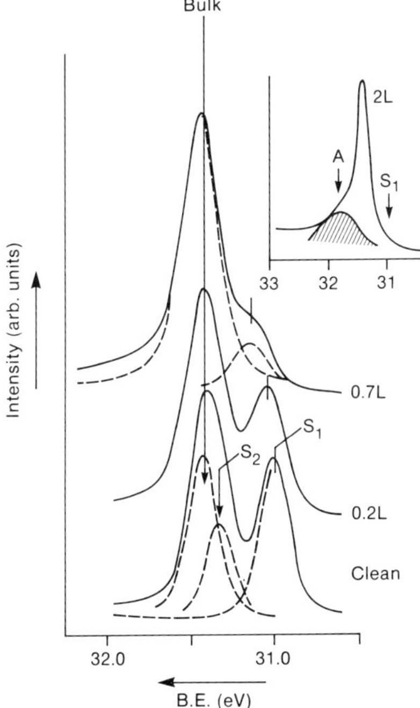

Fig. 31. W(4$f_{7/2}$) XPS spectra as a function of oxygen exposure using $h\nu = 70$ eV photon energy for W{111}/O [140].

authors apparently believe, and that the O coverage at this exposure is probably $\leqslant 4 \times 10^{14}$ atoms cm^{-2}. The strong implication given by the authors is that O_{ad} is attached to W first-row atoms and that the remaining S_1 shoulder represents some first-row W atoms which might still be uncovered. This could certainly be true at such a coverage, but an equally plausible suggestion would seem to be that O_{ad} is in the "well" of second- and third-row atoms, the +0.4 eV shifted W(4f) peak thus representing the second- and third-row atoms, while the perturbing effect is still pronounced enough on first-row atoms to shift S_1 much closer to the bulk. It is not clear, then, that observation of the electronic effects resulting in a chemical shift of the W(4f) levels really helps distinguish the location of O_{ad} between "on-top" first-row and "well" second- and third-row. In the light of other known O_{ad} geometries on a variety of metals, an initial on-top position when higher coordination positions are available would be surprising.

The third study is the angle-resolved photon-stimulated desorption work of Madey et al. [118]. In this work, discussed more fully in Sect. 2.8.3, the O^+ ion yield and angular distribution were measured as a function of incident photon energy for 300 K adsorption. Two pieces of information are relevant here. First the ion yield at $\theta = 0.5\theta_{sat}$ is about 1/5th of that at θ_{sat} and about 1/100th that of a heavily oxidized W{111} surface. Second the ion yield versus photon energy curves are practically identical for $\theta = 0.5\theta_{sat}$ and a heavily oxidized surface. The curves are interpretable only in terms of the Knotek–Feibelman (KF) model for photon- or electron-stimulated desorption [179]. This model involves a W core-hole Auger decay mechanism and apparently requires W to be in the maximal valency form (i.e. W^{6+} as in WO_3). The conclusion, therefore, was that all the PSD data represented minority oxide-like species and that these species were formed on the surface well below θ_{sat}. Thus ESD and PSD yield versus photon energy distributions apparently give little information on the electronic structure of the major adsorbed species, but do apparently provide information on the nature of the minor oxide species. Note that this conclusion is at least partly based on the conviction that the maximal valency rule for the KF mechanism is inviolate, and therefore the W species giving rise to an O^+ ion must be the W^{6+} of WO_3. We have reservations about the maximum valency rule, as discussed in Sect. 2.8, particularly for W{100} surfaces where it is clear that not all the O^+ ESD active states derive from minority oxide species.

2.7.4 Adsorption geometries for W{111}/O

As should be apparent from the previous sections nothing definitive exists concerning the geometric arrangement of O adatoms on W{111}. There is no long-range order at low T, and at high T (> 850 K) facetting occurs to {110} surfaces on which the O adatoms presumably have the same long-range order and adsorption geometries as on flat W{110} surfaces. At 300 K and below the only real attempts at determining the adatoms' geometric arrange-

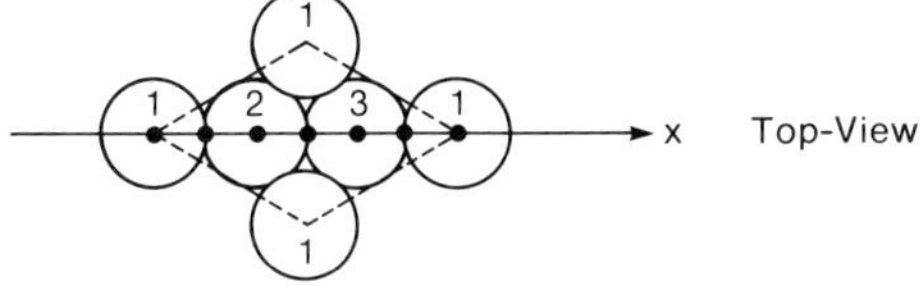

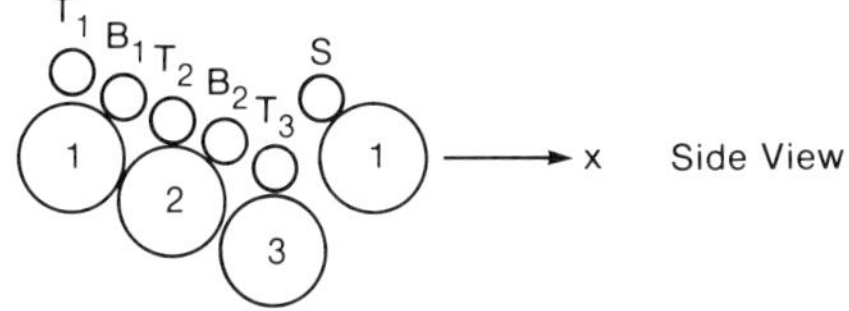

Fig. 32. Unit cell of W{111} showing a variety of possible adsorption sites, using the site terminology of Preuss [91]. Large open circles are W atoms, the numeral referring to which plane the atom belongs. Possible oxygen positions are marked with small open circles in the side view only. For clarity, the sites are marked only with black dots in the top view.

ment have been from the ESDIAD [83] data combined with mathematical simulations of the ESDIAD process [91, 131]. These are discussed in Sect. 2.8.3; here we merely report the general conclusion drawn from the experimental data compared with simulations. Readers should keep in mind that we consider these conclusions highly speculative, both because of minority/majority species questions and ambiguities in correlating the simulations with the data. The unit cell of a W[111] surface and the cross-section are shown in Fig. 32. Possible adsorption sites are marked using the terminology of Preuss [91]. Niehus [83] concluded, from ESDIAD data compared with simulation, that O atoms initially occupied B_2 sites, moved to T_2 and T_3 sites at higher coverage, and finally, at saturation, occupied S sites. Preuss [91] favored the sequence B_1, T_2, S as a function of coverage.

2.7.5 Summary

It is clear that the characterization of the W{111}/O interaction is only in a preliminary stage, compared with the comprehensive work on W{110} and W{100}. The nature of the surface offers the best possibility for significant changes in bonding locations, nature, etc., as a function of O coverage and temperature. It should be perhaps the best of the three surfaces for observing such distinctions, and therefore may turn out to be even more complex in its complete range of interaction than the {100} surface. Painstaking work correlating $\Delta\phi$, AES and XPS effects against coverage, θ, and T is going to be required to bring the state of knowledge up to that of the {100} surface. Regarding the use of ESDIAD as an alternative to LEED in this system, experience in the other surfaces would suggest that the question of minority versus majority species must first be answered before there is any further attempt to interpret the ESDIAD data in terms of the adsorption site geometries.

In the section on W{111}/O, it was necessary to summarize the work on ESDIAD because of the paucity of alternative information. In this section, we present a brief review of the work on electron- and photon-stimulated desorption studies for W/O surfaces in general. We have left this discussion until after the separate sections on the interaction of oxygen with W{100}, W{110} and W{111} because it is our opinion that at present the primary emphasis in this work should be on using the known information concerning W/O surfaces as a guide to understanding the ESD and PSD desorption processes rather than trying to use ESD and PSD to delineate details of the W–O interaction. Certainly, if used in isolation, ESD and PSD do not seem to be sufficiently well understood to provide definitive information on W/O surfaces, though, as will become obvious in this section, there are cases where, in combination with other techniques, additional useful information is obtained.

We start with a brief description of the theory of ESD and the ways in which it has been used in an attempt to extract information concerning the W–O interface.

The one-dimensional Franck–Condon excitation model originally proposed by Redhead [180] and Menzel and Gomer [181] (usually referred to as the RMG model) to explain experimental data existing at that time is summarized in Fig. 33(a). The process is considered to involve Franck–Condon

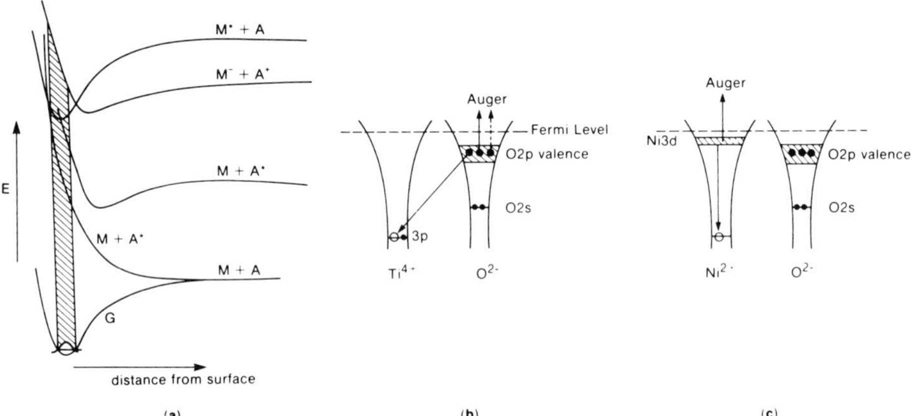

Fig. 33. (a) Schematic potential energy diagram for interaction between absorbate, A, and surface, M. G is the ground state. The other curves represent possible excited states reached by electron-impact excitation. The Franck–Condon envelope within which the electronic transitions must occur (fast compared with vibrational time scales) is indicated by shading [114, 181]. (b) Schematic energy level diagram illustrating the KF ESD process for TiO_2. A Ti 3*p* electron is removed from Ti^{4+} by electron impact and the hole filled by an O(2*p*) electron, releasing sufficient energy to eject Auger electrons and convert O^{2-} to O or O^+ [179]. (c) Equivalent diagram to (b) for NiO. In this case the intra-atomic hole-filling process will dominate, leaving Ni as Ni^{4+}.

electronic excitation from the ground state, M + A, to a variety of excited state potential wells.

If the excited state potential is such that the surface complex is now in a repulsive situation, the adparticle starts to move away from the surface. Its kinetic energy will depend on how far up the repulsive curve the excitation process has taken it. As the adparticles move away, de-excitation can occur by electron tunneling from the substrate resulting in neutralization and possibly recapture. Experimental observations which support the general model are [114] the observation of desorbed ions and neutrals, the range of KEs observed for desorbed ions, the fact that ESD cross-sections are several orders of magnitude smaller than gas-phase dissociation cross-sections (de-excitational capture process) and the observation of large isotope effects on ESD cross-sections (slower moving isotopes more effectively recaptured).

Usually, only the desorbing ions are studied because of the difficulty of detecting neutrals. The observations consist of determining the mass of the ion, measuring the intensities of the ion signal, and sometimes measuring its KE distribution (ESDIED) or the angular distributions (ESDIAD). Very rarely are all these parameters measured. Often only the mass and relative intensities are measured and interpretations based on such restricted data are, at best, ambiguous. On the basis of the assumed correctness of the RMG model, two qualitative correlations are often invoked. (1) Higher KE values and a narrower KE distribution of a desorbing ion indicate closer starting proximities of the adsorbate to the surface. (2) The closer to the surface is the starting position of an ion attempting desorption, the less its chances of survival without recapture. Thus, high ion yields and low KEs should go together and indicate a ground state bonding equilibrium metal–adsorbate distance which is large. Such experimental observations are therefore often ascribed to, or considered to be consistent with, on-top bonding positions, whereas higher KE, low yield, ion desorption states are ascribed to shorter bond lengths, reconstruction situations, or adsorption in "well" sites. The correlations are, of course, dependent on the applicability of the RGM model, and in addition, the KE correlation with adsorption bond distance depends on there being no complicated final state effects. The intensity correlation is only likely to be viable if the total excitation probability from ground to final states is comparable for two different ground states (i.e. yield changes are associated only with recapture probabilities). A situation where neither of the above conditions is likely to hold is in a comparison between two completely different electron bonding situations, such as O_{ads} compared with a W oxide nuclei. This brings us to the main difficulty of ESD interpretation when considering metal/oxygen systems. It is known that ion yields can be several orders of magnitude larger for oxide surfaces than for some chemisorbed O situations, and that some chemisorbed O situations are completely inactive to ESD [e.g. the p(2 × 1) W{110}/O case [68]]. We are therefore always faced with the situation that the ESD signal observed (always O^+ for W/O systems), may represent a minority surface species of high ESD yield (such as oxide nuclei), which is present together with an ESD inactive (or low activity) majority species. This situation has been suggested in many cases by monitoring the depletion of the ESD active state by the desorption process together with the depletion of the total O coverage by AES. Frequently, there is little change in the AES intensity during the time that the ESD state is depleted. The most obvious interpretation of such results is that the ESD active state is a minority species. This is not an unambiguous interpretation, however. The active state could be a majority O species which, on electron impact, largely converts to an inactive state with a minority desorbing as the O^+ ESD signal.

The final problem in interpreting the yield and KE of desorbing ions in terms of bonding characteristics is the validity of the RMG model itself. It has recently been demonstrated that an alternative excitation mechanism to the valence level excitation implied in Fig. 33(a) is at least partially responsible for O^+ ESD intensities observed for W/O situations. This is the Knotek–Feibelman mechanism [KF] [179]. The principle of this mechanism is illustrated in Fig. 33(b) for TiO_2 in which the bonding is assumed to be completely ionic, i.e. Ti^{4+}, O^{2-}. If an electron is removed from the Ti(3*p*) level by electron (or photon) impact, the dominant decay model will be the inter-atomic Auger process depicted, in which the energy released in the O(2*p*) → Ti(3*p*) transfer (31 eV) is sufficient to cause the emission of one or two Auger electrons from the O(2*p*) state. What was initially an O^{2-} species is now either O^0 or O^+. In the O^+ case, ejection will definitely occur owing to Coulombic repulsion with Ti^{4+}. It has been demonstrated that this mechanism is indeed dominant for production of O^+ from TiO_2, WO_3, and heavily oxidized surfaces of Ti and W. This was proven by scanning the electron or photon impact energies, and observing that thresholds for O^+ emission came at the core-level BEs for Ti [179] and W [118]. The necessary conditions proposed for the process are (a) that the core-hole on the metal atom is deep enough for the energy released in the decay step to be sufficient for enough anion Auger electrons to be subsequently ejected such that the anion becomes a positive ion; and (b) that there are no higher lying occupied metal levels which would result in an intra-atomic decay process dominating the inter-atomic decay process necessary for the KF mechanism. Such a process is illustrated for NiO in Fig. 33(c), where Ni^{2+} has the valence region $(3d)^8$ electrons available for the intra-atomic decay to a Ni(3*p*) core-hole. The most probable final state after Auger emission is Ni^{4+} O^{2-} for which no ejection of O^+ should occur. None is observed experimentally [179]. Thus only maximal valency cations, i.e. ones which have completely empty valence shells, are considered able to produce O^+ desorption by the KF mechanism. Such ions include Ti^{4+}, W^{6+}, Va^{5+}, etc. but not Ni^{2+}, Cr^{3+}, Ti^{3+}, W^{4+}, etc. Therefore, it is supposed for W surfaces plus O that if the KF mechanism is confirmed for any O^+ ESD state (by observation of O^+ thresholds at W core level BEs) that state must be a maximal valency, WO_3

complex. In fact, for both W{100}/O [182] and W{111}/O [118], the KF mechanism has been confirmed (by PSD: the thresholds are easier to detect in PSD) and the authors conclusions were that the O^+ desorption states concerned, which were high O coverage high O^+ yield states in both cases, therefore involved maximal valence W ions, i.e. W^{6+} in oxide nuclei, existing as minority species on the surface. While the evidence seems rather good, we would caution that, whereas it has been demonstrated that the majority O_{ads} states on W [118, 182] and even, apparently, Ti_2O_3 [179], do not give KF O^+ desorption, in agreement with the maximal valency requirement it has not been demonstrated that formal oxidation states of W lower than W^{6+} cannot give KF O^+ desorption. It would after all, seem feasible that the inter-atomic decay channel for a W(4*f*) hole could carry some intensity resulting in KF O^+ desorption even if the dominant decay was intra-atomic from higher-lying W electrons which did not lead to KF desorption. KF O^+ desorption with a reduced yield might result and we therefore do not consider it definitively proven that observation of W core-level thresholds in O^+ desorption mandate that the species originates from WO_3 nuclei. What happens in the KF process to the simple predictions of RMG theory concerning the relationship of ion kinetic energies and relative intensities to bonding geometry is unclear, but since their validity was already questionable they should probably be discarded completely.

Finally , it should be mentioned that at least one other theory [183] for ion desorption has been proposed. In it, the first step involves ionization of the adparticle and movement *towards* the metal surface. Neutralization takes place by tunnelling *from* the substrate and re-ionization by tunnelling to the substrate can leave the adatom in an ionized state with high KE, leading to ejection. There is no experimental evidence available to test this mechanism but as the authors point out, it is capable of accounting for neutral and negative ion desorption, while the KF mechanism is not.

Summarizing so far then, wherever the KF mechanism is demonstrated for W/O situations, we can see that (a) any argument relating ion kinetic energies or relative yields to adatom proximity to the surface is unsupportable, and (b) it is likely, but not conclusively proven, that oxide nuclei are involved and therefore the ESD state represents minority species with very high yield. For those cases where ESD and PSD studies do not show a KF threshold behavior, the RMG valence level excitation theory may be valid and therefore the correlation of KEs and relative yields with adatom distance above the surface may be more meaningful. We should stress that the two W/O cases where the minority O species dominated the O^+ ESD and the KF mechanism was found to be operative are the only cases for which the KF mechanism has been tested [118, 182]. Thus, it is at present unknown to what extent any O^+ ESD active states are formed by the RMG mechanism as opposed to the KF mechanism. We now turn to the individual ESD work on W{110}, {100} and {111}.

2.8.1 W{110}

All the evidence available suggests that none of the majority species oxygen on flat {110} surfaces up to a coverage of $\theta \approx 0.6$ mL is ESD active and therefore any observation of O^+ desorption in this range is indicative of adsorption either at sites other than {110} (i.e. defects or steps) or of local high coverage regions (i.e. $\theta \geqslant 1$ mL) or even localized W oxide nuclei. Engel et al. [68] first addressed this question in sufficient detail to draw some definitive conclusions. Working in the O coverage range of $\theta = 0$–0.55 mL at 300 K, they showed that no O^+ desorption occurred below $\theta \approx 0.33$ mL and that there was a linear increase in O^+ intensity between $\theta = 0.33$ and 0.55 mL. Annealing a 300 K exposed surface to ~1600 K (well below oxygen desorption temperatures) reduced the ESD O^+ signal to zero but further exposure at 300 K completely regenerated the signal. Their conclusions from these results and other tests were that the ESD active species constituted less than 4% of the total oxygen present and therefore did not represent the majority species at least up to coverages of $\theta \approx 0.55$ mL and had no relationship to the p(2 × 1) O ordered LEED structure. Madey and Yates [114] looked at the O^+ ESD and ESDIAD characteristics from a deliberately stepped W{110} surface consisting of ~50 Å terraces (Fig. 34). Again, for O coverages where the p(2 × 1) was sharpest, no ESD was observed. At higher exposures, the ESDIAD pattern of Fig. 34 was observed which increased in strength with increased exposure up to 30 L [note the total θ does not increase much with increased exposure beyond the p(2 × 1) coverage, see Sect. 2.5.2]. Clearly, the O^+ desorption in the off-normal spots is related to oxygen located at the steps only. As in the Engel et al. study, annealing (to ~500 K) depressed the ESD signal drastically, even though the p(2 × 1) LEED pattern did not change. Clearly, the annealing either removes O from

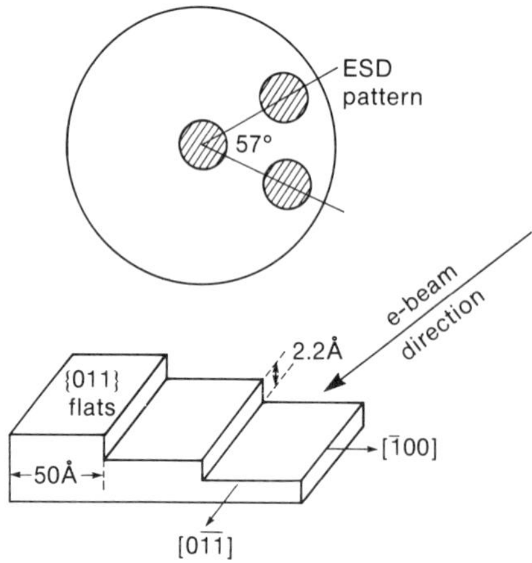

Fig. 34. Stepped W{110} surface and O^+ ESD pattern observed from it [114]. The ESD pattern is viewed on a screen parallel to the W{110} flats.

the steps to terraces or alters the adatom bonding at the steps to an ESD inactive state. The origin of the central spot in the ESDIAD pattern is less obvious. It appears and disappears together with the off-axis spots and is therefore presumably also associated with adsorption at the steps, even though the O^+ emission is normal to the surface.

Steinbrüchel and Gomer [110] also demonstrated that if a $\theta = 0.5\,mL$ 300 K surface was heated to ~600 K, the ESD signal disappeared. They, however, continued the heating process and reported that for $T > 1500\,K$ the O^+ signal returned, but the formation of this state was relatively slow (minutes). Cooling to 600 K, however, quenched the O^+ signal instantaneously. The cycling effect could be repeated many times without variations. The authors speculated that the room-temperature W/O_{ad} underwent a reconstruction at ~700 K and that this phase in turn decomposed slowly at ~1500 K to give the active ESD state. We know now, however, that surface reconstruction at $\theta = 0.5\,mL$ does not take place until 1500–1600 K, just below desorption temperatures (see Sect. 2.5.3). The reconstruction is irreversible though (i.e. it is stable on re-cooling) so the reconstructed surface cannot be directly responsible for the ESD O^+ production, since the signal disappears on cooling. In the light of later work and interpretations on the W{100} surface (see below) we suggest a similar interpretation to those systems. This involves a reconstructed state, formed at ~1500 K, which at that temperature has high mobility of both W and O atoms resulting in some O atoms occupying temporarily "high" positions and/or a high local O/W ratio. These are the ESD-active O adatoms. On recooling to 600 K, the reconstructed surface has no O adatoms occupying "high" positions or a high local O/W ratio and is therefore ESD inactive.

Madey [115] further investigated the ESDIAD on stepped W{110} surfaces by using the polyhedron crystal described in Sect. 2.5.14 and shown in Fig. 18(a). At 300 K, no ESD was observed on any of the facets until $\theta \gtrsim 0.45\,mL$ but by 4 L exposure the ESDIAD patterns of Fig. 18(c) were observed for the stepped surfaces. Further exposure simply increased their intensity. Clearly, the ESD is again associated with oxygen located at the steps and in addition the angular patterns are related to the type of step. Madey suggested O_{ad} located in the positions 4, 3 and 3′ shown in Fig. 18(b) would account for the off-normal angular patterns. "On-top" sites at steps (e.g. site 2) would account for the normal emission (the angular distributions are assumed to arise by the simple bond geometry arguments discussed later for the ESDIAD of W{111} surfaces). The majority p(2 × 1) O atoms have no ESD activity. As in the previous work, heating to ~500 K eliminated the ESD signal. Adsorption at 100 K (where there is no long-range LEED order) was also examined. Patterns of the same symmetry as at 300 K were observed on the stepped surfaces, though the details were different [Fig. 18(c)]. In addition, they appeared at a lower coverage, $\theta \approx 0.2\,mL$, than at 300 K. Warming to 300 K caused the patterns to change in detail irreversibly to those found for adsorption directly at 300 K. Thus the pattern on the flat surface disappeared completely. The interpretation of the differences observed between the stepped surfaces and the flat surface, and the changes with temperature can all be explained within the general scheme proposed for dissociative oxygen adsorption on W{110} surfaces put forward by us in Sect. 2.5.14 at 300 K; the oxygen molecule precursor state dissociates easily at steps or defects in addition to having a low probability of dissociating directly on the flat {110} terraces (small activation energy barrier). The product O_{ad} diffuses away from the steps to the terraces and eventually becomes immobile in the stable p(2 × 1) ordered structure. This oxygen produces no ESD activity. Only when the regions in the vicinity (~20 Å) of the steps or defects become highly populated do O adatoms remain at their dissociating locations (because there are no available hopping sites left). These are the active ESD adatoms. This sequence explains why the ESD signals do not occur until after the p(2 × 1) LEED structures form, even though we know from the increased S_0 values for stepped surfaces that the O_2 molecules do dissociate preferentially at the steps. It also explains why signals are observed to appear at lower coverages at 100 K on the stepped surfaces. O adatoms cannot diffuse away so easily from the steps so the average coverage at which the local coverage near steps reaches the critical value necessary to confine O atoms to ESD active sites is greatly decreased. When the 100 K adsorbed situation is warmed to 300 K, adatoms diffuse to the terraces and the 300 K ESD situation is obtained. Annealing to 500 K causes complete loss of the ESD signal because diffusion is enhanced, allowing all atoms to move to terraces and into p(2 × 1) ordered structures. Further exposure at lower temperatures repopulates the step sites and regenerates the ESD signal.

For the flat {110} surface of the polyhedron, saturation adsorption at 100 K produced a weak ESDIAD pattern [Fig. 18(c)] which seemed to have the combined symmetry of the patterns of the up and down steps of the T and B stepped surfaces. This disappears on warming to 300 K. Our suggestion would be that either or both of the following factors give rise to this effect: (i) restricted adatom mobility increases the concentration of O_{ad} near the (few) steps on the flat surface, forcing atoms that dissociate at these defects to remain there in active ESD sites, and (ii) at 100 K there is no p(2 × 1) ordering, and therefore, the local coverages in regions on the terraces can become much greater than $\theta = 0.5$ mL, a condition which may give rise to ESD activity. This last hypothesis introduces the questions of what is the special attribute of a site which gives rise to O^+ ESD activity? Madey [115] has suggested that "single coordination sites" are required. By this, he means sites where O_{ad} is not in a multiple W bond situation, such as two-fold bridge, three- or fourfold hollow sites, etc. A single W–O "on-top" bond might qualify and clearly WO_2 or WO_3 units would. The most favorable of all might be O_{ad} adsorbed on top of tungsten oxide nuclei giving essentially a WO_4^{2-} unit. (It is known that tungsten oxide surfaces give very strong ESDIAD patterns, and in the case of the stepped surfaces used by Madey the same threefold "down-step" symmetry as for the T and B faces was obtained when the stepped surfaces were deliberately heavily oxidized.) Thus, the general statement can probably be made that to get O^+ ESD activity on W{110} surfaces a high local O/W ratio must be achieved. All the positions suggested by Madey in Fig. 18(b) qualify, as do W oxide nuclei, O_{ads} on W oxide nuclei and a high local coverage (O/W $\approx$ 1) on a {110} flat surface.

Weng [184] has discussed the mechanism and binding sites for O^+ yield species on a flat W{110} surface. In his paper, he refers to unpublished data indicating that the KF mechanism is a major contributor to the O^+ desorption observed at high coverage. The suggested maximal valency rule for the KF mechanism would then indicate that the W atoms involved are formally W^{6+}, i.e. the species involved are WO_3 nuclei. (Weng does not draw this conclusion. He does not refer to the maximal valency rule, but simply suggests that evidence of a KF mechanism indicates a "direct binding" between O_{ad} and W atom.) For the stepped surfaces, the suggestion that oxide nuclei are responsible for the ESDIAD patterns implies that the oxide nuclei grow at the steps with the right orientation for the observed ESDIAD symmetries. It also implies that heating to 800 K "dissolves" these nuclei in favor of ordered p(2 $\times$ 1) formations. This seems highly unlikely and therefore casts doubt on the suggestion that oxide nuclei are involved and therefore on the validity of the maximal valency rule for the KF mechanism. Weng has presented some arguments to support the suggestion that the basic requirement for O^+ desorption is simply a singly coordinated binding site and that therefore the ESD signals observed on flat W{110} surfaces at high coverage at 300 K and lower coverage at 100 K do not necessarily originate from steps or defects but simply represent those (minority) O atoms located in "atop" positions. This suggestion is not in disagreement with our suggestion made above that on the flat {110} surface, a high local coverage (O/W $\approx$ 1) must be achieved for O^+ ESD activity.

2.8.2 W{100}

The O^+ ESD from W{100}/O surfaces is extremely complex and only a carefully correlated ESDIAD/ESDIED study over the full range of O coverages and temperature is likely to have any chance of fully understanding the processes and species involved in generating the observed ESD active states. Keeping in mind that we already know that the surface reconstruction process occurring at ~650 K involves domain growth of reconstructed nuclei with massive atomic transport (Sects. 2.6.2 and 2.6.3) it is also clear that annealing times at any given T may strongly affect the ESD. Bauer and co-workers [90, 106] have become very aware of these problems and have made some progress in understanding these phenomena. A complete understanding of all the ESD features is still some way off, however.

We will start the discussion from the summary given by Madey et al. [85] of the situation in 1975. For adsorption of less than $\sim 8 \times 10^{14}$ atoms cm^{-2} ($\theta \approx 0.7$) for $T \geqslant 77$ K a low ion yield ($< 5 \times 10^{-9}$ ions electron^{-1}) ESD O^+

state is formed, termed the β_2 state. For higher oxygen coverages, a high yield ($\gtrsim 10^{-6}$ ions electron^{-1}) state is formed, termed the β_1 state. It had been realized that the β_1 state arose from a minority surface species and had been attributed to either molecular oxygen or tungsten oxide nuclei [185]. It was thought that the β_2 state was directly associated with the majority species [85]. The later ESDIED work of Bauer and co-workers showed that things are very much more complex than this brief summary might imply, but in 1975 Madey et al. [85] carried out an ESDIAD study on the β_2 and β_1 states. A weak diffusive cross pattern was observed for the β_2 state [Fig. 35(a)] which disappeared on heating to $\sim$700 K. The speculative conclusion made by the authors for this pattern was that it represented O in the fourfold hollow sites. Simple bond direction arguments, however, would predict only normal emission for such an adsorption geometry, and indeed this is what the computer simulations of Gersten et al. give [130] (Fig. 36). In a later review, Madey and Yates [114] get around this problem by suggesting that

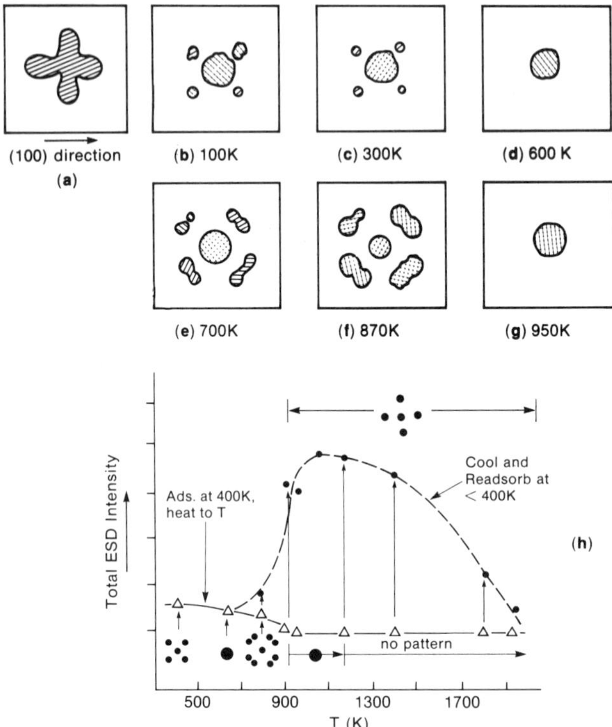

Fig. 35. ESDIAD patterns produced for W{100}/O_2 [85]. In (a)–(g), strong intensity is represented by cross-hatching, weak intensity by hatching. (a) β_2 low-yield ESD state; adsorption at $77\,K < T < 700\,K$ to $\theta \lesssim 0.8$ mL. (b)–(g) β_1 high-yield ESD state. (b) represents saturation adsorption at 100 K. (c)–(g) represent successive 10 s anneals followed by recooling to 100 K and remeasurement. (h) Total ESD intensities and ESDIAD patterns for saturation adsorption at 400 K as a function of annealing, recooling and re-adsorbing (2 L further exposure).

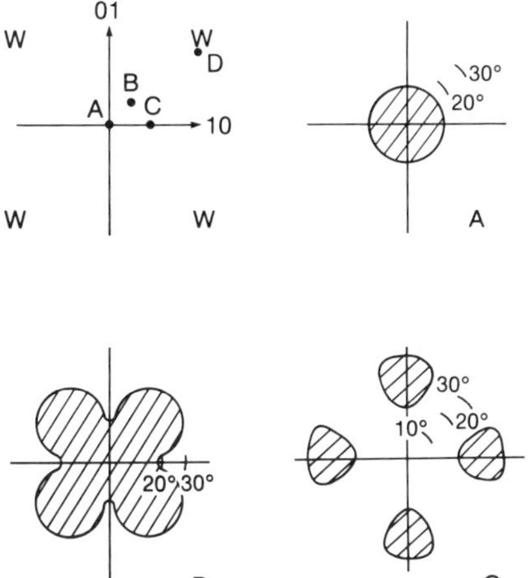

Fig. 36. Predicted ESDIAD patterns from simulations [130] of an adsorbate atom adsorbed in positions A, B and C above a {100} W lattice. An O_2 molecule adsorbed at D could also give pattern B or C (see text).

the cross-pattern results from the anisotropy in the ground state potential for an O adsorbed in the fourfold site. (This arbitrary deviation from previous arguments relating adatom geometry to ESDIAD patterns seems rather like having your cake and eating it.) The suggestion so far then is that, below reconstruction temperatures, the β_2 ESD is characteristic of the majority O_{ads}. Why, then, does the ESDIAD pattern disappear on heating to ~ 700 K? Very likely it does not disappear but just gets weaker and too diffuse to show up as an ESDIAD pattern. Certainly the detailed ESDIED studies of Prigge et al. [106] show that, over the range $\theta = 0.5$–0.75 mL, the total normal emission ESD is not much different after annealing to 700 K. Another possibility is that the surface has already passed through the reconstruction T (~ 650 K), though, as will be seen later, this should lead to changes in ESDIAD patterns and energy distributions, but not to a complete loss of ESD activity. One other major problem remains in associating the β_2 ESD with the majority adsorbed species; at 300 K, the intensity does not increase in proportion with θ but goes through a maximum somewhat near $\theta = 0.5$ mL and reaches a minimum at $\theta \approx 0.75$ mL, just before the β_1 state starts to be populated [85]. On the other hand, Bellard and Williams [186] have shown the depletion of the β_2 state by ESD is approximately paralleled by a decrease in the O Auger signal, indicating association between total coverage and β_2 population. Not much else can be said about the β_2 state without further ESDIAD/ESDIED versus θ measurement being made and so the questions of whether it represents the majority adsorbed species and whether the "hazy cross" ESDIAD pattern is indicative of fourfold hollow site adsorption must remain open.

The "β_1" state, or rather range of ESD states (since a variety of ESD ion energies and ESDIAD patterns are formed), formed at higher coverage and/or after the reconstruction process has started have been studied in more detail than the "β_2" state. For 300–400 K adsorption, the total saturation intensity of "β_1" is at least two orders of magnitude greater than β_2. It grows rapidly over the coverage range $\sim$0.8–1.1 mL, is totally uncorrelated with θ and when depleted by the ESD process produces only small decreases in total θ, i.e. there is no doubt that much of the O^+ ESD activity is from minority species. The three most informative studies are the ESDIAD study of Madey et al. [85], the ESDIED study of Prigge et al. [106] and the ESDIED study by Kramer and Bauer [90] on the $\theta = 0.47$–0.57 mL surface during the surface reconstruction process. Unfortunately, the two ESDIED studies deal primarily with normal emission O^+, and the ESDIAD study has no detail concerning ion energies so cross-correlations are not easy. Nevertheless, a lot of progress in understanding the O^+ ESD has been made through these studies. We start with the ESDIAD study. The sequence of ESDIAD patterns observed under various conditions of annealing is schematically represented in Fig. 35(b)–(g). The starting point is saturation exposure (40 L) at 100 K [Fig. 35(b)] and in all cases the ESDIAD measurements were made after recooling to 100 K. Presumably the saturation exposure corresponds to $\theta \approx 1.4$ mL (see Sect. 2.6.1). The annealing procedure should lead to the onset of reconstruction with excess O somewhere on top of the reconstructed surface in the 600–800 K range (Sect. 2.6.3). This is exactly the region where there is a pronounced transition from an intense normal-only emission O^+ ESD to one which has equal intensity in each of the four lobed off-normal spots [Fig. 35(d)–(f)]. Some $\sim$8/9 of the total intensity is in the off-normal spots in Fig. 35(f). In all other conditions of the annealing sequence, the majority intensity is from the normal emission. We therefore associate the fourfold patterns of Fig. 35(f) in some manner with reconstruction. We note, however, that Madey et al. [85] followed the LEED during their annealing sequences and that at 865 K, the (weak) (1×4) LEED patterns persisted, changing to the reconstructed (1×2) only at $T > 1100$ K. Thus, it would seem that the strongly lobed part of the ESDIAD pattern of Fig. 35(f) is representative of some low concentration intermediate situation during reconstruction and that the final fully reconstructed surface gives only normal emission. This would be in accord with the knowledge that at least part of the "β_1" ESD state represents minority O species. If the O-saturated surface which has been annealed to temperatures well above the reconstruction temperature and then recooled to 400 K is re-exposed to a few L of oxygen, the increase in total oxygen coverage is trivial, but the total intensity of the "β_1" state increases enormously, and the symmetry of the pattern changes, as indicated in Fig. 35(h). Madey et al. [85] suggested that this state represented minority O adsorbed on top of an already reconstructed surface, which had undergone facetting. Since high θ, W{100} surfaces apparently facet to {110} planes (Sect. 2.6.3) and O on W{110} gives either no ESD activity or two- or threefold symmetry patterns associated with steps or defects, it is hard to reconcile the strong four-lobed pattern produced by re-adsorption after heating [Fig. 35(h)] with facetting. It certainly seems true, however, that it does represent minority O formed on a W{100}/O surface which has completed the reconstruction process. The actual local geometric position of an O atom with respect to the W{100} lattice which could give rise to the fourfold lobes of Fig. 35(b) and (h) are shown in Fig. 36, as found by the simulations of Gersten et al. [130]. Position B could signify a WO_2 local bonding unit, as suggested by Madey and Yates [114], but it should be pointed out (a) that there is not a tremendous local difference between positions A, B or C, even though they give rise to drastically different ESDIAD simulation patterns, (b) the hazy cross of the β_2 adsorption state has the same symmetry as that of the high-intensity re-adsorption component of the β_1 state [Fig. 35(h) and position C of the simulation] even though Madey and Yates argue that the β_2 hazy cross represents adsorption at point A (see earlier), and (c) an O_2 molecule adsorbed parallel to the W{100} surface at position D is also capable of generating in the simulations the same symmetries as atomic adsorption at B or C, depending on the rotational orientation of the molecule [130]. Clearly, the ESDIAD patterns are at present of little genuine use in determining the local geometries on the W{100}/O surface, and we must, for the moment, settle for their finger-printing capabilities and their detection capability for rather specific (but unidentified) minority species being formed during the adsorption, annealing and surface reconstruction processes.

We now turn our attention to the two ESDIED papers. These provide rather more insight than the ESDIAD paper, probably because the ESDIED data were taken in conjunction with information from several other techniques.

In the study by Prigge et al. [106], the first important clarification point concerns the very high yield component of the "β_1" state. By choosing the pre-coverages $\theta = 5/8\,\mathrm{mL}$ and $\theta = 1.25\,\mathrm{mL}$ prior to annealing at 1100 K to cause reconstruction, and then choosing a temperature below and above 800 K (the reconstruction temperature) for the subsequent re-adsorption step, they were able to show that the required condition to produce the minority high yield component of the "β_1" state is that reconstruction must be achieved with a total θ of $\gtrsim 5/8\,\mathrm{mL}$. Also since a pre-coverage of $\theta = 5/8\,\mathrm{mL}$ followed by annealing to 1100 K and re-exposure at 800 K does produce the high-yield state, facetting can be excluded as the cause of the high-yield state. This is so because facetting will not occur at any annealing T for a θ of 5/8 mL, and re-exposure at 800 K is below the facetting temperature for any θ. The high-yield component has a KE of 7.4 eV and is considered by Prigge et al. to represent minority W oxide crystallites. It forms on a reconstructed surface below facetting temperatures because the reconstructed surface is saturated and further O can only be taken up by growing oxide nuclei. If the surface is subsequently annealed above $\sim 1050\,\mathrm{K}$, the state disappears because facetting occurs and the excess oxygen of the W oxide crystallites can be incorporated into the increased area of the facetted surface. In a comparison with the ESDIAD work of Madey et al., we therefore conclude that the final parallel-orientation fourfold pattern of Fig. 35(h), produced after annealing and re-adsorption, represents this minority oxide species.

The rest of the Prigge et al. paper mostly concerns classifying the different components of the β_1 state in terms of ion energies, and their yield behavior as a function of (a) pre-coverage and (b) annealing temperature. Both the pre-coverage and the remeasurement after annealing to temperature T were performed at 300 K. No redosing is performed after annealing, thereby excluding the very high-yield 7.4 eV W oxide minority species from this set of measurements. The results are complex and presented in a complicated manner in the paper. We attempt to summarize them as simply as possible here. For coverages between $\theta \approx 0.6$ mL and saturation, the only ion desorption for exposure at 300 K without any annealing is at an ion energy of ~7.5 eV. There is an indication of a slight decrease in the energy (~7.5 → ~7.3 eV) at the $\theta \approx 0.75$ mL point. Prigge et al. attribute the 7.5/7.3 eV state to the majority species adsorbed overlayer oxygen. They suggest that the shift from 7.5 to 7.3 eV occurs because of the attempts to form a more hexagonal overlayer with increased coverage which will force the O_{ads} to more "on-top" positions and should therefore decrease the ion energies. There is also a strong increase in yield accompanying the shift in KE to 7.3 eV which could also be a result of moving into the slightly more on-top position. The implication is that this unannealed and non-re-exposed 7.3–7.5 eV state is entirely unrelated to the very high yield 7.4 eV state ascribed to W oxide above. It is therefore unfortunate that both are usually given the "β_1" state classification. After annealing, the detailed results depend on the pre-coverage but the general pattern is similar at least over the θ = ~0.6–1 mL range. The 7.3/7.5 eV desorption intensity decreases rapidly as annealing temperatures approach 500–600 K (lower T for higher coverage). It has disappeared by 800–900 K and has been replaced entirely by an 8 eV peak with low intensity. Since surface reconstruction is complete by this temperature, the 8 eV peak must be associated directly with O in the reconstructed layer. The increase in the KE and decrease in yield is in accord with the expectation of the RGM model for O moving in towards the surface. During the reconstruction process, another ESD active state is formed with an energy of 5.5 eV. Its intensity and its duration as a function of increased annealing T is a very sensitive function of the exact pre-coverage, θ. From the Prigge et al. data, the authors concluded that the 5.5 eV peak represented excess chemisorbed O located in high on-top positions on top of the W/O reconstructed layer. The high position explains the low KE. Note that this state is different from the W oxide species which give rise to the very high yield 7.4 eV re-exposure state. Clearly, there should only be excess O if $\theta > 0.5$ mL since 0.5 mL can be incorporated into the surface in the reconstruction process (see Sect. 2.6.3) and therefore the 5.5 eV ESD state should only be formed for $\theta > 0.5$ mL. In addition, the supposed high on-top position of the 5.5 eV state should imply a very high yield (RGM model) and hence the O concentration giving rise to the 5.5 eV peak could be very small (i.e. we have a technique for looking effectively at a minority species). So far, then, we have four different "β_1" O^+ ESD active states, two of them minority states (one an oxide, one chemisorbed O following reconstruction) and two considered to be more representative of majority oxygen (for one of which the yield increases dramatically as its overlayer structure changes with coverage). These results are summarized in Table 7.

We now turn briefly to the ESDIED study of Kramer and Bauer [90]. This paper was extensively referred to in Sect. 2.6 since it was the most important piece of work delineating the detailed mechanism of the W{100}/O reconstruction process at $\theta = 0.5$ mL. The results discussed there were $\Delta\phi$, LEED, and AES. Here we discuss their detailed investigation of the nature of the 5.5 eV O^+ ESD peak, which Prigge et al. had already determined originated from excess O chemisorbed on-top of the reconstructed surface after annealing an O covered surface of $\theta > 0.5$ mL through the reconstruction temperature. Kramer and Bauer calibrated their pre-coverages using AES and assigning the first discontinuity in AES O intensity versus exposure to $\theta = 0.5$ mL (see Sect. 2.6.1). The ESD signals for this nominal 0.5 mL situation as a function of annealing temperature are shown in Fig. 37. The behavior is, in general, in agreement with that of Prigge et al. Also shown is the effect of varying θ slightly below and above a nominal $\theta = 0.5$ mL on

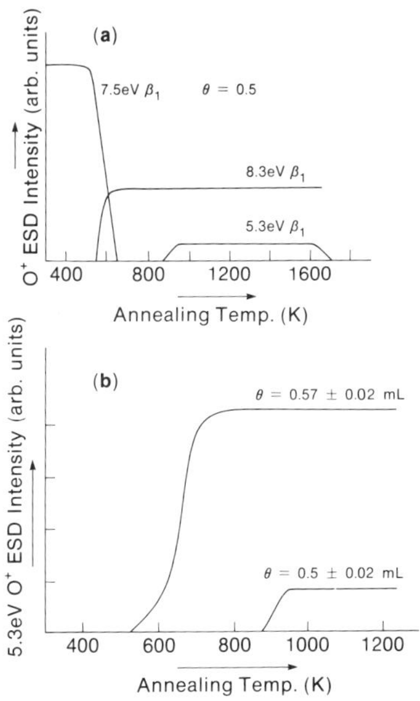

Fig. 37. (a) Effect on O^+ ESD states of annealing a 300 K 0.5 mL W{100}/O surface. (b) Effect of coverage on the 5.5 eV O^+ state formed during the reconstruction process. For a coverage of $\theta \leqslant 0.47 \pm 0.02$ mL, no 5.5 eV state intensity was observed [90].

TABLE 7

Summary of the characteristics of the ESD active states for W{100}/O

Conventional designation	Coverage range (ml)	Temperature conditions	Intensity	Type of species	Ion energy (eV)	ESDIAD pattern
β_2	0–0.6	Ads. between 77 and 600 K	Very weak	?	7.5	Rotated hazy cross
β_1	0.6–0.75	Ads. 300–600 K	Medium	Majority species; overlayer	7.5	Rotated fourfold
	0.75–saturation	Ads. 300–600 K	Strong	A more hexagonal overlayer	7.3	Rotated fourfold
β_1	0.6–1	Annealed $>$ 600–700 K	Medium	Majority; reconstructed W/O surface	8.0–8.3	Central spot?
"Fast" β_1	$>5/8$	Annealed $> \sim 650$ K, re-exposed at >400 K	Very strong	Minority; oxide on reconstructed W/O surface	7.4	Parallel fourfold
β_1	>0.5	Anneal above 600–900 K, depending on θ	Variable f(θ, anneal T)	Minority; excess O on reconstructed W/O surface	5.3–5.5	Rotated fourfold?
	$\theta = 0.5$ only studied	$\gtrsim 900$ K, measured at temperature	Variable, higher at higher T	Minority; O atom of reconstructed W/O surface occupying temporary high positions	6.3	

the 5.5 eV ESD state. This conclusively demonstrates that the 5.5 eV state cannot be formed at a θ significantly below 0.5 mL. Referring to the model derived by Kramer and Bauer for the place exchange and domain growth of the reconstructed surface [Fig. 26(c)], it can be seen that, for the limiting small domain size [the single place exchange "unit" shown in Fig. 26(c)], O atoms could be accommodated in the reconstructed surface up to $\theta = 0.66$ mL. In the opposite limit of a single domain covering the whole surface, the limit of O accommodation is $\theta = 0.5$ mL. The reconstructed domains grow as a function of annealing time for any given T above the reconstruction temperature. Therefore, if the θ precoverage is > 0.66 mL, the 5.5 eV peak must appear as soon as the reconstruction temperature is reached, but for $0.5\,\text{mL} < \theta < 0.66\,\text{mL}$, the 5.5 eV peak will appear at higher and higher T the lower is θ, since only as the reconstructed domains grow will the excess O ($\theta > 0.5$ mL) be forced to take up a position outside the reconstructed domain. For θ precoverage $\leqslant 0.5$ mL, no 5.5 eV peak should occur, e.g. the $\theta = 0.47 \pm 0.02$ mL case. From these arguments, it follows that the $\theta = 0.5 \pm 0.02$ mL case must actually be slightly greater than $\theta = 0.5$ mL. Kramer and Bauer point out that the 5.5 eV peak saturation intensity at $\theta = 0.57 \pm 0.02$ mL is five times that of the $\theta = 0.50 \pm 0.02$ mL case. If the 5.5 eV O^+ ESD intensity is directly proportional to the concentration of the excess O, this then implies that the true coverage for the $\theta = 0.5$ mL case is $0.5 + (0.57 \pm 0.02 - 0.50 \pm 0.02)/5 = 0.514 \pm 0.004$ mL. Thus the intensity of the 5.5 eV O + ESD state is an accurate measure of the oxygen coverage in the $\theta = 0.5$–0.6 mL range.

We conclude, therefore, that the 5.5 eV state is representative of excess O adsorbed on top of a reconstructed W/O surface. It has a high yield allowing the coverages of the excess oxygen to be determined rather accurately between a total θ of ~ 0.5 and ~ 0.66 mL.

One other important result came out of the Kramer and Bauer W{100}/O ESDIED work. They performed ESDIED measurements actually at T, in addition to annealing to T and remeasuring at low T. They found, in agreement with earlier work [77], that the total ion flux increased with T above ~ 800 K. They showed, however, that all the extra intensity went into a new state at 6.3 eV. Their explanation of this state is that it represents those atoms which occupy high positions with respect to the surface after passing through the order–disorder transitions of the already reconstructed surface at ~ 850 K (see Sect. 2.6). They estimated the activation energy of the process as ~ 0.35 eV from an Arrhenius plot of the 6.3 eV emission intensity. This implies that the observed signal must originate from only a few O atoms (5% at ~ 1350 K would occupy the high positions for an activation energy of 0.35 eV).

Summarizing the ESD results we note that, though much has been learned about fingerprinting the ESD active states, and the nature of some of these states is understood (e.g. oxide nuclei, excess O on reconstructed domains) nothing has yet been established about the W/O adsorption geometries, despite the ESDIAD patterns, and there are still questions concerning whether some of the states are or are not directly related to the majority species O as monitored by other techniques.

Finally, in this section on W{100}/O ESD we make a brief comment about the PSD study [182] which confirmed the KF mechanism of O^+ production. The sample was a 100 L exposed surface at 300 K. The observation was that the O^+ PDS thresholds coincided with W(4*f*) and (5*p*) core level BEs. The conclusion was that the KF mechanism had been demonstrated for O^+ desorption. The relevance to metal substrates is ambiguous, however, since the observed signal could all be from WO_3 oxide sites. From the preceding discussions, it should be obvious that a lot more effort is needed in preparing the W{100}/O surfaces and characterizing (by ion energies and intensities) the many ESD O^+ active state prior to performing the threshold measurements, which can distinguish RGM from KF mechanisms. It is also clear that, whatever the conclusion concerning the mechanism, the active ESD states do not all originate from WO_3 nuclei. It would be very interesting to see whether those which definitely do not originate from WO_3 nuclei do or do not have a KF mechanism of production, since this would be a definitive test of the maximal valency rule of KF.

2.8.3 W{111}

ESDIAD data are available from Niehus [83] and from Madey et al. [117] with computer simulations of the ESDIAD process from Janow and Tzoar [131], and from Preuss [91]. No substantial body of ESDIED data exists. We will refer to the data of Niehus since it is the more complete. At 300 K, and low exposures (~0.5 L) a three-spot pattern of orientation, I (Fig. 38), is observed; at higher exposures this fades and is replaced by a single normal spot of comparable intensity, II. Near saturation a three-spot orientation, III (Fig. 38), is formed with 10–20 times the intensity of the two earlier patterns. The computer simulations [91, 131] involve calculating the ejected ion trajectories by classical dynamics following a quantum mechanical treatment to describe the electronic excitation from the ground state (MA) to a repulsive final state ($M + A^+$) from which the ionized adatom, A^+, may desorb. In addition to the many assumptions that must be made concerning the potentials used to describe the initial and final states, a major problem in the simulation is neglect of the re-neutralization process during the ejection. Crude attempts have been made to overcome this problem and give

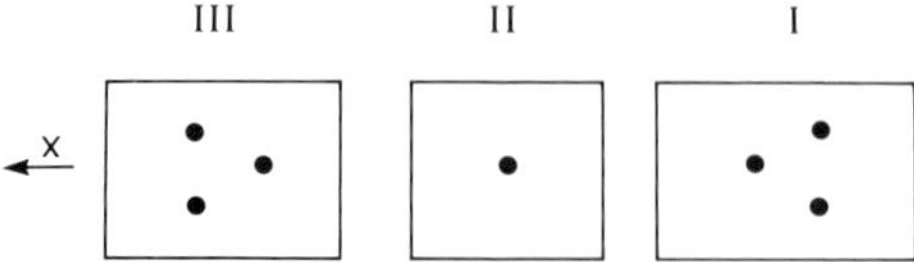

Fig. 38. ESDIAD patterns observed for adsorption of oxygen on W{111} at 300 K [83]. I, Low exposure (> 1 L); II, intermediate exposure (~2–5 L); III, high exposure (> 7 L). The orientation of the patterns with respect to the W{111} substrate azimuthal orientation can be determined by referring to the unit cell of Fig. 32. Direction, *x*, is the same in both figures.

some feel for the expected relative intensities of ESDIAD patterns, in addition to ESDIAD geometries for different adatom configurations. In these attempts, the dominating factor in the re-neutralization process is assumed to be simply the starting distance from the surface; adatoms in high overlayer positions give rise to ions which have a much better survival chance than adatoms closer to or below the surface plane. The general results which are obtained within the above simulation framework can be summarized as follows. Each adsorption site candidate produces a single emission spot. In many cases, the dominant contribution to the ejection ion energies comes from interaction with a single substrate atom and to the extent that this is true the emission spot geometry then reflects the chemisorption bond direction, i.e. ions are ejected approximately along substrate adatom bond directions (clearly this concept breaks down when delocalization on multi-centered bonds between adatoms and substrate atoms are involved. In these cases, rather diffuse angular patterns will be expected). Unfortunately, there are usually several choices of adsorbate site for which the simulations give ion desorption patterns consistent with the ESDIAD geometric patterns actually observed. The details of the exact angle of emission, the mean ion energies, and the desorption and re-neutralization cross-sections are what are needed to distinguish one acceptable geometry from another. It is exactly these parameters which are strongly dependent on the assumptions and parameterized features involved in the simulations. It is not surprising, therefore, that different authors [83, 91] arrive at different conclusions as to the most likely adsorption sites based on a comparison of their simulations with experiment.

The two simulations of the W{111}/O system differ in the mathematical details of the calculations and in addition Preuss [91] attempted to calculate ion survival probabilities to compare with experimental ESD intensities. Referring to the unit cell diagram in Fig. 32, the possible sites for O_{ads} considered in the simulations are marked using the terminology of Preuss. From the general features of these types of simulation discussed above, it is clear that sites T_1, T_2 and T_3 would give rise to single-spot normal emission, sites B_1 and B_2 will give rise to a threefold pattern orientation I, and site S will give a threefold pattern orientation, III. The patterns have threefold symmetry simply because there are three equivalent sites oriented 120° to each other, each site giving rise to an off-normal spot (see Fig. 2). Niehus [83] compared his data with the spot patterns produced by the simulations of Janow and Tzoar and concluded that initial exposure, giving pattern orientation I, was due to O atoms adsorbed at B_2 sites. At intermediate coverages they suggest O_{ad} has moved to sites T_2 and T_3, replacing the three-spot pattern by normal emission, and at saturation the ESDIAD is dominated by emission of O_{ad} located at S sites, giving rise to orientation III. Preuss, on the other hand, compared his later simulations with the data of Neihus and favored the sequence B_1, T_2, S on the basis of a closer agreement between his calculated ion survival probabilities for these sites and the experimentally determined ESD intensities for the three patterns. In Preuss's simulations, the B_2 site, proposed by Niehus for initial adsorption, has an extremely low ion survival probability (as does T_3), and therefore was excluded as a possible ESD active site by Preuss in favor of B_1. It should be noted, however, that the ion survival probability calculated for site S is not greater than for B_1, whereas experimentally a 10–20 fold increase in ESD intensity is observed for pattern III compared with I.

One sees then that even with the dubious assumptions that the patterns represent majority species, it is only at present really possible to extract the symmetry of site responsible for a particular symmetry ESDIAD pattern. Distinction between different sites of the same symmetry requires a better understanding of the details of both the simulated and experimental intensities and the angular distributions, rather than just the symmetries of the patterns.

What about the minority/majority species argument for this system? Niehus [83] based his claim that ESDIAD patterns are representative of majority species on his observed correlation between the decay of ESD O^+ intensity with the decay of O AES intensity using the same beam current. Unfortunately, only the high coverage extreme (120 L exposure) was monitored (i.e. the orientation III pattern) using a very high AES current. It is known that high O coverage W surfaces are susceptible to electron beam desorption at high incident currents, and Niehus's monitoring of the decrease of the O AES intensity certainly verifies this. It is by no means established, however, that the major species being desorbed by the electron beam is the O^+ species giving rise to the ESDIAD pattern. Madey et al. believe that the O^+ ESD [117] and PSD [118] for the system represents only minority oxide species. This is based on the facts that (a) the ion yield versus electron (or photon) energy curve is the same as for a heavily oxidized surface and this curve is compatible only with the KF ESD mechanism which requires a maximal valency W^{6+} species, and (b) the O^+ ESD intensity at $\theta \approx 0.5\theta_{sat}$ is only about one hundredth that of the oxidized surface.

Both Madey et al. and Niehus have examined the ESD/ESDIAD for surfaces exposed at 300 K and annealed to high T. Under these conditions, facetting to $\{112\}$ and $\{110\}$ planes is, of course, expected (see Sect. 2.6.3). A rich sequence of ESDIAD patterns is observed, but they all consist of three-spot patterns of orientation either I or III, though the emission angle and intensities vary. In this case both authors agree that no correlation between the ESD intensities and surface coverages exist. The ESDIAD patterns may well be representative of desorption normal to the facetted planes on the surface.

2.8.4 Conclusions

This review of ESD for W/O surfaces has illustrated both the unresolved problems with the ESD technique and, in specific cases, the definitive additional information that can be obtained. In the problem category, the question of whether the ESD signal arises from the majority O species or not is uppermost. For W{111} there is complete disagreement on whether the 300 K and below adsorption ESDIAD patterns represent majority species. Our opinion, in agreement with that of Madey et al. [117, 118], is that they do not. For W{111}/O surfaces annealed to high T it seems to be agreed that the ESDIAD patterns are not from majority species. For W{110} it seems clear that at $\theta \lesssim 0.5$ all ESD is from minority species [68]. Work on deliberately stepped W{110} surfaces confirms this in that all the ESDIAD patterns observed can be clearly traced to the activity of the steps [114, 115]. For W{110} for $\theta > 0.5$ and for W{100} at all coverages it is possible that some fraction of the ESD signals observed originate from majority species. For W{100} the detailed ESDIED and ESDIAD studies done in conjunction with other techniques have shown that many different ESD active states can be identified, some of which are majority and some minority species [90]. Two of these states have definitively been linked to oxygen species on top of the reconstructed surface. One is thought to be W oxide nuclei and the other excess O which could not be incorporated into the reconstructed surface during reconstruction. The latter O^+ ESD state intensity can be used quantitatively to determine the amount of this type of oxygen present. The second problem, accepting that some majority O species and some minority O species are O^+ ESD active, is establishing what type of bonding or geometry configuration must be achieved to produce activity. There seems to be consensus that a high local O/W ratio is required, but not on exactly what this means. Certainly oxide nuclei and oxygen on top of oxide nuclei qualify. Very likely any "single" O–W bond qualifies, such as in an on-top site, but multicentered bonds (e.g. O in two- or fourfold sites) do not. High θ, such as the $\theta > 1$ overlayer compression stages where the O/W ratio is high, probably also qualifies even though there is no direct O–W "single-bond" to speak of. The third problem is the reliable extraction of adatom geometries from ESDIAD patterns. The empirical nature of the ESDIAD simulations so far, plus the lack of sensible ion yield calculations to go with the geometry pattern simulation, results in a situation where, at best, only the symmetry of adsorbate sites can be extracted from the experimental data. Unfortunately there are usually many choices of adatom position which will give the right symmetry in the simulations.

Finally, with respect to the implications of the KF mechanism, though we agree with the conclusion drawn by the authors [118, 182] that in the two studies made, the O^+ ESD activity which showed a KF mechanism comes from WO_3 oxide nuclei, we do not consider it yet proven that only maximal valency oxide species can give a KF mechanism O^+. Thus the observation of a KF O^+ ESD mechanism does not yet prove that the O originates from WO_3 nuclei and is, therefore, always a minority species.

References pp. 381–388

2.9 THERMAL DESORPTION FROM W/O SURFACES

Desorption of a species from a surface can be monitored in several ways. We shall restrict our discussions here to the method most commonly used, both in general and for W/O studies, that of thermally programmed desorption (TPD). In this method, measurement of the rate of desorption of a species is made as T_s is increased in a programmed manner. Assuming that only a single species desorbs (mass determined by a mass spectrometer) from a single adsorption state, such measurements yield the activation energy for the rate-controlling step in desorption. If the adsorption is non-activated, the desorption energy, E_d, is approximately equal to the differential heat of adsorption. Measurements of this type produce most of the experimentally determined heats of adsorption and estimates of metal adatom bond energies. Since desorption is a process involving an exponential energy dependence, the desorption temperature can be very sensitive to changes in substrate–adsorbate bonding which are too subtle to detect by other spectroscopic methods much as electron spectroscopy. Therefore, if multiple adsorption states (of a single species) with different heats of adsorption coexist on the surface, they may appear as multiple peaks in the TPD spectra and a desorption energy can be assigned to each.

Even assuming that the correct interpretation of a multiple-peaked TPD spectrum is the coexistence, at the temperature of adsorption, T_{ads}, of multiple, independent, adsorption states, the analysis of the multiple peaks in terms of a set of E_d values is still ambiguous and non-unique because of the assumptions necessary in deriving an E_d value from the coverage dependence of both position (desorption temperature) and line shape of the TPD peak. In practice, at least three further complications can arise:

(1) A single adsorption state at T_{ads} can give rise to more than one TPD peak through repulsive lateral interactions between the adspecies. This is an example of a desorption spectrum not representing the situation that existed at T_{ads} prior to the flash. As T is raised and desorption begins the remaining adspecies redistribute themselves on the surface and subsequently desorb at a different T_d.

(2) A single adsorption state, or multiple adsorption states existing at T_{ads}, may convert to other adsorption states as T is raised prior to any desorption. T_d is then only representative of the state populations existing at desorption temperatures.

(3) More than one species may be desorbed, e.g. dissociation or recombination of already dissociated molecular fragments may occur, or species involving combinations of fragments of the adsorbate and the substrate may occur in the desorption spectrum.

In the case of W/O surfaces, it is this last situation which has led to great confusion and originally erroneous conclusions concerning the structural nature of the W/O surfaces, as determined from TPD. At certain coverages, the desorption of WO_2, WO_3 and WO species has been observed and their presence, for one reason or another, has been erroneously taken to imply the presence on the surface, at T_{ads}, of three-dimensional W oxide structures.

Most of what turns out to be the relevant information on desorption spectra for W{110}/O and W{100}/O is contained in the papers of Bauer and co-workers. It is instructive, however, to see how the misconceptions concerning the interpretation of TPD of W/O systems arise.

In 1971, King, et al. [187] studied the desorption spectra of oxygen adsorbed on polycrystalline W filaments. For a T_{ads} of 300 K, they observed O atom desorption only (at T_d = 2000–2400 K) for estimated coverages of up to either 6.4×10^{14} atom cm^{-2} (0.45 ml), 8×10^{14} atoms cm^{-2} (0.56 ml) or 1.2×10^{14} atoms cm^{-2} (0.85 ml). All these figures are mentioned in the paper at various places. Above this coverage most of the additional oxygen desorbs in the form of W oxide molecules (WO_3, WO_2 and WO mainly) at a lower T_d (1400–1800 K). The limiting uptake at 300 K was given as 30.4×10^{14} atoms cm^{-2} (equivalent to 2.15 ml). The interpretation of King et al. was that, at O coverages below the onset of W oxides in desorption, the adsorbate was atomic oxygen and above the coverage of the onset of W oxides in desorption, the additional O was present on the surface in the form of either a reconstructed W/O surface or, at higher coverages, three-dimensional oxide. The main reason for this conclusion was not the trivial (but erroneous) one that desorption of oxides must signify the presence of oxides on the surface, but the absolute coverage determination. The coverage determination was made directly from measurements of the filament geometric area, initial pressure, steady-state pressure, and pumping speeds. The authors' estimate of their accuracy for sticking probabilities was $\pm$ 50%. Keeping in mind that no correction for a roughness factor of the filament was made, the error in absolute coverage determination could be even greater than this. Thus, the estimate of saturation uptake at 300 K of 30.4×10^{14} atoms cm^{-2} could easily be compatible with a true value of 14×10^{14} atoms cm^{-2}, the figure we now know to be appropriate for a close-packed O atomic overlayer for {100} and {110} single crystal surfaces. The compelling reason for interpreting the desorption of W oxides as implying oxide formation on the surface is thus removed. A second seductive reason considered by the authors was that the O coverage onset for W oxide desorption coincided roughly with the onset of high-yield β_1 ESD state discussed in the previous section. We now know that this ESD state can actually comprise many states and those which do indeed arise from oxide are from oxide present only as minority species on the surface for coverages of $\lesssim 1.4 \times 10^{15}$ atoms cm^{-2}.

The subsequent TPD work on the individual W{110} and W{100} faces established very precisely the relationships between coverage and onset of W oxides in desorption and demonstrated that the onset of oxides in the TPD did not imply oxides on the surface at T_{ads}. In addition, they provided the values for E_d which serve as estimates for chemisorption bond energies. We refer briefly to each face in turn below.

2.9.1 W{110}

The TPD work of Engel et al. [68] in the $\theta = 0$–0.5 mL coverage range correlated with the $\Delta\phi$, LEED and AES measurements definitely showed that, up to 0.5 mL, only O atoms desorb from the W{110}/O surface. E_d was determined as ~ 10 eV at zero coverage dropping rapidly to ~ 5 eV at $\theta = 0.25$ mL and staying sensibly constant up to 0.5 mL. Since our conclusions from Sect. 2.5 were that, at low θ, the W{110}/O system reconstructs just below the desorption temperature, the ~ 10 eV E_d value does not relate to the original surface structure for a T_{ads} of 300 K. At higher θ, the O desorption occurred before reconstruction could take place so the ~ 5 eV E_d value is probably more genuinely characteristic of the O overlayer desorption energy at $\theta \approx 0.5$ mL. Note that, even though we know that reconstruction occurs during the TPD process for low θ, no W oxide species are observed in the TPD. This demonstrates that the presence or absence of W oxides in desorption is not related to whether O is in an overlayer or reconstructed form, but rather to the coverage concerned.

In the paper of Bauer and Engel discussing the $\theta > 0.5$ mL regime [69], a direct correlation of all TPD species signals versus θ determined from AES is given. It is reproduced in Fig. 39. Remember that the higher coverages (above 0.5 mL) were obtained by deposition of W oxides followed by 800 K annealing. For values of θ between 0.5 and 1 mL this procedure gives an O_{ads} overlayer. Above 1 mL a three-dimensional oxide is formed. Examination of Fig. 39 shows that the amount of O desorbing increases monotonically up to about 0.8 mL and that the onset of oxide desorption just above 0.5 mL is unconnected with the formation of oxide on the surface during oxygen adsorption. The authors conclude that the various oxide desorption products are determined by local configurational changes with coverage. The phenomenological argument they give for the onset above $\theta \approx 0.5$ is that, as

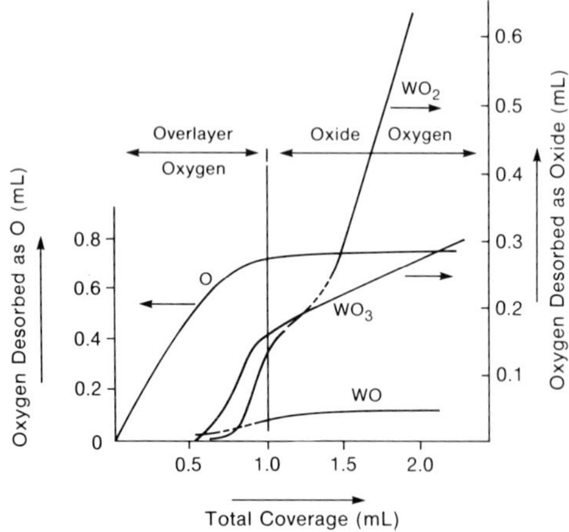

Fig. 39. Absolute amounts of O and W oxides desorbing as a function of oxygen coverage on W{110} [69].

the ratio of O/W increases, W/W bonding is weakened in favor of W–O bonding, thus increasing the probability of W/O species being desorbed together. They also point out that the distribution of W oxides observed in desorption is quite different from that expected from bulk W oxide during thermodynamic quasi-equilibrium considerations. Note, however, for the true 3D-oxide region above 1 mL all oxygen in excess of 1 mL does come off as oxide, with the WO_2 species dominating.

2.9.2 W{100}

The work of Bauer et al. [77] is again the most useful because of the careful coverage calibrations. It is not nearly as extensive as the {110} work, however. Only atomic O desorbed up to $\theta \approx 0.65$ mL. Above this coverage, WO_2 species start and at $\theta \approx 0.85$ mL, WO_3 species. Earlier, Russian work placed the onset of oxide desorption at $\theta \approx 1$ mL [188]. An E_d of 5.7 eV at low coverage and 5.1 eV at $\theta \approx 0.5$ mL was obtained [77]. More extensive attempts to extract E_d values elsewhere yielded 7.4 eV at $\theta \approx 0.15$ mL and 5.2 eV at $\theta \approx 0.5$ mL.

Since we now know that a reconstruction process occurs at ~650 K for all O coverages, but that this process is slow, it is not clear what meaning these E_d values have. Again, since oxygen adsorption at 300 K apparently gives majority overlayer adsorption right up to $\theta \approx 1.4$ mL, the onset of W oxides in the TPD is not correlated with the onset of surface oxidation. Adsorption at 1000 K produces an increased percentage of oxide product in the TPD. Even here, an interpretation that implies enhanced oxidation at 1000 K would be wrong. The results are entirely compatible with the known behavior of W{100} on heating in oxygen at 1000 K, namely that an increased surface area of {110} facets covered with a saturated overlayer of O is formed (Sect. 2.6.3).

2.9.3 W{111}

The only significant TPD work on W{111}/O is that of Vasko et al. [177] referred to earlier in Sect. 2.7.2. Assuming that their coverage calibrations are correct, O only is desorbed up to a coverage of $\sim 3.5 \times 10^{14}$ atoms cm^{-2}. Above this coverage, the O desorption intensity increases to a saturation value of 5.2×10^{14} atom cm^{-2} while the oxide contributions constitutes $\sim 5 \times 10^{14}$ atoms cm^{-2} at the highest exposure achieved. It is not clear that saturation O coverage has yet been reached, however, since the oxide desorption intensities still seem to be growing. The similarity to the W{100} and W{110} cases is the observation of only O desorption initially. The percentage of oxygen desorbing as oxide at high θ seems greater than for W{110} or {100}, however, and θ for the onset of oxide desorption seems lower. Whether any significance can be attached to this is a question that must wait for detailed TPD studies in conjunction with other analytical techniques. Vasko et al. did not, this time, attribute the onset of oxide desorption to oxide on the surface at T_{ads}. The reason is that the oxide desorption spectra followed second- rather than first-order kinetics. They therefore correctly concluded that the oxide desorption products were formed during the heating, and not related to the structural arrangement at T_{ads}. In the low θ regime, where only O desorbed, they derived an E_d value of 6.2 eV.

2.10 SECONDARY ION MASS SPECTROSCOPY (SIMS) FROM W/O SURFACES

The use of SIMS as a qualitative analytical tool to detect which atomic species are present at surfaces is well established [189]. The two main problems with the technique used in this manner are its intrinsic destructive nature and the difficulty in providing a quantitative analysis. The former problem can be overcome by using very low ion beam current densities and high-sensitivity mass detection schemes. The time for sputter removal of a monolayer should be large compared with the analysis time. Even then it should be remembered that the damaged area on a surface may be much larger than might be indicated by the sputter yield. When the conditions are such that one can be sure that the secondary ions detected come from a surface area undamaged by a previous Ar^+ impact, then the technique is known as static SIMS, and it is suitable for fundamental investigation of adsorption systems [190]. The second problem, quantification, is more difficult to overcome. Generally, the ionized species detected constitute only a small fraction of the total sputtered yield. This fraction can change drastically with chemical or electronic changes and in addition if there is the possibility of a range of ionized species (i.e. W^+, WO^+, WO_2, WO_2^-, etc. from W/O surfaces) the distributions of sputtered species throughout this range can also change drastically. The result is that the SIMS yields for any given species may change by several orders of magnitude with changes in the chemical nature of the surface. Quantitative analysis can therefore only be obtained if the SIMS yields are calibrated against some other measurement of species concentration. The principal area where quantification has been routinely achieved is, in fact, not in the use of static SIMS for surface studies, but in its use for depth analysis into the bulk for doped semiconductors [191]. Here analysis is done during sputtering (dynamic SIMS) and the main advantage of the technique is its high absolute sensitivity for the element of concern and the reasonable spatial resolution of the technique. Calibration is achieved through use of ion-implantation standards. For static SIMS for adsorbate/single crystal surfaces very little has been done by way of calibrating the observed SIMS ion intensities against surface adsorbate coverage. For the W/O systems, only W{100}/O has been investigated. Since SIMS yields can change so drastically with chemistry this lack of coverage calibration leads to the same old problem, the inability to distinguish between a minority state with a high yield and a majority species with a low yield (cf. ESD). This problem can make any attempted interpretation of the intensity patterns of the various SIMS positive and negative cluster ion emitted very ambiguous. In the case of W/O (and Mo/O) it is exactly this type of correlation between emitted cluster intensities and the nature of the W–O surface which has been attempted. Though it has been established that this can be done successfully for heavy oxidation (essentially bulk oxide films) and that even the oxidation state of the metal can be determined from the positive and negative cluster ion intensity distributions [191], it is not clear that the same approach can be taken in the adsorption and two-dimensional reconstruction stages.

The first attempt at relating static SIMS ion distributions to the adsorption situation at the surface for single crystal W/O surfaces were by Benninghoven et al. [192]. They used a W ribbon consisting largely of {100} orientation crystallites, and combined the SIMS measurements with ESD measurements. No other method of coverage determination was used and the experimental procedure was to plot the intensities of the secondary ions (2 kV Ar^+ primary beam) observed as a function of oxygen exposure. Owing to a problem with a background signal the O^+ signal at low oxygen exposure was not representative of the surface. This was unfortunate because, as demonstrated later by Yu [193], only the O^+ signal is directly proportional to coverage for θ between 0 and 1 mL. The observed signal intensities of the major ions are plotted as a function of exposure in Fig. 40.

Benninghoven et al. made the following general conclusions based on the above data.

(i) At exposures of < 1 L adsorption occurred without any oxidation. It is not stated in the paper, but from the figures one would conclude that this stage gives rise only to W^+, O^-, and possibly O^+ signals (unsure because of the background O^+ signal).

(ii) At exposures of > 1 L an oxidation process is occurring. Up to 10 L exposure "a mono-molecular W–O structure" is formed, which is characterized by WO_2^- emission.

(iii) Above 10 L a second stage in the oxidation process starts in which the structure becomes three-dimensional. At the same time the oxidation state of the W atoms in the top layer increases. This stage is characterized by WO^+ emission.

The authors were careful not to associate the observation of any specific molecular ion directly with the presence of W atoms in the surface in that configuration, e.g. the observation of WO_3^- was not taken to imply that WO_3 existed on the surface but simply that a configuration existed at the surface from which the emission of WO_3^- becomes possible. We now know, of course, that the sequence (i)–(iii) proposed above is quite incorrect for a W{100} surface. Up to exposures of 20 L, where the SIMS signals start to saturate (Fig. 40), our belief is that all the majority O species are in overlayer positions and the saturation coverage is $\theta \approx 1.4$ mL. The association of specific SIMS ion signals with the specific stages suggested is therefore either quite incorrect, or it is correct but the oriented ribbon is not characteristic of a flat {100} surface, or it is correct but some of the stages referred to represent minority O states, not the majority of O on W{100}.

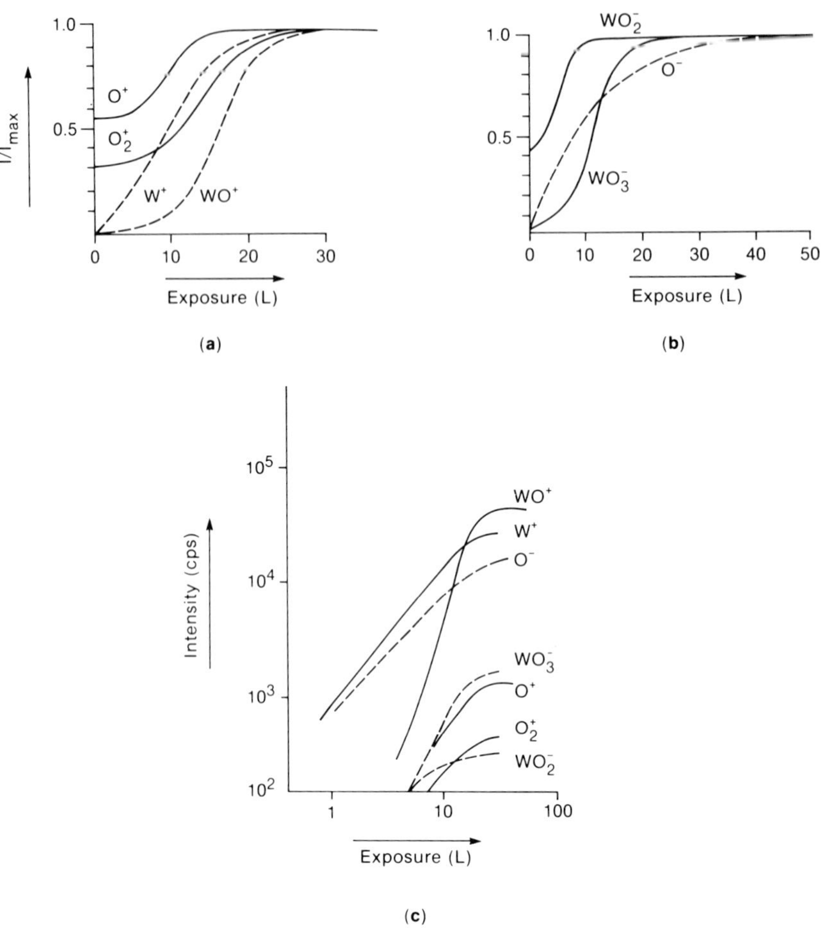

Fig. 40. SIMS intensities as a function of exposure for W{100}/O at 300 K [192]. (a), (b) Intensities relative to saturation values as a function of exposure. (c) absolute intensities as a function of exposure on a log/log scale.

Yu re-examined the W{100}/O system in greater detail [193] using AES as a method of O-coverage determination. He used an Ar^+ beam energy of 500 eV, which restricts the SIMS signal more to the top layer than does a 2 keV beam. He also had the advantage of having the definitive work of Bauer et al. [77] on the W{100}/O system available to him by that time. The first important point he established was that the O^+ intensity and only the O^+ intensity was proportional to O coverage over the whole adsorption range (usually it is not clear whether full coverage in his definition means 1 or 1.4×10^{15} atoms cm^{-2}. Remember that the saturation coverage on W{100} is actually 1.4×10^{15} atoms cm^{-2}, or $\theta = 1.4$ mL). In Fig. 41(a) we show his plot of O^+ against O Auger. The maximum O Auger intensity is taken to be $\theta = 1.0$, thereby establishing the coverage range. The rest of his

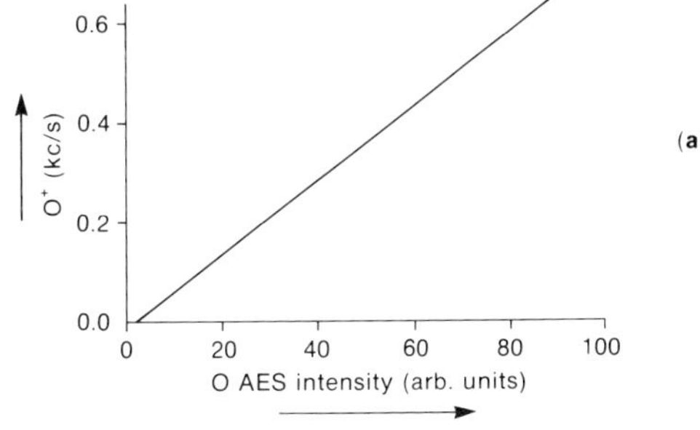

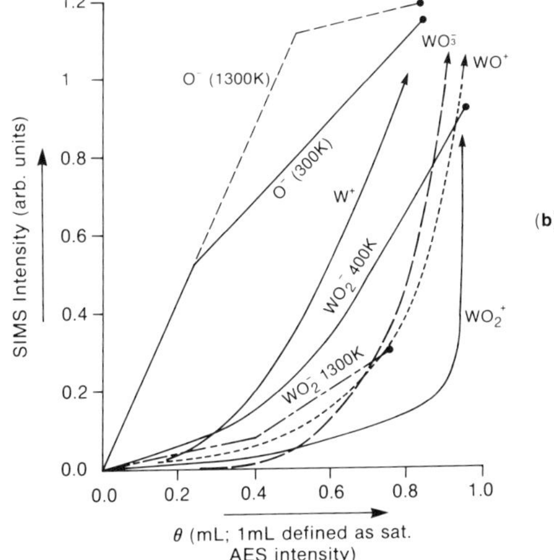

Fig. 41. (a) O^+ SIMS intensity versus O AES intensity for W{100}/O at 300 K [193]. (b) Intensities of all major SIMS signals as a function of θ for W{100}/O. Plots derived by us from the original data of ref. 193. Where not marked, the curves refer to ~ 300 K adsorption. Curves ending in ● imply that the SIMS signal saturates and the saturation θ is obtained. Curves ending in → signify that the signals are still increasing rapidly even though θ, as determined by AES, has essentially saturated.

data are plotted in the paper as the intensity of the SIMS signal concerned against time of exposure, or against O^+ intensity. We have attempted to convert all this to plots of intensity versus coverage [Fig. 41(b)]. The O^- signal is proportional to θ over the whole coverage range, but there is a sharp change in the proportionality constant at $\theta \approx 0.25$ mL. As pointed out by Yu this is just the coverage where the p(4 × 1) LEED structure is obtained. He associates the break in the O^- curve with the onset of W/O ensembles on the surface where the O/W ratio is increased, with the explanation that the probability of O^- ejection at such sites might be expected to be decreased because there is now competition between O atoms to acquire the electrons from bonding W atoms to generate O^-. He suggests that these ensembles are the double and triple rows of the Froitzheim model of the p(4 × 1) structure [133]. We, however, have already pointed out that the Froitzheim model is unrealistic since "p(4 × 1)" is actually a mixture of structures (Sect. 2.6.3). A less ambitious interpretation would be simply to say that any ensemble increasing the local O/W ratio will reduce the O^- yield. Thus up to $\theta \approx 0.25$ mL the local coverage never exceeds 0.25 mL, but above 0.25 mL local coverage must be greater than 0.25 mL in some regions. The annealing experiments are of interest here. Annealing a 300 K dosed surface to successively higher temperatures to cause desorption extended the first linear segment of the O^- curve to $\theta \approx 0.4$ mL and for adsorption at $T \approx 1300$ K to $\theta \approx 0.5$ mL [see Fig. 41(b)]. Note that, though the O^+ yield is apparently unaffected by temperature, the growth in O^- nearly ceases above 0.5 mL at high temperature. We know reconstruction occurs at all coverages for $T > 600$–800 K, so the O^+ and O^- intensities after annealing or from adsorption at 1300 K are characteristic of the reconstructed surface. Several important points can be deduced from comparison of the O^+/O^- intensities below and above the reconstruction temperature. The first is that O^+ (presumably) is still directly proportional to θ. Thus, whether O atoms are present as overlayer or reconstructed O coplanar with W is of no significance as far as O^+ emission probability is concerned. The unchanging yield of O^+ to $\theta = 1$ mL also implies that there is no decrease in O^+ emission probability with increased local O/W ratio (cf. O^-). The fact that the O^- yield is the same for 0–0.5 mL *after* reconstruction as it was for 0–0.25 mL *before* reconstruction [Fig. 41(b)] suggests two things. First, as was the case for O^+, whether O is in an overlayer or a reconstructed layer is not important in determining the O^- yield. Second, since this same high O^- yield can be maintained to $\theta = 0.5$ mL for the reconstructed layer, by which point O atoms are quite close together, the probability of O^- emission does not seem to be significantly affected by this close approach. However, for that fraction of O above $\theta = 0.5$ mL in the reconstructed surface the O^- yield appears to be close to zero. Since the p(2 × 1) reconstruction can only accommodate 0.5 mL O atoms it would seem that the additional O atoms (up to $\theta \approx 0.8$ mL) go into sites where the O^- yield is zero but the O^+ yield remains unchanged.

Yu [193] noted that there is an inverse correlation between the O^- signal behavior and the very weak intensity WO_2^- behavior. The WO_2^- ion grows from zero exposure and saturates at $\theta \approx 1$ mL. (It is not, therefore, characteristic of a "mono-molecular W–O structure" as suggested by Benninghoven et al. [192]). For 400 K adsorption, the WO_2^- versus θ curve shows two linear segments with a break at θ between 0.25 and 0.4 mL. The second segment shows the higher yield, the inverse situation of the O^- behavior. On annealing above reconstruction temperatures the yield for WO_2^- drops dramatically, indicating that O in the p(2 × 1) reconstructed arrangement has only a low probability of generating a WO_2^- secondary ion. There is still a break in the curve at $\theta \approx 0.4$ mL, however, which suggests that any O not associated with the p(2 × 1) reconstruction arrangement has a higher probability of being emitted as WO_2^-. Thus though one can certainly not say that WO_2^- means that WO_2 units can exist on the surface it does seem that a decreased O^- yield and an increased WO_2^- yield signify an increased O/W local ensemble ratio. We have to keep in mind, however, that we have no way of knowing whether his conclusion is representative of the majority O or of minority species. Thus it is quite possible that over the whole adsorption range the presence of WO_2^- signifies a minority species, such as O trapped at defects, and that there is simply an increase in the rate of population of such states for $\theta > \sim 0.4$ mL.

The ions W^+, WO^+ and WO_3^- are also observed from zero coverage, but there is no linear relationship between their intensity and coverage [Fig. 41(b)]. In each case the yield is low at low coverage and rises dramatically as θ increases. Arrows at the end of curves indicate that, though coverage is no longer increasing according to AES, the SIMS signals continue to increase rapidly with further exposure. The onset of this rise is delayed further as one goes from W^+ to WO^+ to WO_3^-. Unlike O^+, O^-, and WO_2^- the signals do not saturate at $\theta \approx 1$ mL but keep increasing. (The WO_2^+ behavior is also plotted in the figure. Although Yu claims it appears only above $\theta = 1$ mL, it is actually quite hard to distinguish its behavior qualitatively from that of WO_3^-.) In each case annealing above reconstruction temperatures greatly reduces the yields at high θ (not shown). Again there are two possible general interpretations. These are (i) that high local O/W coverage ensembles increase the yields of W^+, WO^+, WO_3^-, and WO_2^+, or (ii) that these species are characteristic of minority states of O on the surface which become more populated at higher coverage and are annealed out during reconstruction. Explanation (ii) does not, of course, rule out (i). Even if minority species are represented the reason for their high SIMS yields could still be that the O/W ratio is high (such as defect oxide crystallites). According to Yu, the onset of WO_2^+ represents oxidation because it occurs at $\theta \approx 1$ mL. If it does, then it is certainly a minority state since, as we have already noted, saturation coverage can be achieved with the majority O in overlayer configurations. A sudden large increase in any of the signals W^+, WO^+, WO_2^+, WO_3^- with exposure cannot be taken to indicate a large increase in a different majority state of O because of the possibility that these SIMS signals represent minority states.

We have gone through these few existing SIMS studies on W{100}/O in some detail to try and get several things in perspective. They are:

(a) that quantitative coverage determination by SIMS is difficult and requires proving by calibration.

(b) that even if one SIMS signal intensity is shown to be proportional to coverage (such as O^+ here) it is hard to establish whether changes in intensities of others are due to changes in geometry and/or changes in electronic structure of that majority O, or whether they represent minority O with high SIMS yields.

(c) the association of certain cluster species with certain stages in the oxygen interaction may be possible, but is complicated by (a) and (b) above.

Nevertheless, the very strong changes in yields observed with changes in exposure, annealing, etc. must imply that SIMS can be a very good characterization tool, perhaps for minority species rather than majority, but this remains to be investigated more thoroughly.

2.11 SUMMARY OF TUNGSTEN DATA

A vast amount of data exists for the W{110} and {100}/oxygen systems, with much less for W{111} oxygen. Only for W{110} is there data at a sufficiently low substrate temperature (20 K) for a chemisorbed molecular oxygen state to have been detected (the α state) [13]. Even here, except possibly at very low substrate and gas temperature, it is only formed on top of an atomic oxygen covered surface and is very similar in nature to the subsequent layers of physisorbed, and multilayer condensed oxygen [13, 113].

2.11.1 300 K and below

Contemporary analysis for all (flat) faces at 300 K indicates dissociative adsorption into overlayer positions without reconstruction. One possible exception is the 0–0.1 monolayer range for W{110}, but this might be the result of defects. The ordered structures observed on W{110} at 300 K by LEED are based on the p(2 $\times$ 1) structure up to a half-monolayer. Above a half-monolayer there seems to be a progressive transition with increasing coverage towards a more close-packed pseudo-hexagonal overlayer structure [69]. These structures are more definitely established between 800 and 1250 K. The situation on W{100} is far more complex, with overlayer–substrate registration and distances not known. Between one-quarter and one-half monolayers the structures are best understood as co-existing metastable phases which again attempt to reach hexagonal close-packing by saturation coverage [77]. In this regard the saturation coverages on W{110} and W{100} are identical at 1.43×10^{15} atoms cm^{-2}. No ordered structures are observed for W{111} and it is our opinion that the local geometries implied by ESDIAD studies [83] represent minority species, probably oxide crystallites. In fact at very high exposures at room temperature synchrotron photoemission data and ESD evidence suggest that minority amounts of oxide are formed on all three faces. Kinetic data taken at room temperature and below indicate that the initial dissociative adsorption process goes via a weakly bound precursor molecular state, which is mobile, on the surface of all three faces. Only on the W{110} surface at $T_s = 20$ K and $T_g \leqslant 100$ K does this state ever remain undissociated on bare W surface [108]. At 300 K on W{110}, completion of the ordered p(2 $\times$ 1) O structure blocks all the easy dissociation sites by $\theta = 0.5$ (where easy indicates only weakly activated), and therefore further adsorption to saturation coverage is highly activated with a very low S [107]. W{100} is much more reactive. S_0 is close to unity compared with about 0.3 for W{110}, indicating that the initial dissociative adsorption is completely unactivated. S falls rapidly as $\theta = 0.5$ is approached with the ordered O structures blocking dissociated sites, as is the case for W{110}, but unlike W{110} it does not reduce to near zero, only to $S \approx 0.5$. Thus, the second stage of dissociative adsorption to saturation is still a very fast, only weakly activated, process [107]. For W{111} the kinetic data are much more sparse, but it is clear that S_0 is close to unity, like W{100}, and that it falls off to a low value intermediate between the drastic fall for W{110} and the very mild fall for W{100}, somewhere roughly in the middle of the coverage range, which itself is not very well established [83, 141, 177, 178]. Thus, the process can be divided into an initial unactivated stage and an activated stage once all easy sites are blocked by an O_{ads} layer.

For all the experimental effort (UPS, XPS, ELS) that has gone into learning something about the electronic structure of the W/O interaction, remarkably little definitive information has resulted. The W{110}/O case is the most extensively studied. The results are all compatible with a single electronic species of $O^{\delta-}$, with a character close to that in an oxide. Movement in the O(2p) UPS features with coverage changes [71] can be attributed to dispersion effects. Splittings at high coverages can be attributed to lateral interactions. The W(4f) chemical shift of ~0.5 eV for the W atoms bound to the overlayer O atoms [137, 160] is roughly what one would expect for a situation where each surface W has many W neighbors but only one $O^{\delta-}$ neighbor. For W{100}, the main O species has electronic characteristics very similar to the single species found for W{110}, but there is quite strong evidence for a second, electronically distinct, O species at high coverage [76, 174]. Since the evidence for geometrically distinct co-existing species in the O overlayer at any coverage is ambiguous at best, the question arises as to whether the second electronic state represents O that has penetrated the lattice. For W{111}, the results suggest a dominant single electronic species [141], again rather similar to that for W{110}, though there is some evidence [141] of a minor second species, which again could represent O penetration of the lattice.

2.11.2 High temperature

Adsorption at high temperatures, or annealing oxygen overlayer surfaces to high temperatures, can cause reconstruction or penetration of the surface for all faces. The basic difference between the behavior of the different faces is the temperature at which this occurs and the stability of the product surfaces with respect to facetting. There are also some complicated coverage-dependent effects. The overlayer coverage situation on the least reactive W{110} face is stable with respect to reconstruction at all coverages up to a temperature close to desorption (~1700 K) [68]. Above that temperature reconstruction occurs. The more reactive W{100} surface undergoes reconstruction at ~600 K for all coverages, with the 0.5 mL case having been thoroughly studied [77]. There is also evidence for a higher temperature ($\sim$ 900 K) reversible phase transition of the reconstructed surface where the O atoms move outward somewhat [77]. Facetting to W{110} surfaces occurs at high temperatures and high coverages ($\theta \geqslant 1.25$ mL), the resulting facets having the normal W{110} stable saturated oxygen overlayer up to desorption temperatures [77, 105, 110].

The interpretation of the data concerning reconstruction of W{111}/O surfaces is very tenuous, but it is clear that something drastic happens at around two-thirds saturation coverage and $\sim$1000 K [92, 146]. Most likely this is reconstruction and/or penetration. At higher coverages, facetting to W{110}/O again eventually occurs. Under some coverage and temperature conditions, W{112} facets develop first followed by W(110) facets. Under heavy oxidation conditions (high T and high P), WO_3 is the oxide formed on all faces, and since the W{100} and W{111} surfaces facet to W{110} the epitaxy of the oxide is always {110}-derived.

The ordered structures observed by LEED for high-temperature adsorption consist of overlayer, reconstructed, and facetted situations, depending on adsorption temperature and coverage, as described above. The overlayer structures were mentioned in Sect. 2.11.1. For W{110} and W{100} the reconstructed surfaces are both p(2 $\times$ 1) at half-monolayer coverage. The (110) p(2 $\times$ 1) is probably a missing W row structure and the (100) case has been definitively assigned as such. All the remaining LEED structures observed under high-temperature and high-exposure conditions on the W{100} and W{111} surfaces are representative of the facetting processes mentioned above.

Kinetic data at high temperature are sparse; for W{111} it is non-existent. For W{110} it seems from the S, θ data available that the surface residence time of any molecular precursor has become so small by 1000 K [107] that the precursor mechanism contribution to the dissociative adsorption process is negligible in the 0–0.5 mL coverage range where it is so effective at low temperatures. In the activated stage above 0.5 mL higher temperatures strongly increase S, however, as would be expected for an activated process. The absolute values of S are still very low, however. Remember that for W{110} these 1000 K results still refer to overlayer adsorption. For W{100} the S, θ studies at high T (1000 K and therefore representing the reconstruction process) have resulted in quantitatively different results being reported by different authors [77, 107]. One can see the decrease in effectiveness of the precursor mechanism compared with 300 K in one study, whereas the other shows remarkably little difference between the two temperatures.

Information on any electronic structure effects specific to the reconstruction process is limited to W{100}. There have been no significant studies on W{111} under such conditions and for W{110} the overlayer structure stability up to near desorption temperatures means that no available UPS, XPS data, etc. are relevant to reconstruction. For W{100} the most significant result is the clear identification of effects relating to the well-documented reconstruction process at 0.5 mL and 600 K. The XPS $W(4f)_{7/2}$ chemical shift [160] is quite appropriate to a transition of top W atoms from a chemisorption to more 2-D oxide-like situations and UPS data supports this. ELS data also shows pronounced effects at high T and coverage [168] but they are harder to interpret.

Owing to the macroscopic nature of most surface diffusion measurements, most diffusion data are taken at high temperature. For W{110} such data in the ~900–1600 K regime yield E_{diff} of around 25 kcal for θ greater than 0.3 mL [142, 143]. On W{100} the low reconstruction temperature precludes macroscopic determinations of E_{diff} for the unreconstructed surface. There is one FEM microscopic study which yields 130 kcal at high coverage [169]. The same method yields values of 14 kcal and 95 kcal, respectively, for W{110} [112] and W{100} [169] at low coverages (less than 0.3 mL). For both surfaces the lower values for lower coverages are considered to represent isolated O adatom diffusion compared with diffusion in the presence of ordered overlayer at the higher coverage. The much larger values for W{100} compared with W{110} reflect the much rougher W{100} surface. No diffusion data are available for the roughest W{111} surface.

2.11.3 Steps and defects

An extensive amount of data exists for the stepped W{110} surface, but virtually none for W{100} and W{111} surfaces. The major effects of steps and defects on W{110} are to increase S_0 from ~0.3 to ~1 [115, 149]. The exception is close-packed step orientations which have little effect on S_0 [104]. For regular stepped surfaces, changes in the LEED structures which are related to registration effects with the steps are produced [104]. A comparison of the "flat" surface results and stepped surface results imposes some severe restrictions on the mechanism of O_2 adsorption on W{110} surfaces [115, 149]. The "classical" precursor model, in which a molecule is either instantaneously energy-accommodated into the precursor state or reflects into the gas phase, seems to be inapplicable. The results imply that molecules are always partly energy-accommodated into the precursor state and can always travel significant distances across the surface before becoming fully accommodated or returning to the gas phase. Dissociation probability is greatly enhanced on encountering steps or defects, but the dissociative products move away from the steps leaving them still active.

2.12 GENERAL COMPARISON OF TRENDS FOR Cr, Mo, W

Despite the tremendous variation and richness of detail in the interaction of oxygen with the {110}, {100} and {111} faces of Cr, Mo, and W, some basic trends are clear. The consequences of these trends are collected together in Table 5 (p. 150). The surfaces are more easily reconstructed under the influence of oxygen, and more easily oxidized, in the order Cr > Mo > W. This follows the reverse order of metal–metal bond energies (Table 1). This means that the breaking of the metal–metal bonds is the dominant factor, with the metal–oxygen bond strengths being much less decisive. This is not surprising when one notices that the metal–oxygen bond strengths do not differ a great deal across the series, whereas the metal–metal differences are huge. For a given metal the ease of reconstruction/oxidation generally goes as $\{111\} > \{100\} > \{110\}$, as might be expected from geometries of these surfaces: the {110} is close packed with the highest coordination number and therefore the most metal–metal bonds to break. Overall, then, W{110} is the hardest surface to reconstruct or oxidize, whereas Cr{111} and {100} oxidize easily at room temperature with the oxide on Cr{111} being thicker. Owing to these great differences in propensity to oxidize, the LEED sequences on the equivalent faces of the different metals are not closely related. The greatest similarity is between the most stable pair, W{110} and Mo{110}, at room temperature.

The initial sticking probability, S_0, for dissociative adsorption on all the faces is high, generally greater than 0.1 and usually closer to 1. This, of course, reflects the strength of the M–O bond formed in all the cases and has relatively little to do with M–M strengths. Since W{110} is the most resistant to oxidation and the W–O bond strength is at least as strong as for Mo–O or Cr–O, the W{110} surface forms the most stable overlayer oxygen adatom structure of the series; in fact it is stable nearly up to the desorption temperature of 1700 K.

The stoichiometries of the epitaxial oxides formed under oxidative conditions also show a trend: oxygen content increases with atomic number, i.e. the oxides formed are Cr_2O_3, MoO_2, and WO_3. These are also the oxides generally found as air corrosion products under non-UHV conditions. The growth of epitaxial oxide is generally very temperature- and pressure-dependent because the oxides of the metals are volatile at the temperatures required to produce them on a laboratory time scale. An overpressure is therefore required for them to grow. The oxide desorption temperatures are also generally lower than those for the oxygen-covered metal surfaces.

3. Nickel

The volume of work existing on the interaction of oxygen with single-crystal Ni surfaces is comparable with that for W. Papers reporting on adsorption geometries are actually more numerous for Ni. They include LEED [194–199], SEXAFS [200], He scattering [201–204], ion scattering [205–215], normal photoelectron diffraction (NPD) [216], angle-resolved XPS measurements [217], HRELS [218–226], RHEED [227, 228], angle-resolved UPS [229], and INS [230]. The work on Ni{100} is the most extensive, that on Ni{111} the least. Unlike W, the onset of oxygen penetration into the Ni subsurface to form oxide is quite facile, occurring at $\theta \leqslant 0.5$ mL at 300 K for all low index faces [231–233]. Also in contrast to W, where reconstructed mixed W/O phases first form when penetration occurs rather than bulk oxide nuclei, the Ni/O system forms thin NiO oxide nuclei immediately. The balance between overlayer chemisorption and oxide nucleation is delicate and affected not only by coverage, but also by the substrate perfection, the degree of perfection in the overlayer, and the amount of oxygen that may already exist in the subsurface. Atomic oxygen can diffuse into the crystal bulk at rather low temperatures (450–600 K, depending on surface and coverage [312–314]). This O dissolution is not to be confused with oxide formation. Owing to the possibility of having a mixture of chemisorbed oxygen, NiO, and dissolved oxygen one has to be very circumspect in assessing the various structural determinations performed. The different techniques have different sensitivities to the different types of oxygen and to long-range order compared with local order. It is also now clear that the formation of minor OH [236] and CO_3 [237, 238] contaminants on the surface, wrongly identified as distinct O_{ads} species, can cause problems. It is therefore necessary to delineate fully the experimental conditions under which the measurements were made, and to spell out the limitations of the technique concerned. Lack of appreciation of these aspects has led to confusion concerning the relationship of the structural determinations to other studies such as the vibrational spectra and electronic structure calculations [224, 239, 240].

There is quite a large body of data on the kinetics of the oxygen interaction. Sticking probability measurements versus coverage have been measured using AES [231–233, 241, 242], XPS [243–245], medium energy ion scattering [209–213], and X-ray fluorescence [227, 228]. They cover both the chemisorption and the oxide nucleation and growth stages. The range of T_s and T_g values considered is much more limited than for W, however. Only $T_g = 300$ K has been used and T_s has been varied from 80 to 700 K. Only one piece of work on stepped surfaces has been reported [241], though the effect of disordering the substrate surface by ion bombardment has been more studied [237, 246]. As was the case for W, it is possible from the S versus θ measurements to derive models for the dissociative adsorption mechanism in terms of the roles of surface order, adsorbate order and admolecule and adatom surface mobility [245, 247]. In addition, for Ni{111} [235], and to a less extent Ni{100} [234], attempts have been made to establish the 2-D Ni/O phase diagram using LEED angular profiles (cf. W{110}).

The third major area of work concerns electronic structure investigations. These include experimental studies using photoemission [229, 236, 243–251, 253, 254, 287, 288], SEXAFS [200] and INS [230], and theoretical studies [239, 240, 255, 257–265, 291, 299]. Reliable theoretical calculations of electronic structure are, in principle, much more feasible than for W because of the smaller number of electrons in the Ni atom. All the calculations have attempted to model the {100}/O chemisorbed structures, either by a cluster or a slab calculation approach. The objectives have been to establish the nature of the Ni–O bond, to determine theoretically equilibrium bond distances, and to calculate force constants which can be used to interpret the HRELS vibrational spectra. Unfortunately, though there has been much computational activity, there is as yet no consensus concerning the reliability of the results. The points of concern are the effect of cluster size in modelling the {100}/O interaction, and the theoretical approximations required to handle many Ni atoms. These concerns have meant that predictions and interpretations have varied considerably [239, 240, 257, 258, 264]. Since the theoretical calculations are often considered with respect to experimental data where there may also be some interpretational doubt, the margin for increasing confusion rather than knowledge has been great. We believe, however, that by now both the experimental and theoretical areas are sufficiently advanced for a cogent picture to emerge.

We have organized the discussion for Ni/O along different lines from W. Each face is taken in turn and the work is discussed in the following order. First we briefly review the data necessary to establish reliable distinctions between the different interaction regions: overlayer dissociative adsorption, oxide nucleation and lateral growth, and oxide-layer thickening. This involves data from a wide variety of techniques which are all considered together. We then consider the geometric structure studies and the electronic structure studies for the well-defined chemisorption structures and, where appropriate, for the oxide layer. Next, having obtained an understanding of the geometric and electronic structure aspects, we consider the implications of the kinetics studies in terms of models for the mechanism of the dissociative oxygen adsorption and subsequent oxidation. Having done this for all three low-index faces, we conclude with a brief section comparing the three faces.

For W we made a point of leaving TPD, ESD, and SIMS data until after the main body of discussion on the grounds that, without an existing understanding of the W/O system, the interpretation of these techniques were not sufficiently well understood to contribute much to the knowledge of the W/O interaction. In the case of Ni this proves necessary only for SIMS [236, 265–268]. TPD plays no role at all, since O does not desorb from Ni but diffuses into the bulk. There are no ESD studies of the Ni/O system.

3.1 Ni{100}

3.1.1 General reaction scheme. 300 K and above

The key paper in providing a general understanding of the Ni{100}/O_2 reaction scheme was the combined AES, LEED, $\Delta\phi$ study of Holloway and Hudson [231]. They monitored the uptake of oxygen using the OKLL Auger intensity, while simultaneously monitoring the LEED and $\Delta\phi$ changes. A summary of their results at 300 K is provided in Fig. 42. Some similar non-monotonic oxygen uptake rates had been observed earlier for Ni films [270] for which an S_0 of unity had also been established by molecular beam measurements. The original interpretation from the film work had been that the fast initial stage corresponded to a molecular adsorption which eventu-

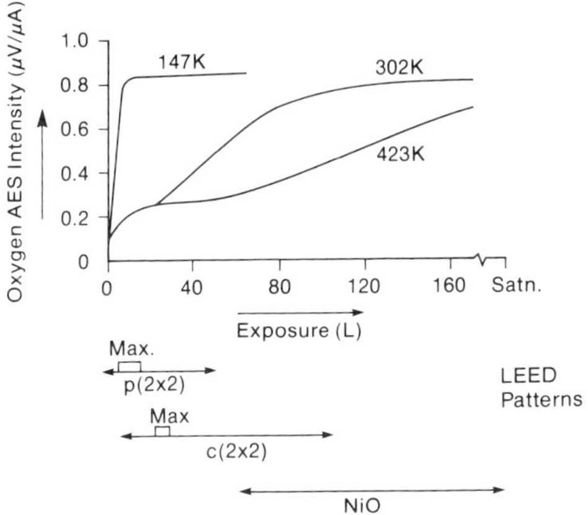

Fig. 42. Oxygen Auger intensity versus exposure for Ni{100} [231] at various temperatures. Individual data points are not shown. The ranges over which the LEED patterns are observed are marked.

ally saturated, and the second fast stage represented dissociative penetration of oxygen into the lattice. Holloway and Hudson established a coverage scale for the {100} surface by assuming that S_0 was unity (cf. the film work) and relating the O Auger intensity to coverage through the measured exposures. Using this calibration the LEED p(2 × 2) adsorbate pattern reached maximum intensity at an atomic coverage of $\theta \approx 0.28\,\mathrm{mL}$, the c(2 × 2) at $\theta \approx 0.40\,\mathrm{mL}$, and the minimum in the oxygen uptake rate was at $\theta \approx 0.32\,\mathrm{mL}$. Though there is no reason to expect this calibration procedure to be highly accurate, subsequent, more accurate, measurements are in good agreement with the Holloway and Hudson values (see Sect. 3.1.7). Since the determined p(2 × 2) oxygen coverage, 0.28 mL, was very close to that expected for a perfect p(2 × 2) O lattice, and only half that expected for a perfect p(2 × 2) O_2 lattice, Holloway and Hudson concluded that the p(2 × 2) adsorbate structure was atomic, not molecular. Again, this argument was not foolproof since the p(2 × 2) might not have covered the whole surface, but the atomic nature of the p(2 × 2) and also c(2 × 2) structures has since been confirmed from UPS and XPS measurements where the O(2*p*) and O(1*s*) BE values are characteristic of atomic rather than molecular oxygen [236, 244, 247]. The rise in sticking probability after the minimum at $\theta > 0.32\,\mathrm{mL}$ was associated by Holloway and Hudson with the onset of NiO nucleation, oxygen incorporation taking place at the perimeter of oxide islands, thus allowing S to increase as the island sizes increased until saturation occurred when the islands coalesced. The merits and limitations of the proposed island growth mechanism are discussed in Sect. 3.1.9. Here we just want to establish that NiO does indeed start to form at less than ~0.5 mL, i.e. before the c(2 × 2) O structure covers the surface. The presence of NiO at less than 0.5 mL has been confirmed by many techniques, though their sensitivity, and therefore the coverage at which NiO is observed, is variable. In LEED [231, 271, 272] the NiO{100} pattern is observed directly once the island sizes are sufficiently large (~20 Å); XPS and UPS detects it by the onset of a Ni^{2+} electronic signature [234, 236, 247]; RHEED [227] detects it directly as in LEED, but at an earlier stage; SIMS detects it as a change in positive and negative ion yields at the nucleation point [236, 266]; and medium-energy ion scattering detects it by the onset of large displacements of surface Ni atoms from their Ni{100} lattice sites [209, 210]. The parameter relating all these studies is θ and there is reasonable agreement among those studies where a θ calibration was made that the oxidation onset is around 0.35–0.40 monolayer.

The NiO fast growth stage saturates at 100–200 L exposure. Though this exposure value for saturation can be very variable [231, 247], depending on surface perfection, θ_{sat} is reasonably constant. Holloway and Hudson [231] estimated the O atom coverage at saturation as ~1.4 mL of atomic oxygen using their O Auger versus coverage calibration. This assumed no attenuation of the OKLL Auger electrons as they escape though the oxide layer. They estimated an attenuation correction of ~14%, using a mean free path length of 9.5 Å for the 507 eV OKLL Auger electrons, which increased the coverage to 1.6 mL. Expressed in terms of NiO{100} layers this translates to two layers of NiO{100} and on the basis of this result Holloway and Hudson concluded that the NiO nuclei grew as bilayers over the surface until all the islands had coalesced. Subsequent, more accurate calibration, indicates that Holloway and Hudson's attenuation correction was too small and that the true "saturation" thickness expressed in the above manner is nearer three NiO layers. The situation is actually more complex than this because more than one NiO epitaxy is observed and the "mix" of these depends on substrate perfection, exposure conditions and temperature. The three-layer NiO value is derived from a determined θ value of 2.4 mL and assumes that the oxygen is all present as NiO{100}.

The sticking probability at the three-layer NiO "saturation" coverage is not zero, of course. A very slow oxide thickening process continues ("tarnishing" regime). At 300 K, however, exposures of up to 10^5 L are insufficient to produce a noticeable increase in the OKLL Auger signal. They can produce continuing changes in the SIMS signals, as discussed in Sect. 3.5, which implies that, although the total oxygen content within the Auger escape depth is not changing significantly with additional exposure, the chemical nature of the surface may be changing somewhat.

Holloway and Hudson monitored $\Delta\phi$ throughout the oxygen interaction [Fig. 43(a)]. For the first measuring point, at $\theta \approx 0.23$ mL, $\Delta\phi$ was $+0.3$ eV; thereafter it decreased monotonically to a value of -0.6 eV at saturation. They assumed, as we have in our discussions on Cr, Mo, and W, that a positive $\Delta\phi$ indicated overlayer oxygen and a negative $\Delta\phi$ indicated oxygen penetration/reconstruction of the surface. They noted that $\Delta\phi$ was decreasing during the second fast reaction stage [Figs. 42 and 43(a)] supporting their

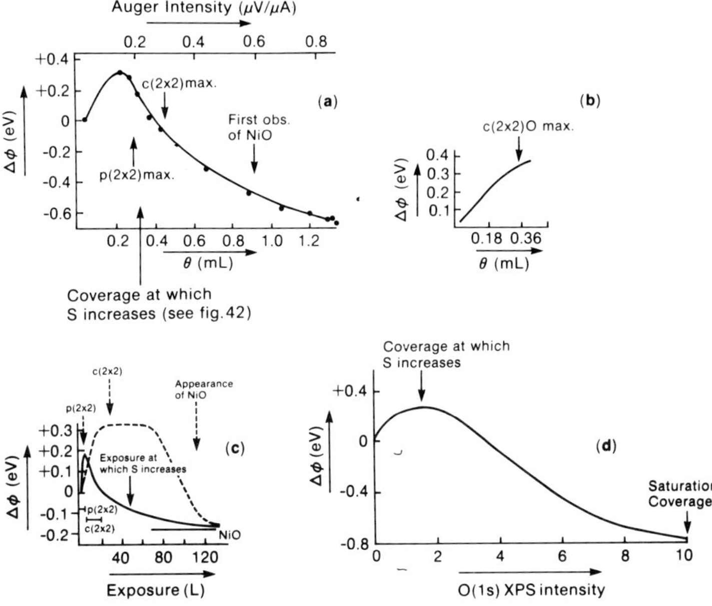

Fig. 43. Work function changes, $\Delta\phi$, during oxygen exposure for Ni{100}/O. (a) 300 K as a function of coverage, from Holloway and Hudson [231]. Auger intensity scale is identical to that of Fig. 42. (b) 300 K as function of coverage, from Demuth and Rhodin [195]. The coverage scale has been rescaled to make the c(2 × 2) coverage 0.35 mL. (c) 300 K as function of exposure. Full curve, Akimoto et al. [273]; dotted curve, Papageorgopoulos and Chen [274]. (d) 300 K as function of coverage, from Krishnan et al. [275].

claim that O penetration and NiO formation was occurring. The implicit assumption for the p(2 × 2) and c(2 × 2) O structures was that they represented overlayer adsorbate structures, but it will be noticed [Fig. 43(a)] that $\Delta\phi$ has actually started to decrease well before the maximum in the c(2 × 2) intensity, though at the latter point $\Delta\phi$ is still positive with respect to the clean surface value. This might be taken as an indication that NiO nucleation actually starts even earlier than 0.32 mL. Alternatively, it could be taken to imply that p(2 × 2) O represented overlayer oxygen and c(2 × 2) O represented a reconstructed/penetrated surface. Our final conclusion is going to be that p(2 × 2) and c(2 × 2) O are both overlayer structures of nearly identical nature [i.e. the difference just represents filling in the empty c(2 × 2) sites], and that the suggestion of early NiO nucleation to explain the $\Delta\phi$ behavior is correct. However, since there has been considerable argument in the literature as to the relative electronic natures and adatom distances, $d_\perp$, above the surface for these two structures [198, 199, 224, 239, 240, 264] we must consider the $\Delta\phi$ data carefully. In Fig. 43 we compare $\Delta\phi$ measurements from five sets of authors [195, 231, 273–275]. It is not possible to compare them all as $\Delta\phi$ versus θ because in some cases only $\Delta\phi$ versus exposure is given, and in others only discrete values corresponding to the best-formed p(2 × 2) and c(2 × 2) structures are given. Though the results of Holloway and Hudson [231], and of Akimoto et al. [273] both show a decreasing $\Delta\phi$ before the c(2 × 2) O structure, the other three are in agreement that $\Delta\phi$ increases through both the p(2 × 2) and c(2 × 2) structures. Demuth and Rhodin [195] note that their data indicate that the change in ϕ per O atom is constant over the whole exposure range up to the well-formed c(2 × 2) structure. This is approximately true for the other two results also. In some of these experiments, the surfaces had either been exposed at up to 550 K or annealed after formation of the LEED structures. This was done to sharpen the LEED structures, but it had been noted that this did not affect the LEED IV characteristics or, apparently, the $\Delta\phi$ values. We therefore conclude that, although the perfection of the long-range adsorbate ordering is affected by the higher temperature, the local bonding characteristics are unchanged. Though the results of Holloway and Hudson and Akimoto et al. are different from the other three with ϕ starting to decrease before the c(2 × 2) O structure forms, the magnitude of the $\Delta\phi$ excursions are quite different in the two studies. Now the kinetics of the oxygen uptake in the Akimoto et al. study does not show a very flat plateau region, and the length of the plateau region in the Holloway and Hudson data is very short (Fig. 44). Other data, particularly those of Hopster and Brundle [236] (also shown in Fig. 44), are different in that the plateau region is flatter and longer, i.e. the onset of NiO formation is delayed. A similar flattening and delay in oxidation growth is observed in the Holloway and Hudson study if the adsorption temperature is raised to ~450 K (Fig. 42). Since we know that raising T in this manner sharpens the p(2 × 2) and c(2 × 2) LEED spots, we suggest that the delayed onset of NiO nucleation is related to the increased perfection in the p(2 × 2) and c(2 × 2) chemisorbed structures. We conclude, therefore, that the differences in the 300 K kinetics between authors are also related to variations in adsorbate overlayer perfection, probably caused by a variation in initial Ni{100} perfection. The latter suggestion is fully supported by the data of Hopster and Brundle [236] where they were able to produce kinetics at 300 K more like Holloway and Hudson [231] by starting with a poorly annealed Ni{100} surface or by deliberately damaging the surface by a brief Ar^+ bombardment (see Fig. 44). Miranda et al. [246] studied the kinetics of oxidation as a function of Ar^+ sputter dose. They were able to reduce the length of the plateau at 300 K from ~100 to ~20 L and increase the oxidation rate after the plateau by a factor of ~2 by damaging the {100} surface. It therefore seems that the differences in $\Delta\phi$ versus θ or exposure behavior, which parallel the differences in kinetics, are also related to differing degrees of perfection in the ordered overlayer. If the ordering is poor at 300 K, then the onset of NiO nucleation occurs at lower exposure than for a well-ordered superstructure and this is reflected in an early decrease in $\Delta\phi$, even though the NiO nucleation cannot be observed by LEED until there are sufficient oxide islands of diameter > 20 Å (oxide islands are observed at an earlier stage by RHEED). There are also questions

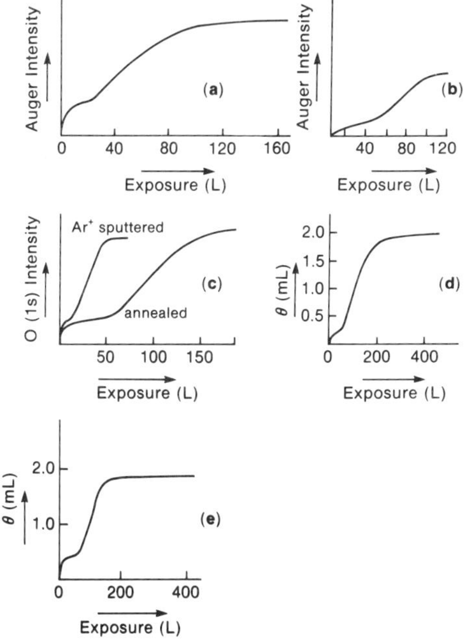

Fig. 44. Comparison of the kinetics of oxygen adsorption by different authors for Ni{100}/O. (a) Holloway and Hudson [231], 300 K. (b) Akimoto et al. [273], 300 K. (c) Hopster and Brundle [236], 300 K. (d) Mitchell et al. [227], 315 K. The vertical axis is directly proportional to coverage (scale marked) since there are no attenuation factors to consider when converting oxygen X-ray fluorescence signal intensity to coverage. (e) Smeenk et al. [209], 325 K. The vertical axis is directly proportional to coverage, as determined by RBS oxygen backscattering intensity. The scale has been corrected by 21%, however (see ref. 20 and text).

about how the number and size of the NiO nuclei change with T and surface perfections during the nucleation process, but we will not discuss them until Sect. 3.1.9. All we wish to establish here is that a surface showing the well-ordered c(2 × 2) O structure and having no NiO nucleation co-existing has the same $+\Delta\phi$ per O atom as does the p(2 × 2) O structure, and therefore the $\Delta\phi$ data are supportive of the idea that both structures are overlayers with very similar electronic and geometric characteristics.

3.1.2 General reaction scheme. Low temperature

The behavior of the Ni{100}/O_2 system at low T has not been so well studied as at $\geqslant$300 K, but it is clear from work in the 77–150 K range that there are major differences in kinetics, if not in chemical structure and compositions at low temperature. The original 147 K data of Holloway and Hudson are shown in Fig. 42. Similar kinetics have been reported in XPS studies by Norton et al. [243] and Hopster and Brundle at 77 K [237]. In comparing with 300 K, the following major points are clear.

(1) The saturation coverage appears to be similar, i.e. equivalent to 2–3

layers of NiO assuming no drastic differences in depth distribution and therefore attenuation factor for the oxygen signal.

(2) The reaction sequence does not consist of a fast initial chemisorption more or less separated from the subsequent oxidation by a period of low sticking probability. The oxide growth is complete within a few Langmuir exposure.

(3) No ordered LEED structures are observed below 200 K, either for the initial chemisorption or for the saturation oxide product.

Holloway and Hudson did not consider the low T data in detail, but the three points above can all be phenomenologically explained using a model where both the lack of order in the oxygen adsorbate layer and the increased residence time of any precursor molecular state are considered (see Sect. 3.1.9 for details). The expected effect of the lack of adsorbate order should be obvious from the above 300 K discussion; oxide nucleation should occur quickly, which is the case. For any discussion of whether the final saturation product at low T is the same as at 300 K or not, a $\Delta\phi$ value would be useful. This has not been measured. Holloway and Hudson report that an O-saturated surface subsequently cooled to 147 K and exposed to oxygen will reversibly adsorb about another 20% of a monolayer and increase ϕ by 0.8 eV. They suggest this may be due to molecular oxygen. The subsequent UPS and XPS work found no evidence of any measurable coverage of molecular oxygen at least down to 77 K [243, 247], and in fact did not detect further oxygen uptake after cooling a 300 K-saturated surface to 77 K. It may be that the Holloway and Hudson additional reversible uptake is electron-beam-induced or possibly represents H_2O or CO_2 contamination at low T. The pick-up of minor contaminants is a problem, particularly at low T, as discussed in the next section, and can confuse the issue of whether the oxide product is the same at 77 K as at 300 K. Since, in the situation where we are sure there are no contamination products, Ni^{2+} is detected by XPS at 77 K in similar quantities to 300 K [238] we conclude that there are no major differences in chemistry, only in long-range order.

3.1.3 Problems of co-adsorption and reaction

One of the early points of contention for the Ni/O_2 system was the question of whether it was possible to have more than one type of chemisorbed oxygen species. Originally there seemed to be the possibility of a $\theta = 0.25$ mL molecular chemisorbed species below a certain exposure (the film work of Horgan and King [270]). When this was disproved for Ni{100} (Holloway and Hudson [231] and subsequent work), a possible electronic distinction between p(2 × 2) oxygen and c(2 × 2) oxygen had to be considered. Early discussion in this area centered around the interpretation of the O(1s) XPS spectrum for $Ni\{100\}/O_2$ [275]. A typical spectrum is shown in Fig. 45(b). It represents a 200 L 300 K exposure at 5×10^{-9} Torr. The base vacuum of the UHV system used is $\sim 1 \times 10^{-10}$ Torr. In interpreting the two O(1s) features observed, the possibilities are (a) that one represents chemi-

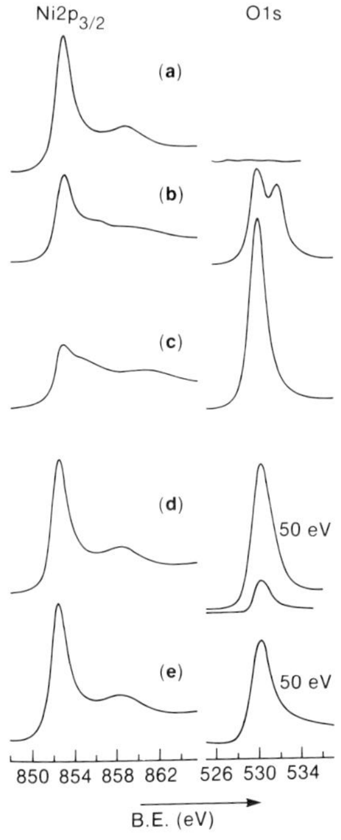

Fig. 45. O(1*s*) and Ni2*p* XPS spectra for the oxygen exposure on Ni{100} at 300 K under various conditions. (a) Clean Ni{100} surface. (b) 200 L exposure at 2×10^{-9} Torr at 300 K [286]. (c) 500 L exposure at 1×10^{-6} Torr at 300 K [286]. (d) 20 L exposure at 1×10^{-6} Torr at 300 K; sharp c(2 × 2) [286]. The O(1*s*) spectrum is also shown at lower resolution (50 eV band pass). (e) 18 L exposure at 1×10^{-7} Torr at 300 K; sharp p(2 × 2) [286]. O(1*s*) at low resolution only.

sorbed O and one NiO [275]; (b) one represents p(2 × 2) O and one c(2 × 2) O; (c) one represents NiO and the other a defect "Ni_2O_3" species [276] or (d) that one is atomic and one molecular. In fact, we can state, unambiguously that the shoulder at ~531.4 eV for the particular spectrum shown is not representative of p(2 × 2) O, c(2 × 2) O, nucleated NiO, "Ni_2O_3", or molecular oxygen. It is unambiguously attributable to OH_{ads} formed by the reaction, on the surface, between oxygen and contaminant water. We do not wish to discuss the nature of this reaction here, but note that the reaction takes place predominantly around the plateau stage of the kinetics. Here the relative importance of residual H_2O, either as a component of the base pressure or generated by interaction of the introduced oxygen on the chamber walls, will be greatest since the sticking probability of O_2 is so low, i.e. a long exposure is required. We are sure that the 531.4 eV peak of Fig. 45(b) is OH_{ads} because in this particular experiment, and in others where H_2O was deliberately introduced with oxygen, the 531.4 eV O(1*s*) peak intensity correlated directly with the OH^- intensity observed in SIMS [236]. In addition, OH_{ads} has a characteristic UPS signature, two O(2*p*) derived features separated by approximately 4 eV. If stringent precautions are taken to minimize the H_2O problem, e.g by "seasoning" the chamber walls with oxygen and then working with a high oxygen-dosing pressure, the O(1s) spectrum shows only one peak [Fig. 45(c)]. In addition, this peak is at the same BE, within ± 0.3 eV, for p(2 × 2) O, c(2 × 2) O and NiO. The fact that OH contamination can cause problems has been recognized before [243], but it has been suggested that an O(1s) BE of 531.4 eV in the Ni/O_2 system cannot always be ascribed to OH. For example, under some conditions, the species giving rise to the 541.4 eV feature is stable to moderate heating, whereas OH_{ads} deliberately produced will disproportionate back to O_{ads} and H_2O, which desorbs. For adsorption at 77 K an O(1s) feature was found at 531.5 eV which did not give a SIMS OH^- signal and therefore could not be OH_{ads}. Hopster and Brundle originally and erroneously interpreted it as being atomic oxygen which had penetrated the Ni lattice but had not formed an ordered NiO nucleus [237]. Subsequently it was found that it represented the oxygen of a CO_3 contaminant formed by reaction between oxygen and residual CO_2 which is adsorbed at 77 K [238]. The CO_3 species so formed is stable to 600 K, at which temperature it starts to decompose back to CO_2 and O_{ads}.

Our belief, then, is that at this point there is no proven case of atomic oxygen on Ni{100} or of NiO having an O(1s) BE of other than ~530 eV, and that in many cases OH and CO_3 contaminants explain the 531.4 eV shoulder. There is still a belief that a "defect Ni_2O_3" structure could contribute to this shoulder [243]. Defect Ni_2O_3 has been claimed in XPS studies many times. In most cases the conditions were such that OH or CO_3 are the likely explanations, but we do not rule out the possibility of some genuine cases.

We have gone into these co-reaction aspects on Ni{100} in some detail because they are well-documented and there have been many arguments on the interpretation of "high BE" O(1s) features for Ni/O_2. We want to make the point, however, that we believe these to be quite general problems and that there are many other metal/O_2 studies where such co-reactions have been misidentified as pure O_{ads} situations. We also note that, in addition to OH and CO_3 giving O(1s) features at ~531.5 eV, CO, CO_2 and H_2O all give features between 531 and 534 eV BE.

3.1.4 Geometric structure of p(2 × 2) O and c(2 × 2) O

From the previous sections we know that it is possible to generate well-ordered p(2 × 2) and c(2 × 2) oxygen-induced structures where the oxygen is in the atomic state and there are no significant contributions from nucleated NiO. The conditions for producing the sharpest LEED patterns are adsorption on a well-annealed surface at up to ~500 K or adsorption at 300 K followed by subsequent annealing to up to 500 K. If the temperature is too high, dissolution into the bulk will occur. The temperature for dissolution is coverage-dependent, being ~550 K for the $\theta = 0.25$ mL p(2 × 2) structure and ~700 K for the $\theta = 0.35$ mL c(2 × 2) O structure [233]. At higher coverages the c(2 × 2) O structure co-exists with surface oxide and the bulk dissolution temperature is even higher [233]. If the ordered LEED structures are prepared by heavy oxidation at 300 K and subsequent heating until sharp LEED patterns are observed (the practice in some work), what actually happens is that the excess surface oxygen of the oxide formed at 300 K dissolves into the top ~35 Å of subsurface leaving c(2 × 2) or p(2 × 2) structures on the surface. The O Auger intensities may be nearly identical to the pure overlayer cases, but the large amounts of subsurface oxygen show up easily in medium-energy ion-scattering measurements, as has been fully documented for the Ni{110}/O system [212]. The behavior of overlayer plus subsurface oxygen (reactivity, bulk dissolution temperatures and rates) will be different from the pure overlayer case. The geometric structure of an overlayer in the presence of subsurface oxygen may or may not vary from the pure overlayer case and the different techniques used to characterize the overlayer structure may show differences which are either genuinely reflecting variations in an overlayer geometry, or are reflecting a sensitivity of the technique to the subsurface oxygen. Clearly, for unambiguous answers concerning overlayer geometries, both the presence of nucleated NiO and dissolved subsurface oxygen should be avoided. Keeping this in mind we will review the structural data.

LEED IV profile measurements compared with scattering calculations as a function of geometry have been the most extensively used methods of attempting to determine the p(2 × 2) and c(2 × 2) O geometries [194–199]. Unfortunately both the data, the calculations, and the claims of what constitutes agreement between the two, have changed significantly over the years. Except for the most recent work, all the calculations have assumed that the O atom sits at the center of the fourfold hollow site (Fig. 46) and simply varied the vertical distance above the surface, $d_\perp$, to search for a fit to the data. This assumption was based on early trials with different locations of atoms (head-on; bridge site) which apparently proved totally untenable. The most recent work varies the lateral position of the O within the fourfold hollow site [199].

The first data and calculations of any respectability were those of Andersson et al. [194]. They looked at four beams, including the (1/2, 1/2) beam, for the c(2 × 2) O structure (formed in a manner likely to give a pure overlayer) over the energy range 20–60 eV. Comparison of the IV profiles with the dynamical calculations led to a proposed best-fit of $d_\perp$ equal to 1.5 Å. By today's standards, however, the fit is abysmal, and calculation below 60 eV would not be considered reliable. In any case, trial fits for $d_\perp$ < 1 Å were not attempted. The next measurements were made by Demuth and Rhodin [195] with calculations by Marcus et al. [196]. Both p(2 × 2) and c(2 × 2) structures were considered and trial fits between 0.815 and 1.760 Å were tried over an energy range up to 240 eV. Where the results overlap and those of Andersson et al. (the 20–60 eV range for the 1/2 1/2 beam of the c(2 × 2) structure), there are differences both in the experimental data and the calculations [224, 239, 240]. Marcus et al. [196] claimed a reasonable and unique fit in their data at a $d_\perp$ value of 0.92 Å for both p(2 × 2) and c(2 × 2) overlayers, based

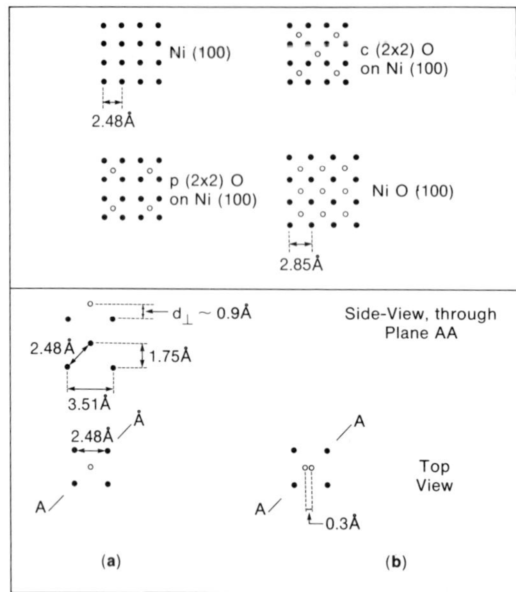

Fig. 46. Geometry of O_{ads} on Ni{100}. Upper panel: schematic of Ni{100} lattice (full circles), O_{ads} chemisorbed superstructures, and the NiO{100} lattice. It is assumed that O_{ads} (open circles) occupies the fourfold hollow sites. For a schematic of the equivalent LEED patterns, see Fig. 50. Lower panel: top and side views of Ni{100} plus O_{ads}. (a) $d_\perp$ as determined by Marcus et al. [196] assuming symmetric location of O_{ads} in fourfold site; (b) lateral displacement of O_{ads}, as suggested by Demuth et al. [199]. $d_\perp$ was found to be ~0.8 Å.

largely on agreement of peak positions in the 20–80 eV range for fractional-order beams. The quality of the fit was certainly better than that of Andersson et al. but it was by no means perfect. The value of 0.92 Å was accepted as correct for several years, however, and calculations by Van Hove and Tong [277] compared with the same experimental data for p(2 × 2) O also produced best fits at ~0.9 Å. Pendry [278] extended his original calculations [194] to $d_\perp$ values of < 1 Å at a later date and found an improved fit at 0.9 Å compared with the original best fit of 1.5 Å. The value is also supported by a low-energy ion-scattering study on the c(2 × 2) structure by Brongersma and Theeton [206], which showed that the oxygen had to be present as an overlayer rather than being in the plane of the surface, and gave an estimated $d_\perp$ of 0.9 ± 0.2 Å. Since, however, in this study the Ni surface was cleaned by heating an O-covered surface to 800 K, the c(2 × 2) O structure certainly had a lot of subsurface oxygen present, even though the ISS measurements would not detect this.

At a later date the range of trial fits to the LEED data was extended to include variation in the distance of the top Ni layer from the bulk [279]. The quality of the fit did not improve. In 1980, Hanke et al. [197] repeated the LEED measurements for p(2 × 2) and c(2 × 2) structures using a fast scanning LEED system. The range was extended to ~450 eV. Comparison with

Demuth and Rhodin [195] over the 20–240 eV range showed that the new data had small but significant differences from the old data and generally showed a sharper structure. The experimental data must be highly suspect, however, since the p(2 × 2) and c(2 × 2) structures were annealed at 1100 K for up to 15 min! In the absence of substantial subsurface oxygen, all oxygen from the overlayer structures would be rapidly dissolved into the bulk at this temperature. Hanke et al. still conclude, however, that ~0.9 Å is the correct answer.

Around this time, a lot of activity developed using structural methods other than LEED, including ab initio electronic structure calculations on clusters. They are discussed later, but the impetus for LEED was to provoke remeasurement and recalculation for the c(2 × 2) structure. Tong and Lau [198] recalculated the IV profiles for the c(2 × 2) structure, extending the trial structures down to $d_\perp \sim 0$ Å, since it had been suggested from an interpretation of HRELS data that this structure had a very short $d_\perp$ value. They concluded that no new $d_\perp$ value gave a better fit to the Demuth and Rhodin data than the originally proposed ~0.9 Å, but that $d_\perp \leqslant 0.1$ Å gave a roughly comparable value and therefore the existing LEED data could not exclude such a value. They also pointed out that a more extensive experimental data set on the oxygen-induced fractional order beams was what was required to help differentiate between the two $d_\perp$ values. Demuth et al. [199] then remeasured the IV curves for the c(2 × 2) O structure. The first premise in his new work was to discard all comparison between experiment and theory below 60 eV on the grounds that the calculations in this region are too sensitive to the uncertainties in the surface barrier and the scattering potentials used. This is equivalent to saying that nearly all the previous LEED work, including their own, should be discarded as unreliable! The c(2 × 2) structure was prepared in a manner which should give rise to a pure overlayer, but the authors noted that the O Auger intensity was 1.8–2 times that for the p(2 × 2), which would suggest a θ of ~0.5 mL, whereas we have suggested earlier that oxide nucleation should occur around $\theta = 0.35$ mL. EELS measurements, however, apparently did not reveal the presence of any nucleated NiO, though this has been questioned recently by Ibach [280]. The data was for the (00), (1/2 1/2) and (1/2 3/2) beams up to ~260 eV and it again showed small variations compared with the earlier data by Demuth and Rhodin [195]. They performed a wide search procedure in the trial fits to theory for the data above 60 eV, varying not only $d_\perp$, but also the registry of the O with the {100} surface. While the $d_\perp$ best fit was still nearly identical to the original, at 0.80 ± 0.025 Å, it was claimed that a considerably better fit was obtained if the O atom was also moved 0.3 ± 0.1 Å from the fourfold hollow site along the {110} direction [Fig. 46(b)]. It is our opinion that the variations in data over the years are comparable with the improvements in fit obtained for the single (last) set of data using a 0.3 Å off-center O position, compared with the center position, so we still consider it an open question, from LEED, as to the location of the O atom within the fourfold hollow. We note, however, that, after all these measurements and recalculations, a $d_{\perp}$ value of ~ 0.9 Å is still the preferred LEED value.

The other structural methods that have been applied to the p(2 × 2) and c(2 × 2) structures are EXAFS, angle-resolved XPS, normal photoelectron diffraction (NPD), and He diffraction. In addition, vibrational spectra, recorded by HRELS, and angular-resolved UPS data have been interpreted in terms of support for particular geometries. Since the most dramatic claims, subsequently shown to be quite wrong, have been made from the vibrational studies, we will consider these first. The vibrational spectra for the p(2 × 2) and c(2 × 2) O structures by Lewald and Ibach [222(a)] are shown in Fig. 47(a) and (b). The original spectra of Andersson [219, 220] show less fine structure, either because of lower resolution or because some of the structure was genuinely missing. Demuth [281] recently remeasured the vibrational spectra at comparable resolution to the Lewald and Ibach study, paying particular attention to the possibility of incipient oxide nucleation,

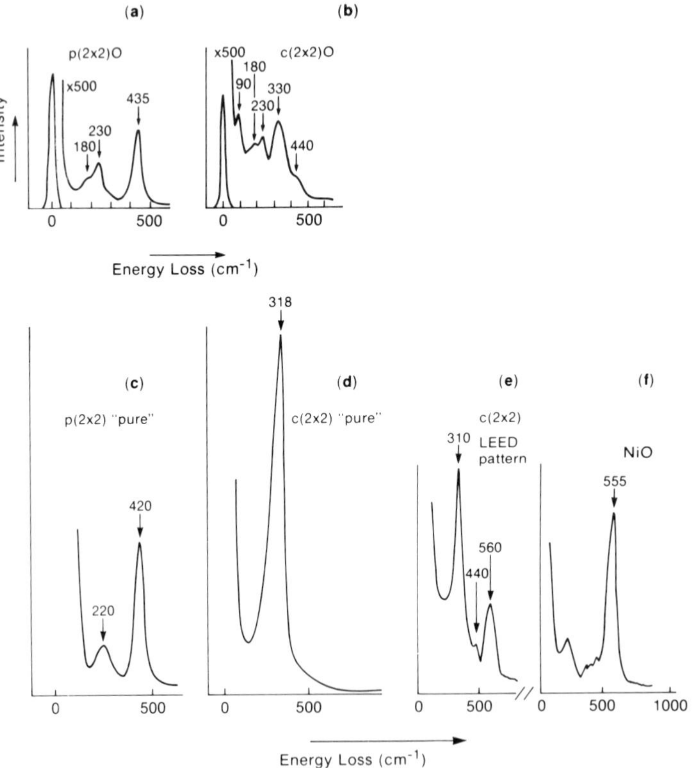

Fig. 47. (a), (b) HRELS spectra of p(2 × 2) O and c(2 × 2) O on Ni{100} at 300 K [222(a)]. (c), (d), HRELS spectra of "pure" p(2 × 2) O and c(2 × 2) O [222(b)]. (e), (f), HRELS for a surface showing a c(2 × 2) O LEED pattern, but which in fact contains [in (e)] admixtures of p(2 × 2) O and NiO or [in (f)] a large amount of NiO [222(b)].

oxygen dissolution, etc. in the p(2 × 2) and c(2 × 2) structures. He obtained identical results, which seemed to rule out the possibility that the spectra were not representative of pure p(2 × 2) and c(2 × 2) O structures. There are very striking differences between the p(2 × 2) and c(2 × 2) vibrational spectra, with the main Ni surface–O vibrational frequency decreasing from $\sim 435\ cm^{-1}$ for the p(2 × 2) structure to $\sim 330\ cm^{-1}$ for the c(2 × 2) (Table 8). Ignoring, for the moment, the additional fine structure of the spectra, one has to be able to explain this large decrease in terms of a difference in force constant for the O–Ni surface in the two structures. Andersson [219, 220] originally gave an explanation in terms of a decreased barrier for O penetration into the lattice for the c(2 × 2) structure (0.8 eV) compared with the p(2 × 2) (1.5 eV), both of them being small compared with the O heat of

TABLE 8

Experimental and theoretical $d_\perp$ values and vibrational frequencies, ω_e, for Ni{100}

Ref.	System	Electronic description of state	$d_\perp$ (Å)	ω_e (cm^{-1})	ΔE (eV)[a]
196, 222	p(2 × 2) O		0.86 ± 0.07	430	
196, 222	c(2 × 2) O		0.86 ± 0.07	310	
257, 258	Ni_5O cluster using pseudo-potential	Oxide	0.65		
	Ni_5O^+ cluster using pseudo-potential	Radical	0.96		
239	$Ni_{20}O$ cluster using pseudo-potential	Oxide	0.55	218 (290)[b]	0.34
		Radical	0.88	371 (427)[b]	
239	Ni_2O^+ cluster using pseudo-potential	Oxide	0.26	266	1.03
		Radical	0.83	355 (419)[b]	
224	Lattice dynamical fit to $Ni_{20}O$ and $Ni_{20}O^+$ results	Oxide	0.27	307	
		Radical	0.88	445	
240	Ni_5O using pseudo-potential[c] (no $4p$ functions)	Oxide	0.57	254	0.42
		Radical	0.87	374	
	+ $4p$ functions	Oxide	0.68	290	0.1
		Radical	0.98	358	
	+ $4p$ + CI	Oxide	0.74	320	0.89
		Radical	0.91	317	
264[d]	Ni_5O electron	Oxide	1.05	380	
	Corrected pseudo-potential	Oxide	0.93	400	
	$Ni_{25}O$ corrected pseudo-potential	Oxide	0.83	400	
	$Ni_{25}O_5$ corrected pseudo-potential	Oxide	0.72	315	

[a] Energy separation between the "oxide" state and "radical" state. In each case the oxide is found to be the ground state.
[b] Empirically adjusted to account for Ni lattice interaction (see ref. 239).
[c] An identical pseudo-potential to that used in ref. 239 for the $Ni_{20}O$ calculation was used.
[d] Ground state (oxide) results only reported.

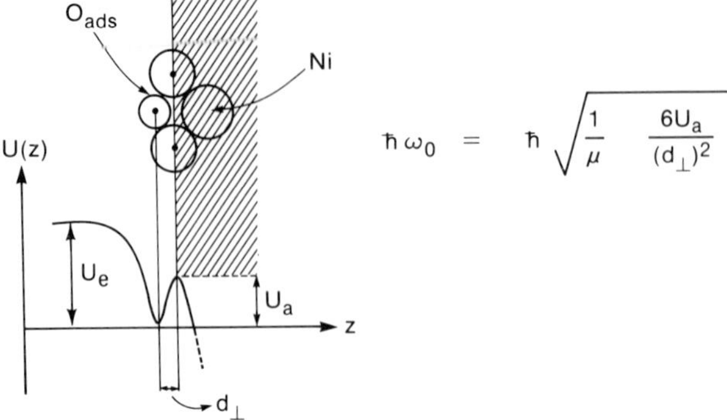

Fig. 48. One-dimensional potential energy representation of O on Ni{100} [219, 220] and the derived formula for the Ni–O stretching frequency. μ is the reduced mass of the vibrating system, $d_\perp$ is the distance of O_{ads} above the Ni surface, U_a is the potential barrier to O_{ads} penetration of the lattice. For a $d_\perp$ of 0.9 Å, U_a values of 1.5 and 0.8 eV for the p(2 × 2) and c(2 × 2) structures reproduce the observed $\hbar\omega$ values (see Fig. 47).

adsorption which may be estimated as 4–5 eV from molecular and solid NiO measurements. The one-dimensional potential energy diagram is sketched in Fig. 48. For the particular anharmonic potential Andersson used to model the system, the values of U_a and U_e were chosen to reproduce the observed vibrational frequencies of the p(2 × 2) and c(2 × 2) O structures for a $d_\perp$ of 0.9 Å in each case. As pointed out by Andersson, this interpretation is supported by the knowledge that O dissolves into the bulk easily, indicating a low potential barrier. In addition we know that the oxide nucleates from the c(2 × 2) structure, suggesting that the barrier might be lower for the c(2 × 2) structure. Armond and Theeten [218] performed lattice dynamical calculations on the p(2 × 2) and c(2 × 2) structures, showing that a fit was obtained if the force constant reduced by a factor of 2 in going from one to the other. They suggested a reason basically similar to that of Andersson, namely that the force constant decreases because the change in the O coordination number reduced the potential barrier. Andersson also pointed out that the lack of any large decrease in vibrational frequency between the p(2 × 2) and c(2 × 2) S structures on Ni{100} was perfectly compatible with the known behavior of S, compared with O, on Ni surfaces; namely that it does not diffuse into the lattice but tends to segregate out to an overlayer position. This suggests that, in contrast to O, the value for U_e for S is comparable with U_a and therefore small differences between U_e p(2 × 2) S and U_e c(2 × 2) S will not cause major changes in vibrational frequency, in agreement with experiment.

All the above arguments for the rational interpretation of $\hbar\omega$ for p(2 × 2) and c(2 × 2) O in terms of an identical $d_\perp$ of 0.9 Å for each structure were subsequently ignored by Upton and Goddard [239] and by Rahman et al. [224] in their interpretations of the vibrational data. Upton and Goddard performed electronic structure calculations on a $Ni_{20}O$ cluster as representative of the Ni{100}/O surface. For the $Ni_{20}O$ cluster, in addition to the ground state, which had a normal closed-shell O(2*p*) structure (termed the "oxide" state by Upton and Goddard) an open shell, "radical" excited state was found with an energy of only 0.3 eV higher than the ground state at its equilibrium bonding distance ($d_{\perp}$ = 0.88 Å for the radical state; 0.55 Å for the oxide state). The calculated Ni–O vibrational frequencies for the oxide and radical state were 218 and 371 cm^{-1}, respectively (Table 8). Upton and Goddard corrected these empirically to 290 and 427 cm^{-1} to account for interaction with the Ni lattice phonons. On the basis of the $d_{\perp}$ value of 0.88 Å and the frequency of 427 cm^{-1} for the radical state, Upton and Goddard concluded that the p(2 × 2) O structure must represent O atoms in this open-shell radical state. They then performed a similar set of calculations on a $Ni_{20}O^+$ cluster, which they claim is more likely to be representative of the higher coverage c(2 × 2) structure because of possible charge depletion in the Ni layer at high O coverage. In the $Ni_{20}O^+$ cluster the oxide state becomes greatly stabilized, compared with the radical state, and had a calculated $d_{\perp}$ of 0.26 Å and frequency of 266 cm^{-1}. Upton and Goddard therefore concluded that the c(2 × 2) O structure represented the closed-shell oxide state, and that $d_{\perp}$ for this state was not 0.9 Å, as believed from LEED, but 0.26 Å. Their explanation for the reduced vibrational frequency for c(2 × 2) O versus p(2 × 2) O was therefore a large reduction in $d_{\perp}$ rather than a decrease in U_a. It was the claims of these calculations which triggered the new round of experimental determinations of $d_{\perp}$. Unfortunately all the measurements subsequent to their calculations have confirmed that $d_{\perp}$ is ~0.9 Å for both p(2 × 2) and c(2 × 2) structures, and no evidence for a difference in electronic structure between the two structures, as would be expected for "radical" versus "oxide" states, has been found. In addition, almost every conclusion concerning the Upton–Goddard (UG) calculations, except the actual existence of a higher-lying radical state, has been challenged in subsequent calculations by Bauschlicher et al. [240, 264]. We will return to this in the section on electronic structure. Here we just want to note that the calculation by a lattice dynamical approach of the vibrational spectra by Rahman et al. [224] of the p(2 × 2) and c(2 × 2) O structures in terms of "normal" (0.9 Å) and "short" (0.26 Å) $d_{\perp}$ values, respectively (Table 8), was entirely dependent on the force constants supplied from the UG calculations and the unjustified belief that a difference in these force-constants implied a difference in $d_{\perp}$. Recently this belief has been abandoned by the authors [282] since, when they empirically adjusted the force constants used, a perfectly good fit to $d_{\perp}$ ~ 0.9 Å was obtained for the c(2 × 2) O structure. In addition, the dispersion of the vibrational frequencies in ELS for c(2 × 2) O has now been measured and apparently the calculated dispersion can only be fit with a $d_{\perp}$ of 0.9 Å, not 0.26 Å [282]. Finally, one should note that the latest round of experimental spectra for "pure" p(2 × 2) O and c(2 × 2) O [222b], shown in Figs. 47(c) and (d), do not show any of the fine structure of the original spectra [Fig. 47(a) and (b)] which apparently was

References pp. 381–388

due to co-existing oxide. The end conclusion, then, from the ELS data is that, far from providing any support for the Upton and Goddard short $d_\perp$ structure, as originally claimed, a normal $d_\perp$ of 0.9 Å is supported for p(2 × 2) and c(2 × 2) and a short $d_\perp$ for c(2 × 2) O is completely untenable.

Angle-resolved XPS measurements of the Ni{100}/O system were reported by Petersson et al. [217] prior to the Upton and Goddard calculations. Azimuthal anisotropies in the O(1*s*) photoemission are caused by elastic scattering of the photoelectrons from neighboring atoms. The photoelectron energy (~ 10^3 eV) is sufficiently high such that the scattering is strongly forward peaked and reasonably well represented by a single-scattering model. The experimental anisotropies can be used to establish both the symmetry of the adsorbate site and $d_\perp$ by comparison with single-scattering calculations. Petersson et al. measured the O(1*s*) azimuthal angular behavior for various oxygen exposures at 300 K. For the p(2 × 2) regime they found a best fit between experiment and calculation at $d_\perp \approx$ 0.8–0.9 Å. For exposures where the c(2 × 2) O LEED pattern was strongest (~ 100 L) a $d_\perp$ value of ~ 0 Å was favored. Upton and Goddard quoted these results in support of their suggestion that the c(2 × 2) O structure had a very short $d_\perp$. Petersson et al., however, were quite careful to qualify their results by two important statements. First, at 100 L exposure the O coverage is approximately unity, i.e. well in excess of that present in the c(2 × 2) structure. We conclude, therefore, that the c(2 × 2) is coexisting with a lot of oxide at this exposure. Second, they point out that, because of the strong forward scattering involved, the XPS anisotropies will be much more strongly affected by O in or below the Ni plane (e.g. oxide) than O in an overlayer position [e.g. the p(2 × 2) structure]. At 15 L exposure, for which a weak c(2 × 2) is observed, a somewhat similar anisotropy to the 100 L exposure was found even though the O coverage was about a factor of two lower (the experimental statistics were also poor). Thus, either the c(2 × 2) O exhibits the same anisotropy as oxide or, even at 15 L exposure, there is sufficient oxide present to affect the observed anisotropy. It has been estimated that a 15% oxide contribution would be sufficient to cause the effect [283]. We conclude, then, that the likely explanation of the short $d_\perp$ derived from the angular XPS for > 15 L exposure is the presence of small amounts of incipient oxide, and in fact angular XPS may be a good way to detect such minority species.

Normal photoelectron diffraction (NPD) measurements [216] for c(2 × 2) O formed at 300 K (20 L exposure) were reported after the angular XPS measurements. In NPD the O(1*s*) intensity is measured, normal to the surface, as a function of photon energy and is compared with scattering calculations for trial values of $d_\perp$. The results gave a best fit of 0.9 ± 0.04 Å. A later, more refined analysis gave 0.85(4) Å. One should note that a $d_\perp$ value of $\leqslant$ 0.5 Å was not tried so $d_\perp$ values below this value cannot be explicitly excluded by the NPD measurement, though Rosenblatt et al. [216] do give reasons why such distances would not be expected to fit the data.

Following the claims of Upton and Goddard [239] and of Rahman et al.

[224] of dramatic differences in $d_\perp$ between p(2 × 2) O and c(2 × 2) O, Stöhr et al. [200] examined the SEXAFS of both systems. By examining both the O–Ni distances obtained from the Fourier transform of the EXAFS signal and the polarization dependence of the amplitude ratios for different angles of incidence they were able to show that the p(2 × 2) and c(2 × 2) structures both occupied fourfold hollow sites with identical $d_\perp$ values of 0.86 ± 0.07 Å, in complete disagreement with Upton and Goddard and Rahman et al. The resolution of the EXAFS structure is such that the presence of a small amount of oxide together with c(2 × 2) O oxygen would not affect the determined $d_\perp$ value, making the EXAFS results compatible with the angular XPS where we noted above that 15% of incipient oxide could affect the angular XPS results. In addition, Demuth et al. [199] reanalyzed the Stöhr et al. data in the light of their LEED based claim of a ~0.3 Å lateral displacement from the fourfold hollow position. Such a displacement leads to two different nearest-neighbor O–Ni distances and correspondingly two closely spaced peaks in the Fourier transformed EXAFS. Demuth et al. showed that the Stöhr et al. data would not be able to resolve these peaks so the EXAFS data, while firmly establishing identical $d_\perp$ values for p(2 × 2) and c(2 × 2) O, does not resolve any question concerning lateral displacements of the O atom in the c(2 × 2)O structure.

In addition to recording the EXAFS data, which can be analyzed directly by Fourier transform, Stöhr et al. [200] also recorded the X-ray absorption near-edge structure (XANES) for the O(1*s*) level for the p(2 × 2) and c(2 × 2) O structures. Multiple scattering is important for the near-edge structure and the data must be compared with a full multiple scattering calculation. Norman et al. [284] performed these calculations for the c(2 × 2) O structure using a 30 atom cluster with $d_\perp$ varied. The calculated XANES fits the experimental XANES for a $d_\perp$ value of 0.9 Å, whereas for $d_\perp$ = 0.2 Å the fit is qualitatively incorrect.

The most recent structural studies of the p(2 × 2) and c(2 × 2) O structures have been by He atom diffraction [201, 202]. The scattering occurs off the valence electron density tails, i.e. 2–4 Å out from the surface nuclei. It does not, and cannot, determine atomic positions directly. They must be inferred, or better calculated, from the variations in the charge density sensed by the He atoms at these large distances from the surface. At the level of sophistication currently used, the scattering data are converted to an experimental corrugation function assuming a hard-wall potential for the He–surface interaction. For the p(2 × 2) and c(2 × 2) structures on Ni{100} identical corrugations of 0.55 Å (Fig. 49) are found, compared with a nearly flat corrugation (0.01 Å) for the clean surface [201, 202]. (The structures were formed at 600 K, so it is likely that some subsurface O is present.) The fact that the two experimental corrugations are identical is, in itself, strong evidence that $d_\perp$ is the same for both structures. As Rieder [201, 202] points out, however, it is conceivable that both $d_\perp$ and the Ni–O charge distribution could change simultaneously leaving an unchanged corrugation. Unfor-

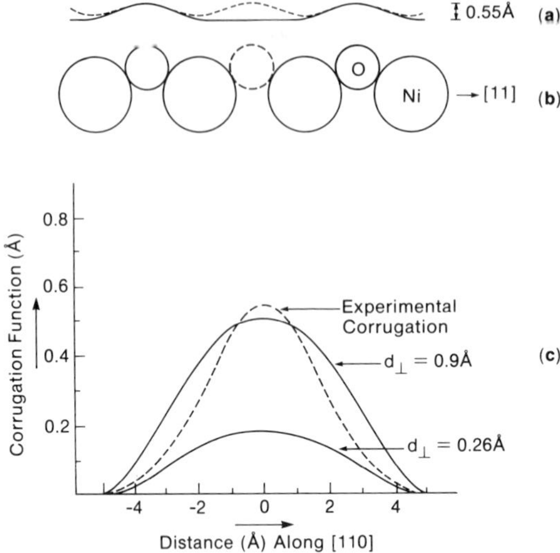

Fig. 49. He scattering corrugation function results for Ni{100}/O, plus calculated corrugation functions. (a) Best-fit corrugation functions [201, 202] along the ⟨11⟩ direction (along line AA in Fig. 46) for the p(2 × 2) O structure (solid line) and the c(2 × 2) O structure (broken line). Identical corrugations are found along the {110} direction. (b) Hard-sphere model along the ⟨11⟩ direction for the p(2 × 2) O structure. The broken circle represents the additional O adatom of the c(2 × 2) structure. (c) Calculated corrugation function [203] along the {110} direction for $d_\perp$ = 0.9 and 0.26 Å for a charge transfer to O_{ads} of unity, compared with the experimental corrugation function [201, 202], for p(2 × 2) O on Ni{100}. Nearly identical results are obtained for c(2 × 2) O.

tunately this attempt to connect hard-sphere radii to He atom scattering corrugation has little validity since, as discussed above, the He scattering occurs much further out than hard sphere radii. We will not discuss this approach further. Barker and Batra [203] took the approach of calculating the charge density (by the superimposition of self-consistent Hartree–Fock atomic charge densities) for a slab model of the Ni{100}/p(2 × 2) and c(2 × 2) O structures. Qualitatively it is immediately clear from these calculations why the Ni{100} surface appears flat to the He atom whereas the O overlayer structures are high corrugated. The Ni(4*s*) wave functions die off slowly with distance compared with the distance between nuclei, leading to a flat charge density surface. The O(2*p*) wave functions are much more localized, such that at p(2 × 2) or c(2 × 2) nuclei separation values the superimposition of the Hartree–Fock charge densities leads to a pronounced hill-and-valley structure across the surface. Since the amount of charge transfer, $O^{\delta-}$, is in question, Barker and Batra performed the charge density superimposition for $\delta-$ between 0.68 and 1.32*e* (the values suggested by various electronic structure calculations, see later). They then calculated the corrugation function that a He atom would see for the calculated charge density distribution using an He–surface interaction potential which has been shown to be appropriate for the clean surface case. The results are shown in Fig. 49(c) for a $\delta-$ value of unity and $d_\perp$ values of 0.9 and 0.26 Å for the p(2 × 2) O structure. The c(2 × 2) results are practically identical. Varying $\delta-$ between 0.68 and 1.32e varied the corrugation height for the 0.26 Å distance between 0.1 and 0.25 Å and for the 0.9 Å distance between 0.3 and 0.6 Å. These results seem unambiguously to rule out the possibility that c(2 × 2) O $d_\perp$ is 0.26 Å. Even for the highest charge transfer the corrugation is a factor of two smaller than the experimental value, whereas for a $d_\perp$ value of 0.9 Å a good fit with the experimental corrugation amplitude is achieved.

In the above discussion on the p(2 × 2) and c(2 × 2) O geometries we have not addressed the question of whether there are any small displacements of the Ni substrate atoms, except to note that trial LEED fits varying the Ni interlayer spacing apparently did not improve agreement with the data [279]. The very recent medium energy RBS work of Frenken et al. [214] in which careful shadowing and blocking experiments using 50–100 keV H^+ ions were performed, has shown that, on going from the clean Ni{100} surface to the c(2 × 2) O covered surface, the first interlayer Ni spacing changes from being ~0.6 Å contracted, compared with bulk, to ~0.1 Å expanded, compared with bulk.

Summarizing this section on the geometric structures of p(2 × 2) and c(2 × 2) O on Ni{100} we can say with certainty that, within ~0.15 Å, both structures have the same $d_\perp$ value of ~0.9 Å. The fine detail of the most recent LEED results on the c(2 × 2) O have been interpreted by the authors [199] as implying a lateral displacement of ~0.3 Å from the fourfold site. It was also shown that the present quality of SEXAFS results would not resolve this lateral displacement. For most of the other structural techniques attempts at accommodating such a lateral displacement have not yet been made. In HRELS, however, it has been pointed out that the symmetry on an off-center O adatom is incompatible with the number of vibrational modes actually observed [280] and, for the He scattering results, a displacement of 0.3 Å is apparently too large to be compatible with the data [285]. A small outward relaxation of ~0.1 Å of the top Ni layer occurs for the c(2 × 2) O structure [213].

3.1.5 Thin oxide epitaxy and geometry

For 300 K and above the NiO growth that follows the c(2 × 2) O stage can be observed by LEED. In the original Holloway and Hudson work [231] two independent patterns were observed (Fig. 50). Note that the patterns only become recognizable for $\theta > 0.7\theta_{sat}$ (~80 L exposure in their work), i.e. well beyond the start of nucleation as judged by the kinetics of Fig. 42 and indicative of poor ordering or small microcrystallite sizes. The two patterns represent NiO in a parallel orientation; Ni{100} ‖ NiO{100}, and a {111} orientation Ni{100} ‖ NiO{111}. The later was not identified as such until later work [221, 271]. Within the accuracy of the LEED measurements the NiO LEED pattern conforms to the normal bulk lattice parameter

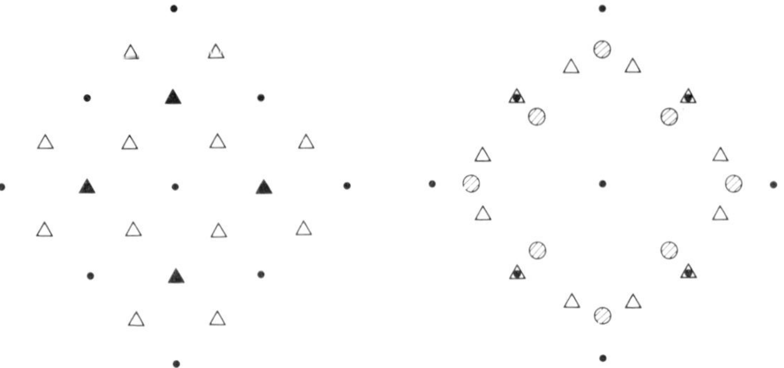

Fig. 50. LEED patterns observed during the interaction of oxygen with Ni{100}. Left, chemisorption pattern. ●, Original Ni{100} spots; ▲ and △ extra spots of the p(2 × 2) O pattern. The △-marked spots disappear in the c(2 × 2) O stage. Right, oxide regime. ●, Original Ni{100} spot positions. ⊘ NiO with {100} planes and ⟨011⟩ directions of NiO and substrate Ni parallel [231], △ {111} orientation of NiO [221, 271].

(a_0 = 4.18 Å, cf. Ni a_0 = 3.24 Å), but subsequent RHEED studies showed that the {100} orientation was compressed 2 ± 0.2% compared with bulk, whereas the {111} orientation was compressed 0.7 ± 0.1% [227]. The oxide is observed much earlier in the RHEED study, at $\theta \sim 0.6$ mL (or $\theta \sim 0.3\theta_{sat}$) presumably because of the greater sensitivity of RHEED to small island structures on surfaces. The parallel orientation gives very diffuse LEED and RHEED patterns, whereas the {111} orientation is quite sharp. Mitchell et al. [227] put the microcrystallite sizes as 7–9 Å for the {100} and > 30 Å for the {111}. They note, however, that if large inhomogeneous strain contributes to the {100} RHEED broadening, the crystallite size need not be as small as 7–9 Å. Matsudaira and Onchi [272] showed that the angular-resolved Ni $M_{2,3}VV$ AES pattern for the Ni{100}/NiO{100} surface was identical to that of cleaved bulk NiO{100}, from which they concluded the NiO domain size must be ≳ 10 Å and the thickness at least 5 Å. Mitchell et al. report that, at 320 K, the {111} epitaxy forms slightly earlier in exposure than does the {100} and that the epitaxies and apparent crystallite sizes do not change up to saturation. Holloway and Hudson [231], however, claim that the {100} orientation grows first with the {111} appearing only near saturation. Other LEED work agrees more with Mitchell et al. In our experience the "mix" of {100} and {111} orientations is not very reproducible, and depends on the details of the surface during the nucleation regime (disorder and contaminant levels) and the pressure at which the oxide is grown. At temperatures above the dissolution temperature of O (~ 500 K) the oxide can be grown with a maintained epitaxy up to about 18 Å thick ($\theta \approx 6$ mL), provided the exposure pressure is high enough [231]. For low pressures, misfits develop. If a 320 K exposed surface is heated to 500 K in vacuum the {111} orientation is converted to the {100}, but the domain size of the latter does not increase (no sharpening of RHEED lines).

In the original Holloway and Hudson LEED/AES work, and in most subsequent studies, it has been suggested that the NiO grows by the expansion of islands of NiO of terminal thickness (2–3 layers) across the surface, rather than by a homogeneous thickening of the O-covered surface. Thus the initial thickening of the oxide nuclei is fast compared with the island growth. The evidence for this was largely based on a fit to the kinetic data using this island growth model [231] plus the fact that the c(2 × 2) O and NiO structures coexist over quite a wide exposure range. The island growth model has recently been supported at 415 K by medium-energy ion scattering measurements where, by using channeling and blocking techniques, a depth resolution of ~4 Å is achieved [209, 210]. At 415 K the backscattered H^+ energy distributions as a function of oxide growth clearly show that islands of terminal thickness exist from near the start of oxidation [210]. The experimental spectra during oxidation can be simulated (Fig. 51) by a linear mixture of c(2 × 2) O and terminal oxide thickness spectra, whereas a fit on the basis of homogeneous overlayers of increasing thickness fails. At lower temperatures (325 K; faster oxidation) the island growth mechanism does not fit the backscattering data quite so well, leading to the suggestion that the island growth is now sufficiently fast that it takes place simultaneously with the nuclei oxide thickening. Finally, the medium energy ion-scattering data

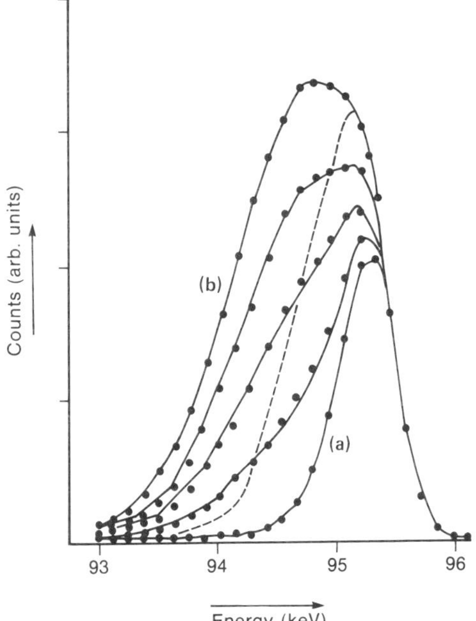

Fig. 51. 100 keV H^+ backscattering energy distributions from Ni during the oxidation of Ni{100} at 415 K [209, 210]. Spectrum (a) is taken during chemisorption just prior to the start of oxidation; (b) represents the passivated surface. The data points are marked individually. The full lines are the calculated profiles using linear combinations of spectra (a) and (b). The dotted line, for comparison, is the simulated energy distribution for a one-monolayer homogeneous NiO layer.

should be able to pin-point accurately the coverage where oxidation initiates by observing the coverage at which the surface Ni atoms are observed to start moving. The value given in the authors' first publication [209] is $\theta = 0.32$ mL. In their most recent paper [214], however, the authors have suggested that their coverage calibration was 21% too low, which would place the oxide nucleation point at $\theta \approx 0.39$ mL rather than 0.32 mL.

Adsorption at low temperature (77 K [237, 247] or 147 K [231]) into the oxidation range produces no ordered LEED structures. The final saturation coverage is the same, and, in the absence of problems with contaminant OH and CO_3 species [238], the electronic structure aspects appear the same (judged by XPS and UPS). We suggest, therefore, that the main difference from higher temperatures is simply the lack of long-range order, the low-temperature oxide being amorphous or small-grain polycrystalline.

Summarizing this section, we can say that the epitaxial relationships of the NiO formed are well documented; the terminal thickness is 2–3 layers of NiO at 300 K, but can be up to 18 Å at 700 K; the {100} epitaxy has very small domain size and/or large inhomogeneous strain, and the layer clearly grows by an island growth mechanism at a slightly elevated temperature (415 K). At room temperature the evidence for island growth is less clear. At low temperature no ordered oxide is formed. The confirmation of an island growth mechanism does not necessarily imply agreement with the model proposed by Holloway and Hudson for the mechanism of the island growth. This area is contentious, as discussed in Sect. 3.1.7.

3.1.6 Electronic structure aspects

There are a large number of publications concerning the electronic structure of the chemisorbed O phases on Ni{100}. The experimental studies mostly involve UV photoemission [229, 243, 249, 251, 253, 254]. Some of the studies involve various combinations of angle-resolved, variable photon energy (synchrotron radiation), and polarization-dependent data. These variations allow a better chance of delineating, for instance, the O(2p) bonding characteristics (Oπ as in O2p_{xy} and Oσ as in O2p_z) or the band structure nature of the interactions (O–O and O–Ni) leading to dispersion in the spectra. Unfortunately the experimental situation is complicated by doubts concerning the "purity" of the spectra, i.e. is only chemisorbed O present or are there contributions from OH or O dissolved beneath the surface, or in the form of oxide? To extract the full electronic structure information from the photoemission spectra requires comparison with theoretical calculations. One approach is to compare the angle-integrated photoelectron spectra with calculated densities-of-states using some model structure to represent the Ni{100}/O surface. Clusters, such as Ni_5O or $Ni_{20}O$ may have their electronic structures calculated in great detail, including determining equilibrium geometry positions. There is always some doubt, however, concerning the effectiveness of a small cluster in representing a real metal surface, and, of course, band-like properties and O–O interactions of the real p(2 × 2) or c(2 × 2) structures cannot be included. Slab models can represent the overlayer structures more appropriately, including the band structure effects, but the quality of the electronic structure determination may be rather crude. Whatever the appropriateness of the physical model, and the accuracy of the electronic structure calculation, in order to properly relate to the photoemission data, particularly the angle, photon and polarization dependent data, requires additional calculation concerning the actual photoemission process itself; i.e. the scattering properties of the outgoing electrons and the electronic structure of the final states.

Other experimental data, such as INS [230], core-level XPS [243, 247], vibrational spectra [219–222], and XANES [284] are also relevant to electronic structure discussions. As indicated in the last section one of the recent controversies concerns whether the p(2 × 2) O and c(2 × 2) O phases have different electronic structures, and what the effects are on the vibrational spectra. Despite the fact that cluster calculations with only one O atom cannot directly consider differences between p(2 × 2) and c(2 × 2) O structures, many of the theoretical electronic structure calculations in the controversy concerning the electronic structure are cluster calculations with one O atom using ab initio techniques and the arguments relate to the proper consideration of cluster size, basis set description, inclusion of correlation effects, etc. on the description of the electronic structure and, in turn, on properties such as vibrational frequencies. Recently a direct consideration of the lateral interaction effects in a cluster calculation has been achieved using a $Ni_{20}O_5$ cluster [264].

Another area of some controversy, or confusion, concerns the discussion of charge transfer in terms of covalency and ionicity, and the way in which these notions affect one's expectations for properties such as work function changes on adsorption.

Since the largest body of data concerns the UV photoemission and comparisons of various electronic structure calculations to the photoemission, we review this area first.

Aspects of the UV photoemission of Ni/O have been reported by many groups, but the number of studies on well-defined p(2 × 2) O and c(2 × 2) O structures on Ni{100} are actually very limited. HeI and HeII angular-integrated spectra at the p(2 × 2) and c(2 × 2) maximum intensity coverages are shown in the top right panel of Fig. 52 [286]. The only significant difference between the p(2 × 2) and c(2 × 2) angle-integrated spectra seems to be the expected increased intensity of the oxygen-induced feature for the higher coverage c(2 × 2) case. Both structures show a depression in the Ni(3*d*) band intensity compared with a clean Ni{100} surface but no evidence for a shifted *d*-band component. There are no significant differences from spectra for oxygen on other Ni faces [243, 250] or on a polycrystalline surface [249]. In discussing the many calculations of Ni{100}/O we will deal first with those concerned primarily with comparisons of calculated densities-of-states, DOS, with the angle-integrated spectra. An early cluster

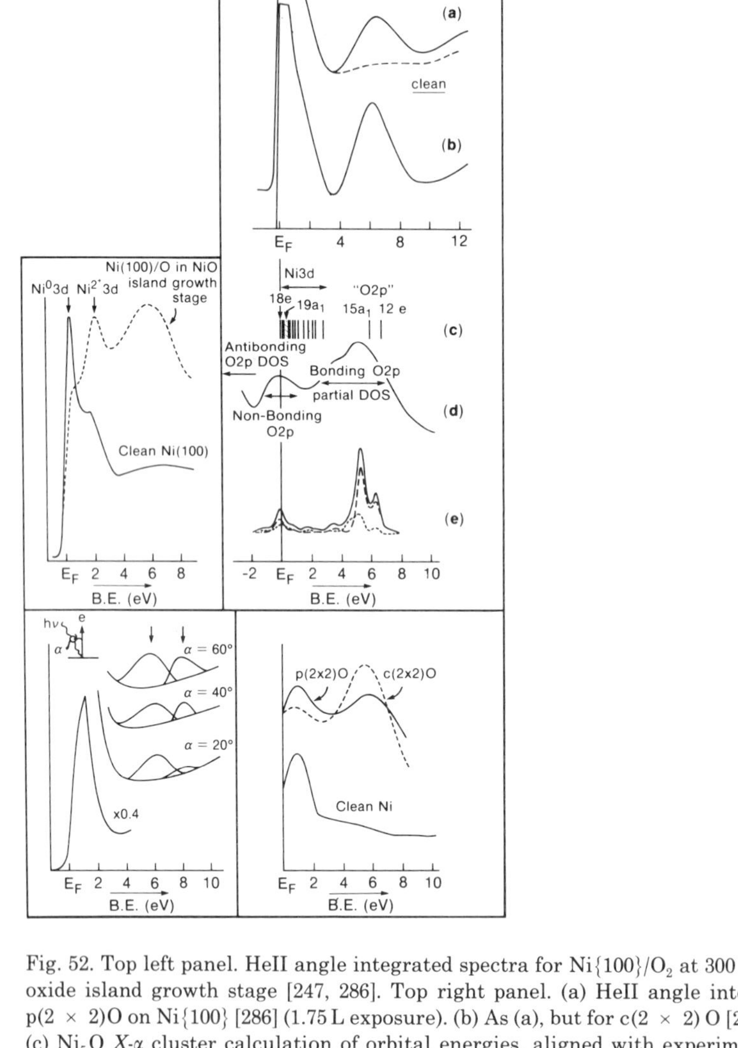

Fig. 52. Top left panel. HeII angle integrated spectra for Ni{100}/O_2 at 300 K; clean and in the oxide island growth stage [247, 286]. Top right panel. (a) HeII angle integrated spectra for p(2 × 2)O on Ni{100} [286] (1.75 L exposure). (b) As (a), but for c(2 × 2) O [286] (30 L exposure). (c) Ni_5O *X-α* cluster calculation of orbital energies, aligned with experimental spectra at E_F [259]. (d) O 2*p* partial density of states for Ni_5O *X-α* cluster calculation [261]. (e) O 2*p* partial density of states from the ab initio band structure calculation of Wang and Freeman [256] for c(2 × 2) O on Ni{100}. The full curve is the total O(2*p*) DOS. The individual (e) and (a) symmetry components are shown as broken and dotted curves, repectively. Bottom left panel. Angle-resolved HeII spectra for p(2 × 2) O on Ni{100} from Jacobi et al. [253]. The geometry is defined in the inset. The curves represent the authors' curve fitting and background subtraction. For the actual data points, see Fig. 1 of ref. 253. Bottom right panel. Unfold functions from the INS data for clean Ni{100}, p(2 × 2) O, and c(2 × 2) O on Ni{100} from Hagstrum and Becker [230].

calculation on Ni_5O [259] provides a good qualitative understanding of the basic features in the spectrum. The calculation is an SCF–X_α spherical-wave calculation using overlapping atomic spheres. The oxygen is fixed in the fourfold site at a $d_\perp$ value of 0.88 Å. This type of calculation using muffin-tin potentials will not give reliable total energies (in fact the Ni_5–O cluster is not bound energetically in the calculation), but the cluster orbital energies calculated, within the transition-state approximation, are expected to be comparable with actual IPs. A best fit to the data is obtained if the calculated orbital energies are rigidly shifted 3 eV lower [Fig. 52(c)]. The $15a_1$ and $12e$ orbitals which formally represent the $O2p_z$ and $O2p_{xy}$ adsorbate levels actually contain about 40–50% Ni *s*, *p*, and *d* character from all five Ni atoms, indicating that the O(2*p*) levels interact with Ni *s*, *p*, and *d* levels to produce bonding orbitals. In the Ni *d*-band region the theoretical changes from those of a bare Ni_5 cluster involve some O(2*p*) contributions to orbitals having *e* and a_1 symmetry, but they cannot be properly compared with experiment because the experimental spectrum in this region is weighted by Ni metallic character from those Ni atoms which are within the escape-depth of the measurement but too deep to be bonded to surface O atoms. Nevertheless, one cannot see any evidence for a shifted Ni(3*d*) component in the chemisorbed spectra, which is a good indication that the Ni(3*d*) levels are not heavily involved in the bonding. This point has similarly been raised for the Ni{111}/O situation [229]. The Ni_5O cluster orbital energies, being discrete, cannot account for the experimental width of the O2*p*-related peak, but within that width there is no evidence for the separation of $O2p_{xy}$ and $O2p_z$ orbitals found in the Ni_5O cluster calculation.

Some further insight concerning the bonding character of the O(2*p*) orbitals and their effect on the UPS spectrum can be obtained from the later Ni_5O cluster calculation of Ellis et al. [261]. They performed a further variation of the MS-Xα type calculation, using the HFS model in a self-consistent-charge approximation. They converted their discrete orbital energies to density of states plots by assigning an arbitrary 1 eV Lorenzian width to each level and folding this with the atomic population. A partial density of states curve for O(2*p*) is shown in Fig. 52(d). It splits into three regions; bonding, non-bonding, and antibonding, in accordance with simple classical expectations for interaction with the Ni levels. Tracking the changes in relative positions of calculated orbital levels as a function of $d_\perp$ (Fig. 53) shows that, at the experimental $d_\perp$ value, the Ellis et al. [261] calculation predicts a reversed order of $12e$ ($O2p_{xy}$) and $15a_1$ ($O2p_z$) compared with Batra and Robaux [259]. Since the $12e$ value varies so strongly with $d_\perp$, obtaining an accurate separation between $12e$ and $15a_1$ is likely to be beyond the capabilities of either calculation.

In 1977, Li and Connolly refined the MS-Xα calculations on Ni_5O by including non-muffin-tin corrections to the potentials [260]. This allows a better calculation of total energy, and now produces a bound cluster. Total energy as a function of $d_\perp$ was calculated and a minimum (0.33 Ry bond

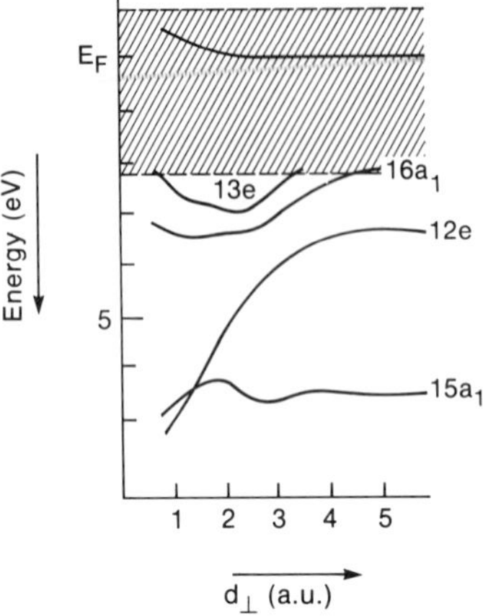

Fig. 53. Variation of Ni_5O cluster valence MO energies with oxygen height, $d_{\perp}$, above the cluster surface. The densely occupied region is indicated by shading. Some important adatom levels are marked individually, in particular the 12*e* (O $2p_{xy}$) and $15a_1$ (O $2p_z$) levels. Taken from Ellis et al. [261].

energy) found at 0.7 Å, in reasonable agreement with the experimental value of 0.9 Å.

Bullett and Cohen performed an early slab calculation of c(2 × 2) O structure on Ni{100} [262]. They used a three-layer Ni slab with fixed bulk geometry and a fixed Ni–O bond length of 1.98 Å. The electronic interactions between atoms were treated within an LCAO framework. On-top, bridge and fourfold sites were tried. The fourfold site was found to be the most stable (Ni–O bond energy 4.7 eV) and partial DOS for this site were presented. They showed a decrease in Ni3*d* states between 0 and 2 eV below E_F, and $O2p_{xy}$ and $O2p_z$ states of about 1.5 eV width located at an energy of 3 eV below E_F. Compared with the Xα cluster calculations described above it is clear that this calculation had introduced O(2*p*) band-widths more appropriate to a real two-dimensional c(2 × 2) structure, but at the expense of accuracy in treating the actual electronic interactions between atoms. Thus the O(2*p*) levels are calculated to be at ~3 eV below E_F instead of the experimentally observed 5.5 eV.

In 1979 Gallagher et al. [263] summarized the description of the chemisorption O–Ni bond from calculations of O on Ni{100} in the fourfold site at a $d_{\perp}$ of 0.9 Å. The calculation involved elements of both cluster and band structure approaches. Simple explanations for several observables were suggested. The calculations showed that the interaction between O_{ad} and the Ni surface leads to a transfer of electron density from the delocalized Ni4*s*4*p* band to a localized O(2*p*) orbital of closed shell, $O2p^6$, configuration. The radius of this decoupled O(2*p*) orbital is large, however, so that the O(2*p*) electrons spatially extend over the nearest-neighbor Ni atoms (Fig. 54). This simple analysis of the electronic situation, which is fully consistent with the

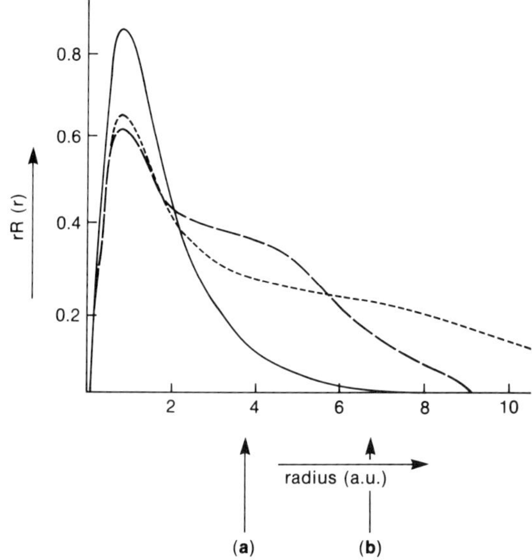

Fig. 54. Radial distribution of O(2p) orbitals. Full curve, atomic oxygen; broken curve, O(2p) for O atom chemisorbed on Ni{100}, spherical approximation to potential; dotted curve, O(2p) for O atom chemisorbed on Ni{100}, improved spherical approximation. The positions marked (a) and (b) are the distances of the first Ni atom and the first c(2 × 2) O atom from the O atom considered. Taken from Gallagher et al. [263].

earlier Xα and later ab initio cluster descriptions, which also find a closed O(2p) configuration, immediately points to the futility of arguments concerning the degree of ionicity versus covalency in the chemisorption bond. The oxygen atom is formally O^{2-}, but it extends spatially over Ni sites. Ionicity/covalency prescriptions depend on dividing up charge in space or considering transfer of charge from the orbitals of one atom to another or some combination of the two. In the present situation these alternative prescriptions cannot give similar results. The large spatial extent of the O(2p) orbitals also explains why attempts to estimate $\Delta\phi$ using a point charge model are futile. Any such attempt (often made!) would have to conclude either that $d_\perp$ was much shorter than 0.9 Å, or that the chemisorption bond was entirely "covalent" to account for the small $\Delta\phi$ ($\sim$0.5 eV) actually observed.

Gallagher et al. [263] also point out that their description of the O(2p) orbital explains why a (1 × 1) O structure is never formed. At such a "short" O–O separation (4.72 a.u.) there is closed shell repulsion between O adatoms, whereas at the p(2 × 2) separation of 9.43 a.u. there is an attractive Van der Waal's tail to the potential, which explains the early island formations of p(2 × 2) O. Finally, the closed-shell description of O_{ad} may explain why NiO nucleates before 0.5 mL. At 0.5 mL there is insufficient electron density in the surface Ni 4s4p bond to form an O(2p)^6 configuration. Electrons must transfer from Ni3d, as occurs for bulk NiO.

In the case of bulk NiO, or of the 2–3 layer NiO on Ni{100}, the problem of unambiguously describing the bonding in terms of ionicity and covalency still exists to some extent, but it is more straightforward than the chemisorption case exactly because of the heavy involvement of Ni3d electrons. For the chemisorption case, though Ni3d electrons may be slightly involved the dominant interaction is between Ni4s4p and O2p electrons and since they occupy similar regions in space it becomes hard to assign electrons in an unprejudiced way. Ni3d electrons in a Ni atom are, however, well-localized around the atom (i.e. more point charge-like) and therefore any bond between Ni3d and O2p orbitals is more easily described in terms of a population analysis where a fraction of the charge is assigned to the Ni atom and a fraction to the O atom. That the Ni3d electrons are substantially involved in the bonding for the 2–3 layer oxidation stage is quite clear from the angle-integrated UPS spectrum for Ni{100}/O at saturation coverage [286] [Fig. 52, top left panel where a shifted Ni(3d) band is evident]. It is also evident in the Ni2p XPS [Fig. 45(c)] where the chemically shifted Ni^{2+} feature is evident at saturation coverage [236, 246]. The oxide components of the UPS and XPS spectra of the 2–3 layer NiO on Ni{100} are identical to those of bulk NiO, from which we conclude that the electronic structures are also very similar.

We now turn to angular-resolved photoemission which should have two advantages compared with integrated spectra for electronic structure information. First, overlapping features may become identifiable by their differing angular/polarization behavior. In the present case the hope would be to distinguish O2p_{xy} (π symmetry) from O2p_z (σ symmetry). The second advantage is that experimental band dispersion measurements allow comparison with E versus k plots from band calculations. Thus direct information on the band properties of the real periodic overlayer O structures is obtained. Any dispersion of O(2p) features would relate to the strength of lateral interactions between O adatoms (both direct and through substrate) in the p(2 × 2) and c(2 × 2) O structures.

In practice the experimental results for Ni{100}/O are rather confusing. In some cases clear separation of O2p_{xy} and O2p_z features are claimed, with agreement with predicted angular-polarization behavior. In other cases essentially only one O(2p) feature is observed which is hard to separate into components by angular/polarization behavior. The discrepancies seem to relate to the old problems of surface preparation and the possibility of OH, oxide nuclei and subsurface O contamination.

Weeks and Plummer [251] reported angular-resolved measurements for c(2 × 2) O (possibly containing subsurface O) using unpolarized HeI radiation (21.2 eV). No spectra were shown, but the intensity of the broad oxygen-induced feature at ~6 eV was plotted as a function of polar emission angle for normal incidence, 50° incidence parallel to the {100} collection plane, and 50° incidence perpendicular to the {100} collection plane. Tong et al. [252] performed Xα scattering calculations on an Ni_5 O cluster to obtain the initial states and then calculated the transition matrix elements to the appropriate final photoemission states (frozen-core approximation). The appropriate summation of $O2p_x$, $O2p_y$ and $O2p_z$ calculated intensity as a function of polar collection angle showed some similarity to the experimental behavior of the unresolved O(2p) band reported by Weeks and Plummer, but little real information concerning the electronic nature of c(2 × 2) O on Ni{100} resulted.

Jacobi et al. [253] reported angular-resolved UPS using polarized HeII (40.8 eV) radiation for the p(2 × 2) O structure. An example of the observed angular variations is shown in Fig. 52 (bottom left panel). The intensity variation of the 6 and 8 eV features apparently agree with model calculations for orbitals of π and σ symmetry, respectively, and so the 6 eV feature was assigned to $O2p_{xy}$ and the 8 eV feature to $O2p_z$. No dispersion in either peak was reported. Guillot et al. [287] reported results on the p(2 × 2) O structure using p-polarized 44 eV photons (Orsay synchrotron) in which clear 6 and 8 eV features were also observed with maximum dispersions of $\leqslant$0.2 eV. The intensity variations as a function of incidence/emission angle could not be explained, however, by the assignment of the 6 eV feature to $O2p_{xy}$ and the 8 eV feature to $O2p_z$. In a subsequent paper [254], involving some of the same authors, an equivalent set of data for c(2 × 2) O was reported. Here the 8 eV feature was observed only as a weak shoulder at close to normal emission. The authors were able to establish that this feature was a bulk Ni feature, unconnected to c(2 × 2) O. They do not comment whether the same holds for their previous p(2 × 2) results, but they do quote Jacobi (private communication) as re-interpreting his data for p(2 × 2) O in terms of the 8 eV feature being subsurface oxygen (the method of preparation of the overlayer makes this likely). It seems, then, that only the 6 eV feature can be considered as representative of p(2 × 2) and c(2 × 2) O. Jepsen et al. [254] report a dispersion of 0.6 eV in this band for the c(2 × 2) O case, at least a factor of 3 larger than for the p(2 × 2) case, as might be expected because of the shorter O–O distance. They also report a band structure calculation of the c(2 × 2) O structure which shows strong mixing of O(2p) and Ni4sp levels, which makes it difficult to separate $O2p_{xy}$ and $O2p_z$ components in the spectra. The total dispersion of these bands is about 0.6 eV, in agreement with their experiment, but the direction of calculated dispersion as a function of increasing θ, the emission angle relative to normal emission, causes them to conclude that the major component of the 6 eV experimental peak is $O2p_z$, in disagreement with the Jacobi et al. assignment [253]. It also leaves the unsatisfactory situation that, with this interpretation, the $O2p_{xy}$ contributions are apparently missing from the experimental spectra.

Simultaneously with the measurements and calculations of Jepsen et al. [254] and Jacobi et al. [253], Liebsch [255] independently reported a detailed theoretical study of the c(2 × 2) O structure using the layer multiple scattering technique on a semi-infinite Ni{100} substrate. The geometry was fixed with O adatoms in the fourfold site at a $d_\perp$ of 0.9 Å and the calculation was not self consistent: i.e. a "reasonable" choice of one-electron potentials was made and the objective was to illustrate the main features qualitatively. The main conclusions were as follows.

Interaction with the Ni4*s*4*p* and 3*d* levels causes the O(2*p*) levels to split into a bonding pair O2p_z and O2p_{xy} about 3 eV below E_F, and an equivalent antibonding set 4–5 eV above E_F. The "non-bonding" O(2*p*) region within the Ni3*d* band carries little weight. This description is roughly equivalent to the cluster calculation results, except for the apparently rather larger involvement of the Ni3*d* levels in addition to Ni4*s*4*p*. The new information, not available from the cluster calculations, concerns the dispersion of the O(2*p*) levels. An isolated c(2 × 2) O overlayer already shows a ~1 eV dispersion in the 2p_{xy} bands. The additional interaction of the O layer with the Ni{100} substrate modifies the dispersion and imparts a *k*-dependent width to the individual levels (maximum value ~0.5 eV). The 2p_{xy} and 2p_z levels now both have dispersions of ~1 eV, and their relative positions reverse as a function of $k_\parallel$. Thus the relative energy position of any identifiable O2p_{xy} and O2p_z character in an angle-resolved photoemission experiment should cross with appropriate variation of angle. Unpublished data of Lapeyre is quoted in support of this. Using the appropriate incidence/emission symmetry and polarization (pure *s* or an *sp* mixture) Lapeyre [288] does find that, within the region ~5–7 eV. The adsorbate-induced structure shows 2p_z and 2p_{xy} character whose energy position varies with $k_\parallel$ (p_z deeper at $k_\parallel = 0$, p_{xy} deeper at $k_\parallel \sim 0.5$). There does not seem to be any suggestion of resolved peaks within this 5–7 eV experimental O-induced density, which may imply that the experimental widths of the individual *p* levels is much broader than the calculated 0.5 eV maximum. There is also no report of any ~8 eV feature. Leibsch also notes that, in his earlier equivalent calculations on the hypothetical p(1 × 1) O structure [289], the individual O(2*p*) widths were much broader (~1 eV), as might be expected from the closer-packed structure.

More recently, Wang and Freeman [256] have performed an ab initio self-consistent band-structure calculation for c(2 × 2) O on Ni{100}. They therefore lay greater claims to accuracy than Liebsch concerning bonding description and the energy positions of the chemisorption induced states. Their O(2*p*) local DOS is plotted in Fig. 52(e). The main distinctions from the Liebsch results seems to be (a) the O(2*p*) assigned levels are centered at ~ −6 eV, in agreement with the UPS data, whereas Liebsch has them at ~ −3 eV; (b) the total width of the O(2*p*) DOS is greater than Liebsch and more in accord with experiment. Like Liebsch, they find that the relative ordering of *e* and *a* O(2*p*) levels reverses on traversing the Broullion zone. We note, however, from Fig. 52(e), that the two-peaked structure of the O(2*p*) DOS is not primarily an *e*/*a* separation, but is a splitting of the *e* states. This implies that any resolved structure in the O(2*p*) photoemission should probably not be interpreted as an *e*/*a* separation.

Goddard and co-workers [239, 257, 258] have performed ab initio cluster calculations for up to $Ni_{20}O$ in an attempt to provide sufficient accuracy to predict Ni/O properties; in particular, total energies and therefore equilibrium geometries. Walch and Goddard [257, 258] reported calculations on Ni_5O. The generalized valence bond (GVB) method was used. This is different from the self-consistent one-electron approximation (Hartree–Fock) in that some electron correlation effects are explicitly included. Only the 10 valence electrons were explicitly included in the self-consistent calculations, however. The 18-electron Ni core was represented by a pseudopotential. The Ni atom valence electron basis set consisted of two *s*, one *p*, and one *d* function contraction. Calculations as a function of $d_\perp$ with O in the fourfold site were performed for Ni_5O and Ni_5O^+ and it was suggested that the latter cluster is the more representative of a real Ni{100}/O surface. The main argument for this claim was that the $d_\perp$ equilibrium value was found to be 0.96 Å for the Ni_5O^+ cluster, close to the then accepted experimental value of 0.9 Å, but only 0.65 Å for the Ni_5O cluster. The difference comes from the increased electronegativity of the Ni_5 unit in the Ni_5O^+ cluster which withdraws charge from the O adatom compared with the situation for the neutral cluster. This leads to a more covalent bond, a longer $d_\perp$, and an Ni–O bond strength increase from 1.9 to 3.1 eV, closer to the estimated value for ΔH of O on polycrystalline Ni ($\sim$5 eV) [290]. It was also noted that, for the Ni_5O^+ cluster, the $d_\perp$ increase from 0.65 to 0.95 Å caused an orbital configuration change so that the ground state at the equilibrium distance had an open shell "radical" oxygen 2*p* structure instead of the normal closed shell configuration.

Upton and Goddard [239] subsequently reported calculations on $Ni_{20}O$ and $Ni_{20}O^+$. The comparison with Ni_5O is confused, however since, in addition to increasing the cluster size, it proved necessary to incorporate all the Ni electrons except the 4*s* into the effective potential. Thus a Ni atom is represented by only *one* valence electron in a basis set of two 4*s* functions and no 4*p* functions (cf. the Walch and Goddard Ni_5O basis set of two *s*, one *p*, and one *d* functions) [257, 258]. Two low-lying states, the closed shell "oxide" state and the open shell "radical" state, were again found. Figure 55 plots the total energy of each state as a function of $d_\perp$ for $Ni_{20}O$ and $Ni_{20}O^+$. Note that here the oxide state, with a very short $d_\perp$ (0.26 Å) is found to be the ground state for the ionized cluster, which is the opposite of the behavior found for Ni_5O^+ where the claim was that the radical state represented the ground state for c(2 × 2) O on Ni with a $d_\perp$ of 0.96 Å. Without making any reference to the contradictory previous results, Upton and Goddard [239] suggested that the Ni_{20} atom cluster results could be interpreted as implying that the lower-coverage p(2 × 2) O structure was represented by the $Ni_{20}O$ calculation with O in the "radical" state at $d_\perp$ = 0.9 Å while the $Ni_{20}O^+$ calculation represents the higher-coverage c(2 × 2) O which, therefore, has the "oxide" state electronic structure and a very short $d_\perp$. The motivation for this suggestion of drastic differences in $d_\perp$ and electronic structure for

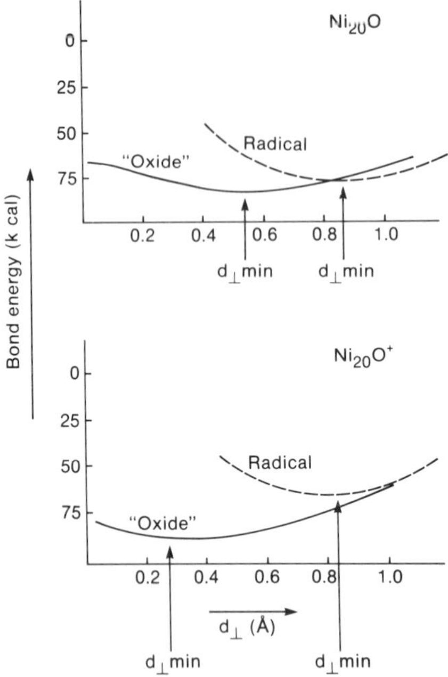

Fig. 55. Potential energy curves for the two low-lying states of $Ni_{20}O$ and $Ni_{20}O^+$. The former cluster was considered to be representative of p(2 × 2) O on Ni{100}, the latter of c(2 × 2) O on Ni{100} by the authors, Upton and Goddard [239].

p(2 × 2) O and c(2 × 2) O comes from the apparent agreement of the calculated Ni–O vibrational frequencies from the neutral and ionized clusters with the experimental values (Table 8). As discussed earlier in Sect. 3.1.4, however, there is no supporting evidence for a $d_\perp$ of 0.26 Å for the c(2 × 2) O and no spectroscopic evidence to suggest a different electronic structure for p(2 × 2) O and c(2 × 2) O. In addition, the HRELS vibrational frequencies have been well-explained using a $d_\perp$ value of 0.9 Å for both structures [219, 220, 264, 282]. Since the whole aim of the calculations was to demonstrate sufficient accuracy for reliable geometric determination, something was clearly rather amiss and it tempted others to ask how such misleading claims came about.

Bauschlicher and co-workers [240] performed a series of Hartree–Fock and Hartree–Fock plus CI calculations on a Ni_5O cluster in an attempt to answer this question. They first used the same Ni effective potential and Ni basis functions as Upton and Goddard used for their $Ni_{20}O$ calculation and, with an HF calculation, obtained very similar $d_\perp$ and w_e values and "oxide"–"radical" energy separation (Table 8). This demonstrated (a) that increasing the cluster size from 5 to 20 is not having a significant effect, and (b) that the correlation included by the GVB treatment of Upton and Goddard is not having a significant effect. They then put the 4p function into the Ni basis set. Equilibrium $d_\perp$ values, w_e values, and "oxide"–"radical" energy separation all changed considerably and became similar to the values found by Walch and Goddard for Ni_5O using the GVB treatment (Table 8). In the authors' opinion this demonstrated that the difference observed between the Walch and Goddard Ni_5O [257, 258] and Upton and Goddard $Ni_{20}O$ [239] calculations was not due to cluster size effects, but came about because the Ni4*p* function, known to be very important, had been left out of the $Ni_{20}O$ calculation. Bauschlicher et al. [240] then went beyond the HF level and included electron correlation by CI calculations. Substantial changes were again observed (Table 8), in particular the "oxide" state become greatly stabilized compared with the "radical" state. Further improvement in correlation treatment would be expected to stabilize the oxide state further. The end conclusions from this series of calculations were that an Ni_5O calculation, which includes all the necessary physics (namely adequate representation of Ni4*p* character and adequate treatment of correlation effects), leads to a $d_\perp$ value in reasonable agreement with experiment (Table 8) with the normal oxide state being the ground state. It is not necessary to attempt artificial high coverage effects by invoking ionized clusters. As far as vibrational frequencies are concerned it is clear that they vary significantly with the level of calculation, and since it is, in addition, clear that lateral adatom interactions also play a role, a cluster with an isolated O atom cannot be expected to do well in calculating w_e for a high-coverage state.

Subsequently Bauschlicher and Bagus completed an all-electron Ni_5O calculation and also performed $Ni_{25}O$ and $Ni_{25}O_5$ calculations using a pseudopotential [264]. The all-electron Ni_5O calculation (Table 1) shows $d_\perp$ increased from 0.74 to 1.05 Å and w_e from 320 to 380 cm^{-1} for the "oxide" ground state compared with their "best" Ni_5O pseudopotential calculation, which included 4*p* functions and CI (Table 8). Both values are now closer to experiment and the full calculation also allows a check on the suitability of the pseudopotential used earlier. The pseudopotential can then be modified to bring the Ni_5O pseudopotential calculation into agreement with the all-electron results (Table 8). The modified pseudopotential was then used for the Ni_{25} calculations where an all-electron calculation cannot yet be performed. The comparison between $Ni_{25}O$ and $Ni_{25}O_5$ now represents a reliable cluster calculation direct simulation of increasing coverage. The $Ni_{25}O$ and $Ni_{25}O_5$ calculations simulate p(2 × 2) O and c(2 × 2) O environments, respectively. The results (Table 8) indicate (a) that the "oxide" state is the ground state in both cases, (b) that $d_\perp$ is nearly the same, and that the w_e decrease from p(2 × 2) O to c(2 × 2) O is largely due to the O–O interaction since 70% of the experimentally observed shift is reproduced by adding the extra four O atoms to the cluster. The remaining shift can apparently be ascribed to a change in the coupling strength to the substrate phonon modes*. In terms of the charge transfer or ionicity/covalency involved in the

*According to Andersson et al. [226] for a $d_\perp$ of 0.9 Å for both structures, ~43% of the experimental shift can be ascribed to the stronger coupling of the p(2 × 2) O structure to the substrate phonon modes, as determined by lattice dynamical calculations.

bonding, all the calculations suggest about a one-electron charge transfer from Ni to O, judged by a Mulliken population analysis of orbital occupancies, and that delocalization of O(2*p*) levels leads to significant covalent contributions to the bonding. Ni4*s* and 4*p* levels are primarily involved, but Ni3*d* levels do make some contributions. The importance of this *d* involvement is still not fully clear and may require even further calculations before being fully delineated. Remember that in the UPS no strong effects other than attenuation were experimentally observed on the Ni(3*d*) band, indicating that Ni3*d* involvement cannot be large.

Although most of the comparison between electronic structure theory and experiment have involved valence level photoemission, other useful comparisons are possible. The example of surface vibrational frequencies has just been discussed above. Differences in electronic structure should also show up in XPS O(1*s*) BE values [244] and the O(1*s*) X-ray absorption near-edge fine structure (XANES) [284]. In fact, identical results for p(2 × 2) and c(2 × 2) O are found in both cases. The ab initio cluster calculations [291] predict a ~ 1.1 eV separation between O(1*s*) BE values for the "radical" versus the "oxide" state, which would be easily resolvable. This is additional evidence, then, that both geometric structures have the same electronic structure. For the XANES result, a calculation of the edge structure as a function of $d_\perp$ produces fair agreement with the c(2 × 2) O XANES at 0.9 Å but qualitative disagreement at 0.2 Å, further proof of the normality of the $d_\perp$ value for c(2 × 2) O [284].

The original experiments addressing the question of electronic structure for Ni{100}/O were the INS experiments of Hagstrum and Becker [230]. In INS, which involves the neutralization of a slow He^+ ion at the surface by an electron from the surface and the simultaneous ejection of a valence electron from the surface (a two-electron Auger process), similar information to UPS is provided with two important differences. The first is that the neutralization mechanism restricts the energy range of the DOS information obtained to ~ 10 eV below E_F. The second is that the incoming He^+ ion does not penetrate the surface and therefore the DOS obtained is more surface-sensitive than UPS which may sample several layers. The INS results for clean Ni{100}; p(2 × 2) O; and c(2 × 2) O are shown in Fig. 52, bottom right panel, for comparison with the UPS data. The O(2*p*) feature falls in the same energy region as in UPS, but the Ni3*d* band suppression is greater than in UPS. Likewise, the O2*p*/Ni3*d* intensity changes in going from p(2 × 2) to c(2 × 2) are larger than for UPS, but there are no significant shape or position changes. The original conclusion of Hagstrum and Becker from the similarity of the p(2 × 2) O and c(2 × 2) O energy distributions was that the local electronic structure must be very similar; a statement which has stood the test of time.

Summarizing this section we can say that, owing to the enormous theoretical effort applied, most features of the electronic structures of p(2 × 2) and c(2 × 2) O are qualitatively and semi-quantitatively understood. Both structures have the same closed shell, $O2p^6$, ground state. The primary bonding interaction is between Ni4*s*, 4*p* and O(2*p*) levels. There is about unit charge transfer from Ni to O, but there is also a strong covalent element to the bonding. Owing to the high degree of delocalization of the O(2*p*) levels it is not possible to be more specific concerning the relative ionicity/covalency involved. Simple arguments involving hard-sphere or ionic radii, or $\Delta\phi$ behavior are useless and misleading. If one wants to compare with NiO as a reference, the major differences are a slightly higher charge transfer and a much stronger Ni3*d* involvement for the oxide. Though the Ni3*d* involvement is certainly much less for the chemisorbed case its significance compared with the Ni4*s*4*p* involvement has yet to be fully established. The main difference between p(2 × 2) and c(2 × 2) structures involves O–O interactions. Whether these are considered direct or through-substrate seems a semantic question to us. Since it is the interaction between O2*p*s which are concerned and the nature of the O2*p*s is strongly modified by the Ni, the best description is perhaps "substrate-modified adatom interaction." This interaction leads to dispersion in the O(2*p*) levels, which is greater for the more closely packed c(2 × 2) O case. The complexity of the dispersion and width behavior, coupled with the propensity in the experimental UPS to produce spectra which contain some artifacts, have so far made it difficult to properly identify the $O2p_{xy}$ and $O2p_z$ components in the angular/polarization resolved spectra. From the calculations it seems that any genuine energy-resolved structure within the O(2*p*) band (not yet experimentally established) is likely to have both $O2p_{xy}$ and $O2p_z$ character spread over all features and should not be thought of as a separation into *a* and *e* symmetry bands. The O–O interactions are also responsible for a large fraction of the difference observed in Ni–O vibrational frequencies between p(2 × 2) O and c(2 × 2) O [264].

3.1.7 Kinetics and mechanisms. General

Having reviewed the literature concerning the geometric location of the O atoms and the nature of the Ni–O bonding as a function of θ and T we are in a position to consider more carefully the implications of the kinetics as a function of θ and T in terms of the processes controlling dissociation, formation of ordered structures, oxide nucleation, etc. In attempting to review the kinetics data from different laboratories the key factor is to establish reliable, interrelatable θ scales, as was the case for W. For Ni{100}, however, the problem is greater because of the very variable propensity for oxygen penetration into the surface. Merely a total θ measurement, by O(1*s*) XPS intensity, for instance, may not be sufficient, therefore, since this could be partitioned differently between θ(overlayer) and θ(elsewhere). On the other hand using $\Delta\phi$ to indicate total θ may not be reliable since the sign of $\Delta\phi$ depends on whether O is above or below the surface. This problem was alluded to in Sect. 3.1.1 where it was shown that, whereas some authors reported a linear relationship between $\Delta\phi$ and θ_{total}, as determined by AES, up to the completion of the c(2 × 2) O overlayer structure, others reported a decrease in $\Delta\phi$ before the c(2 × 2) O overlayer was completed. Our interpretation, once we were sure that the c(2 × 2) O structure was an overlayer with $d_{\perp}$ the same as the p(2 × 2) O structure, was that the $\Delta\phi$ measurement senses different rates of incipient incorporation of O atoms into the surface in the $\theta = 0.25$–0.39 mL range prior to the point where formation of NiO could be inferred from RHEED, LEED, Ni2*p* XPS, or whatever technique was being used. In any comparison of one set of kinetic data with another it is, therefore, very useful to have both $\Delta\phi$ measurement and independent θ(total) measurement. Unfortunately these two are not usually routinely measured together, which reduces the reliability of comparison between experiments from different laboratories. Keeping all this in mind we will review the kinetic data. We will attempt to separate it into two sections, the "purely chemisorptive overlayer" regime and the "oxide nucleation and island growth regime" while still being aware of the overlap and interaction between these regimes.

3.1.8 Kinetics and mechanism of chemisorption

Examples of the kinetic uptake data were given in Figs. 42 and 44. The low exposure regime from the Holloway and Hudson data [231] is shown in more detail in Fig. 56(a) and the conversion to S versus θ curves in Fig. 56(b). Note that this conversion is based on the assumption that $S = 1$ between zero and ~0.4 L exposure at 147 K, and that the exposure scale, determined by ion gauge pressure readings, is accurate. This allows the coverage at the 0.4 L exposure point in Fig. 56(a) to be calculated from the gas impingement equation and therefore the AES O KLL intensity scale to be converted to a θ scale, provided there are no attenuation effects in the AES data. The latter would be expected to be true for submonolayer overlayer adsorption. If the assumption of $S_0 = 1$ were incorrect, the θ scale would also be wrong. We

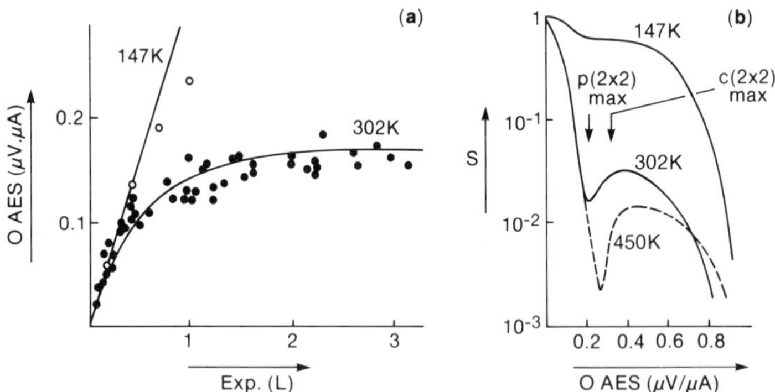

Fig. 56. (a) O AES intensity versus oxygen exposure at 302 and 147 K for Ni{100}/O in the chemisorption region [231]. The straight line fit through the origin and first two points at 147 K defines the coverage since S is assumed to be unity for the fit. (b) Plot of sticking probability versus O AES intensity derived from Figs. 42 and 56(a) [231].

have replotted Holloway and Hudson's data on a linear instead of logarithmic scale in Fig. 57(a). We wish to consider the kinetics up to the completion of the c(2 × 2) O overlayer. We have to keep in mind, however, the possibility that surface penetration and incipient oxidation competes with the c(2 × 2) O overlayer formation. Holloway and Hudson state that NiO nucleation occurs between $\theta = 0.3$ and 0.4 mL, i.e. before the maximum of their c(2 × 2) O LEED spots ($\theta = 0.44$ mL). This statement was based on their interpretation of the increasing S beyond $\theta \sim 0.35$ mL rather than the appearance of NiO in the LEED, which was not observed until $\theta \sim 1$ mL. In fact their $\Delta\phi$ data, which we have replotted on the same θ scale in Fig. 57(a), suggest that O penetration occurs as early as $\theta \approx 0.25$ mL. In the numerous redeterminations of O uptake versus exposure curves since the Holloway and Hudson work the general shape of the curves in Fig. 42 are usually reproduced, but the details of the "plateau" differ. This is due to differing oxide nucleation behavior, as discussed earlier. Only one other group, Brundle et al. [244, 245, 247, 286], have, however, made accurate measurements in the early stages and also converted these to S versus θ curves. They did not rely on a θ scale calibrated from the assumption of $S_0 = 1$ and

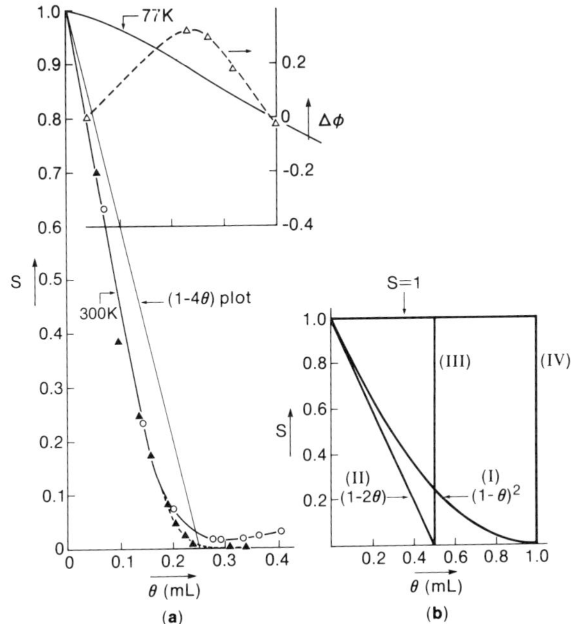

Fig. 57. (a) Sticking probability, S, versus coverage, θ, for Ni{100}/O_2 at 300 K. ○, Holloway and Hudson data [231]; ▲, Brundle [286]. An approximate curve for 77 K is also shown. Inset shows $\Delta\phi$ versus θ (data from Holloway and Hudson [231]). (b) Model behavior for S versus θ [245]. (I) Langmuir kinetics for chemisorption with no chemisorbed ordering; (II) Langmuir kinetics for perfect c(2 × 2) ordering on a square lattice; (III) infinite precursor lifetime and c(2 × 2) adatom ordering; (IV) infinite precursor lifetime and no adatom ordering.

accurate exposure determination, but calibrated the θ scale directly by comparison of the O(1*s*) XPS intensity for Ni{100}/O_2 with Ni{100}/CO at 77 K. θ_{sat}CO in the latter case is well-established as 0.68 [292], thereby allowing a direct conversion of O(1*s*) XPS intensities into O coverage in the Ni{100}/O_2 case. A representative 300 K S versus θ curve, taken from many reproducible runs, is also shown in Fig. 57(a). There is extremely good agreement with Holloway and Hudson between $\theta = 0$ and ~ 0.19 mL, then the Brundle et al. data starts to deviate, dropping much lower than the Holloway and Hudson data and not starting to rise until a somewhat later coverage. We associate these differences with a decreased incipient nucleation process in the Brundle et al. data, as discussed earlier. The significance, for the interpretation of the kinetics, is that the Brundle et al. 300 K data can be used for modelling the dissociative chemisorption process to higher θ, in fact right up to the maximum of the c(2 × 2) O structure, whereas the Holloway and Hudson data cannot be used beyond $\theta = 0.25$ mL because of the oxide nucleation interference.

Holloway and Hudson concentrated on the mechanism of oxide nucleation stage in their original paper, but they did note that in the chemisorption range up to $\theta \approx 0.25$ mL the kinetics fit the relationship

$$1 - \theta' = \exp\left(-\frac{2I}{N_0}t\right) \tag{2}$$

where θ' is the coverage relative to that at $\theta = 0.25$ mL, N_0 is the number of filled adsorption sites at $\theta = 0.25$ mL, which equals $4 \times 10^{14}\,\mathrm{cm}^{-2}$, I is the molecular impingement rate and t is the time of exposure. Brundle and Hopster [244] later noted that their data gave an S versus θ relationship close to

$$S = (1 - 4\theta) \tag{3}$$

and Holloway pointed out in a review paper [293] that eqn. (2) reduced to $S = (1 - 4\theta)$ when differentiated. In reality both sets of data fall well below the $S = (1 - 4\theta)$ curve, as indicated in Fig. 57(a), and this fact has real importance when we try to understand the significance of the curves in terms of the dissociation process. Before considering the interpretation we note that, though no other detailed S versus θ curves have been reported, both X-ray fluorescence/RHEED [227] and medium-energy ion scattering measurements [209, 214] have been used to calibrate the θ scale independently and they agree with the AES and XPS data given here in the sense that S_{min} falls between $\theta = 0.33$ and 0.39 mL and that NiO nucleation is observed beyond S_{min}.

A suitable starting point for the interpretation of the S versus θ data is the paper by King and Wells on the kinetics of adsorption of N_2 on W{100} [14]. They modelled the kinetics starting with the basic assumption that it is necessary for an N_2 molecule to have access to two nearest neighbor empty sites, nn, (Fig. 58, upper left) for dissociation to be possible. This is a

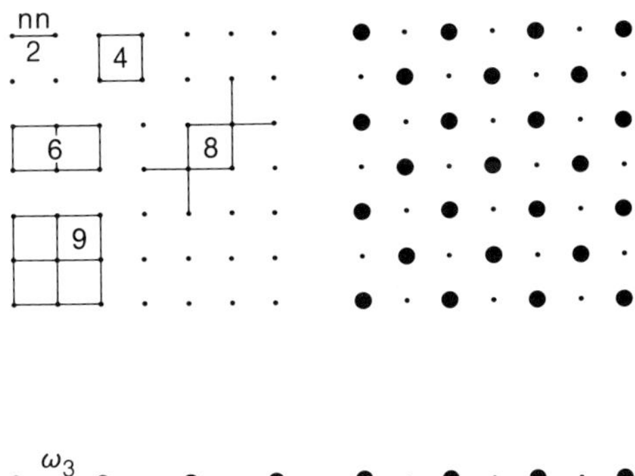

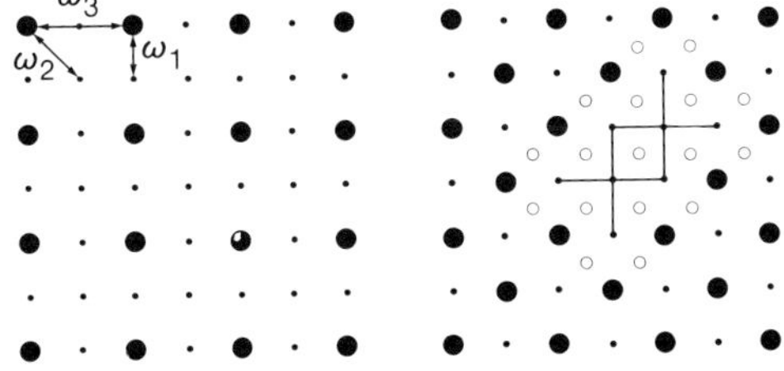

Fig. 58. Upper left, possible empty cluster site arrangements necessary for dissociative adsorption of a diatomic on a square lattice (see text). Upper right, perfect c(2 × 2) structure (large full circles) on a square lattice. Lower left, perfect p(2 × 2) structure (large full circles) on a square lattice. Lower right, c(2 × 2) structure with two adjacent vacancies revealing an 8 empty site cluster. Substrate Ni{100} atom positions are indicated in the cluster site region for reference (open circles).

commonly held assumption based on the belief that to break the strong N_2 triple bond requires a transition state of bonding between the N_2 molecule and the two sites that the dissociated N atoms will occupy after the dissociation process. Since we know the latter are the fourfold hollow sites, it is assumed that an N_2 molecule cannot straddle a distance greater than the nearest neighbor separation distance in achieving the transition state (see Fig. 58), hence the requirement of nn sites for dissociation. Whether this is a valid assumption should emerge from a comparison of the model's predictions for the kinetics compared with the experimental data. In addition to the nn requirement, King and Wells included the effects of a mobile N_2 state existing on the surface prior to dissociation and of lateral interaction and resultant ordering in the final chemisorbed N layer. If N_2 mobility prior to dissociation is negligible, the situation is referred to as *Langmuir adsorption*. It implies that the only possibility for N_2 to dissociate is by direct impingement from the gas phase in the vicinity of an available dissociation site. Molecular N_2 striking an already-covered region reflects back into the gas phase. $S(\theta)$ is then proportional to the number of dissociation sites available, in this case to θ_{nn}, the fractional coverage of nn empty pairs. If N atoms occupy the chemisorption sites randomly after dissociation (equilibrium achieved by a few hops to dissipate energy) then $\theta_{nn} \propto (1 - \theta)^2$. This leads to the simple Langmuir relationship for dissociative adsorption

$$S = S_0\theta_{nn} = S_0(1 - \theta)^2 \qquad (4)$$

and for $S_0 = 1$

$$S = (1 - \theta)^2 \qquad (5)$$

Such a theoretical plot is shown in Fig. 57(b), curve (I).

If adatom ordering occurs at the surface, instead of random occupation of sites, θ_{nn} will deviate from $(1 - \theta)^2$. In the limit of infinite pair-wise repulsion between adatoms on nearest neighbor sites, a c(2 × 2) structure will be established statistically as θ increases. In this case [14]

$$\theta_{nn} = (1 - 2\theta) \qquad (6)$$

$$S = S_0(1 - 2\theta) \qquad (7)$$

and dissociative adsorption must cease at $\theta = 0.5$ (Fig. 58) where θ_{nn} becomes zero for a perfect c(2 × 2) lattice. The theoretical S versus (θ) curve for this situation is shown in Fig. 57(b), curve (II).

For less than infinite repulsion, and therefore less than perfect equilibrium adatom ordering, nn(θ) can be established from the Guggenheim quasi-chemical approximation, as discussed by King and Wells. The important parameter is exp (w/kT_s) where w is the pair-wise repulsion. As exp (w/kT_s) goes from ∞ [large w; perfect c(2 × 2) order] to 0 (zero w, random adatom distribution) nn(θ) varies between the two limits of $(1 - 2\theta)$ and $(1 - \theta)^2$, and therefore so does S.

Having established the limits of the effects of adatom order of the kinetics (within the assumption of only pair-wise repulsive interactions between adatoms on nearest-neighbor sites), we must consider the additional effect on any mobility of N_2 trapped in a precursor state on the surface. Qualitatively the effect is obvious; S will decrease more slowly than θ_{nn}, the no-precursor Langmuir kinetics limit. The more hops the trapped N_2 can make before desorbing, the slower will be the decrease of S compared with the decreasing θ_{nn}. In the limit of infinite precursor lifetime (i.e. zero desorption probability), the kinetics reduces to

$$S = S_0 \qquad \text{for } \theta_{nn} > 0 \qquad (8)$$

and

$$S = 0 \qquad \text{at } \theta_{nn} = 0 \qquad (9)$$

Thus curves (III) and (IV) in Fig. 57(b) are obtained for the perfect c(2 × 2) adatom ordering case and the random adatom distribution case, respectively. A finite precursor lifetime will lead to curves lying between (II) and (III) [Fig. 57(b)] for perfect adatom ordering, and (I) and (IV) for random adatom distribution. In reality, of course, both the adatom ordering parameter and the precursor residence time parameter might take intermediate values, in which case a curve anywhere between (II) and (IV) is possible. King and Wells computed such sets of curves and compared them with the experimental $S/S_0(\theta)$ curves for W{100}/N_2. At $T_s = 300$ K, they concluded from this comparison that both the adatom ordering effect and the precursor residence time exerted strong influences on the adsorption kinetics. Switching back to Ni{100}/O_2 we can see that the situation is completely different by comparing the experimental 300 K data in Fig. 57(a) with the model curves in Fig. 57(b). Two things are immediately obvious. One, there is a complete absence of any precursor residence time effect at 300 K; and two, $S(\theta)$ falls much too rapidly to be explained even by the perfect c(2 × 2) ordering model [curve (II), Fig. 57(b)]. Now, the latter curve is based on two assumptions; the need for nn sites for dissociation and the presence of only pair-wise nearest-neighbor repulsions for determining the adatom ordering. We know the latter assumption is incorrect for Ni{100}/O_2 from LEED, since a p(2 × 2) pattern forms prior to the c(2 × 2), and in fact is observed to grow in intensity from low coverage. This implies the formation of p(2 × 2) islands and therefore an attractive interaction, w_3, between third nearest-neighbor O adatoms (Fig. 58). As c(2 × 2) islands are not observed at low θ, the second nearest neighbor adatom interactions, w_2 (Fig. 58), must be repulsive, but only weakly because the c(2 × 2) does form as soon as the p(2 × 2) is completed. The nearest-neighbor adatom repulsion, w_1, is of course large, as assumed in the model, since no p(1 × 1) structure is ever formed. Any number of pair-wise interactions can, in principle, be treated using Monte Carlo procedures (see also Sect. 2.5.5 for W surfaces). As it happens, for a strong nearest-neighbor repulsion, i.e. w_1/kT large, θ_{nn} is always independent of w_2 and w_3; that is to say, θ_{nn} is always given by eqn. (6), when no nearest-neighbor sites are occupied whether or not a p(2 × 2) structure precedes the c(2 × 2) and whether or not islands form. Curve II [Fig. 57(b)] is therefore still the limiting case for the basic assumption that nn sites are required for dissociation. Since the experimental curve is actually nearer $1 - 4\theta$ than $1 - 2\theta$ one might be tempted to assume that a block of 4 empty sites, rather than 2 is actually required for dissociation. In the Monte Carlo approach one can follow the concentration of any configuration of empty sites, $\theta_{cluster}$ as a function of θ. In Fig. 59 we have plotted $\theta_{cluster}$ for a number of different configurations as a function of θ for an optimum set of w values [245]. The shapes of the empty cluster configurations are shown in Fig. 58. We can see that, independent of the exact choice of w values, a comparison with the experimental $S(\theta)$ data excludes not only nn sites but also the 4 and 6 site configurations. The appropriate candidates seem to be the 8 or 9 site configurations and a distinction on the basis of the experimental data cannot be made over the coverage range up to completion of the p(2 × 2) structure. If, however, the Monte Carlo calculations are extended into the c(2 × 2) region a strong distinction in their behavior occurs. This is shown in Fig. 60 where we have replotted $\theta_{cluster}$ on a log scale so that the behavior into the c(2 × 2) region is observable. θ_9 continues to decrease rapidly after $\theta = 0.25$ mL, whereas θ_8 stays almost constant. This is related to the fact that, for $\theta > 0.25$ mL for which c(2 × 2) regions start to grow, vacancies of

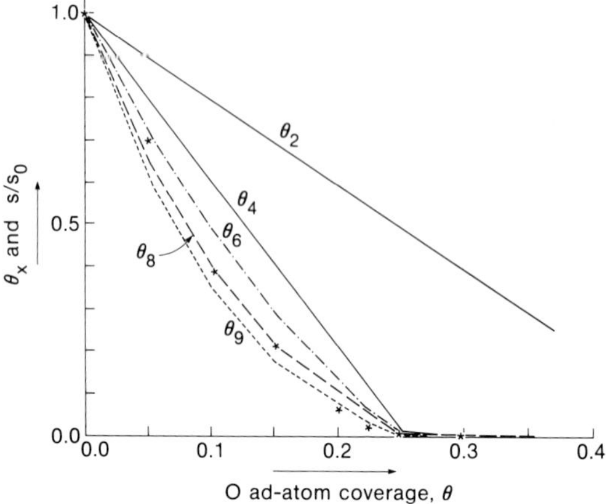

Fig. 59. Plots of the fractional coverage, θ_x, of the empty clusters of configuration shown in Fig. 58 as function of oxygen adatom coverage as determined by Monte Carlo calculation using the pairwise interaction potentials $w_1 = \infty$, $w_2 = kT$, $w_3 = -1.2kT$ [245]. The experimental S values, $*$, are included for comparison [245].

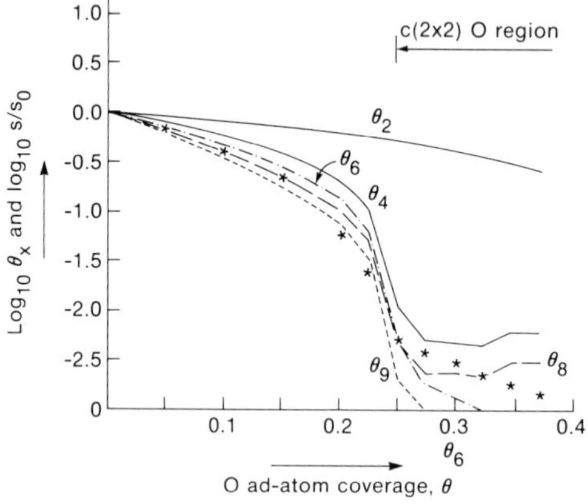

Fig. 60. As Fig. 59 except on a logarithmic scale so that the behavior in the c(2 × 2) O growth region can be seen [245].

2 adjacent c(2 × 2) sites reveal 8 site empty configurations, whereas vacancies on 3 or 4 adjacent sites are required to produce 6 or 9 site empty configurations, respectively (Fig. 58). Thus as θ increases above 0.25 mL the relative availability of 8 site empty configurations compared with 6 or 9 increases. The experimental $S(\theta)$ data, clearly more closely follows the 8 site behavior. It is tempting therefore, to conclude that the 8-site configuration is the minimum requirement for dissociation. Some caution is necessary, since any early NiO nucleation will artificially increase S and confuse the comparison. Since, however, we believe that nucleation does not occur until $\theta \approx 0.35$ mL, in the data presented we *do* conclude that the 8 sites are

required for dissociation. The significance of this conclusion is that it negates the view that the adsorption step and ordering of O_{ads} under the influence of lateral interaction can be regarded as separate steps. An O_2 molecule approaching the Ni{100} surface will not dissociate unless the two final adjacent c(2 × 2) O_{ad} sites are both vacant. The molecule clearly feels the Ni substrate induced O_{ads} lateral interactions as it approaches and enters any "transition state" involving the molecule and the surface. Interestingly, because of the awareness of the molecule of the lateral interactions this "transition state" must be thought of as the O_2 molecule bridging across a Ni atom to two fourfold sites, i.e. two next-nearest empty sites rather than the originally suggested transition state to nn sites which does not straddle a Ni atom (Fig. 58). Of course, once the molecule has dissociated into two O_{ad} on c(2 × 2) sites there is a high probability that O_{ad} will diffuse, provided empty sites are available and T is sufficiently high to overcome any hopping barrier. At 300 K the latter is clearly the case, since the p(2 × 2) structure dictated by the attractive w_3 and weakly repulsive w_2 is formed at low coverages.

We now consider the kinetics measured at low T. The original data of Holloway and Hudson (Fig. 42) clearly showed a drastic difference at 147 K compared with 300 K, though no discussion of the origin of the difference was given. The later data of Brundle and co-workers [237, 247] at 77 K are converted to S versus θ in Fig. 57 for comparison with the 300 K case. Other authors find a qualitatively similar effect at 77 K [243]. The data are qualitative only, partly because the increased rate of reaction does not allow much measuring time in a dynamic experiment. No ordered LEED structures are observed at 147 or 77 K, but Ni^{2+} is detected in XPS at least as early in coverage as was the case at 300 K [237, 247]. Problems with reaction with contaminant CO_2 to give a carbonate invalidate an earlier suggestion [237] that an intermediate oxide state formed prior to NiO nucleation [238]. All we can really say is that Ni oxide is formed at $\theta \leqslant 0.35$ mL, but the long-range order is too poor to give an NiO LEED pattern. Though only a qualitative consideration of the data is justified, therefore, it is quite clear that at low T the precursor O_2 lifetime now significantly affects the kinetics. $S \approx S_0$ up to $\theta \approx 0.2$ mL, whereas at 300 K it was 50 times smaller at $\theta \approx 0.2$ mL. Since there is no reason to suspect that the site requirement for dissociation changes with T, we must conclude that either the availability of θ_8 sites with θ changes (adatom order effect) or the ability to reach them (precursor effect) changes, or both. Thus S may be kept high by a combination of an increased availability of θ_8 sites as a function of θ, and an increased probability of an incoming O_2 molecule being transported to such a site via a precursor mechanism. If oxidation did not occur, θ_8 and S should approach zero at $\theta = 0.5$ mL. Clearly it does not; $S \approx 0.5S_0$ at 77 K, and since we know that Ni^{2+} forms at least as low as a θ of 0.35 mL at 77 K it becomes futile to try to interpret the data between $\theta \approx 0.35$ and 0.5 mL in terms of a chemisorption only model.

3.1.9 Kinetics and mechanism of oxide nucleation and island growth

Holloway and Hudson associated the increase in S at $\theta \approx 0.35\,\text{mL}$ for 300 K adsorption with dissociation and capture of oxygen molecules at NiO islands. Thus, as the islands grow the perimeters increase and S increases until coalescence of the islands occurs and the whole surface becomes covered by oxide, at which point S falls towards zero. Though they observed NiO in LEED, this was not until $\theta \approx 1\,\text{mL}$ and the suggestion of NiO islands growing laterally at $\theta \approx 0.35\,\text{mL}$ was entirely based on the kinetics and the inversion of $\Delta\phi$, which suggested oxygen incorporation. Such island models had been proposed before [294, 295] and there is no doubt that Holloway and Hudson were qualitatively correct since, in addition, LEED, RHEED [227] and RBS data [209, 214] confirm the existence of NiO islands for $\theta > 0.35\,\text{mL}$. Ni^{2+} character was also detected for $\theta > 0.35\,\text{mL}$ by XPS [236].

Within the general model there can be considerable latitude, however, and it is not obvious that the form of the kinetics can enable one to distinguish between different possibilities. Questions that can be asked are; what is the nucleation site density and does it change with temperature and surface condition, what is the thickness of the NiO islands and does it remain constant with island lateral growth; what is the mechanism of O capture at the island perimeters; and finally what causes the initial nucleation sites to form?

Holloway and Hudson developed the island model within a specific set of assumptions as follows.

(1) Physisorbed (precursor) oxygen exists on the surface with mean residence time τ_p and diffusion constant D_S. No distinction is made between O chemisorbed regions and NiO island regions.

(2) In the island growth region oxygen is dissociated at NiO perimeter sites only.

(3) Oxide islands are circular.

(4) All oxide islands are initially nucleated at time zero (i.e. short compared with island lateral growth time).

(5) Islands have a fixed thickness.

Based on earlier treatments for three-dimensional oxide growth [296], they derived the kinetic equation

$$(1 - \theta) = e^{-K_i N_0 P^2 t^2} \tag{10}$$

where θ is the oxide island coverage, N_0 is the nucleation site density, P is the pressure, t is the exposure time, and K_i is a rate constant which has the form

$$K_i = \text{const}\,[2B_i e^{E_i/RT} + h] \tag{11}$$

where h is the island height. B_i and E_i are pre-exponential and activation energy terms which have limiting forms depending on the dominant mechanism for oxygen capture. The schematic model suggested for the oxide growth process is shown in Fig. 61. If oxygen capture were limited by precursor diffusion to the island perimeters, E_i would equal $1/2(E_a - E_d)$; if it is limited

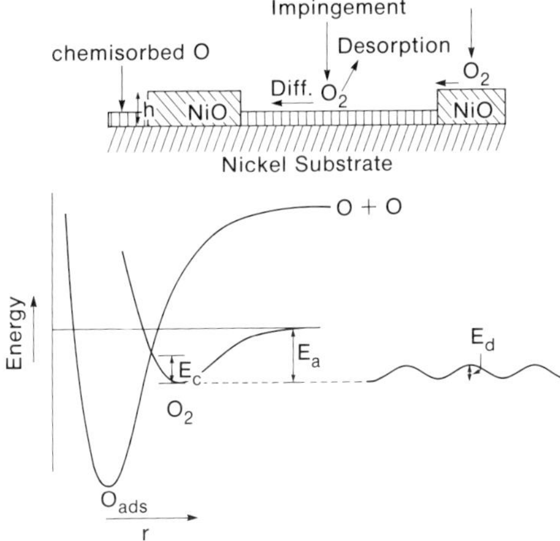

Fig. 61. Schematic model of oxide nuclei growth process on Ni{100}, as proposed by Holloway and Hudson [231]. The assumed schematic one-dimensional energy diagrams for molecular adsorption, dissociation and diffusion are also shown.

by the capture rate at the perimeter, E_i would equal $(E_a - E_c)$. In practice, Holloway and Hudson used $K_i N_0$ as a fitting parameter to eqn. (10) for each temperature measured. Values of 3×10^8 to $8.6 \times 10^7\,\mathrm{Torr^{-2}\,s^{-2}}$ were found to fit the data well for temperatures between 300 and 423 K (Table 9). From these they calculated $E_i = -1.5 \pm 0.3\,\mathrm{kcal\,mol^{-1}}$, assuming that N_0 was constant with T. They suggested that this value was consistent with either a surface diffusion-limited reaction or a surface diffusion and oxygen incorporation-limited reaction. From out discussion earlier we find it very hard to understand how the process can in any way depend on precursor diffusion over this T range since in the chemisorption regime the effect of the precursor on the kinetics was zero; i.e. τ_p is very short and the O_2 reaching the island perimeters is restricted to that impinging in the vicinity directly from gas phase.

While using eqn. (10) to model the kinetics with $K_i N_0$ as the fitting parameter produces a good fit, we do not believe this proves that the proposed mechanism for the island growth is correct. As stated above, we believe the precursor diffusion effect to be negligible at 300 K. In fact over the region $\theta \approx 0.35$ mL to the point where S starts to decrease again it is possible to fit the S versus θ behavior reasonably well simply assuming $S \propto \theta^{1/2}$, where θ is the oxide island area. This form of kinetics would result if oxygen was captured by direct impingement from the gas phase at perimeter sites with no activation energy for incorporation into the oxide. A further problem in the Holloway and Hudson treatment arises from their derived value of N_0 ($6 \times 10^9\,\mathrm{cm^{-2}}$) and the assumption that it is constant with temperature. For a nucleation site density of only $\sim 10^{10}\,\mathrm{cm^{-2}}$ the oxide islands would have to reach a size of the order of 1000 Å diameter by the time they start impinging and the surface becomes passivated with the oxide

TABLE 9

Comparison of oxide island growth parameter, K_iN_0 (Torr $^{-2}$ s^{-2}), in the Holloway and Hudson kinetic equation for the three low-index Ni surfaces

Face	Authors	Temperature (K)	K_iN_0
Ni(100)	Holloway and Hudson [231]	302	3×10^8
		373	1.2×10^8
		423	8.6×10^7
	Mitchell et al. [227]	313	1×10^8
		425	1.4×10^7
		473	2×10^6
	Smeenk [5]	325	3.0×10^8
		415	6.0×10^6
Ni(110)	Holloway and Outlaw [233]	296	7×10^9
		356	2×10^9
		373	7×10^8
		423	4×10^8
	Mitchell et al. [228]	313	6.6×10^9
		373	6×10^8
		423	1.6×10^8
	Rieder [269]	296	1.1×10^{10}
	Smeenk [5]	296	6.3×10^9
		448	7.1×10^7
Ni(111)	Rieder [269]	296	3×10^9
	Holloway and Hudson [232]	147	3.9×10^{11}
		302	8.7×10^9
		346	2.5×10^9
		423	2.0×10^9

layer. The RHEED data, however, strongly suggests very small island sizes, at least for the {100}-oriented NiO, of the order of 10 Å [227]. Clearly an N_0 value of $\sim 10^{12}$ cm^{-2} is required to fit with this information. Questions concerning the constancy of N_0 and also the nature of the activation energy term in K_i have been addressed by temperature-jump experiments. Before considering these we should perhaps review the medium-energy ion-scattering work on Ni{100}/O_2 as far as it relates to the kinetics of oxide growth. As mentioned in Sect. 3.1.4 on geometric structure, the H^+ backscattering work [209, 210] confirmed the existence of NiO islands of a fixed thickness during oxidation at 415 K. At lower temperature (325 K), however, the H^+ backscattering energy distribution did not fit this model so well. Smeenk et al. [210] actually suggest that the backscattering data are consistent with the presence, early on, of a thin uniform oxide layer which then grows in depth until saturation. Despite this difference in mechanism, they could still fit in their kinetic data at both temperatures using the Holloway and Hudson rate equation, which resulted in values of K_iN_0 of 3×10^8 and 6×10^6 Torr^{-2} s^{-2}

at 325 and 415 K, respectively (Table 9). If the Smeenk et al. interpretation of the backscattered energy distribution is correct this clearly shows that being able to fit the kinetic data empirically with $K_i N_0$ as a fitting parameter is not sufficient to validate the details of the Holloway and Hudson model. It is not clear to us, however, that the backscattering interpretation is as certain as claimed. It is probable that at 325 K a model which has many small islands which grow laterally and in depth simultaneously can also fit the data. One of the great advantages of the ion scattering when used in the channeling and blocking mode at high depth resolution, is that the displacement of surface Ni atoms from their {100} positions can be followed while monitoring the oxygen uptake. Figure 62 (upper panel) shows both O uptake and the number of Ni atoms displaced as a function of exposure [209] for both 325 and 415 K. It is obvious that the movement of Ni atoms is strongly correlated with the increase in S. The data, replotted in Fig. 62 (lower panel) as the number of displaced Ni atoms, Ni_d, as a function of θ, show an apparently linear relationship beyond the nucleation point. At 325 K the final stoichiometry is close to NiO, whereas at 415 K a total excess of $\sim$0.5 Ni atoms move. The authors attribute this to Ni displacement at the NiO/Ni{100} interface because of lattice mismatch as NiO tries to grow epitaxially. At 325 K the temperature is too low to displace the Ni metal atoms and the resultant NiO epitaxy is much poorer. The linear relationship between θ and Ni_d is compatible with an island growth mechanism, as discussed more fully by the authors for Ni{110}/O_2 [212]. Smeenk et al. originally used this compatibility to support the island growth mechanism at both 415 and 325 K [209]. When they later suggested, on the basis of the H^+ backscattering

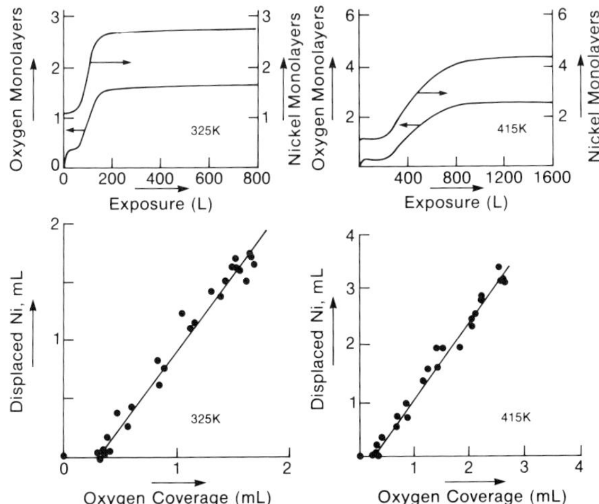

Fig. 62. RBS results for the adsorption of oxygen on Ni{100} at 325 and 415 K (Smeenk et al. [209]). Top, oxygen coverage, as determined by RBS, and the number of visible Ni monolayers as a function of exposure at 325 and 415 K. Bottom, above data replotted as the number of displaced Ni atoms as a function of oxygen coverage.

H^+ backscattering energy distributions, that at 325 K growth actually occurred by a uniform thickening of a thin layer, no further mention of the linear Ni_d versus θ curve was made [210].

The data of Fig. 21 also well illustrate that there are actually three separate factors one has to keep in mind when describing oxidation as "faster" or "slower". The first is the exposure at which oxide nucleation initiates; clearly greater at 415 than 325 K here. The second is the rate of oxide growth after the nucleation onset; clearly much faster at 325 than 415 K here. The third is the final thickness; significantly thicker at 415 K here. The first two factors might be expected to go together, as they do here, since they are both connected with the propensity to form oxide nuclei. The final thickness is more diffusion-related, however, and would be expected to be greater at higher temperature.

Let us now return to the implications of N_0 and the activation energy term in K_i. Two groups have attempted temperature-jump experiments for the Ni{100} surface. Mitchell et al., using low angle X-ray fluorescence to monitor the kinetics, studied the 313 to 473 K range [227]. Using eqn. (10), they were able to fit the kinetic uptake curves very well, with typical $K_i N_0$ values varying from 1×10^8 Torr^{-2} s^{-2} at 313 K to 2×10^6 Torr^{-2} s^{-2} at 473 K (Table 9). These values are 3–6 times smaller than those of Holloway and Hudson. Mitchell et al. note that the determined values of $K_i N_0$ are more reproducible for repeat experiments on a particular surface than for different surfaces or a surface heavily sputtered and re-annealed between experiments. They ascribe this behavior to a variation in nucleation site concentration, N_0, as a function of surface re-preparation. They also relate the presence of the two different oxide crystallite orientations to a variable nucleation site effect. As we have pointed out earlier, our own observations are that the relative proportions of the different oxide orientations change from experiment to experiment. Though Mitchell et al. could fit the kinetics using the Holloway and Hudson equation, they made a basically different assumption concerning the nature of the fitting parameter $K_i N_0$. Holloway and Hudson assume that N_0 is a constant, fixed at time zero. The variation of $K_i N_0(T)$ is then ascribed to the activation energy term in K_i. Mitchell et al. assumed that the variation of $K_i N_0$ with T was due to a variation in N_0. Thus the slower oxidation growth at higher T is ascribed to a decrease in N_0, rather than to the negative activation term, E_i. Of the three temperature-jump experiments they performed, two do not distinguish the two mechanisms: raising T from 317 to 473 K during the chemisorption stage produced subsequent oxidation kinetics appropriate to 473 K and annealing from 317 to 473 K and returning to 317 K in the oxide region produced subsequent oxidation kinetics appropriate to 317 K. The third experiment, raising T from 317 to 473 K in the oxide region and continuing at that temperature, produced a startling result, however. Oxidation continued at the (faster) 317 K rate. In the Holloway and Hudson model the rate should decrease owing to the negative activation energy term in K_i. Mitchell et al. explained their result by suggesting that

N_0 varied with T up to the end of the chemisorption region, then it became fixed as soon as oxide islands started to grow. Though they do not explicitly mention it, the implication of the lack of a temperature effect after oxide nucleation is initiated is that there is no activation energy term in "$K_i N_0$", i.e. it supports our suggestion made a few pages earlier that the oxygen-capture process during oxidation is by direct impingement and non-activated dissociation at perimeter sites.

Hopster and Brundle performed similar types of temperature jump experiments, but between 77 and 300 K [237]. They found that if T is reduced from 300 to 77 K just after oxidation is initiated the oxidation rate remains that at 300 K, except for a small initial fast uptake. Remembering that at 77 K the molecular precursor becomes effective the small initial increase in S can be ascribed to the precursor effectively finding all remaining chemisorption sites. The fact that the oxidation kinetics then remains at the 300 K rate rather than the much faster 77 K rate supports the Mitchell et al. data and the interpretation we have offered above. If T is raised from 77 to 300 K during the oxidation region, Hopster and Brundle found that the oxidation rate decreased from the 77 K value but is still twice as fast as the 300 K value. Some adsorbate long-range ordering is likely to occur on raising T from 77 to 300 K, and the precursor effect disappears. Both these effects will tend to reduce S. The fact that, despite this, the oxidation kinetics is still twice as fast as for oxidation at 300 K also supports the idea that N_0 is determined by T in the chemisorption and nucleation stage and then assumes a fixed value.

Holloway [293] argued against the above interpretation on the basis of his own temperature-jump experiments on Ni{110} rather than Ni{100} which apparently do not show the effects described above [233]. We will consider this data in the section on Ni{110} since the behavior of this surface is not the same as Ni{100}. His position, based on his Ni{110} data, is that N_0 is fixed at time zero and is therefore associated entirely with a substrate phenomena, possibly defect sites, and is not related to conditions in the chemisorbing overlayer. He makes the point that $K_i N_0$, as determined by several authors, does not vary by more than a factor of 3 for a given temperature, which he finds remarkable and supportive of the suggestion that N_0 is fixed at time zero. Our comment is that the actual spread of data is more like a factor of 6 (Table 9) and that, because of the form of the model kinetic equation used, such variations corresponds to enormous oxidation rate changes (see Fig. 44). We believe that the number of potential nucleation sites which become active nucleation sites depends on both the perfection of the substrate and the perfection of the O chemisorbed overlayer being formed. We also believe that Ni{110}/O is rather different from Ni{100}/O (see later). Can we say anything further to characterize these sites and the mechanism by which they become active? Miranda et al. [246] and Hopster and Brundle [236] attempted to answer this question by deliberately creating substrate defects by ion bombardment and following the effects on the kinetics. Representative results from Hopster and Brundle were shown in

Fig. 44. Miranda et al. noted that the maximum in the sticking probability, S_{max}, during the oxide nucleation step increased rapidly with Ar^+ sputtering dose and then saturated. They assumed that the saturation point corresponded to the situation where maximum surface damage had been achieved and, on the basis of the rapid increase at lower doses, speculated that Ni substrate defects may be responsible for the oxidation process on the unsputtered annealed surface too. The maximum increase in S_{max} is only ~ 2.5 on sputtering. This 2.5 increase would suggest a $\sim 2.5^2$ increase in defects if $S \propto \theta^{1/2}$. This seems a relatively small increase for such a drastic treatment and may suggest that not all defects introduced by Ar^+ bombardment can become active oxide nucleation sites. Another point that is observable in the sputtering data is that, in addition to S_{max} increasing, the exposure at which nucleation appears to start also decreases. This suggests, in keeping with the proposal of Mitchell et al., that whether defect sites on the Ni substrate become active nuclei depends on competing events: O capture by the defects and O loss to the chemisorbed phases forming on defect-free regions. Whether or not the defect plus O grows into an oxide island nucleus which then expands irreversibly will depend on conditions, including those of the ordered overlayer; i.e. we arrive at the idea of an incipient site activated by the local behavior of the chemisorbed phase. Hopster and Brundle [237] have suggested that the critical situation is a local coverage $\theta > 0.5$ mL, i.e. once one or a few O adatoms are added to a local c(2 × 2) O region by "mistakes" either originating from substrate defects, overlayer defects, or a combination of both, this local region becomes unstable with respect to NiO formation. Thus the driving force for the Mott–Fehlner Ni–O place-exchange process [15] may be regarded as a repulsive $O^{\delta-}$–$O^{\delta-}$ interaction for chemisorption above $\theta_{local} = 0.5$ mL and an attractive dipole arrangement for a bilayer of NiO. Hopster and Brundle suggest that at low T (77 K) many more "mistakes" in the overlayer ordering occur because of the lack of O_{ad} mobility and therefore a local θ of > 0.5 mL is more easily achieved, leading to the higher oxidation rate. At this temperature the diffusion of precursor oxygen to island perimeters is also effective, of course.

Gallagher et al. [263] made a somewhat similar suggestion on the basis of electronic structure effects. Their calculations suggested that, at $\theta \sim 0.35$ mL, since all available Ni *sp* electrons have become involved in the Ni–O bonding, Ni *d* electrons start to become involved. This makes the bonding more like bulk NiO, explaining the desire to convert to an NiO geometric arrangement. This is a simple and appealing idea, though it lacks the local aspects to explain small island nucleation.

Summarizing this section then, we can say that, though it is certain that NiO on Ni{100} forms at least partly by an island nucleation process, a detailed understanding is lacking and at the lower temperatures (less than 300 K) it is not even clear that the idea of fast island nucleation to a depth of 2–3 layers, followed by a slower lateral island growth, is correct. It may be that the lateral growth is on a time scale comparable with the in-depth 2–3 layer thickening. The original kinetic derivation assuming lateral growth only, fixed island thickness, fixed N_0, an effective oxygen precursor feeding the island perimeters and an activation energy term in the "rate constant," K_i, can *always* be made to fit the data. We have shown, however, that, because of the exponential form of the fitting procedure, such a fit does not prove much. Certainly other models can be made to fit in the significant range immediately beyond nucleation, the simplest of which is direct O_2 impingement on island perimeters for which $S \propto N_0^{1/2}\theta^{1/2}$. There is other evidence to suggest that a molecular precursor plays no role above 300 K, that there is no Arrhenius activation-energy-like behavior, and that N_0 depends on both substrate and overlayer defects, varying with T in the chemisorption region but becoming fixed as soon as a critical nucleation size is reached. The exact nature of the sites is unclear as is the way in which they become fixed oxide nuclei, but suggestions based on the idea of a change in the relative stabilities of chemisorbed overlayer versus NiO bilayer at $\theta = 0.35$ mL or $\theta_{local} > 0.5$ mL have been suggested. Clearly to go further requires more studies over a wide temperature range with controlled defect concentration and character, i.e. stepped surfaces.

3.2 Ni{110}

The reaction scheme of Ni{110} is basically similar to that for Ni{100} so the reader should also consult the relevant sections on Ni{100} where some pains were taken to substantiate statements which are also made here without further justification. There are, however, complications present for Ni{110} that are not present for Ni{100} or {111} surfaces. These are concerned with the propensity for the open structure of the {110} surface to reconstruct under the influence of oxygen in the chemisorption stages prior to oxidation.

The amount of data available for Ni{110}/O_2 is comparable with that for Ni{100}, except in the area of electronic structure studies where there are no theoretical studies attempting to simulate specific Ni{110}/O geometries.

3.2.1 General reaction scheme. 300 K and above

The reaction of oxygen with Ni{110} has been followed up to the passivation stage by LEED [195, 225], AES [233], XPS [243], X-ray fluorescence [228], SIMS [269], He diffraction [204], and ion scattering [207, 208, 211–213]. The general kinetic behavior is very similar to Ni{100} and, by analogy, it can be considered as involving three stages: rapid dissociative chemisorption during which time S drops from ~ 1 to a low minimum value; NiO nucleation and oxide growth, during which S increases, until the passivation oxide thickness corresponding to $\theta \approx 2$ mL is reached and S drops towards zero; and finally a slow diffusion-limited oxide thickening process requiring large exposures (10^5 L) to produce observable changes [228]. Like Ni{100}, dissolution of O into the subsurface occurs for temperatures above ~ 500 K

[233], the exact values depending on coverage and existing subsurface O content. The kinetics in the fast oxide growth region can be fit by the lateral island growth model [233], and, as was the case for Ni{100}, such a fit does not necessarily provide much insight into the true mechanisms operating.

On closer comparison with Ni{100} some striking differences emerge. A comparison of the two kinetics curves, as determined by medium-energy ion scattering [209], is shown in Fig. 63. The plateau length for Ni{110} is much reduced and the NiO oxidation rate after initiation is much faster. Within the island growth model of Holloway and Hudson this implies either a larger number of oxide nucleation sites for Ni{110}, or that there is a smaller activation energy for oxygen incorporation into the growing oxide islands. The shorter plateau length implies either an easier initial nucleation process from the chemisorbed overlayer, or, more likely, simply that the kinetics during completion of the chemisorption stage are more rapid for Ni{110}. That the latter is the real reason is supported by the fact that, though oxide nucleation initiates earlier on Ni{110} in terms of exposure, the O atom coverage at this point is similar for both faces, $\theta \approx 0.35$ mL. Actually, for Ni{110} some authors claim a value as high as $\theta = 0.5$. In terms of absolute O atom concentration, a θ value of 0.35 mL on Ni{100} is equivalent to 0.5 mL on Ni{110} because of the different Ni atom concentration of the two surfaces (1.6×10^{15} atoms cm^{-2} for Ni{100} and 1.4×10^{15} atoms cm^{-2} for Ni{110}). In the chemisorption stage there is strong evidence that the Ni{110} surface reconstructs [204, 207, 208, 211–213]. For the p(2 × 1) LEED pattern observed for coverages at the plateau, it is certain that the pattern is generated by a surface with missing Ni close-packed rows in the top Ni atomic layer. This effect has been confirmed by He scattering [204], low-energy ion scattering [207, 208], and medium-energy ion scattering [211–213]. It is not surprising, then, that with such a facile rearrangement of Ni atoms under the influence of chemisorbed O atoms, oxide nucleation and growth should occur very easily on Ni{110}. In addition one might expect that the

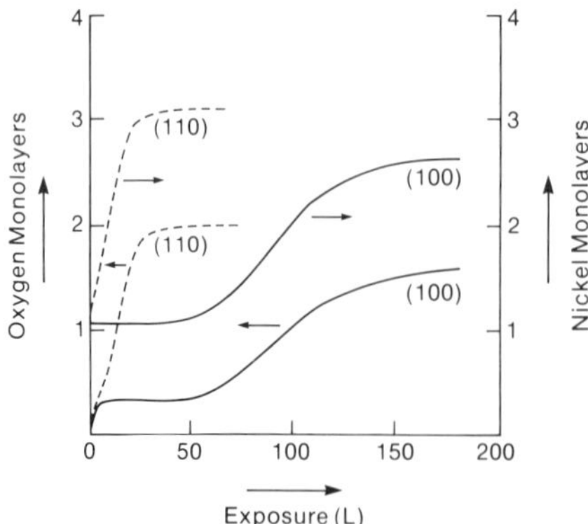

Fig. 63. Comparison of RBS data at room temperature for Ni{100} and Ni{110}. The correlation between oxygen uptake kinetics and the number of visible Ni monolayers is plotted [209].

exact form of $S(\theta)$ during the chemisorption would be rather different from Ni{100} where such reconstructions do not occur. Though there is not much detailed $S(\theta)$ data for Ni{110} over the 0–0.5 mL range, what there is does suggest drastic differences from Ni{100}. The O atom locations and geometric structures of the chemisorbed phases are not well established, in contrast to the Ni{100} case. The $\Delta\phi$ behavior is similar to Ni{100} in that it is positive in the chemisorption regime (with a small oscillation associated with Ni reconstruction) and starts to decrease as oxide nucleation occurs [225].

3.2.2 General reaction scheme. Low temperature

Adsorption at ~80 K is qualitatively the same as on Ni{100}; the fast dissociative chemisorption stage merges into the NiO formation region without much of a decrease in S and without the formation of any ordered LEED structures. The data at 80 K for Ni{110} are actually somewhat more extensive than for Ni{100}, there being an AES study [297], an XPS/UPS study [243], and UPS studies at 80 and 20 K [298–300]. One difference from Ni{100} is that apparently the saturation coverage at 80 K is only ~0.7 that at 300 K [243, 297]. For Ni{100} the value is ~0.95. Benndorf et al. [297] have suggested that a full monolayer of molecular O_2 is formed at 80 K (with little or no oxygen present in any other form). This is wrong, as is clearly apparent from the UPS data of Norton et al. [243], and of Jacobi et al. [298–300] where, at 80 K, the oxygen-induced feature is that of atomic oxygen. In fact, Jacobi et al. found that, even at 20 K, molecular O_2 features are not observed in UPS until O_2 physisorbs on top of a complete first layer of dissociative oxygen.

3.2.3 Geometric structure of chemisorbed species. General

In the discussion of the Ni{100}/O chemisorbed structures the final conclusion, after the confusion and over-interpretation of the quantum chemistry and vibrational spectroscopy, was that the structures were rather simple with the O atoms placed where one would expect them in the {100} fourfold hollow sites of the unreconstructed top Ni layer. The only remaining doubt was whether the O atoms were always exactly centrally located in the fourfold sites. For Ni{110}/O the situation is much more complex, as might be anticipated for the more open and asymmetric {110} Ni arrangement. Four different LEED structures have been identified; p(3 × 1), initial; p(2 × 1); p(3 × 1), final; and p(9 × 4), in addition to highly streaked patterns in intermediate situations. It is quite certain from He scattering [204] and ion scattering [207, 208, 211–213] that reconstruction of at least part of the top Ni layer is involved for the p(2 × 1) structure, and strongly suggested that this is true also for the two p(3 × 1) patterns. There has been a suggestion from ELS vibrational data [225] that the p(2 × 1) and p(3 × 1), final, patterns each represent two surface structures, one being reconstructed and one not. Later we offer a different interpretation for the ELS data which does not involve such complexity. The p(9 × 4) pattern is almost certainly a pseudo-NiO pattern (compressed structure compared with bulk NiO). It appears well after oxide nucleation has started according to the kinetic data [212, 225, 228, 269]. No NiO (1 × 1) is formed at 300 K, in contrast to the Ni{100} results. All the general points made for Ni{100} concerning complications due to oxide nucleation or subsurface or bulk O dissolution at elevated temperatures, and impurity products such as OH or CO_3 also apply here.

We will begin by describing the LEED patterns and their coverage, temperature and work function behavior, and then go on to consider the geometry of each structure in detail.

At 300 K the sequence of LEED (and RHEED) patterns described below is observed as a function of exposure. A typical correlation between LEED patterns and exposure [225] and coverage is shown in Fig. 64. Between ~0.2 and 0.8 L exposure a poorly ordered (3 × 1) pattern is observed by some [195, 204, 225, 269, 301], but not all [212, 228], authors. The pattern is sharper on exposure at 400 K or with subsequent annealing. At exposure between 0.3 and 0.8 L, depending on the authors, a p(2 × 1) structure emerges and persists to 1.2–3 L. Above this exposure, or sometimes overlapping with it, the second p(3 × 1) pattern is observed. In some cases streaked patterns are observed between the end of the p(2 × 1) and the start of the second p(3 × 1). The p(3 × 1) persists up to 4–8 L. At much higher exposures (12–15 L) the p(9 × 4) pattern develops. The discrepancy in exposures between authors is due either to variable kinetics or absolute exposure calibration problems, since when converted to a *coverage* scale, there is good agreement between authors. Originally it was assumed that the coverages concerned were 1/3, 1/2 and 2/3, in the p(3 × 1), p(2 × 1), and p(3 × 1) structures [195, 301]. When careful coverage determinations were made, however, by medium-energy RBS [212] or X-ray fluorescence [228], it was clear that the p(2 × 1) formed at 0.25–0.3 monolayer, and that the second p(3 × 1) structure was complete by, at the most, 0.5 monolayer and probably by 0.4 mono-

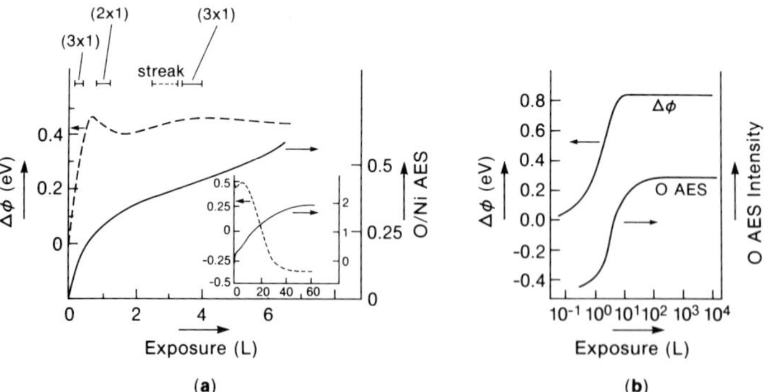

Fig. 64. Kinetics, LEED pattern and work function changes for Ni{110}/O_2 (a) at 300 K [225] and (b) at 77 K [297].

layer. Above 0.4–0.5 monolayer, oxide nucleation and growth occurs, as detected directly by RBS [212] or low energy ion scattering [207, 208] or indirectly from the sudden increase in S after the plateau (Figs. 63 and 64). Other coverage calibration procedures used the NiO passivation O coverage of $\theta = 2$ mL (as determined by RBS or X-ray fluorescence) as a calibration point for AES and XPS in order to reference the LEED coverages and the kinetic plateau coverage. In these studies the reported oxide nucleation point varies from $\theta = 0.4$ [272] to 0.55 mL [225] and the three LEED patterns precede this coverage. The first p(3×1) pattern occurs at $\theta \sim 0.15$ mL. It is clear, then, that simple LEED-compatible coverages are incorrect and that the surfaces are O deficient compared with such coverages, either because of surface patching/statistical vacancies, or because the patterns are generated by the Ni atoms rather than the O atoms. It is also clear that the coverage difference between the p(2×1) and second p(3×1) patterns is rather small, and that the distinction between the end of the chemisorption stage and the start of the oxidation stage is less clear than for Ni{100} (Figs. 64 and 65).

The general form of $\Delta\phi$ as a function of exposure at 300 K is similar to that for Ni{100}. There is an initial rise in the chemisorption regime followed by a strong decrease associated with NiO nucleation and oxidation. A typical example of the correlation between $\Delta\phi$, exposure and coverage is included in Fig. 64 [225]. Other authors have reported identical behavior [297]. The interpretation of these changes in terms of O atom location is more complex than for Ni{100}, however, as might be expected when reconstruction of the top Ni layer is occurring. For Ni{100} we were able to conclude that, provided there was no interference from early oxidation, the increase in $\Delta\phi$ was linear with coverage through both p(2×2) and c(2×2) regions, reinforcing our belief that both structures contained essentially similar O_{ad}-atom locations. This linearity is clearly not observed for Ni{110} over the whole of the chemisorption range. The initial sharp rise (Fig. 64), which is approximately linear with coverage, is certainly indicative of overlayer O_{ad} even though at least part of the Ni surface, that associated with initial p(3×1) formation, is thought to reconstruct during this exposure regime. During the coverage increase from ~ 0.2 to ~ 0.4 mL, during which time the p(2×1) and final p(3×1) patterns are completed, ϕ barely changes except for a slight oscillation which seems to be associated with the completion of the p(2×1) structure. A plausible explanation for the sharp increase from 0 to 0.2 mL followed by little change between 0.2 and 0.4 mL is as follows. During the first 0.2 mL, a significant fraction of the O_{ad} occupies disordered overlayer positions on unreconstructed regions of the Ni{110} surface. This fraction gives a strong $+\Delta\phi$ and the surface regions giving rise to the weak (3×1) pattern contribute much less strongly to $+\Delta\phi$. Between 0.2 and 0.4 mL all additional oxygen becomes involved with the p(2×1) and final (3×1) structures and some or all of the disordered O_{ad} oxygen also converts. Since the $+\Delta\phi$, per O_{ad} atom, of the ordered structures is smaller than that

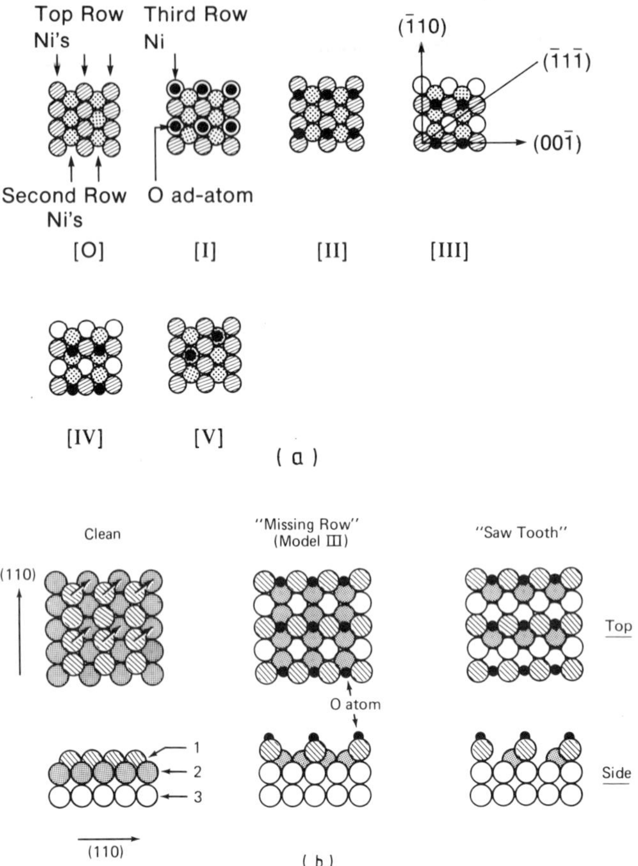

Fig. 65. Proposed models for the p(2 × 1) O structure on Ni{110}. (a) [0] Clean Ni; [I] mixed Ni/O reconstructed layer [301]; [II] short-bridge overlayer O_{ads} [304]; [III] reconstructed Ni top layer, O_{ads} in long-bridge overlayer positions [207] (known as "missing row" model); [IV] modified version of [III]. O_{ads} is rotated towards second layer Ni atom [207]; [V] disordered overlayer O_{ads} at low θ (see text). (b) Comparison of "missing row" model [207] and "saw-tooth" model [310] showing both top view and cross-sectional view. First, second and third row atoms are marked 1, 2 and 3, respectively. The clean Ni{110} surface is shown to demonstrate how the "saw-tooth" structure might be obtained.

of the disordered structures, an approximately constant ϕ is maintained over this region. Also, since ϕ does not start its monotonic decrease towards a final NiO-like value until $\theta \geqslant 0.4$ mL, it is likely that though there is strong evidence that the p(2 × 1) and final p(3 × 1) structures have Ni reconstruction, they do not involve Ni and O place exchange, i.e. they are not NiO-like structures with O beneath a Ni surface giving rise to a negative $\Delta\phi$ term.

Adsorption at temperatures greater than 300 K, or subsequent annealing, produces only a sharpening of the chemisorption LEED patterns, without any change in the coverage relationships, provided the O dissolution temperatures are not exceeded [204, 272]. For the higher exposure oxidation stage, increasing T does cause changes, of course, the kinetics being much slower and, according to Smeenk et al. [212], at 475 K a (9 × 14) LEED pattern follows the final p(3 × 1) chemisorption pattern. It persists up to $\theta \approx 0.6$ mL (70 L exposure) and at ~200 L a NiO{100} structure is formed. These data should be treated with caution, however, since in a subsequent paper [210] the authors suggest that the true temperature may have been much higher, in which case O dissolution would occur. Holloway and Outlaw report that O dissolves from chemisorbed coverages above ~450 K, and forms a saturated NiO layer at ~650 K [233]. A number of the chemisorption structural determinations discussed below were, however, performed at up to 500 K without any significant changes, other than LEED spot sharpening, being observed compared with 300 K. The dissolution temperature may, therefore, be quite dependent on substrate surface perfection. The practice, however, of deliberately exposing to saturation at 300 K (i.e. NiO formation) and then annealing to ~800 K to produce the ordered LEED structures certainly puts a lot of oxygen in the subsurface. Smeenk et al. [212] were able to show, by RBS, that a perfect p(2 × 1) pattern generated in this manner also had $\sim 7 \times 10^{15}$ Ni atoms cm^{-2} displaced in a ~35 Å deep subsurface region owing to dissolved oxygen. Thus annealing LEED structures to 400–500 K is acceptable, provided no dissolution oxygen loss is observed, but applying high exposures and heating to 800 K to generate "surface" LEED patterns is not.

Adsorption at 80 K produces no ordered LEED structures. As mentioned in Sect. 3.2.1, and discussed again in Sects. 3.2.10 and 3.2.11, the kinetics are quite different from 300 K with no plateau region (Fig. 64). $\Delta\phi$ also increases monotonically to its final value of +0.85 eV [297]. The suggestion [297] that this indicates the presence of a chemisorbed molecular oxygen overlayer is clearly wrong, since XPS and UPS data show that the oxygen is atomic. Over the first ~0.2 monolayer adsorption, $\Delta\phi$ is very similar at 85 K to that at 300 K, from which we might conclude that, though no ordered overlayers are formed, the local positions of O_{ad} may be similar in both cases. The question arises as to why $\Delta\phi$ continues to increase up to saturation coverage at 80 K. All our previous arguments concerning $\Delta\phi$ effects would suggest that the continued increase in $\Delta\phi$ implies continued overlayer adsorption, i.e. no NiO formation. Other factors, such as the saturation coverage (1.3 mL), the results of annealing to 400 K and the available electron spectroscopic information lead us to doubt this, but we reserve further discussion to Sect. 3.2.8 on electronic and chemical structure effects.

3.2.4 Geometry of the p(2 × 1) structure. $\theta \approx 0.25$–0.3 mL

Though the p(2 × 1) structure is not the lowest-coverage structure, it is the best studied and so we consider it first. There have been many LEED studies [195, 269, 301–304], the original model proposed [301] being that of Fig. 65 (1). In this reconstructed model every other close-packed (110) row is missing and replaced by O atom rows. Subsequently, a full dynamical analysis of the structure was reported [304] to show a best fit to model II (Fig. 65) (i.e. an unreconstructed surface with O in the short-bridge position $d_\perp \approx 1.5$ Å). By today's standards the fit between experiment and the theory for model II is very poor. The result was accepted for several years, however, and tentatively supported by some low-energy ion scattering data [305, 306]. In 1979, a detailed low-energy ion scattering study of Ni{110} plus 1 L O_2 at ~520 K was reported [207]. No LEED was performed but under these conditions the p(2 × 1) should be at maximum intensity. The study was performed along several scattering azimuths and in the low-angle scattering mode to ensure that the multiple scattering events, which contain much of the structural information, dominated. The qualitative conclusions were that a large decrease in Ni atom density along the close-packed (110) row occurred and that O adatoms were strongly shadowed by Ni atoms along the (001) direction. Model II, the unreconstructed short-bridge site model with $d_\perp \approx 1.5$ Å, is completely incompatible with these conclusions. Model I is partly compatible since the "missing row" Ni reconstruction explains the decreased Ni density in the close-packed directions. It cannot, however, account for the strong O shadowing by Ni along the (100) direction. Verheij et al. [207] proposed model III (Fig. 65) to account for the scattering data, i.e. a reconstructed Ni surface but with O_{ad} in true overlayer positions on the long-bridge sites. Computer simulations of the scattering process were made with this model to see if a quantitative fit could be obtained. These suggested two things. First the O atoms cannot be more than 0.5 Å below the plane of the reconstructed top layer Ni atoms, and probably are not more than 0.1 Å below it. Second, it was not possible to simulate the scattering in the (001) direction correctly with model III. The authors suggested [207] a modification (model IV, Fig. 65) in which the O atoms are moved slightly off the Ni(001) rows, as a possibility for accounting for the discrepancy between simulation and experiment. A lateral displacement of 0.5 Å and a $d_\perp$ value of 0.5 Å would maintain a reasonable NiO bond length (1.9 Å). Finally the authors point out that the tentative support for model II from the earlier ion scattering results [305, 306] could equally well be considered as support for model III or IV.

At a similar time to the low-energy ion scattering work, the first of a series of papers using medium-energy RBS appeared [210–213]. 100 keV H^+ ions were used in a channeling and blocking mode. The main objective of the first paper [211] was to determine any change in $d_\perp$ for the top Ni layer of the clean surface. This was done by channeling along the $(3\bar{1}4)$ bulk directions and blocking along the (011) bulk direction. Determining the exact angle of maximum blocking with respect to the bulk (011) direction allows a determination of $d_\perp$. The result indicated a 4% contraction, which agrees with the LEED dynamical results for the clean Ni{110} surface [307]. For an O-covered surface (2 L 300 K exposure, 0.35 mL) the surface Ni layer was shown to expand back to within 1% of its bulk position [211]. The O_{ad} atoms do not contribute significantly to the blocking so their positions could not be determined by channeling and blocking. It was noted, however, that the H^+ backscattering minimum yield in the (011) direction was identical for the clean and O-covered surfaces. This was taken to imply a non-reconstructed Ni surface for the p(2 × 1) O structure (i.e. excluding models I, III and IV, Fig. 65). In a later paper [213] it was realized that the (314), (011) geometry used could not distinguish a missing-row surface from a non-reconstructed surface, provided all Ni atoms are on bulk lattice sites, since in the former case the yield from Ni rows 1 plus 2 equals the yield from row 1 of the latter case. The scattering measurements were therefore repeated for p(2 × 1) surfaces prepared at 470 K and checked by LEED. Several scattering geometries were used where a reconstructed surface would give more blocking than an unreconstructed one. A lower yield was found in all the directions, as predicted by the missing-row model and firmly excluding model II. The measurement also indicated that the remaining surface Ni atoms are located within 0.1 Å of bulk lattice site positions. It does not, of course, distinguish between models I, III and IV. The O coverage for the study was determined as 0.36 ± 0.04 mL. From quantitative aspects of the minimum yield values it seems that only ~ 50% of the surface is reconstructed. Together with the less than $\theta = 0.5$ mL coverage, this implies that the p(2 × 1) structure covers only part of the surface.

Masuda et al. [225] have reported HRELS data on the Ni{110}/O system under a wide range of adsorption conditions. The data are correlated with simultaneous LEED, AES and $\Delta\phi$ measurements. The vibrational frequency changes observed as a function of exposure (300 K), coverage, and LEED patterns are shown in Fig. 66. Three frequencies can be discerned: a 480 cm^{-1}

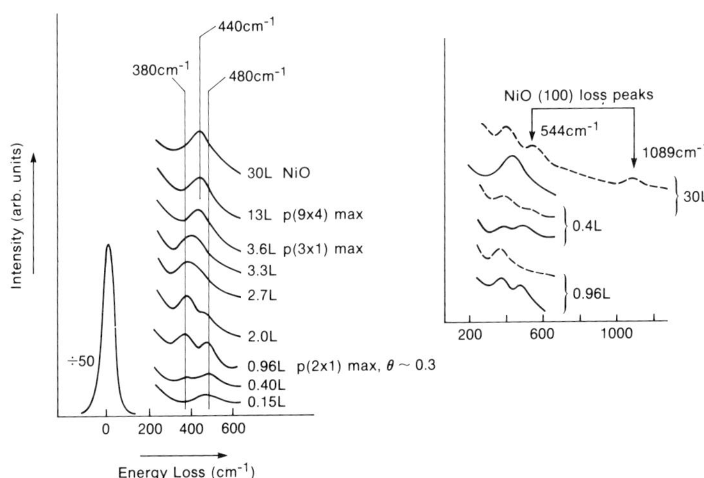

Fig. 66. HRELS vibrational spectra of Ni{110}/O_2 as a function of oxygen exposure at 300 K [225]. Inset shows the effects of annealing to high temperature (broken curves) after various exposures (see text).

initial frequency, a 380 cm^{-1} frequency growing in above 0.2 mL and having maximum intensity between 0.27 and 0.35 mL, and a 444 cm^{-1} frequency emerging at high coverage [exposures marked in Fig. 66 can be converted to θ using Fig. 64 and the knowledge that the p(2 × 1) maximum is at 0.25–0.3 mL]. The assignment of the final 444 cm^{-1} frequency at θ = 2 mL to disordered NiO is straightforward. Between the NiO nucleation point (~ 0.4 mL) and saturation coverage the resolution is insufficient to determine whether there are only superimpositions of the 380 and 444 cm^{-1} frequencies, or whether intermediate values representing, for instance, the (9 × 4) "pseudo-oxide" are present. The maximum intensity of the 380 cm^{-1} peak coincides with the p(2 × 1) O formation and we therefore assign it to O in the p(2 × 1) missing-row Ni reconstructed phase. Annealing lower coverage situations causes the 480 cm^{-1} frequency to convert to the 380 cm^{-1} (see inset, Fig. 66) together with the observation of a sharpening of the p(2 × 1) structure or, at ⩽0.2 mL, the emergence of the p(3 × 1) initial pattern which was not observed for 300 K exposure. Thus the 380 cm^{-1} frequency seems characteristic of O on the reconstructed surface rather than being specific to a particular LEED pattern. This suggests that the reconstructed p(3 × 1) and p(2 × 1) patterns have basically the same missing-row local structure. So far we are in agreement with the authors' own interpretations. We diverge, however, concerning the origin of the 480 cm^{-1} frequency. They suggest it also represents p(2 × 1) and p(3 × 1) O and that these structures are therefore mixtures of reconstructed and unreconstructed ordered O. They suggest a mixture of models II and III for the p(2 × 1). We believe the 480 cm^{-1} frequency represents overlayer disordered O, probably in the center of the rectangular mesh (Fig. 65, V). The fact that this frequency converts to the 380 cm^{-1} frequency on annealing as the p(3 × 1) appears or, at higher coverage, as the p(2 × 1) sharpens, seems to support this suggestion. Masuda et al. [225] dismiss this possibility without any reasonable explanation. We note that the 480 cm^{-1} frequency is higher than the 423 cm^{-1} value for p(2 × 2) O in the fourfold Ni{100} hollow. This might be expected if it corresponded to O in the rectangular Ni{110} hollow since the more open nature of the hollow should drive the frequency more towards an on-top site value. However, as has been demonstrated for Ni{100}/O, the interpretation of Ni–O frequencies in terms of O location is so unsure, because of possible lateral interactions and lattice phonon coupling, that we do not claim the 480 cm^{-1} frequency as a definitive determination of the O_{ad} atom location in the disordered phase. Likewise though the 380 cm^{-1} frequency is characteristic of O involved in the reconstructed missing-row Ni phase, we do not believe the short bridge site, model III, for the O_{ad}-atom position can be proven from the 380 cm^{-1} value.

Engel and Rieder [204] have reported He diffraction studies on the p(2 × 1) and p(3 × 1) structures for adsorption at 500 K. The one-dimensional electron density corrugation function in the ($\bar{1}$10) azimuth derived from the diffraction data using the hard-wall formalism (see Ni{100} for a discus-

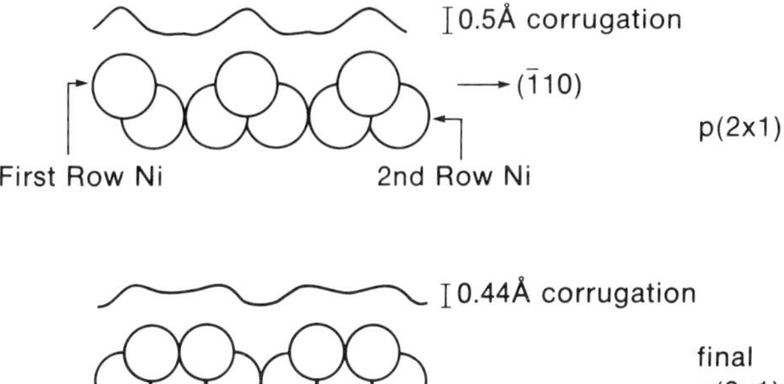

Fig. 67. One-dimensional electron density corrugation function in the (1̄10) azimuth from He diffraction for the p(2 × 1) phase (upper) and the final p(3 × 1) O phase for Ni{110}/O. The Ni layer reconstructions consistent with the data are shown schematically within a hard-sphere model (Engel and Rieder [204]).

sion) is shown in Fig. 67 for the p(2 × 1) structure. It is compared with the Ni atom structure of the missing-row model and the comparison gives clear and unequivocal support to the model. In the (001) azimuth the corrugation amplitude is negligible compared with its value for the clean surface. This is a strong indication that the O_{ad} atoms are located between Ni atoms along these rows (i.e. model III or IV), smoothing out the electron density in that direction. If the O_{ad} atoms were as in model III with a $d_{\perp}$ of 0.9 Å, the value found for p(2 × 2) O on Ni{100}, one would expect localized maxima from O_{ad} to be observable in p(2 × 2) O/Ni{100}. Since they are not observable, $d_{\perp}$ is presumed to be much shorter than 0.9 Å. This could most easily be achieved by the lateral displacement of O_{ad} from the rows, as in model IV. In the absence of full calculations of the electron density contours for various $d_{\perp}$ values and lateral displacements it is not yet possible to know what ranges of $d_{\perp}$ and lateral displacements are compatible with the scattering data and derived one-dimensional corrugation functions. At this higher level of sophistication it would also be more appropriate to calculate the scattering directly and compare it with the experimental data, rather than using a hard-wall one-dimensional corrugation function interpretation of the data.

Summing up the information on the p(2 × 1) structure we can say with certainty that the pattern is caused by the Ni atoms themselves in a "missing-row" reconstruction arrangement. In addition the remaining Ni rows have the Ni atoms within 0.1 Å of their ideal bulk lattice positions. Both the low-energy ion scattering and RBS work indicate that the p(2 × 1) reconstruction does not cover the whole surface, but we do not find any evidence to support the suggestion that there are two surface structures giving rise to a p(2 × 1) pattern, i.e. the missing-row reconstruction plus unreconstructed regions with O_{ad} located in the short-bridge sites in a p(2 × 1) arrangement (model II). Our interpretation of the HRELS data that led to such a suggestion [225] is that at 300 K a fraction of the O_{ads} goes into disordered overlayer sites on the unreconstructed surface and does not contribute to the p(2 × 1) pattern, but does give rise to a different vibrational frequency. On annealing to 500 K the long-range order improves considerably in the p(2 × 1) pattern because this disordered oxygen is incorporated into the p(2 × 1) reconstruction regions.

The question of the location of the O_{ad} atoms in the reconstructed regions is not settled. The low-energy ion scattering [207] and the He scattering [204] data suggest that they must be located in the vicinity of the (001) rows and between Ni atoms with $d_{\perp}$, (the distance between the O_{ad} plane and the top Ni plane) being much less than 0.9 Å. We suggest model IV. Since the O coverage of such an ideal structure would be 0.5 mL and the actual coverage is only ~0.3 mL, it was originally considered that the O_{ad}-atom distribution in sites along the (001) row must be random. If it were a regular structure such as one O atom per every two sites ($\theta = 0.25$ mL), a doubled periodicity should be observed in the (001) direction. This is not the case. However, random O_{ad} occupation of sites along the (001) direction should produce a high scattered background, which is also not observed. Keeping in mind, however, that only part of the Ni surface is reconstructed, a total θ of 0.3 mL is enough to have a local θ of 0.5 mL in the reconstructed regions. The periodicity is then unchanged compared with the clean surface and no high background due to random site occupation would be expected.

Very recently a further development has taken place concerning the nature of the p(2 × 1) structure. Schuster and Varelas [310(a)] used a novel channeling technique in which ion-induced Ni and O Auger electron intensities were monitored as a function of the ion impact geometry. From the Ni Auger results it was shown that the channeling profile was consistent with a reconstruction model where every other top Ni[001] row was missing in agreement with the conclusion of the medium-energy [213] and low-energy [207] ion scattering results. An additional claim was made, however, that every other second Ni row in the [001] direction was also missing, giving rise to an asymmetric "saw-tooth" reconstruction. The difference between this proposed structure and the simple missing row structure of Fig. 65, III is shown in the bottom panel of Fig. 65. The distinction between the two structures in the ion-induced Auger data depends on a comparison of the quality of fit of computed channeling profiles to the experimental data, and it is not clear that there is significant evidence for the "saw-tooth" model over the "missing-row" model in such a comparison. Neither is it clear whether the "saw-tooth" model can fit the RBS data [213]. The structure is interesting, however, because it provides an easy explanation of how reconstruction can occur without massive transport of Ni. Ni atoms in the original top layer have simply to move 2.5 Å to form a new top layer, as depicted in the bottom panel of Fig. 65.

In addition to examining the Ni substrate structure, Schuster and Varelas also located the site of the adsorbed O atoms from the O Auger data. The long-bridge site was found, in agreement with the low-energy ion scattering data [207], and from a quantitative comparison of experimental channeling data with computer simulation a $d_\perp$ of 0.25 ± 0.125 Å was suggested. No comment is made on whether the O_{ads} might be rotated about the long-bridge bond, such as suggested in Fig. 65, IV for the simple missing-row model.

Even more recently scanning tunneling microscopy (STM) results [310(b)] have been claimed to agree with the saw-tooth model. From the quality of the data actually presented, however, it is impossible to comment on the validity of this claim.

3.2.5 Geometry of the p(3 × 1) initial structure. θ ≈ 0.15 mL

The initial p(3 × 1) structure has not often been observed and when it has, the information obtained on it is meager. Apart from the original LEED observations of the (3 × 1) periodicity [301], the only significant information on the nature of its structure comes from the HRELS work of Masuda et al. [225] and the He scattering data of Engel and Rieder [204]. In the HRELS work a well-ordered (3 × 1) pattern was observed only after briefly annealing a 0.4 L 300 K exposed surface (~0.15 mL coverage) to 630 K [225]. In conjunction with the appearance of the (3 × 1) pattern a 480 cm^{-1} vibrational frequency disappeared leaving only the 380 cm^{-1} frequency. The latter frequency reaches maximum intensity with further exposure at a coverage coincident with the maximum intensity of the p(2 × 1) structure. This is strong evidence that the Ni–O local geometry is similar for the p(2 × 1) and p(3 × 1) structures and that both are "missing-row" Ni reconstructed surfaces. The 480 cm^{-1} vibrational frequency is then associated with O_{ad} in fourfold overlayer sites prior to reconstruction, as discussed in the previous section on the p(2 × 1) structure.

In the He scattering work [204] at 500 K (no well-ordered diffraction data could be obtained at 300 K) no detailed analysis of the initial p(3 × 1) diffraction data was made. It was stated, however, that the similarity of the data for all three phases, p(3 × 1) initial, p(2 × 1) and p(3 × 1) final, indicated similar types of electron corrugation function behavior in each case and therefore similar types of geometric structure in each case. To explain the low coverage (3 × 1) periodicity and the corrugation function periodicity in the $(1\bar{1}0)$ azimuth, a "missing-row" model with two out of every three rows missing is required, i.e. a 7.47 Å spacing between surface Ni atoms in the $(1\bar{1}0)$ direction. As in the p(2 × 1) structure the exact location of the O_{ad} atoms is not known but they must be close to or in the top Ni(100) rows to account for the decreased corrugation in the (001) direction compared with the clean surface. Not much else can be said about the initial (3 × 1) structure except that the broadness of the He diffracted beams suggests that it exists as small islands on the Ni{110} surface.

3.2.6 Geometry of the p(3 × 1) final structure. θ ≈ 0.3–0.6 mL

Work on the final p(3 × 1) LEED structure is somewhat more extensive than on the initial p(3 × 1), but still meager compared with the p(2 × 1) structure. It has been observed in most LEED studies, but the coverage range over which it occurs is variable. For example, the Masuda et al. HRELS study [225] reports its existence (at 300 K) only close to $\theta \approx 0.4$ mL ($\sim$4 L exposure) whereas Rieder in his AES/SIMS study [269] observes it up to 11 L exposure, which extends into the accelerated kinetics region associated with NiO oxide formation. In the medium-energy RBS work of Smeenk et al., it is observed between 0.3 and 0.6 mL at 300 K, but only around 0.3 mL at 475 K [212]. We interpret these variations as implying co-existence of the (3 × 1) with nucleated oxide in much the same manner that the c(2 × 2) O structure on Ni{100} co-exists with NiO formation well into the accelerated kinetics regime.

The best geometric information on the structure comes from the He scattering data [204]. The best-developed (3 × 1) structure corresponded to 3.8 L exposure at 500 K. The corrugation function in the $(1\bar{1}0)$ direction derived from the diffraction data is shown in Fig. 67 together with a hard-sphere model of the "missing-row" structure where every third Ni atom in the (110) direction is removed. The corrugation follows the atomic positions closely and provides convincing evidence that this is the correct structure. The corrugation in the (001) direction is negligible, as was the case for the p(2 × 1) structure, again suggesting that the O_{ad} atoms are located between Ni atoms along, or adjacent to, the (001) rows. In the low-energy ion-scattering study of Van den Berg et al. [208], the changes observed in the exposure range at the chemisorption plateau, which should correspond to the p(3 × 1) structure (500 K exposure), imply a surface structure with both Ni and O atoms more closely packed than for the preceding p(2 × 1) structure in the $(1\bar{1}0)$ direction. This can be interpreted in terms of the double Ni rows of Fig. 67; indeed, these authors [207, 208] were the first to suggest such a model though the subsequent He diffraction data [204] are more direct evidence than the low-energy ion-scattering data.

The HRELS data show a gradual shift of the 380 cm^{-1} peak associated with O in the initial p(3 × 1) and the p(2 × 1) structures to 444 cm^{-1} during final p(3 × 1) formation [225]. This is also the value for the p(9 × 4) "pseudo-oxide" structure which follows at higher exposures and the final saturation-covered NiO. The authors were puzzled by the similarity of the vibrational frequency for all three situations. We note that the movement of the vibrational frequency of the p(3 × 1) final structure to $\sim$444 cm^{-1} is back towards the 480 cm^{-1} value which we assigned as O in the rectangular site of the unreconstructed surface (Fig. 65, V). In fact, if the p(3 × 1) final structure is a paired Ni row structure (Fig. 67) then this rectangular site is reformed and if oxygen occupies it the only difference from O on the unreconstructed surface is the fact that each Ni rectangle is isolated from the next in the (110) direction by a missing Ni row. It is reasonable then that the vibrational frequency should be similar, but not identical, in the two cases. It is also reasonable that the p(3 × 1) value should be very similar to that for the "oxide" (9 × 4) structure that follows at higher coverage, and the very small NiO islands following that, because the essential difference between these structures is small movements of the Ni atoms with the O atoms remaining in the hollow site. Eventually these Ni movements convert the original Ni{110} rectangle to the {100} square lattice of NiO.

The channeling and blocking medium-energy RBS measurements which confirmed the missing row model for the p(2 × 1) O structure [213] were not carried out for the p(3 × 1) O structure. In another paper [212], however, the adsorption kinetics over the complete adsorption range were correlated in RBS with the displacement of Ni atoms from their ideal lattice sites, which is a signature of NiO formation. We note that at 300 K this correlation shows that NiO grows in a coverage regime ($\geqslant 0.35$ mL) which overlaps the coverage existence range of the p(3 × 1) structure. Since the He scattering data [204] indicate that the p(3 × 1) structure has its surface Ni atoms at bulk lattice sites (Fig. 67), we interpret the RBS data as supporting our suggestion that at 300 K NiO nucleates before the p(3 × 1) chemisorbed structure is complete [cf. c(2 × 2) O for Ni{100}]. At 475 K the RBS data show no Ni atom displacement until after completion of the (3 × 1) structure [212]. The Ni atom displacement occurs at a similar coverage, but much higher exposure than at 300 K. This simply indicates that, like Ni{100}, the onset of NiO nucleation depends on the perfection of the substrate and chemisorbed structures formed and that this perfection is better at higher temperatures, thus inhibiting NiO nucleation.

3.2.7 Summary of chemisorbed structures

From the foregoing discussions we think it is established beyond doubt that the three LEED structures observed prior to NiO nucleation all represent Ni "missing-row" arrangements, as indicated in Fig. 67. The LEED patterns of all of them are therefore generated by the Ni substrate arrangement and not the O_{ad} atoms. For the p(2 × 1) structure the Ni atoms are known to remain at lattice sites to within 0.1 Å from the blocking and channeling RBS data. The location of the adatoms has not been definitively determined for any of the structures, but it is fairly certain that they are located along, or close to, the top Ni(00$\bar{1}$) rows with a $d_{\perp}$ of considerably less than 1 Å, but without any Ni/O place exchange process (i.e. not a precursor to NiO). They do not introduce any change in periodicity in this direction in LEED and therefore must have one of the following environments: (a) the same periodicity as the Ni atoms, i.e. one O for every Ni, (b) random occupation of a specific site or (c) occupying disordered sites. Possibilities (b) and (c) should give rise to a higher background in the He scattering data of the p(2 × 1) structure. Since this was not observed we tentatively favor (a). For the paired Ni rows of the p(3 × 1) final structure (Fig. 67), we suggest that the O_{ad} atoms are located between the rows in order to account for the observed Ni–O vibrational frequency difference for this structure compared with the other two.

At very low coverages (< 0.1 mL) at 300 K we believe most of the O_{ad} atoms occupy the fourfold overlayer hollow position (Fig. 65, V) on the unreconstructed surface. This occupation is either statistical if there is one specific position within the hollow or possibly there are many disordered possible sites within the hollow. No LEED pattern results. On increasing the coverage or annealing, this situation converts to the reconstructed Ni phases. Some O_{ad} atoms may remain and coexist with the reconstructed phase, however, depending on coverage and annealing conditions.

One important question has not yet been raised here. By what process are the "missing-row" structures generated? Several of the papers referred to in this section have addressed this point, but we defer discussion until Sect. 3.2.10 on the kinetics of the adsorption process.

3.2.8 Geometric structure of the oxide species

At 300 K oxidation, the start of oxide nucleation can be judged by the three "signatures" of oxidation: an increase in S [233], the decrease in $\Delta\phi$ [225, 297] and the displacement of Ni atoms from ideal bulk Ni lattice positions as determined by RBS [212]. Oxide initiation starts between $\theta = 0.35$ and 0.4 mL according to these measurements.

The structures of the oxides formed have been investigated by LEED, RHEED [228] and HRELS [225]. In early LEED work [308] a (9×4) LEED pattern was observed to develop at exposures beyond those for formation of the final (3×1) structure. Continued exposure finally converted this to a NiO (1×1) structure. In other LEED studies this final NiO structure has normally only been seen after annealing, though the (9×4) preceding pattern is observed in most studies [204, 212, 225, 269]. At temperatures below 300 K no ordered structures are seen in the oxide regime.

The (9×4) structure was termed a "pseudo-oxide" structure because it resembles the face-centered square array of oxygen atoms in the $\{001\}$ plane of the NiO lattice, compressed by $\sim 5\%$ [308]. The implication was that it represented a 2D oxide structure which preceded the thicker NiO growth (which we now know to be ~ 3 layers thick at saturation). Subsequent work, however, tends to be ambiguous as to whether the (9×4) structure should be classified as an oxide or a chemisorbed structure. The reason for the ambiguity is that, in some studies, the exposure and coverage range of the (9×4) existence extends back into the chemisorption plateau region. In defining a structure as oxide-like *we* mean, as usual, that O adatoms have penetrated at least the top Ni layer and the Ni atoms have started to move from Ni lattice positions towards NiO lattice positions.

The HRELS/LEED study of Masuda et al. [225] and the SIMS/LEED study of Rieder [269] both place the LEED observation of the (9×4) structure at 300 K at $\theta \approx 0.5\theta_{sat}$, i.e. at ~ 1 mL coverage and therefore well into the oxidation stage. These results support the contention that the (9×4) structure forms after oxide island nucleation initiates and is essentially the precursor of (1×1) NiO. In the RBS study [212], the LEED observation of the (9×4) structure was at a lower coverage of 0.5–0.7 mL, spanning the end of the plateau stage of the chemisorption kinetics and into the accelerated kinetics range of oxide growth. However, according to the same RBS work, by 0.6 mL O adatom coverage 0.6 mL of Ni atoms have started to move. This was taken to imply that all O adatoms at this point were part of growing oxide regions. Thus the (9 × 4) structure would have to be an oxide structure involving Ni atom displacement.

In the RHEED observation of the (9 × 4) structure [228] the onset is also at the end of the chemisorption plateau, but the structure persists through to ~1 mL, by which time epitaxial NiO reflections are observed. The authors refer to it as a chemisorbed structure but also note the pseudo-oxide classification, so it is not clear whether they view it as a 2D oxide-like structure or not. Since, however, RHEED reflections of epitaxial NiO with the relationship NiO(001)‖Ni(100), NiO($1\bar{1}0$)‖Ni($1\bar{1}0$) develop smoothly after the (9 × 4) structure (at ~1 mL) with a lattice constant compressed ~5% in the ($2\bar{2}0$) direction ($d_{2\bar{2}0}$ = 1.41 Å, whereas $d_{2\bar{2}0}$ for bulk NiO is 1.476 Å) it seems very likely to us that the (9 × 4) structure is the 2D precursor to the 3-layer NiO formation. Between 1 and 1.7 mL the lattice parameter does not change but over the last 0.3 mL O uptake it expands to its bulk NiO value. The widths of the RHEED diffraction maxima suggest a very small island size for the epitaxial NiO ($\leqslant$10 Å), as was the case for the (001) orientation of NiO on Ni{100}. In LEED this small island size is probably responsible for the non-observation of the NiO(1 × 1) structure at 300 K in most studies. On annealing, the NiO(1 × 1) appears or, if the annealing temperature is too high, some O dissolves into the subsurface, producing a sharp p(3 × 1) or p(2 × 1) LEED structure from the O remaining at the surface. The HRELS data for such an anneal is rather interesting [225]. As discussed earlier the disordered NiO produced at saturation at 300 K gives a Ni–O vibrational frequency of 444 cm^{-1}, essentially the same as for the preceding (9 × 4) structure. When annealed to 770 K for 1 min, a sharp p(2 × 1) structure is observed and a vibrational spectrum showing peaks at 400, 544 and 1089 cm^{-1} is produced. Since multiple loss peaks of 544 cm^{-1} spacing are observed for thick NiO{100} films grown on Ni{100}, the 544 cm^{-1} and 1089 cm^{-1} frequencies are assigned to $h\nu$ and $2h\nu$ of an optical phonon mode for semi-infinite NiO, i.e. whereas very small NiO islands give a vibrational frequency of 444 cm^{-1}, which is a very local Ni–O mode, the annealed situation produces optical phonons characteristic of long-range 3D order. The 400 cm^{-1} peak represents the O adatoms of the observed p(2 × 1) LEED structure, which presumably implies that the shift from the normal 380 cm^{-1} value (see earlier) is due to interaction between the p(2 × 1) surface patches and the coexisting 3D NiO.

As was the case for Ni{100}, extensive exposure at temperatures above the O dissolution temperature can produce quite thick NiO layers. In the RHEED/X-ray fluorescence work, θ values up to 5 mL were produced for temperatures between 450 and 600 K with exposures between 10^3 and 10^5 L [228]. At 300 K the oxide can also be made thicker, though it is a very slow process, taking 10^8 L to go from $\theta = 2$ to 3 mL. At this coverage and all higher coverages the (001) NiO epitaxy has broken down and been replaced by a constant, but very complex, epitaxial oxide relationship which has not been interpreted in detail from the observed RHEED reflections [228].

At temperatures much lower than 300 K no ordered oxide structures are observed and the limiting oxygen coverage is apparently reduced (1.4 mL at 80 K compared with 2 mL at 300 K [243, 297]). With the lack of any order the knowledge of the nature of the oxide is restricted to electronic structure information. Such information is available at 20 and 80 K [298–300], as discussed in Sect. 3.2.8. The data are compatible with the idea that at $\leqslant$ 80 K very small oxide islands are formed and the oxygen atom coverage is concentrated more in the top layer. This would reduce the amount of true Ni^{2+} behavior observed, which is the case [243]. In geometric terms a plausible scheme is that local geometries such as the 2D "pseudo-oxide" (9 × 4) are retained to a higher coverage at 80 K at the expense of the growing 2–3 layer-thick NiO islands.

Summarizing this section on the geometric structures of the oxide species formed, we can say that, like Ni{100}, two epitaxial relationships are observed below θ_{sat}. The first (lower coverage) case is the (9 × 4) structure, sometimes referred to as a chemisorbed structure and sometimes as a "pseudo-oxide". We believe it to be a genuine oxide structure in the sense that the Ni atoms have moved significantly from their metallic lattice positions and O atoms have penetrated the first layer. The structure probably corresponds to a 2D {100} NiO layer with a ~5% compressed lattice.

The second epitaxy, observed above 1 mL coverage, is a 3D NiO(1 × 1) structure with a well-developed epitaxial relationship to the {110} substrate and a ~5% contraction in the $(2\bar{2}0)$ direction, the only one determined. At saturation coverage, 2 mL, which corresponds to 3 layers of NiO, the normal bulk NiO lattice parameter is achieved. This NiO appears to form as very small islands ($\leqslant$ 10 Å). In many cases the disorder is sufficient that it is not observed at all in LEED at 300 K. Such disordered NiO has quite a different vibrational spectrum from well-ordered NiO to which it can convert when annealed. At temperatures above the O dissolution temperature, a complex oxide epitaxy with θ up to 5 mL can be achieved. At temperatures well below 300 K, the limiting θ is reduced to ~1.4 mL and the evidence is that the 2D NiO structures are stabilized compared with the subsequent growth of 2–3 layer NiO.

3.2.9 Electronic structure aspects

In marked contrast to the Ni{100}/O system there are no electronic structure calculations available. The reason is clear. Whereas the location of the O atoms on Ni{100} in the fourfold hollow is well established, it should be obvious from the previous section that the location on Ni{110} is a subject of conjecture. The one "determination" by LEED for the p(2 × 1) O structure is definitely wrong. In addition, since it was only recently established that the chemisorbed structures involved reconstructed Ni surfaces, it is just as well that no electronic structure calculations specific to the Ni{110}/O have been performed since an incorrect Ni/O local geometry would almost certainly have been used.

With the lack of theoretical input, our discussion of electronic structure effects becomes limited to a comparison between equivalent experimental data for the Ni{110}/O system and the Ni{100}/O system. A large part of the experimental information for the latter came from angle-integrated, and in particular, angle-resolved UV photoemission. No angle-resolved photoemission exists for the Ni{110}/O system, but one of the angle-integrated studies is instrumental in answering a question that arises from the work of Benndorf et al. [297]. They performed detailed AES/$\Delta\phi$ studies on Ni{110}/O at 300 and 80 K and concluded, mainly on the basis of the $\Delta\phi$ changes (Fig. 64) that, at 80 K, a molecular layer of O_2 ($\theta_O \approx 1.3$ mL; $\theta_{O_2} \approx 0.65$ mL) was adsorbed. The reasoning was that, since $\Delta\phi$ increased monotonically at 80 K and did not go negative in the later stages, oxidation was not occurring. Though this is the simplest explanation, the angle-integrated UPS data of Hsu et al. at 20 K [299] (Fig. 68) clearly show, however, that it does not mean molecular oxygen. Even at 20 K oxygen is dissociatively adsorbed up to ~2.4 L exposure, as indicated by the typical O_{ad} (2*p*) peak at ~6 eV, the spectrum being very similar to that of Ni{100}/O at the p(2 × 2) O or c(2 × 2) O stage. The oxygen is therefore present as atoms and any differences in the local Ni–O electronic structure between the Ni{100}/O chemisorbed case and the Ni{110}/O (20 K) up to 2.4 L exposure are insufficient to cause significant differences in the valence band spectrum. It also seems, from the Hsu et al. data, that, up to 2.4 L exposure, no NiO is formed. NiO should show up as the emergence of a Ni^{2+} 3*d* band peak at ~2.0 eV below E_F. For Ni{100} [286] and {111} [250] the emergence of the Ni^{2+} 3*d* feature in UPS upon oxidation is well documented. There is much less data for Ni{110}, even at 300 K, but the

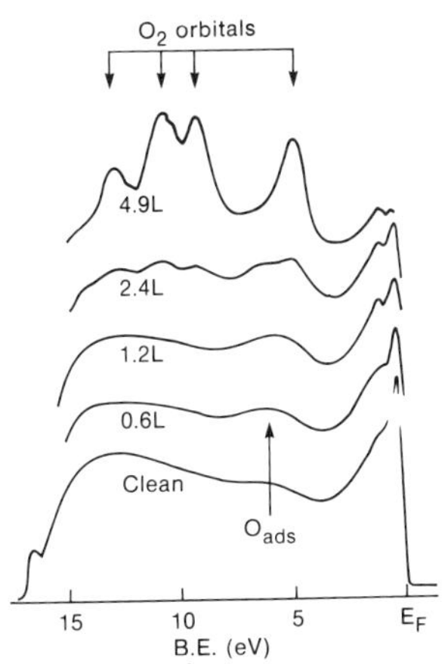

Fig. 68. HeI angle-integrated photoemission spectra for Ni{110} plus O_2 at 20 K (Hsu et al. [299]).

saturation coverage (300 K) spectrum of Norton et al. [243] does show a shifted Ni^{2+} $3d$ peak. Remarkably, though there is UPS data as a function of exposure at 20 K [299], none exists at 300 K over the chemisorption range. Such data should be informative since the Ni substrate reconstruction occurs over this range at 300 K and this may lead to specific changes in the O(2p) and Ni(3d) spectral characteristics. The strong similarity between the Ni{100}/O UPS at 300 K and the Ni{110}/O UPS at 20 K may indicate that the Ni{110} surface is not reconstructed by the O_{ads} at 20 K, but until the UPS is examined in detail at both temperatures as a function of coverage, nothing more can be said. Returning to the 20 K work of Hsu et al., it is clear that, for exposure above 2.4 L, molecular oxygen, recognizable from its molecular orbital structure, condenses on the O_{ad}-covered surface (Fig. 68). Owing to the superimposition of the molecular O_2 in the spectrum it is not really possible to say whether there is any further O_2 dissociation above 2.4 L. Since no chemisorbed molecular O_2 forms on the bare Ni{110} surface at 20 K, it must be so that none is formed at 77 K. At this temperature there will also be no O_2 condensation on top of the dissociated O layer. This then effectively dismisses the idea of a molecular chemisorbed O_2 layer at 80 K but does not answer the question of whether any oxidation occurs at 80 K. Comparison of the $\Delta\phi$ data at 20 K [299] and 77 K [297] is informative here. At 20 K, $\Delta\phi$ rises monotonically to its maximum at +0.85 eV, which is coincident in coverage with the first observation of condensed O_2 in the UPS. The build up of condensed O_2 has no further effect on $\Delta\phi$. At 80 K, $\Delta\phi$ rises monotonically also to +0.85 eV and then remains steady, but the simultaneous O Auger data shows that the +0.85 eV value is reached at $\theta \approx 0.78\theta_{sat}$ and the remainder of the O uptake does not change $\Delta\phi$. Since θ_{sat} at 80 K is estimated by Norton et al. at ~1.3 mL (which agrees with Benndorf et al. [297] if their 300 K saturation value is normalized to $\theta = 2$ mL), $\theta(\Delta\phi_{max})$ is at ~1.0 mL. This suggests that the first monolayer is dissociatively adsorbed and the next 0.3 mL O uptake involves the onset of oxidation, resulting in a reduction in the dipole moment per O atom such that $\Delta\phi$ remains constant. At 300 K, $\Delta\phi$ was approximately +0.4 eV at the end of the chemisorption stage for a θ value of ~0.4 mL. The dipole moment per O_{atom} for 77 K adsorption at the $\Delta\phi_{max}$ point is thus similar to that at 300 K, again suggesting that the same type of O_{ad} overlayer is present in both cases. Since we have no idea whether the Ni surface reconstructs at 80 K during the chemisorption stage it is not really possible to take the arguments further.

Our tentative conclusion, then, concerning the low temperature data is that, on Ni{110}, close to one monolayer of O_{ad} can be reached before any three-dimensional oxidation occurs and that at saturation (~1.3 mL) most of the O present is not in the form of 2–3 layer NiO. As discussed in the previous section, a plausible interpretation is that it is present in the 2D (9 × 4) "pseudo-oxide" geometry which at 300 K precedes 2–3 layer NiO formation. This is supported by XPS data, where the Ni2p spectrum shows little evidence of a Ni^{2+} component of NiO at saturation unlike at 300 K [243]. This situation differs markedly from Ni{100} where the saturation coverages at 80 and 300 K are nearly identical and the Ni2*p* XPS spectra, while not identical, clearly show that an oxide is formed at 80 K. One caution should be considered concerning the low-temperature Ni{110}/O work, however. We know that water and carbon dioxide contaminants initially caused severe problems for the Ni{100}/O system at 80 K, leading to misidentification of new "O_{ads}" species. We do not believe that the efforts made to eliminate these artifacts for the Ni{100}/O system have yet been made for the Ni{110}/O system. It is possible, therefore, that some of the results and conclusions discussed above are influenced by impurity effects. For instance, for Ni{100} in the absence of impurity reactions the XPS O(1*s*) BE does not vary significantly over the whole oxygen adsorption range, or with temperature. For Ni{110}/O, the O(1*s*) spectrum upon which the claim of θ_{sat} (80 K) $\sim$ 1.3 mL is based looks markedly different from that at 300 K [243]. It is much broader, showing evidence of two types of O(1*s*), and is altogether rather similar to what is obtained for Ni{100} when a large amount of carbonate impurity is present [238]. We think, therefore, that the 80 K adsorption work on Ni{110} should be repeated rather than expending further effort on trying to establish the differences from the Ni{100} system based on existing data. For this reason, although Benndorf et al. have reported shifts in O(KLL) Auger position and in Ni(M_{23}VV) positions during adsorption (at both 300 and 80 K) and have attempted to interpret them in terms of electronic structure effects [297], we will not consider them here.

3.2.10 Kinetics and mechanisms. General

As was done for Ni{100} we will separate this section into "chemisorption" and "oxide nucleation and growth" parts, keeping in mind that these two regions can overlap considerably and that the amount of overlap is likely to vary from experiment to experiment because the oxide nucleation is strongly affected by substrate and overlayer defects. As for Ni{100} we note that an accurate coverage scale is the key to a consistent interpretation and comparison between data, and that the combination of a total coverage measurement and a $\Delta\phi$ measurement is particularly useful under these conditions where oxide nucleation might be occurring. The general form of the kinetics for Ni{110}/O has already been contrasted to Ni{100}/O in Fig. 63.

3.2.11 Kinetics and mechanism of chemisorption

The extraction of a model for the dissociative chemisorption of oxygen in the 0–0.4 mL coverage range in which the chemisorption structures are formed (300 K) is much more difficult than for Ni{100}. In addition to problems of accounting for the Ni{110} substrate reconstruction over this range and of not being exactly sure where the O adatoms are, there is not the detailed data of $S(\theta)$ for Ni{110} that there is for Ni{100}. Though $S(\theta)$ at 300 K during the chemisorption regime has been approximately followed in many studies, it was usually obtained during the pursuit of other objectives and is not of sufficiently good quality to draw many conclusions. There are two recent exceptions where the determination of $S(\theta)$ was the main objective. In the work of Winkler et al. [241], the relative θ was determined from AES O(KLL) intensities. This was converted to an absolute coverage scale using a quartz crystal microbalance calibration. The integral mass sensitivity of the microbalance had previously been determined by evaporation of an Ag film of known thickness. The microbalance calibration also showed that there was a linear correlation between O(KLL) AES intensity and θ up to at least one monolayer. Their $S(\theta)$ curve is shown in Fig. 69. It is strikingly different from that of Ni{100} (see Figs. 56 and 57). The constancy of $S \approx S_0 \approx 0.9$ up to $\theta \approx 0.2$ mL strongly suggests a molecular O_2 precursor adsorption effect which was totally absent for the {100} surface at 300 K. The authors briefly discuss their data in terms of a precursor model, to which we will return after introducing further data. One surprising feature of the Winkler et al. data is that there is no evidence for the plateau stage at the end of the chemisorption process. In fact, S has only dropped to ~ 0.1 by $\theta \approx 0.5$ mL. If the coverage scale is accurate this may indicate a poor quality substrate surface so that the oxidation starts well before the end of the chemisorption stage (~ 0.4 mL), thereby obliterating the plateau by keeping S high. In this case the values of S beyond $\theta \approx 0.25$ mL will not be meaningful for any chemisorption model.

Jacobi and Rotermund [300] followed the relative kinetics of Ni{110}/O_2 at 20, 80 and 300 K by monitoring the decrease in the metallic Ni(3d) feature at 0.3 eV below E_F in the UPS. The data are shown in Fig. 70. In the submonolayer stage the decrease should be linear with coverage, provided no oxidation occurs, since the decrease is due to the signal attenuation by inelastic scattering through one atomic thickness of O_{ad}. The 300 K curve shows a region of constant S where the Ni(3d) decrease is linear with

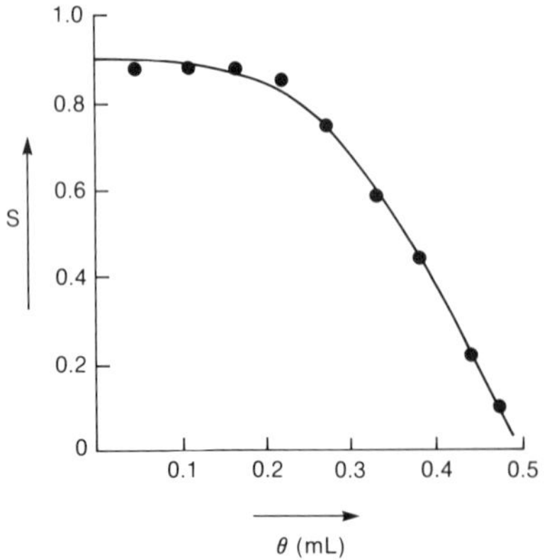

Fig. 69. Sticking probability of O_2 on Ni{110} at 300 K as determined by a microbalance calibrated Auger method (Winkler et al. [241]).

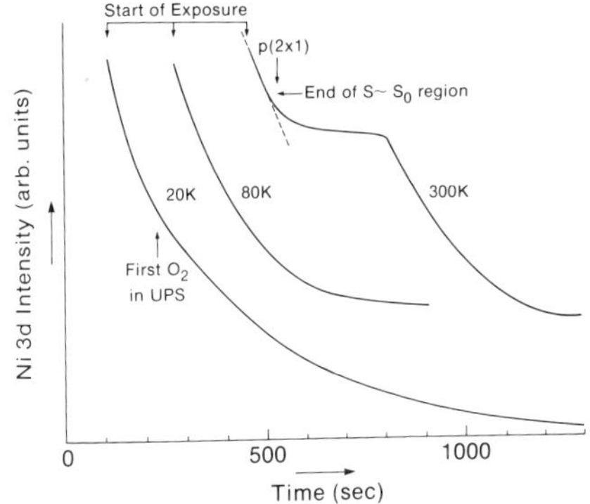

Fig. 70. Ni(3*d*) peak (0.3 eV below E_F) intensity during adsorption on Ni{110} at 20, 80 and 300 K (Jacobi and Rotermund [300]). The sudden rapid decrease at ~800 s in the 300 K curve is due to a pressure increase of a factor of 10.

exposure. The chemisorption plateau stage is then reached. This point should correspond to ~0.4 mL, therefore, which allows us to fix the end of the constant S region at ~0.2 mL, in rough agreement with Winkler et al. [241]. In addition if we assume an S_0 of 1 for 20 K adsorption, the S_0 value at 300 K is ~0.8, again in agreement with Winkler et al. In the Jacobi and Rotermund data, however, S at the end of chemisorption has dropped to $\leqslant$0.03 compared with the ~0.1 value of Winkler et al. at their $\theta \approx 0.4$ mL.

Of the other θ versus exposure plots at 300 K that exist, the AES data of Rieder [269] and of Masuda et al. [225] support the suggestion that $S \approx S_0$ up to $\theta \sim 0.2$ mL. Benndorf et al. [297] claim a different relationship

$$S \propto \left(1 - \frac{\theta}{\theta_{max}}\right)^2 \tag{12}$$

where θ_{max} is the oxygen coverage at the end of the chemisorption stage. It is not clear on what evidence this is based since their published AES uptake data is not extensive enough in the chemisorption regime to derive such a fit. They do give a detailed $\Delta\phi$ versus exposure plot up to the p(2 × 1) LEED region (Fig. 19 of ref. 297). The relationship appears to be linear over the first half of this range (our observation, not theirs) which, if one assumes a constant dipole per O_{ad}, implies a constant S over this range, in agreement with the work discussed above.

In comparison with the Ni{100} kinetics we can say, then, that whereas there was no effect at all of a precursor state at 300 K for Ni{100}, on Ni{110} an O_2 molecule impinging at an O_{ads} occupied site is always able to "diffuse" to an empty dissociation site, rather than desorb, for coverages $\leqslant$0.2 mL. In terms of the potential well diagram of Fig. 61, this implies either that the molecular potential well depth, E_a, for O_2 on O-covered regions is greater relative to the diffusion barrier, E_d, for Ni{110} than for Ni{100} or the dynamics of the trapping process into the molecular state are different in the two cases, even though the well depths may be similar. The former suggestion holds if one can consider the problem in two separate parts; instantaneous energy accommodation (trapping) with unit probability into the molecular state, followed by a competition between diffusion hops to neighboring sites and desorption. This is the standard philosophy, which, as was discussed in some detail with reference to the W{110}/O system is probably far from the truth. The latter suggestion implies that there is more efficient energy accommodation, leading to trapping in the Ni{110} case, compared with the Ni{100} surface where there is no trapping. Since it is the normal component of momentum that must be dissipated for the O_2 molecule to remain on the surface, "trapping" could simply mean transformation of normal to parallel momentum, enabling the molecule to remain on the surface, rather than desorb, and access a wide area quickly. Such a state is not a classical precursor since it is not fully energy-accommodated with the surface (i.e. does not have the surface temperature), but might be termed a "hot precursor" (see the W{110}/O section). It might be expected that such a transformation of normal to parallel momentum would be accomplished much more easily on the very rough Ni{110} surface, especially since impact on O_{ads}-covered regions is involved and we know that the Ni substrate reconstruction occurring there leads to even greater atomic roughness.

Winkler et al. [241] tried to explain their $S(\theta)$ data on the basis of a generalized treatment of the Kisliuk [309] model of precursor adsorption. Instantaneous accommodation on impact into a trapped precursor is assumed and the subsequent possible competing process (dissociation, hopping (diffusion) to a new site or desorption) are assigned probabilities, P_{diss}, P_{diff} and P_{des} for location at a bare site and P^*_{diss}, P^*_{diff} and P^*_{des} at an O_{ad}-covered site. In the original Kisliuk treatment, P^*_{diss} was zero. In the King and Wells [14] treatments of $W\{100\}/N_2$ referred to earlier, $P_{diff} = P^*_{diff}$, $P_{des} = P^*_{des}$ and $P^*_{diss} = 0$ and, of course, ordering effects were properly included. The fully generalized form used by Winkler et al. has six variable parameters and no restrictions related to adatom ordering or to site requirements for dissociation. A good fit to the data was obtained by Winkler et al. but the lack of any sensible way of including ordering effects and geometric site requirements for dissociation, plus the fact that six variable parameters were used, means that any physical insight is lost. In addition some of the parameter values extracted from the fit seem bizzare unless they are misquoted in the paper. For instance, P^*_{diss}/P_{diss} is given as 12.8, which means it is 12.8 times more probable that oxygen dissociates at a covered site than a bare one! Given this unphysical situation, plus the fact that the original assumption of instant accommodation into the precursor state is probably wrong, we conclude that pursuing such a parameterized model further at this point is futile. A general statement about the observed ordering effects on the kinetics, or rather the lack of them, can perhaps be made. The strong influence of the adatom ordering plus the 8-site cluster requirement for dissociation found for Ni{100} clearly has little parallel here. If it were effective, S would have to be much lower than it is at $\theta \approx 0.4$–$0.5\,$mL. It is reasonable to suggest that the ordering and adatom repulsive effects on the kinetics would be less for the rough Ni{110} surface than the smooth Ni{100}. Put differently, the probability for dissociative adsorption could be controlled by more local effects on the rough Ni{110} and more long-range effects on the smooth Ni{100}. Another possibility, however, is simply that the oxide nucleation process is so facile on the Ni{110} surface that $S(\theta)$ is kept high by its early inception and any ordering and dissociative site requirement effects in the chemisorbed layer, which would drive S to a low value by $\theta \approx 0.4$–$0.5\,$mL, are suppressed. A reasonable way to view the comparison between the Ni{100} and Ni{110} behavior is perhaps to note that the Ni{110} 300 K behavior is more like the Ni{100} low-temperature behavior ($S \approx S_0 \approx 1$ up to some fractional coverage and little or no plateau region prior to oxidation). The changes in the Ni{100} behavior on going from 300 K to low temperature, were of course, also ascribed to a mixture of increased precursor lifetime and reduced adatom ordering.

Good ordering of the p(3×1) and (2×1) structures is only observed for slightly elevated temperatures and several of the structural studies of the ordered phases were performed at elevated temperatures. The effect on the kinetics of raising T is qualitatively the same as for Ni{100}; it stretches out the chemisorption plateau considerably such that oxidation is not initiated for several tens of Langmuir exposure. Armour and co-workers [208] studied the kinetics at 500 K by ISS in addition to their structural determination of the p(2×1) geometry [207]. Oxygen coverage was monitored from ISS O peaks in several scattering directions. The decrease of Ni peaks and the growth of a new Ni peak associated with the reconstruction process were also monitored. In the ($\bar{1}1\bar{1}$) and ($\bar{1}10$) directions the O-peak behaviors with exposure are identical, showing a smooth monotonic growth (Fig. 71). It was therefore assumed that in these directions no screening by Ni atoms occurred and the ISS intensity was proportional to coverage. The authors fit the uptake curve to the equation

$$(1 - \theta') = \exp\left(-\frac{2I}{N_0}t\right) \tag{13}$$

where θ' is the coverage relative to that at the completion of the chemisorption stage, N_0 is the number of filled adsorption sites corresponding to that coverage, I is the oxygen flux and t the exposure time. This simply follows the approach of Holloway and Hudson for Ni{100} [231]. Assuming S_0 is 1, the fit gives a value of N_0 corresponding to $\sim 0.2\,$mL, i.e. for $\theta' = 1$, $\theta = 0.2\,$mL. In terms of sticking probabilities eqn. (13) thus translates to

$$S = (1 - 5\theta) \tag{14}$$

The authors themselves suggest that the true N_0 is probably higher than 0.2 mL. If it were 0.25 mL, the equation would become

$$S = (1 - 4\theta) \tag{15}$$

References pp. 381–388

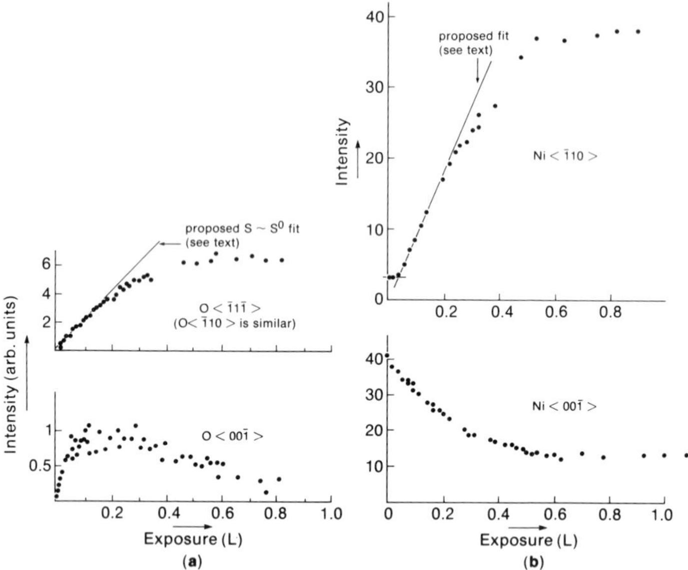

Fig. 71. ISS O and Ni intensities during oxygen chemisorption on Ni{110} at 500 K. (a) O peak heights as functions of exposure along three scattering directions (Van den Berg et al. [208]). (b) Growth of new Ni multiple scattering peak representative of reconstruction (see text), and concomitant decay of single-scattering Ni peak.

Whatever the true value of N_0 (our reading of the literature would place it at ~0.35–0.4 mL as discussed earlier), the functional form claimed at 500 K is different from the $S \approx S_0$ behavior at 300 K. If the 500 K fit is to be believed there is no precursor effect at this temperature. It is not clear, however, that the fit is justified, since a straight line fit in the early stages, corresponding to $S = S_0$, seems equally valid (Fig. 71). More accurate data, and especially an independent assessment of N_0 rather than leaving it as a variable parameter, are required to distinguish the two functional forms. The (001) direction O ISS peak behavior is quite different from the $(\bar{1}1\bar{1})$ and $(\bar{1}10)$ behavior (Fig. 71). It initially rises, but saturates between 0.05 and 0.1 L exposure and then decreases again. The early saturation and subsequent decrease are associated with the onset of the reconstruction process which produces a strong screening of O atoms because they are located between Ni atoms in the (100) direction. The onset of the screening occurs at $\sim\frac{1}{6}$th the saturation chemisorption coverage, or $\theta \approx 0.04$–0.06 mL, suggesting that prior to this coverage O_{ad} atoms are located in unscreened positions such as the rectangular site of Fig. 65, V. This agrees with our interpretation of the ELS data, discussed in Sect. 3.2.3, that initial adsorption is on the unreconstructed surface, but that both p(3 × 1) structures and the p(2 × 1) structure are Ni reconstructed surfaces.

The growth of a new Ni multiple scattering peak in the (110) direction as a function of oxygen exposure [Fig. 71(b)] is associated with the reconstruction process [208]. The peak is at lower scattered energy than the equivalent process peak for the clean surface because the multiple scattering influence has been reduced owing to the density of Ni atoms in the (110) direction being halved (see reconstruction models of Fig. 65). There is a strong correlation between the growth of this Ni reconstruction peak and the O(001) peak behavior of Fig. 71(a). Thus, over the first ~ 0.05 L exposure, virtually no growth of the Ni reconstruction peak is observed and it is over this exposure range that the O(001) peak grows normally. Beyond this exposure there is a sharp rise in the Ni reconstruction peak intensity which coincides with the saturation and then decrease in the O(001) intensity, signifying that both effects represent the onset of the same reconstruction process.

The final mechanistic model proposed by Armour and co-workers [208] for the "missing row" reconstruction process seems very reasonable. They suggest that surface Ni atoms diffuse along (110) channels (facile at 500 K) and are trapped by O_{ad} atoms. Owing to the longer range attractive and repulsive forces, these Ni and O atoms eventually build up added rows in the (001) direction. Obviously to form a high-coverage structure this must require massive Ni transport and it is therefore likely that the reconstructed surfaces are not flat, but consist of steps and terraces rather like the W/O reconstructed surfaces for W{100}/O (Sect. 2.6.3).

The only data on the kinetics at low temperatures for Ni{110}/O is the UPS work of Jacobi and Rotermund [300] and the AES/$\Delta\phi$ study of Benndorf et al. [297]. Both studies indicate that at 80 K, S does not drop significantly as the end of the chemisorption stage is approached, i.e. the precursor effect is extended. This is, of course, what happens for Ni{100} also and is ascribable to an increased precursor lifetime and an elimination of any adatom ordering effects which would otherwise reduce the availability of specific empty site clusters.

Summarizing this section on chemisorption kinetics, it is apparent that our understanding is less complete than for Ni{100}/O, which is not surprising considering the complex Ni reconstruction that occurs. The qualitative conclusions, however, are that the molecular precursor is effective to higher temperatures on Ni{110}, resulting in a much faster approach to saturation chemisorption coverage and the onset of oxidation. This effect is compatible with the greater roughness of the Ni{110} surface. The proposed mechanism for the reconstruction process, trapping of diffusing Ni surface atoms by O_{ads} to build up added Ni/O rows in the (001) direction is compatible with all the existing experimental data.

3.2.12 Kinetics and mechanism of oxide nucleation and island growth

Many of the aspects of oxide nucleation and growth on the {110} surface are similar to those on {100} so the discussion here will concentrate on comparative aspects and on any information available for Ni{110}/O that is not available for Ni{100}/O.

Recapping on the Ni{100} oxidation we had concluded that the phenomenological sequence of 2–3 layer NiO nucleation in small islands, followed by lateral island growth to saturation coverage, as proposed by Holloway and Hudson for 300 K oxidation, was basically correct [231]. We had also concluded, however, that fitting the kinetic data to their model kinetic equation (Sect. 3.1.9) in no way proved that the mechanistic steps of their nucleation and island growth model were correct. Derived values of $K_i N_0$, where K_i is a rate constant term and N_0 a nucleation site density, will not vary greatly even if the oxidation rates vary greatly because of the exponential form of the fitting. The Holloway and Hudson interpretation of the data in terms of their model implies a mechanistic detail for the oxide nucleation growth which is highly questionable. The most contentious point was the conclusion that N_0, the nucleation site density, was fixed at time zero and was independent of temperature. This implied that the way in which nucleation and subsequent island growth occurred depended entirely on the original (time zero) condition of the Ni{100} substrate (number of defects, perhaps) and was independent of the conditions in the chemisorbed O_{ad} layer prior to oxide formation. Variation in $K_i N_0$ with T was then ascribed to a negative activation energy term in K_i representing the difference between the precursor adsorption energy and an energy barrier for oxygen diffusion or capture by the oxide islands at their perimeter. Alternative evidence, however, was presented by us to indicate that N_0 was a function of T during the chemisorption stage and that this was why $K_i N_0$ varied with T rather than because of an activation energy term in K_i. Holloway and Hudson assumed that diffusion of the oxygen precursor to island perimeters was an important contribution to the island growth kinetics. In fact oxide capture was considered by Holloway and Hudson to be rate-limited either by the diffusion process or by an activation barrier at the island perimeter following diffusion. On the basis that, at 300 K, no evidence for an effective precursor could be found during the chemisorption stage, we found this assumption hard to believe and concluded that island growth at 300 K and above occurred only by direct perimeter impingement.

We now review the situation as it applies to Ni{110}. Holloway and Hudson did not study the {110} surface in their original LEED/Auger work, but Mitchell et al. [228] showed by X-ray fluorescence/RHEED that the general form of the kinetics; fast chemisorption, plateau, oxidation also applied to the {110} surface. Since then, of course, many studies have confirmed the main qualitative distinction in the oxidation stage between the two surfaces in that the oxidation growth process starts at an earlier exposure and is more rapid on the {110} surface (see Fig. 63). We suggested in the previous section that the reason for the earlier nucleation start, in terms of exposure, was simply that the chemisorption completion was faster, i.e. oxide nucleation does not start at an earlier *coverage*. The implications of the faster oxidation once nucleation has initiated need exploring in more detail, however. Mitchell et al. fit their data to the usual Holloway and

Hudson island growth equation and extracted $K_i N_0$ values ranging from 6.9×10^9 at 313 K to 1.6×10^8 at 423 K (i.e. the oxidation slows considerably at high temperature. Data at higher than 423 K are not considered because of the possibility of O diffusion into the bulk). Table 9 (p. 316) compares values for the determinations of $K_i N_0$ for Ni{110} with Ni{100}. We note that at 300 K $K_i N_0$ is more than an order of magnitude greater for Ni{110} than for Ni{100}, but that as T is increased this difference is reduced. The variation in $K_i N_0$ with T is due, within the Holloway and Hudson model, to a variation in K_i through the activation energy term, as explained above. N_0 is considered to be constant with T. Unlike their work on Ni{100} Mitchell et al. did not perform any temperature-jump experiments to test the T-dependence of $K_i N_0$. As a test of the claims by Mitchell et al. [227] and Hopster and Brundle [237] for Ni{100}/O that N_0 was T-dependent during the chemisorption stage, Holloway and Outlaw [233] performed temperature-jump experiments for Ni{110}/O using AES to establish the T-dependence of $K_i N_0$. They concluded that the effects of T changes during the oxidation stage were in accordance with their model for oxide growth, i.e. a temperature change after the initiation of oxidation produced an oxidation rate appropriate to the new temperature, as would be expected if K_i contained an activation energy term and N_0 was independent of T. These results are in complete disagreement with the T jump experiments for Ni{100} where it was shown that T changes after oxide initiation had no, or much reduced, effect on the subsequent oxidation rate, i.e. the rate was determined entirely or partly by the temperature during chemisorption, indicating a strong dependence of N_0 on T during that stage and that N_0 only became fixed at the end of the chemisorption stage. The disagreement implies either that there are artifacts in the results for one of the two surfaces, or there are genuine differences in behavior. We do note that the Holloway and Outlaw [233] data at 450 K for Ni{110} seem somewhat atypical in that evidence for a well-defined plateau is missing, whereas it is quite clear in the RHEED/X-ray fluorescence data of Mitchell et al. [227] and the RBS data of Smeenk et al. [212] at elevated temperatures. This, however, does not invalidate their general conclusion that an activation energy-like behavior is found during oxidation. They were even able to demonstrate that the temperature behavior fitted reasonably well to an Arrhenius plot and extract an effective activation energy of 5.5 kcal mol^{-1}. In Sect. 3.2.10 we have already concluded that there is a genuine increase in the precursor effectiveness for Ni{110} compared with Ni{100} during chemisorption. This is exactly what is required to explain the observed differences between Ni{100} and Ni{110} in the oxidation stage as well. If the oxidation is done in a temperature range where the precursor is effective, then the Holloway and Hudson mechanism for island growth is plausible since a central tenet of that mechanism is the diffusion of O_2 to growing islands where it is incorporated into the oxide. The decrease in oxidation rate between 300 and 450 K for Ni{110} is then ascribable to a decrease in lifetime of the precursor and the effective activation energy term extracted from an Arrhenius plot relates predominantly to the O_2 diffusion process. For Ni{100} there is already no precursor effectiveness at 300 K and therefore raising the temperature during the oxidation process to 450 K will have no effect, as was observed. At 77 K the precursor is effective on Ni{100} and therefore raising T to 300 K during oxidation should have an effect, as observed. The effect is, however, smaller than if the entire chemisorption and oxidation had been done at the new temperature, indicating that the value of N_0 was also dependent on the temperature during the chemisorption stage.

Our tentative conclusion, then, is that there is evidence for both an activation energy term in K_i and a T dependence of N_0 during chemisorption. The conditions under which one or the other temperature dependence becomes dominant are different for Ni{100} and Ni{110}. For Ni{100}, T must be low for precursor diffusion to play a role. N_0 appears to vary strongly with T during chemisorption, probably because well-ordered chemisorption on the surface does not cause any movement in the substrate atoms so that nucleation sites have to be generated by disorder in the overlayer leading to a $\theta_{local} > 0.5$ mL, which is unstable with respect to NiO formation. For Ni{110} any variation of N_0 with variation in T during chemisorption is not evident, probably because N_0 is always made high by the facile reconstruction process occurring in the presence of oxygen. Disorder in the overlayer may thus be of minor significance in creating further nucleation sites. In addition, precursor O_2 is more effective on Ni{110} producing the observed Arrhenius behavior in K_i, even above 300 K. If the precursor is effective on Ni{110} at 300 K it will, of course, be more effective at 80 K, as is indicated by the very fast oxide kinetics at 80 K, just as was the case for Ni{100}. Another way of expressing this would be to say that at sufficiently low temperature the kinetic differences between Ni{100} and Ni{110} caused by structural effects either of the substrate or of the adlayer are eliminated. What remains unexplained is why Ni{110}, which has the faster kinetics above 80 K, has a limiting oxygen coverage at 80 K which is apparently only ~70% that at 300 K, whereas Ni{100} has the same limiting coverages at both temperatures. We have suggested earlier that the 2D "pseudo-oxide" geometry observed to precede NiO formation at 300 K on Ni{110} is responsible, being retained at low temperature instead of converting to 2–3 layer NiO. For Ni{100} no such stable 2D "pseudo-oxide" is formed, presumably because the lattice dimensions are unfavorable for the formation of such a structure. The chemisorbed O structures thus pass directly to the 2–3 layer NiO structure by island nucleation. The idea that on Ni{110} lower temperature favors formation of a stabilized 2D oxide structure, whereas this effect is less marked for Ni{100}, is supported by the RBS data [216]. Both the backscattered H^+ energy distributions and the determinations of the ratio of displaced Ni atoms per O atom as a function of θ support the idea that for Ni{110} the basic model of rapid in-depth nucleation of 2–3 layer NiO, followed by lateral growth of the NiO islands starts to break down as T is lowered. It changes towards a model where there is rapid 2D growth followed by an in-depth thickening. This, of course, is qualitatively what would be expected in a situation where lateral growth is accelerated by low T (precursor effect) but in-depth nucleation is suppressed by low T (stable 2D "pseudo-oxide" effect).

3.3 Ni{111}

The reaction scheme of Ni{111}/O_2 is very similar to that of Ni{100} and so we concentrate here on a comparison between the two. Where the same effects and same explanations are appropriate for both surfaces only concise summaries are given for Ni{111} and it is expected that the reader will consult the Ni{100} section for justifying detail. One factor that encourages comparison with the {100} surface, rather than {110}, is that neither the {100} or {111} surfaces suffer the massive reconstruction effects during the ordered chemisorption regime that occur for Ni{110}. This observation is in accord with expectations, given the order of close packing of these surfaces. The amount of data available for Ni{111} is much less than for either of the other two surfaces. This is rather surprising since it is the close-packed surface and might therefore be expected to be the first and best studied. The only explanation we can offer is that things have not proven so controversial for Ni{111} so less volume of material has been generated.

3.3.1 General reaction scheme. 300 K and above

The reaction of oxygen with Ni{111} has been followed up to the passivation stage by LEED [195, 232, 235, 250, 301, 311, 312], AES [195, 232, 241, 269], UPS [229, 243, 250], XPS [243], RBS [215] and SIMS [269]. The general kinetic behavior is the same as for Ni{100}: rapid chemisorption ending with a plateau in the coverage versus exposure plot (Fig. 72), NiO nucleation and oxide growth, during which S increases again, until a passivating film covers the surface, and slow oxide thickening beyond this. The main differences from Ni{100} are (i) the chemisorption kinetics to the plateau are faster, (ii) the plateau region is shorter, (iii) the oxide growth kinetics after oxide nucleation are faster, and (iv) there is a suggestion that the passivating thickness corresponds to $\theta = 3$ mL [215] rather than 2, though the Holloway and Hudson study [232] of Fig. 72, plus at least one other study [243] claim 2 mL in agreement with the Ni{100} situation. The generally faster kinetics on the {111} surface compared with {100} might be thought anomalous since {111} is the more close packed. Plausible explanations are available, however, as discussed later. At elevated temperatures O will dissolve into the bulk, as on the other two surfaces. For Ni{111} the temperature at which this occurs is ~ 500 K, apparently independent of coverage [235]. The phase diagram for coverages between 0.1 and 0.34–0.38 mL (the oxidation onset coverage) and temperatures between 150 and 500 K has been studied in some detail [235] in contrast to Ni{100} [234] where

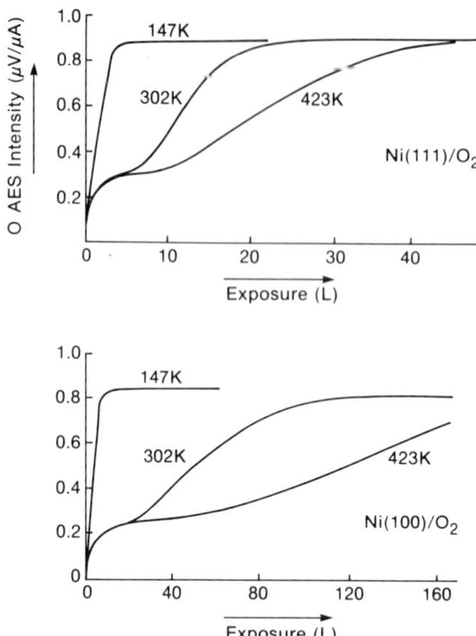

Fig. 72. Comparison of oxygen adsorption kinetics on Ni{111} [232] and Ni{100} [231]. Methods and instrumentation are the same for both cases. Note the difference in exposure scales.

information was sketchy because dissolution into the bulk interfered with attempts to study the chemisorption phase diagram. Two transitions, identified in the phase diagram study, clarify apparent anomalies in previous LEED reports of the chemisorption regime. For temperatures above 300 K the p(2 × 2) O structure is observed only between $\theta \approx 0.21$ and 0.25 mL. For coverages below 0.21 mL and T above 300 K, O_{ad} exists as a lattice gas. At coverages above ~0.28 mL some authors report a $\sqrt{3} \times \sqrt{3}$R30° structure, and some do not. This disagreement apparently comes about because of a sharp disordering transition from the $\sqrt{3} \times \sqrt{3}$R30° structure right at 300 K [235].

Early O_2 adsorption studies on Ni{111} erroneously suggested that a molecular chemisorbed phase might exist at low coverages at 300 K [195]. As is the case for Ni{100} and {110}, only dissociated chemisorbed O_{ad} is formed for temperatures at least as low as 80 K. The chemisorption geometry for the p(2 × 2) O structure has been determined from LEED IV data and theory [196], with support from other techniques [215, 222].

3.3.2 General reaction scheme. Low temperature

Adsorption at low temperatures (down to 80 K) has been studied [232, 243] and the scheme is similar to Ni{100}; fast chemisorption (no long-range order at 80 K), with little decrease in S, followed by rapid NiO formation (no long-range order). The passivation coverage appears to be the same as at 300 K, as was the case for Ni{100} but not for Ni{110}. There is a report [222] that at "low temperature" (not specified) a diffuse LEED (1 × 1) pattern is formed. Since the phase diagram study of Kortan and Park [235] goes down to 150 K and no mention of a (1 × 1) phase is made, it seems likely that the (1 × 1) structure is from the substrate showing through an overlayer which has no LEED pattern.

3.3.3 Phase diagram

Since order–disorder transitions occur right around 300 K where most geometry, electronic structure or kinetic studies have been made, and because a fairly detailed description of the phase diagram is available, we have chosen to discuss the phase diagram prior to dealing with geometric, electronic and kinetic effects. To discuss things sensibly in this sequence it is necessary to anticipate the actual real-space structures to which the observed LEED patterns correspond. The justification of these structures is given in Sect. 3.3.4.

The suggested phase diagram for Ni{111}/O is shown in Fig. 73 [235]. It was assumed, on the basis of LEED IV measurements [195, 196], that the O adatoms in the p(2 × 2) structure occupy threefold hollow sites [Fig. 74(b)], and the coverage scale was established by assuming that the maximum intensity p(2 × 2) pattern corresponded to $\theta = 0.25$ mL, i.e. a perfect p(2 × 2) lattice. Between $\theta = 0.1$ and 0.21 mL the (1/2, 0) LEED spots of the p(2 × 2) O structure are very weak and broad and are only observed below a critical temperature, T_c, which is close to 300 K at $\theta \approx 0.18$ mL. The weakness and breadth of the spots indicates that the structure exists as small islands surrounded by a lattice gas, as does the fact that the p(2 × 2) can be observed at all at $\theta = 0.1$ mL. Islands in turn indicate an attractive third-nearest-neighbor adatom interaction, as discussed earlier in this chapter. The order–disorder transition temperature, T_c, in Fig. 73(a) was experimentally determined by monitoring the intensity of the (1/2, 0) LEED spot as a function of T. The transition corresponds to the dissolution of the p(2 × 2)

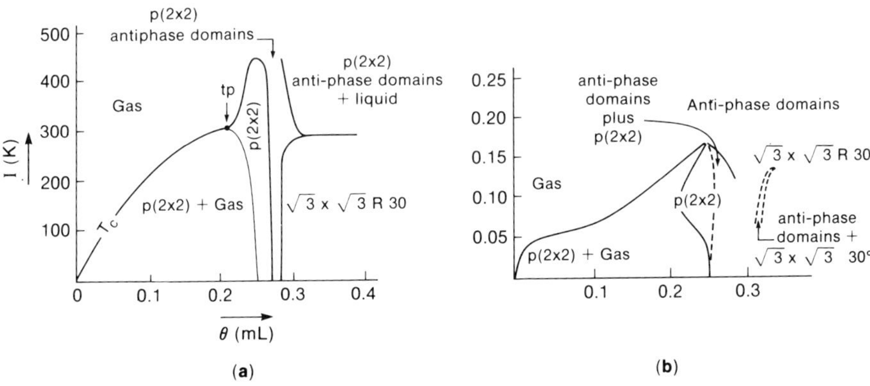

Fig. 73. Phase diagram for Ni{111}/O. (a) Experimentally determined (and extrapolated) from LEED from Kortan and Park [235]; (b) theoretical (Monte Carlo simulation) [313].

islands as the attractive third-nearest-neighbor interaction is overcome by kT. Between $\theta = 0.21$ and 0.25 mL the (1/2, 0) spots below T_c become rapidly very strong and sharp as the p(2 × 2) islands plus lattice gas phase gives way to a single p(2 × 2) plus vacancies phase. At $\theta = 0.25$ mL, T_c is 150 K higher than at $\theta = 0.21$ mL because now all p(2 × 2) sites are filled and the disorder transition requires some O atoms to move into repulsive nearest-neighbor positions. Though the detailed phase diagram of Fig. 73(a) was derived by Kortan and Park [235], the order–disorder transition at and below $\theta = 0.25$ mL was originally observed by Davisson and Germer [311] and studied in more detail by Germer and MacRae [301]. Unfortunately, Germer and MacRae's interpretation of the real-space structure corresponding to the p(2 × 2) LEED pattern was quite incorrect. It was done in the days prior to dynamical LEED calculations and, based on the incorrect assumption that the $\frac{1}{2}$-order LEED beams were too intense to be accounted for by scattering entirely from O atom, they concluded that the p(2 × 2) pattern derived from a mixed Ni–O 2D layer, i.e. the p(2 × 2) pattern was caused by missing Ni atoms in the top substrate layer in a (2 × 2) configuration, with the further assumption that O atoms filled these gaps.

So far the general phase diagram behavior and explanation reported above are quite similar to those of the p(2 × 1) W{110}/O system. Above $\theta = 0.25$ mL complications occur. Between $\theta = 0.25$ and 0.29 mL splitting and streaking of the p(2 × 2) pattern is observed, interpreted by Kortan and Park as due to the formation of anti-phase domains of p(2 × 2) incorporating the excess O above 0.25 at the domain boundaries [Fig. 74(b)]. Between $\theta = 0.29$ and 0.38 mL, a sharp $\sqrt{3} \times \sqrt{3}$R30° structure [Fig. 74(a)] is observed provided T is below ~300 K. The suggested real-space structure is shown in Fig. 74(b). Since a perfect $\sqrt{3} \times \sqrt{3}$ structure corresponds to $\theta = 0.333$ mL the situations between 0.29 and 0.33 mL and between 0.33 and 0.38 mL cannot be a purely $\sqrt{3}$ phase. In fact between 0.29 and 0.33 mL the $\sqrt{3}$ structure coexists with the anti-phase domain p(2 × 2) structure. Above 0.33 mL it is possible that the $\sqrt{3}$ structure coexists with incipient NiO nuclei. Kortan and Park [235] claim that oxidation starts only above $\theta = 0.5$ mL but offer no evidence for this and in fact in an earlier publication the same authors refer to the oxidation onset as being at $\theta = 0.4$ mL [313]. Other estimates (see later) are that $\theta = 0.34$–0.38 mL is the nucleation point.

Raising T above ~300 K in the 0.29–0.38 mL range causes the $\sqrt{3} \times \sqrt{3}$ structure to convert to the split/streaked p(2 × 2). The existence of this phase boundary right at 300 K explains why some authors working at a nominal 300 K see the $\sqrt{3} \times \sqrt{3}$ pattern and some do not. The transition does not seem to correspond to the overcoming of a single attractive term, as was the case below $\theta = 0.25$ mL. Comparison of the real space structures in Fig. 74(b) show that the transition actually corresponds to moving adatoms from less favorable second-nearest-neighbor sites to attractive third-nearest-neighbor positions, but with the excess O so generated located at domain anti-phase boundaries and, supposedly, in a "liquid" phase [235].

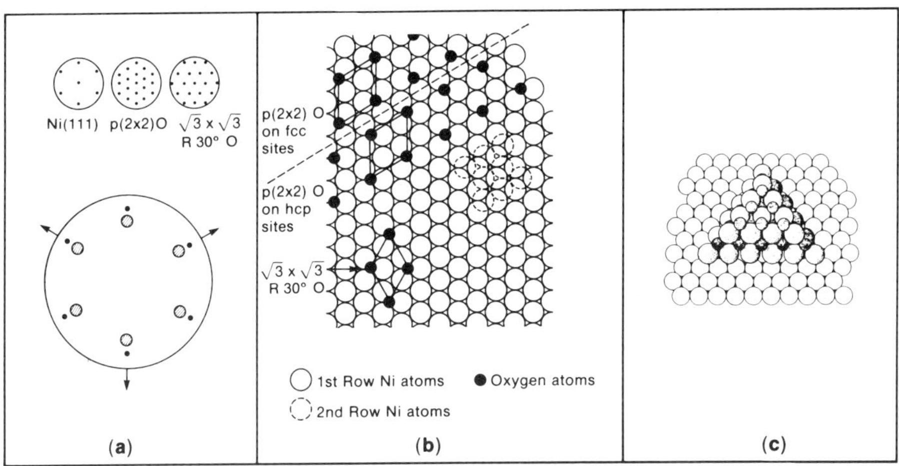

Fig. 74. LEED patterns and real space structure for Ni{111}/O. (a) LEED patterns showing the clean Ni{111} hexagonal pattern, the sharp p(2 × 2) O formed at $\theta = 0.25$ below 450 K, and the sharp $\sqrt{3} \times \sqrt{3}$R30° structure observed for $\theta > 0.29$ below 450 K. At intermediate coverages between 0.25 and 0.29, complicated streaked and split patterns resulting from the formation of anti-phase domains occur [235]. The LEED pattern for adsorption at 500–600 K into the oxidation regime is shown at the bottom [312]. The outer ring contains the original Ni{111} spot, the inner diffuse spots being due to NiO{111}. The same pattern is found for saturation exposure at 300 K [232, 250, 269]. The arrows mark the zones in which three additional NiO spots were observed at 500–600 K and which move away from the origin on decreasing the LEED beam energy [312]. (b) Real space structures. p(2 × 2) O sublattices existing on both h.c.p. and f.c.c. threefold hollow sites are shown. The boundary between them, represented by the broken line, is an anti-phase domain boundary. A $\sqrt{3} \times \sqrt{3}$R30°O unit cell is also shown [235]. (c) The pyramidal NiO with three {100} faces exposed found to grow on Ni{111} by MacRae [312]. Open circles represent the Ni{111} substrate. Shaded circles represent the oxide (small circles Ni^{2+}; large circles O^{2-}).

Summarizing, we can say that up to $\theta = 0.25$ mL the phase diagram behavior is quite clear and well explained. Between $\theta = 0.25$ and 0.38 mL complex phases exist, the boundaries of which have been mapped. Though the proposed geometries for these phases are apparently consistent with the LEED behavior, it is not really clear from the experimental data alone what adatom interactions are causing the transitions. Neither is it certain that NiO nucleation is not interfering above $\theta = 0.34$ mL. Roelofs et al. have attempted to simulate the experimental phase diagram by Monte Carlo [313]. They considered six adatom pair interactions (Fig. 75). E_1, E_2, and E_3 are the first-, second- and third-nearest-neighbor interactions if one breaks the hexagonal lattice into two sublattices. This is necessary because there are two types of threefold site on the surface, h.c.p. sites where there is a Ni atom from the second Ni layer immediately below, and the f.c.c. sites with no second layer Ni atom below [see Fig. 74(b)]. Both sublattices can apparently be occupied, at least in the 0.25–0.29 mL range, since the anti-phase domain p(2 × 2) structure [Fig. 74(b)] could not otherwise exist. Roelofs et al. [313]

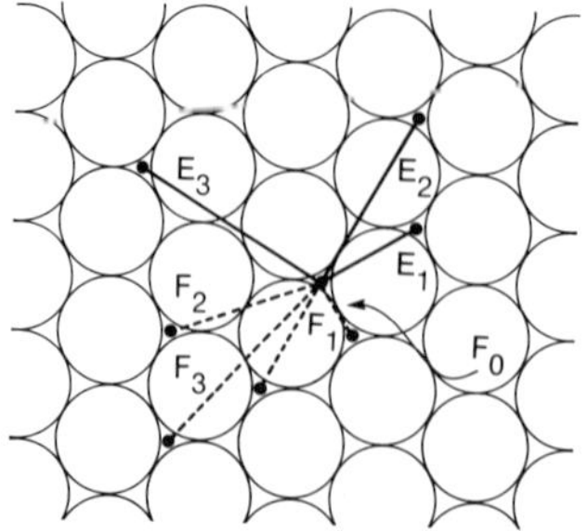

Fig. 75. Pairwise adatom interaction included in the Monte Carlo simulation of the phase diagram for Ni{111}/O (Fig. 73). "E" denotes interaction between sites on the same sublattice, "F" between sites on different sublattices. Taken from Roelofs et al. [313].

therefore also included the three interactions F_1, F_2 and F_3, which are the counterparts of E_1, E_2 and E_3 for two adatoms on different sublattices. F_0 was set to $+\infty$ since the distance, 0.73 Å, is so short. F_3, the longest-range interaction was set to zero to simplify the simulation. In addition to the five remaining pair terms, the energy difference, E_0, between atoms in an h.c.p. sublattice and in an f.c.c. sublattice had also to be included. This term, which represents the difference in O_{ad} bonding to the substrate with and without a second-layer Ni atom directly beneath, is, however, small compared with the pair-interaction terms. The best phase diagram "match" to experiment they were able to produce by judicial choice of interaction terms is shown in Fig. 73(b). It reproduces many of the features of the experimental results. The one anomaly that could not be rectified, however, was the failure to produce the tricritical point, tp, of Fig. 73(a). It was suggested that a possible explanation for this anomaly is that below $\theta = 0.25$ mL only one sublattice can be occupied (i.e. E_0 varies with coverage). A related suggestion, or an explanation for how this could come about, is that the Ni substrate atoms relax from their normal positions under the influence of the oxygen and that the manner of this relaxation is different above and below $\theta = 0.25$ mL. This information, which comes from RBS measurements [215], is discussed in the next section.

3.3.4 Geometric structure of chemisorbed species

Structural determinations for Ni{111}/O are rather sparse compared with Ni{100}/O and Ni{110}/O. The original electron diffraction paper by Davisson and Germer in 1927 [311] reported the p(2 × 2) O structure and its order–disorder phase transition. In the mid-sixties Germer and MacRae [301] reported the p(2 × 2) O and $\sqrt{3} \times \sqrt{3}$R30° structures prior to NiO formation, whereas Park and Farnsworth [314] observed only the p(2 × 2) prior to oxidation. Extensive LEED IV data for the p(2 × 2) structure were reported by Demuth and Rhodin in 1974, together with comparison with such data on the other Ni faces and qualitative suggestions for interpretation [195]. Detailed IV calculations were subsequently reported by Marcus et al., and the interpretation of the p(2 × 2) geometry rests on these results [196]. No further LEED IV work has been reported since 1975, but important RBS [215] and HRELS [222] measurements have been made.

Prior to the IV calculations, Germer and MacRae [301] had originally interpreted their experimental LEED IV data in terms of a missing Ni atom p(2 × 2) structure. As mentioned in the previous section, this interpretation was based on the erroneous belief that the half-order LEED spots were too intense to originate from O atom scattering. Demuth and Rhodin [195] showed that $\Delta\phi$ was positive and directly proportional to O coverage (determined by AES) up to at least completion of the p(2 × 2) structure. A value of + 0.7 V was reported at this coverage. The subsequent work by Kortan and Park [235] confirm this linearity but places the p(2 × 2) value at + 0.5 V. Capehart and Rhodin give a value of + 0.65 V [229]. Neither Demuth and Rhodin nor Kortan and Park comment on the behavior of $\Delta\phi$ at higher coverages. Capehart and Rhodin, however, indicate that $\Delta\phi$ increases from + 0.65 to + 0.75 V at the $\sqrt{3}$ LEED stage. Demuth and Rhodin noted that their 0.7 V value was twice that for similar a coverage on Ni{100} and on this basis and some slight differences in O AES line shapes suggested that the bonding on Ni{111} must be quite different from {100}. Either a molecular phase or a much more weakly bound atomic phase were suggested. As discussed earlier, neither of these suggestions is correct. Kortan and Park pointed out that, since $\Delta\phi$ at 310 K is proportional to O coverage over the whole range up to the well-ordered p(2 × 2) structure, it is clear that no chemical environment changes are occurring for O_{ad} in going from the low-coverage lattice gas arrangement to the long-range ordered p(2 × 2) structure. One can extrapolate from this that no local site changes occur either, i.e. O_{ads} is in the same site throughout with the same $d_\perp$. A brief ISS report [315] provides confirmation of the $\Delta\phi$ interpretation that O_{ads} is in an overlayer environment since strong Ni shadowing effects were observed.

The results of the LEED IV calculations gave a best fit for the p(2 × 2) structure with O in the threefold hollow site at a $d_\perp$ of 1.20 Å [196]. The LEED IV calculations are not capable of distinguishing between the h.c.p. and f.c.c. sites. The $d_\perp$ value of 1.20 Å yields a Ni–O bond length of 1.88 ± 0.06 Å. The 0.9 Å $d_\perp$ value for p(2 × 2) O on Ni{100} corresponds to a Ni–O bond length of 1.98 ± 0.05 Å. Thus it is clear that the more close-packed arrangement of Ni{111} causes O_{ad} to sit higher above the plane of atomic nucleii in the threefold site than the fourfold site on Ni{100}. The longer $d_\perp$, with no significant difference in Ni–O bond length, indicates that the greater $\Delta\phi$ value for p(2 × 2) O/Ni{111} compared with p(2 × 2) O/Ni{100} is simply due to the larger dipole separation caused by $d_\perp$ and not to a significant difference in charge distribution in the Ni–O bond.

No LEED IV data or calculations have been reported for the $\sqrt{3} \times \sqrt{3}$R30° structure. Indeed, the structure drawn in Fig. 74(b) seems to have been suggested merely as a plausible LEED pattern and O-coverage compatible structure. Subsequent RBS [215] and HRELS [222] work do tend to confirm its correctness, however, though no determined value for $d_\perp$ exists, as yet. The high-energy RBS study of Narusawa et al. [215] monitored the Ni surface peaks in the normal (111) and off-normal (110) channeling directions at energies between 0.5 and 2 MeV He^+ as a function of oxygen exposure. Coverage was determined by AES using a calibration against the backscattered O signal. As in the medium-energy RBS work of Saris and co-workers on the Ni{100} and {110} surfaces [209–213], the positions of the O adatom cannot be directly determined but any change in position of the top Ni atoms as a function of oxygen coverage can be monitored very sensitively. This information can then be correlated with the LEED behavior. The coverages of the sharpest p(2 × 2) and $\sqrt{3} \times \sqrt{3}$R30° structures were found to be 0.2 ± 0.04 mL and 0.31 × 0.05 mL, respectively. The corresponding exposures were ~2 and ~5 L. Saturation O coverage, for which NiO was formed (see later) corresponded to 2.8 ± 0.4 mL O at an exposure of ~50 L. For the clean Ni{111} surface the absolute intensity of the (110) channeled He^+ surface peak and its angular variation around the (110) direction both indicated that the surface Ni layer is bulklike with any relaxation or reconstruction limited to < 0.02 Å. On exposure to oxygen at 300 K the off-normal (110) direction surface peak showed increases of intensity as the p(2 × 2) and then the $\sqrt{3} \times \sqrt{3}$R30° structures were formed, whereas the normal direction (111) measured surface peak showed no change. The only explanation compatible with this behavior is that an outward relaxation of top layer Ni atom is occurring. Since the magnitude of the surface peak increase is a function of both the size of the outward relaxation and the fraction of surface Ni atoms involved, the absolute value of the increase does not uniquely determine the Ni top layer geometry. For the $\sqrt{3} \times \sqrt{3}$R30° structure two models were found to give equally good fits to the data. One assumes that one-third of the Ni atoms expand 0.30 Å; the second assumes that all Nis expand to 0.15 Å. An attempt was made to distinguish between these possibilities by determining the angular profile of the (110) scattered beam. A significantly better fit was obtained for the case with all Ni atoms involved in the expansion. This tends to confirm the correctness of the assumed threefold hollow location of O_{ads} for the $\sqrt{3}$ structure, since every Ni is equally bonded to O_{ads} in the arrangement. If the O adatom were in an on-top site, one might expect that one-third of the Ni atoms (those directly below O_{ads}) would be affected differently from the other two thirds.

Similar considerations for the p(2 × 2) O structure indicated a best fit for a model in which three-fourths of the Ni atoms were expanded by 0.15 Å with the remaining one-fourth not affected. This can be rationalized from Fig. 74(b) where it can be seen that one-fourth of the Ni atoms are not involved in bonding to O_{ads}. Thus, without directly monitoring O_{ads}, but rather its effect on the top Ni layer, confirmation of the LEED derived location of O_{ads} is obtained. In addition, as pointed out by the authors [215], for the p(2 × 2)

O structure the change in IV characteristics compared with those of the clean surface are not just due to the O overlayer but also to the reconstruction (outward relaxation) of three-fourths of the Ni atoms. This, of course, casts some doubt on the accuracy of the value of $d_\perp$ of 1.20 Å obtained from the LEED IV calculations, since no Ni relaxation was included in those calculations.

The interpretation of the HRELS-obtained vibrational spectra for Ni{111}/O [222] is somewhat controversial, as was the case for Ni{100}/O. In the Ni{111} case, however, no unusual bond-lengths are required in lattice dynamic calculations to fit the data. The uncertain interpretation refers instead to why certain Ni surface phonon frequencies can be excited in the presence of some adsorbate structures but not others. Allan and Lopez [223] calculated lattice vibrational spectra for O on Ni{111} in the p(2 × 2) and $\sqrt{3} \times \sqrt{3}$R30° structure [and p(2 × 2) O and c(2 × 2) O on Ni{100}], using a 21 layer slab and a tight-binding band structure model which included only Ni *d* and O *p* states. The calculation has 8 parameters, 6 of which were fixed by fitting to bulk Ni properties (phonon dispersion and bulk cohesive energy relationships). The remaining 2, which represent Ni/O hopping integrals and repulsive potential terms, were adjusted to give the best fit, for one pair of parameter values for all observations, to the observed experimental frequencies and $d_\perp$ values for the p(2 × 2) and $\sqrt{3} \times \sqrt{3}$R30° Ni{111} cases and the p(2 × 2) and c(2 × 2) Ni{100} cases. Good fits to the experimental data were obtained for all cases except c(2 × 2) O/Ni{100}, which, as discussed earlier, involves O–O lateral interactions. The fits to the experimental spectra for the p(2 × 2) O and $\sqrt{3} \times \sqrt{3}$R30° O structures on Ni{111} with O_{ad} occupying the threefold sites are shown in Fig. 76 for a $d_\perp$ value of 1.26 Å. There is no difficulty, then, in reproducing, for both LEED structures, the localized Ni–O vibrational frequency and the Ni surface phonon frequencies using a $d_\perp$ value very close to that determined by LEED for the p(2 × 2) O structure.

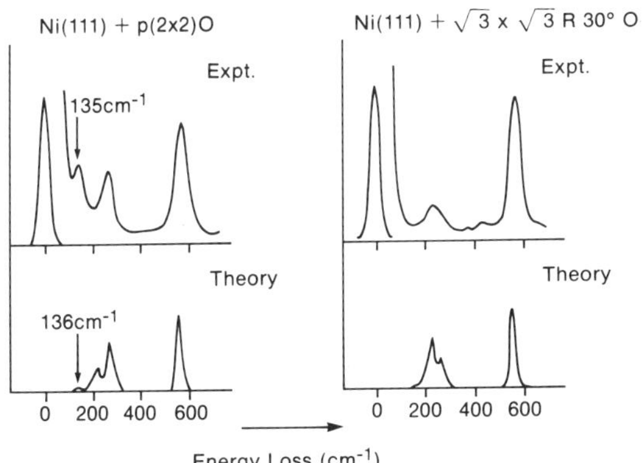

Fig. 76. Comparison of HRELS data for p(2 × 2) O and $\sqrt{3} \times \sqrt{3}$R30° O on Ni{111} [222] with lattice dynamical theory parameterized for bulk Ni. The fit is for an assumed $d_\perp$ of 1.26 Å [223].

This result is the only "determination" of $d_{\perp}$ for the $\sqrt{3} \times \sqrt{3}R30°$ structure. Since the sensitivity of the theory to $d_{\perp}$ is not reported, however, we cannot really comment on the ability of the theory to extract $d_{\perp}$ from the experimental HRELS spectra. Only dipole strengths, not scattering amplitudes, were calculated so no direct comparison with experimental vibrational intensities can be made. In practice it appears that the $\sim 135\,cm^{-1}$ Ni surface phonon mode, which is perpendicular to the surface, is excited only in the presence of a p(2 × 2) overlayer, whereas the $\sim 250\,cm^{-1}$ modes, which are primarily parallel to the surface appear for all adsorbate conditions (isolated adatom, p(2 × 2), $\sqrt{3} \times \sqrt{3}R30°$) [222, 316]. Ibach and Bruchman [316] have interpreted this in terms of symmetry selection rule arguments. Whether these arguments are correct or not the presence of the $135\,cm^{-1}$ mode certainly seems to signal p(2 × 2) formation since it reversibly disappears and reappears when crossing the phase boundary at $\sim$360–410 K [316].

Lewald and Ibach mention that, at 150 K, 2.8 L O_2 exposure does not produce the p(2 × 2) pattern or the $135\,cm^{-1}$ Ni phonon frequency associated with it [222(a)]. The adsorbate is apparently completely disordered since only a diffuse (1 × 1) pattern (presumably the substrate) is observed. Since the phase diagram indicates that at $\theta < 0.25$ mL the p(2 × 2) O overlayer is stable at 150 K, the obvious explanation would be that, at this temperature, the adatom ordering is kinetically limited and thermodynamic equilibrium has not been established. The Ni{111} behavior would then be in line with Ni{100} and Ni{110} where no LEED order is observed at low T either. One caveat is appropriate though. Lewald and Ibach [222(a)] seem to assume the 2.8 L exposure will give $\theta = 0.25$ mL at 150 K because it does so at 300 K (or at least there is no further discussion on this point in their paper). Sticking probability considerations (see later) make this unlikely. The coverage should be higher for a given exposure at the low temperature, just as was the case for Ni{100} and {110}. It is therefore possible that θ is already into the oxide nucleation stage, which at low T would be amorphous.

No other structurally related data seem to exist in the literature for low T adsorption of O_2 on Ni{111}.

3.3.5 Geometric structure of the oxide species

At 300 K NiO formation starts at 0.34 mL according to Holloway and Hudson [232], $\sim$0.45 mL according to Norton et al. [243] and 0.33 mL according to Rieder [269]. All these estimates refer to the coverage at which the sticking probability starts to increase after the chemisorption plateau, rather than to any direct observation of NiO. The Holloway and Hudson [232] AES coverage calibration was identical to that in their Ni{100} study [231], so the reader should see the appropriate sections on Ni{100} for details. The Rieder value is based on fitting a Langmuir dissociative adsorption equation to the kinetics of the chemisorption region. The coverage at the plateau is obtained as a parameter from this fit. Since the basis for the type of fit is questionable, the value cannot be considered as unambiguous. The Norton et al. XPS value is based on an assumption that saturation coverage corresponds to three layers of NiO with parallel {111} orientation [243]. Though a three-layers saturation limit is consistent with the value for the {100} and {110} surfaces, their own nuclear microanalysis measurements give a saturation value of 3.4 layers. An X-ray fluorescence measurement by Mitchell et al. [228] yielded a value of 3.0 layers and the RBS results of Narusawa et al. [215] gave saturation coverage as 2.8 ± 0.4 mL of O which converts to 3.8 ± 0.5 layers of {111} oriented NiO. No report of the coverage at which the oxidation onset occurred was given, though this certainly could have been derived from the He^+ O scattering intensity at the point (~10 L exposure) where Ni atoms began to move because of oxide formation (very clearly defined in Fig. 2 of ref. 215). Thus, although the reported values agree reasonably well, the onset of oxide nucleation perhaps cannot be considered to be as firmly established as in the case of Ni{100}.

The first LEED report of NiO formation on Ni{111} was by Park and Farnsworth [314]. They reported a diffuse NiO pattern with the [1$\bar{1}$1} plane parallel to the {1$\bar{1}$1) plane of the Ni substrate and with the (110) directions parallel. The pattern coexisted with the original Ni{111} pattern, indicating that the oxide was in patches and did not cover the whole surface. These LEED results have since been confirmed in the studies by Holloway and Hudson [232] (NiO observed for exposures greater than a nominal 15 L), by Rieder [269] (NiO observed above a nominal 35 L exposure) and by Conrad et al. [250]. Holloway and Hudson [232] report that at saturation only the NiO {111} pattern remains (i.e. the surface is now completely covered). Rieder [269] reports that the NiO [111] spots become sharper and more intense with increased exposure. None of the papers report the formation of any other orientations. This then differs from both Ni{100} where mixed epitaxial relationships formed and Ni{110} where the long-range order of the oxide was quickly lost as exposure increased. The ion-scattering results of Narusawa et al. [215] are not completely compatible with the above description for Ni{111}. At saturation coverage, which they claim corresponds to $\theta = 2.8$ mL (or 3.8 layers of {111} NiO), they suggest that the interface is very sharp and unstrained (the substrate Ni interface atoms do not move) indicating an amorphous NiO overlayer. The discrepancy may simply be a question of defining how sharp an interface is meant. If one of the layers of the three Ni layers displaced at saturation is actually the Ni substrate interface layer, the other two being in the form of the ordered NiO overlayer (i.e. O coverage closer to 2 rather than 2.8 ± 0.4 mL) the ion scattering results become compatible with the LEED results.

MacRae discussed the epitaxial relationship of NiO on Ni{111} in his early LEED work [312]. He worked at 500–600 K, so some oxygen probably also diffused into the Ni subsurface. The LEED pattern he observed is shown in Fig. 74(a). It represents the same epitaxial relationship of NiO as reported above at 300 K. The NiO lattice has a lattice constant, $a_0 = 4.17$ Å, 18% larger than that of Ni. Thus there is a large mismatch in lattices at the interface, even though an oriented epitaxy forms. This is very similar to the Ni{100} situation where mismatch of the same magnitude occurs for the {100} NiO parallel growth (see Sect. 3.1.5). The existence of this mismatch, the coexistence of the Ni and NiO patterns and the width of the NiO LEED spots all indicated that the oxide grows as small islands, estimated as ~25 Å by MacRae. In addition to this oxide pattern, however, MacRae also observed three additional NiO LEED spots oriented at 120° to each other which moved outward from the origin on decreasing the LEED beam energy [see Fig. 74(a)] instead of the inward movement of the Ni{111} and NiO{111} spots. The energy dependence of these spots is entirely consistent with them originating from the 3 equivalent {100} faces of triangular pyramidal facets of NiO [Fig. 74(c)]. The existence of these stable 3D NiO structures clearly indicates that for $T > 600$ K any model describing the oxide growth purely as lateral island growth is an oversimplification. At least part of the surface is covered by oxide growing normally to the surface plane.

Finally, we should briefly mention oxidation at low temperature. The only relevant studies seem to be the XPS kinetic data of Norton et al. at 77 K [243], and the AES/LEED kinetic data at 147 K by Holloway and Hudson [232]. No mention of any ordered LEED structures in the oxide region is made by Holloway and Hudson. From the saturation values of O AES and O(1*s*) in these studies it appears that the thickness of the presumably amorphous or very fine grained polycrystalline NiO formed at low temperature is similar to that at 300 K.

3.3.6 Electronic structure aspects

The only significant studies dealing specifically with electronic structure aspects of Ni{111}/O are two UPS studies [229, 250] and an early INS study [317]. No cluster or band-structure calculations specifically aimed at O_{ad} in the threefold Ni{111} site have been reported, though rudimentary correlation diagrams for the interaction of Ni_3 cluster orbitals with X 2*p* orbitals for Ni_3X, where X is a chalcogen, have been discussed [250]. Becker and Hagstrum [317] reported the INS spectrum of p(2 × 2) O on Ni{111}. It is extremely similar to that reported for the p(2 × 2) O on Ni{100}, and not much different from the p(2 × 1) O structure on Ni{110}. Our conclusion from these similarities is that the *local* Ni–O bond is electronically very similar in all three cases, despite the geometric differences. Conrad et al. [250] followed the HeII UPS angle-integrated spectrum of the Ni{111} surface as a function of oxygen exposure at 300 K while monitoring the overlayer structure by LEED. Their results are shown in Fig. 77(a). At the p(2 × 2) overlayer stage no significant change in the character of the Ni{3*d*) band was observed, only an attenuation accompanied by the growth of an O 2*p* band at ~5.5 eV. The authors concluded that the Ni(3*d*) states were not significantly perturbed by the p(2 × 2) O adsorbate overlayer, and therefore the bonding primarily involved the Ni 4*s*4*p* electrons. This is, of course, in general agreement with the detailed discussions of electronic structure

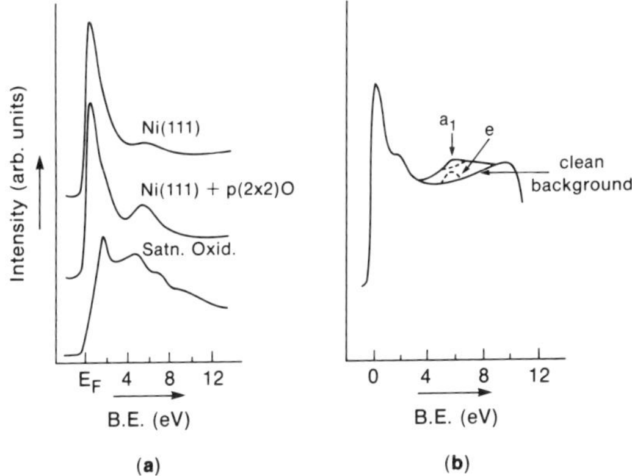

Fig. 77. (a) HeII angle-integrated UPS for Ni{111} + O_2 at 300 K at various stages [250]. (b) $h\nu$ = 16.8 eV angle-resolved UPS for Ni{111} + $\sqrt{3} \times \sqrt{3}$R30°O structure. Data taken at normal emission, 45° incidence angle. The authors suggested decomposition of the O(2*p*) feature into *a* and *e* components is marked [229].

presented for Ni{100}/O earlier. Conrad et al. also concluded that the lack of change in the Ni *d* band region was good evidence that Ni reconstruction did not occur for the p(2 × 2) structure. They were almost certainly thinking of a massive reconstruction, e.g. the formation of a 2D NiO array with Ni atoms laterally displaced, and in this sense they were certainly correct. The RBS results [215] referred to in Sect. 3.3.4, however, showed that there *is* an outward relaxation of ~ 0.15 Å occurring for the 3 Ni atoms bonded to each O_{ad} in the p(2 × 2) structure. Presumably this small relaxation does not perturb the Ni(3*d*) states sufficiently to be observed in the angle-integrated HeII UPS, or to show up in any way in the INS spectra [317].

The HeII spectrum at saturation coverage (10^3 L exposure) [Fig. 77(a)] clearly shows the presence of the shifted Ni^{2+} *d* band at ~ 1.7 eV below E_F for which the LEED pattern was very similar to that shown in Fig. 74(a), i.e. epitaxial NiO{111} was present. The complete spectrum at this point correlated well with the SCF Xα calculations of NiO_6^{10-} orbital energy levels. Messmer et al. [318] had originally thought that this showed that an NiO_6^{10-} cluster was a reasonable representation of chemisorbed O, since earlier Ni(poly)/O UPS spectra [249] which were very similar to the 10^3 L exposure spectrum had been interpreted as representing chemisorption. Conrad et al. pointed out that this type of spectrum was characteristic of oxide and the Xα NiO_6^{10-} calculation therefore simulated an oxide environment, as might have been expected given the environment of the single Ni atom in the cluster.

The angle-resolved UPS behavior of O, S, Se, and Te on Ni{111} in p(2 × 2) and $\sqrt{3}$ environments (monitored by LEED) were studied by Capehart and Rhodin [229] using HeI and NeI energies. A representative spectrum is shown in Fig. 77(b) for the $\sqrt{3}$ structure. The authors' primary interests were to compare the spectra of the chalcogens and see whether the spectra and their differences could be explained by simple Huckel arguments concerning the Ni_3X bonding. For S, Se, and Te, the X p band was shown to be split into two parts, which Capehart and Rhodin ascribed to a splitting into $a_1(p_z)$ and $e(p_xp_y)$ levels due to interaction primarily with the Ni_3 cluster 4s orbitals. The intensity changes of the two orbitals with incidence angle (normal emission) were as predicted for the a_1 and e assignments. For O, however, no clear splitting could be observed either in any individual spectra or from intensity variations as a function of incident angle. Nevertheless a tentative decomposition into a and e components was suggested [Fig. 77(b)]. The difference of the O behavior from that of S, Se or Te, was ascribed to the greatly increased separation between O 2s and O 2p atomic levels compared with the separation between s and p levels for S. This has the effect of reducing the separation between a_1 and e Ni_3 cluster orbital levels, as schematically depicted in Fig. 7 of ref. 229. This description of the bonding and the interpretation of the photoemission data is, of course, very simplified and cognizance of the much more complex discussions for the Ni{100}/O case is advisable. There it was concluded that it was really not possible to consider any splitting of the O 2p band in simple terms of a_1 and e levels, but rather that there was a spread of a_1 and e character over the entire 2p band.

3.3.7 Kinetics and mechanisms. General

As was done for Ni{100} and {110} we separate this section into "chemisorption" and "oxide nucleation and growth" regions, keeping in mind that the two regions can overlap considerably and that the degree of overlap is likely to vary from laboratory to laboratory and experiment to experiment because the oxide nucleation step is strongly affected by substrate and possibly adsorbate overlayer defects. As for the other surfaces we note that an accurate coverage scale is important for consistent interpretation between different sets of data.

In fact the kinetic aspects of oxygen adsorption on Ni{111} have received the least attention of all three surfaces. The original AES/LEED study of Holloway and Hudson [232] offers a good comparison with Ni{100} [231] since they studied both surfaces in an identical fashion using the same equipment. The other studies in which aspects of the kinetics have been studied are the AES/SIMS/LEED study by Rieder [269], an AES study by Liu et al. [242], an XPS study by Norton et al. [243] and an AES/microbalance study by Winkler et al. [241]. Only in the last study was the determination of $S(\theta)$ the primary goal.

Comparison of the kinetics of Ni(111)/O and Ni{100}/O is easier and perhaps more enlightening in terms of trying to establish dissociation mechanisms than comparison between Ni{110}/O and either of the other two surfaces because of the massive reconstruction that occurs only on the Ni{110} surface.

3.3.8 Kinetics and mechanism of chemisorption

The method of coverage calibration in the Holloway and Hudson LEED/AES study [232] is identical to that discussed for Ni{100} earlier. Their raw data were reproduced in Fig. 72 where it is compared with their {100} data. In Fig. 78 we have replotted the data on a linear scale. Holloway and Hudson observed the p(2 × 2) O structure with maximum intensity at $\theta = 0.27$ mL, but did not see the $\sqrt{3}$ structure, presumably because the temperature at which they were working was a few degrees too high (see Sect. 3.3.3). They did note, however, that the minimum in S came after the formation of the p(2 × 2) structure and, in fact, we can see from Fig. 78 that it is probably close to $\theta = 0.33$ mL, the coverage of a perfect $\sqrt{3}$ structure.

Comparing the data on Ni{111} with those of Ni{100}, it is clear that the dissociative adsorption kinetics prior to oxide nucleation is faster for Ni{111} but reaches its minimum S value at a very similar coverage. Comparison of Figs. 57 and 78 shows that while the initial $S\{\theta\}$ curve on Ni{100} is projecting towards a zero value at $\theta = 0.25$ mL; on Ni{111} it projects to zero near $\theta = 0.33$ mL. In addition the minimum S value is lower for the Ni{100} surface (~0.021 at 300 K and 0.002 at 423 K for {100}, ~0.045 at 300 K and ~0.025 at 423 K for {111}). Note at the slightly elevated temperature of 423 K the difference is over one order of magnitude. Before discussing the significance of these differences, determined by Holloway and Hudson [231, 232], we will review the data on the {111} surface from the other authors.

Norton et al. [243] followed the adsorption and oxidation kinetics of all three surfaces by XPS. Their data are not detailed enough in the chemisorption region to derive reliable S values, but it is clear from Fig. 1 of their paper that at 300 K the average rate on the {111} surface is faster. This, then, agrees with the Holloway and Hudson data.

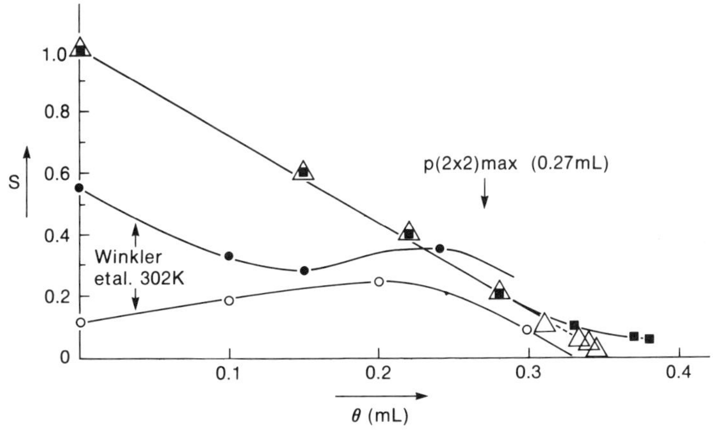

Fig. 78. Sticking probability data for Ni{111}/O_2. ■, △, taken from the 302 and 423 K data of Holloway and Hudson [232]. The other curves are the flat (○) and stepped (●) data of Winkler et al. at 302 K [241].

The AES/LEED/SIMS 300 K comparison by Rieder [269] of the Ni{110} and {111} surfaces also does not show sufficient detail for reliable S values in the chemisorption region. From Figs. 3 and 2 of Rieder's paper, one can conclude that chemisorption saturation is reached in comparable exposures in both cases. Since we believe (Sect. 3.2.9) that S is high for Ni{110} over most of the chemisorption region, it therefore follows that the average S is also high for Ni{111}. The other piece of information that is extractable from the Rieder data is that S_{min} is reached at a coverage $\sim$1.3 times larger than the optimum p(2 × 2) LEED pattern coverage. This is in close agreement with the results of Holloway and Hudson.

Finally Winkler et al. [241] reported $S(\theta)$ data at 300 K for Ni{110} and Ni{111} using AES and a quartz crystal microbalance. LEED was not used so correlations to ordered structures was impossible. Their {110} data was discussed earlier. The {111} data are shown in Fig. 78 for comparison with the Holloway and Hudson data. The two sets of data bear little resemblance to each other. We can offer no logical explanation for the Winkler et al. data on the {111} surface. They also report data for a stepped Ni{111} surface (8 {111} atoms per terrace; {100} steps) (see Fig. 78). The authors explain the differences between the flat and stepped results as being due to a high S value at the steps and a low S value on the terraces. The stepped results look somewhat more like the Holloway and Hudson {111} results [232], but we are quite reluctant to believe that their work, and that of Rieder [269], and of Norton et al. [243] are dominated by massive defect behavior. If the Winkler et al. data were correct, explanation of the dissociation mechanism would have to follow the lines of those discussed in detail for W{110} (Sect. 2.5.14). For the present, however, we choose to believe the majority of authors and will consider the data of Holloway and Hudson as essentially correct and base our interpretations on this.

In their original publication Holloway and Hudson [232] fit their Ni{111}/O kinetics in the chemisorption region to their usual Langmuir equation. A reasonable fit was obtained if the limiting chemisorption coverage of 0.34 mL was assumed. In a later review, however, Holloway [293] describes the kinetics up to the p(2 × 2) formation region as fitting the expression $S = (1 - 4\theta)$, identical to his proposed fit for the {100} surface. This fit, of course, requires a limiting coverage of 0.25 mL. In fact as we showed in Sect. 3.1.8, the {100} data of Holloway and Hudson and of Brundle et al. do not fit a $(1 - 4\theta)$ curve but decrease faster than this functional form. The Ni{111} data of Holloway and Hudson do not fit the $(1 - 4\theta)$ form either. S decreases more slowly than this, approximating

$$S = (1 - 3\theta) \tag{17}$$

It is clear then that the functional forms on the two surfaces are quite different, though in neither case is there evidence for precursor behavior at 300 K. For the square lattice of Ni{100} detailed Monte Carlo simulations led to the proposal [245] that for O_2 to dissociate on this surface two next-nearest-neighbor empty sites each of which in turn had an empty nearest neighbor was required. This arrangement constitutes the 8-site cluster whose availability as a function of coverage was found to parallel the behavior of S as a function of coverage. No such detailed Monte Carlo determinations of empty cluster site availability as a function of θ have been performed for the hexagonal Ni{111} lattice. The 6 pair-wise adatom interaction necessary to simulate the phase diagram (Fig. 75) might make such a determination considerably more complex than for Ni{100}. In addition, not knowing the conditions under which both sublattices can be occupied complicates things further. However, knowing the natures of the ordered O structures on the two surfaces does allow us to suggest reasons for the different behaviors on the two surfaces. Such a comparison was considered by Liu et al. [242] in their AES study of oxidation of a cylindrical Ni crystal exhibiting {100}, {110}, and {111} faces. Most of their comments concerned explanations for the oxide nucleation effects, but consideration of the adsorption sites available may be equally relevant for the chemisorption regime. The {111} lattice with the p(2 × 2) and ($\sqrt{3} \times \sqrt{3}$)R30° O overlayer structures is represented in Fig. 79. It can be compared with the p(2 × 2) and c(2 × 2) O structure on Ni(100) in Fig. 58. Liu et al. [242] pointed out that, whereas transition from p(2 × 2) to c(2 × 2) structures on the {100} surface does not, in principle, require adatom mobility, conversion of the p(2 × 2) to $\sqrt{3}$ structure on the {111} surface does. In addition we know from the phase diagram study that at 300 K such conversion is facile. This, then, explains why there is no obvious break in the $S(\theta)$ curve at $\theta \approx 0.25$ mL for the {111} surface (Fig. 78).

The fact that S approaches zero at $\theta = 0.25$ mL on Ni{100} even though there are plenty of nearest-neighbor empty pairs still available was proof that nearest-neighbor empty pair availability was insufficient for O_2 dissociation. On Ni{111} we can see from Fig. 79 that, for the $\sqrt{3}$ structure, nearest-neighbor empty pairs are still plentiful and therefore, since S does

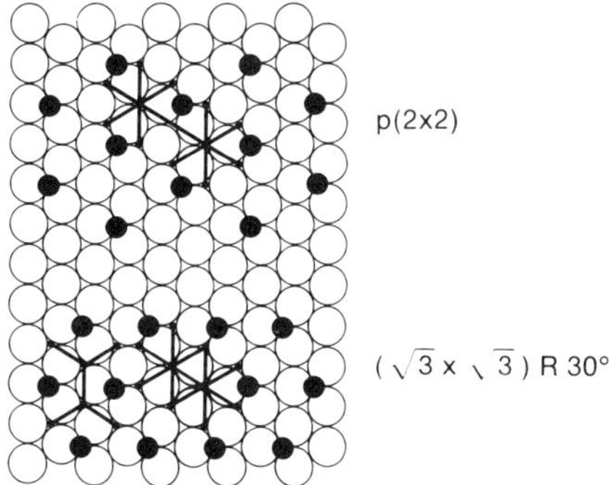

Fig. 79. Comparison of the empty cluster site availability for p(2 × 2) O and ($\sqrt{3} \times \sqrt{3}$)R30°O structures on Ni{111}. For the latter structure, the availability for a perfect lattice and a lattice with one O_{ad}-vacancy are shown (cf. Fig. 58 for Ni{100}).

go towards zero at $\theta = 0.33$ mL, it must also be true that a nearest neighbor availability is insufficient for O_2 dissociation on the {111} surface. What the detailed site requirement is is unknown, but we note, for instance, that for the $\sqrt{3}$ structure the removal of one O_{ad} atom leaves a multi-site cluster arrangement which consists of a nearest-neighbor pair with each site in the pair having no occupied nearest-neighbor sites (Fig. 79). It bears some similarity, then, to the 8-site cluster on the Ni{100} surface, which is available if one O_{ad} is removed from the p(2×2) structure though in the (111) case O_{ad} atoms entering the cluster from an O_2 molecule would go into a nearest-neighbor pair rather than a next-neighbor pair. Whether this cluster availability, or something like it, goes as $(1 - 3\theta)$ must await proper Monte Carlo simulations.

3.3.9 Kinetics and mechanism of oxide nucleation and island growth

The general scheme of oxide nucleation followed by island growth by reaction at the island perimeters is qualitatively similar to that for Ni{100} and so here we will concentrate on a comparison between the two systems.

In comparing the oxidation of Ni{110} with {100}, we concluded that at 300 K Ni{100} showed no evidence for involvement of diffusing molecular O_2 in the oxidation kinetics. This implied that the observed temperature variation of $K_i N_0$ in the Holloway and Hudson oxide kinetics equation was due to a temperature variation of N_0 and not because of an activation energy term in K_i. For Ni{110}, however, molecular oxygen did appear to play a role both in the chemisorption kinetics (precursor effect) and in the oxidation growth stage, such that K_i did incorporate an activation energy term. We suggested that the reason for the difference between the two surfaces was the roughness of the Ni{110} surface, particularly following the reconstruction during chemisorption, which allowed the normal component of the energy of an impinging O_2 molecule to be more easily accommodated with the surface. For Ni{111}, the smoothest of the surfaces, we would expect that if there were no precursor effect on Ni{100} there would be none on Ni{111}. In the chemisorption region this appears to be born out by the $(1 - 3\theta)$ behavior of S. In the oxide nucleation stage, which can be fit easily enough by the standard Holloway and Hudson equation, we would expect that a comparison between $K_i N_0$ values should give information on relative N_0 values and that any T variation in $K_i N_0$ on Ni(111) could be interpreted in a like manner to Ni{100} as due to a variation in N_0.

Holloway and Hudson made a detailed study of both surfaces in the oxide nucleation region using the same apparatus and under the same conditions [232]. The values they extracted for $K_i N_0$ and related parameters are included in Table 9. Rieder [269] also fit the oxidation kinetics, as determined by Auger spectroscopy, to the Holloway and Hudson formula. His 300 K $K_i N_0$ value is also included in Table 9. It agrees within a factor of 3 with the Holloway and Hudson value and both are much larger than the Ni{100} values. In the original uptake data, this shows up in the saturation coverage exposure values, which are 20–40 L for Ni{111} (dependent on author) and 100–200 for Ni{100} (depending on author). From measurements of the oxidation rate at different temperatures and the assumption that N_0 was independent of temperature and that K_i contained an activation energy term, Holloway and Hudson derived negative activation energies of 1.4 kcal mol^{-1} for Ni{111} and 1.5 kcal mol^{-1} for Ni{100}. The respective calculated nucleation site densities were 1×10^{11} cm^{-2} and 6×10^{9} cm^{-2}. Thus, within their model, the activation energy for O_2 diffusion to and/or incorporation in the growing island perimeters (see Fig. 61) is the same at both surfaces, whereas N_0 is 20 times larger for Ni{111}. Our interpretation would be that K_i contains no activation energy term at 300 K and above, since O_2 surface diffusion plays no role, and that the temperature variation of $K_i N_0$ above 300 K in both cases is related only to the variation in N_0, which is established during the chemisorption stage. No temperature-jump experiments have yet been performed to test this hypothesis. The origin of the twenty-fold increase in N_0 on going from Ni{100} to {111} is the interesting fact. As discussed in Sect. 3.1.9 we do not believe that the absolute values are meaningful since they depend on the Holloway and Hudson [231] expression for K_i, which we believe is incorrect. We know, for instance, that a nucleation site density of 6×10^{9} cm^{-2} on Ni{100} is at least a factor of 100 too small to account for the island sizes at coalescence, as estimated from LEED (Sect. 3.1.9). The relative values of $K_i N_0$ should be meaningful, however, which implies that the probability of producing a nucleation site on the Ni{111} surface is far higher than on the Ni{100}. The reason may involve the same structural blocking effects that control the chemisorption kinetics. For Ni{100} the c(2×2) O phase slowly builds up after the (near) completion of p(2×2) O. During this time S continues to drop, indicating that nucleation is not yet occurring (or that the number of nuclei is still small). The c(2×2) phase structurally blocks the surface to further dissociation very effectively, since in perfect regions not only are there no 8-site empty clusters available, there are not even nearest-neighbor empty pairs. Therefore, within the model of Hopster and Brundle [237] where a local θ of > 0.5 mL is the driving force for oxide nucleation because of the repulsive dipole interactions at such coverages, the only nucleation that can occur is at defect sites, either in the overlayer or the substrate. For Ni{111} the situation is rather different since the structural blocking effects of the chemisorbed phase are much weaker. The $\sqrt{3}$ O structure is 5–10 times less effective in stopping further O_2 dissociative adsorption than is the c(2×2) O on Ni{100}, as judged by the difference in S_{min} achieved on both surfaces. Liu et al. [242] argue that this is due to a greater mobility of O_{ads} on the flatter Ni{111} surface. Their reason for proposing this is that to progress from p(2×2) to $\sqrt{3}$ structures on Ni{100} requires O_{ads} to change sites in large numbers, whereas to go from p(2×2) O to c(2×2) O on Ni{100} requires, in principle, no movement of O_{ads} atoms. They argue that this mobility allows a crucial local O coverage to be achieved more easily for Ni{111}, leading to more oxide nuclei. While we agree that O_{ads} mobility is clearly demonstrated on Ni{111}, we cannot see the logical connection to increased oxide nucleation. Rather, we would have anticipated that it would allow more perfect $\sqrt{3}$ O ordering on Ni{111} than c(2 × 2) O ordering on Ni{100}. If the $\sqrt{3}$ structure were structurally blocking in the same way as the c(2 × 2), one would then expect S_{min} to be lower on Ni{111} and the subsequent oxide nucleation density to be lower. That this is counter to experimental observation leads us to believe that the perfect $\sqrt{3}$ structure on Ni{111} does not inhibit oxide nucleation in the manner that the perfect c(2 × 2) does on Ni{100}. The reason is probably that, for the one-third coverage situation of the $\sqrt{3}$ structure, there are still nearest-neighbor empty sites available, and, in fact, each site of such a pair has one-half of its nearest-neighbor sites empty, generating a 6-site empty cluster (see Fig. 79). We propose, then, that dissociative adsorption into such clusters is much more facile than dissociative adsorption into the c(2 × 2) O structure and that this generates regions of high local θ which become oxide nucleation sites. The difference between the blocking power of the two structures, $\sqrt{3}$ and c(2 × 2) is increased with increasing temperature. This is in agreement with the suggestion that O_{ads} is less mobile on Ni{100} since it means that the c(2 × 2) O structure on Ni{100} will require higher temperatures to anneal out overlayer defects than does the $\sqrt{3}$ structure on Ni{111}. Thus at 300 K the $\sqrt{3}$ structure probably already has its full blocking effectiveness, whereas for the c(2 × 2) this is not achieved until higher temperatures. Therefore S_{min} should decrease more strongly with increasing T on the {100} surface, which is the case.

3.4 INTERCOMPARISON OF Ni{100}, {110}, AND {111} SURFACES

In this section we briefly recall and compare the major features of oxygen interaction with the three low-index Ni faces.

3.4.1 General reaction scheme

The general scheme is similar on all faces. At 300 K a fast chemisorption stage uses up active dissociative adsorption sites so that S drops to a low value. Oxide nucleation initiates towards the end of the chemisorption. This causes S to increase because reaction occurs at nuclei perimeters which increase in length as the islands enlarge. When the islands coalesce, reaction effectively ceases under UHV conditions. The oxide is 2–3 layers thick. A very slow transport-limited thickening will occur with continued higher pressure exposure (tarnishing reaction). At higher temperatures, but below the oxygen dissolution temperature (450–700 K depending on face, coverage, and defect concentration), the perfection of the chemisorbed phases can be increased, the oxidation reaction gets slower, and the epitaxial relationship of the oxide improves. Only for temperatures above the dissolution temperature can the oxide be grown significantly thicker, however. At low T (77–150 K) the situation changes drastically. Ordered chemisorption structures which block further reaction are not formed. This, plus the fact that the molecular O_2 residence time increases, results in a rapid reaction right up to the 2–3 oxide layer passivation coverage.

Despite the general similarity of all three surfaces there are distinct and significant differences which manifest themselves in strong variations in chemisorption and oxidation rates. The differing Ni surface geometries affect both the probability for O_2 to dissociate and the molecular O_2 residence time. In addition, the propensity for the substrate to reconstruct is different. Defects influence oxidation rates on all surfaces. The greatest effect is on the Ni{100} because, in the absence of defects, it is the slowest to oxidize. Impurity co-adsorption and reaction must always be considered. Their effects are well-understood and documented for Ni{100} [236, 238], but we believe they are general to all three faces.

3.4.2 Geometric structure of chemisorbed and oxide species

The geometry of the p(2 × 2) and c(2 × 2) O chemisorbed structures on Ni{100} have been very well documented and it is certain that the O adatoms are located in the fourfold hollows at a $d_{\perp}$ value of ~0.9 Å [196–206, 216, 226]. The only remaining doubt is whether there is any lateral displacement from the symmetric fourfold position [199]. Outward relaxations of the top Ni layer have been determined for the p(2 × 2) and c(2 × 2) O structures. The best estimates of these values, ~2 and ~5%, respectively, come from the latest round of medium-energy ion scattering results [214].

For Ni{111} the threefold hollow site is occupied [196] in the two overlayer structures formed, p(2 × 2) O and $\sqrt{3} \times \sqrt{3}$R30°O. There are two types of threefold site, h.c.p. sites and f.c.c. sites, and from the LEED it appears that both can be occupied to form anti-phase domains [235]. Owing to this complication, interpretation of the Ni{111}/O phase diagram, the only Ni surface for which much experimental phase transition data exist, in terms of pairwise adatom interactions is complex [313]. The value of $d_{\perp}$ on the Ni{111} surface is thought to be ~1.2 Å [196], though this is not established with the certainty of the Ni{100} case. The increase over the 0.9 Å Ni{100} value seems to be purely due to the threefold geometry, i.e. the Ni–O bond lengths are similar and there is no evidence for any difference in bonding character, such as degree of charge transfer. There are small (0.15 Å) outward relaxations of the Ni atoms involved in bonding to the O atoms for the p(2 × 2) and $\sqrt{3} \times \sqrt{3}$R30°O phases, as determined by RBS [215].

For Ni{110} the chemisorbed situation is much more complex. Of the three LEED structures formed [low coverage p(3 × 1), p(2 × 1), and higher coverage p(3 × 1)], it is established beyond doubt that the p(2 × 1) involves a Ni top layer massive reconstruction and it is very likely that the other two structures do too [204, 207, 210–213]. The reconstruction process produces missing Ni rows along the ⟨110⟩ direction, which implies that the observed LEED fractional order patterns are generated by the Ni atoms rather than the O atoms. The location of the O adatoms is not settled, but it is somewhere

References pp. 381–388

close to the bridge site on the top Ni rows, probably rotated towards a threefold hollow site.

All faces nucleate oxide starting around $\theta \approx 0.35$–0.4 mL. In the case of the {100} and {110} surfaces this occurs before completion of the preceding chemisorbed structures. For Ni{111} the one-third $\sqrt{3}$ structure appears to be completed before nucleation occurs. Ni{100} oxidation produces two epitaxial relationships, NiO {100} and {111} faces parallel to the {100} substrate [221, 231, 271]. The {100} morphology has very small grains (~10 Å) [227]. At higher temperatures {111} converts to {100} [227]. The mix between the two orientations is irreproducible. The final thickness at 300 K is 2–3 layers NiO [210, 227, 231]. At 700 K the oxide can be grown to ~18 Å with maintained epitaxy if the exposure pressure is high enough [227]. For Ni{111}, NiO islands grow with one morphology only [39, 57, 76, 122]; the $\{1\bar{1}1\}$ plane is parallel to the Ni$\{1\bar{1}1\}$ plane with ⟨110⟩ directions parallel. As for Ni{100} the epitaxy is maintained to 2–3 layer oxide saturation and the oxide islands are small (~25 Å) [312]. At higher temperatures (500–600 K) triangular NiO pyramids grow out of the 2–3 layer epitaxial NiO [312].

The Ni{110} surface is again rather different in behavior. A (9 × 4) 2D "pseudo-oxide" precedes the formation of regular NiO 2–3 layer thick [307]. The epitaxial NiO develops smoothly from the 2D structure, initially with a 5% lattice compression but later with normal NiO lattice dimensions. The island sizes of the oxide are again small (~10 Å) [307] and the epitaxy with the substrate is often completely destroyed at saturation coverage. The difference of the Ni{110} oxidation from the other two surfaces is, then, that the thin 2–3 layer NiO oxide islands grow laterally across the surface either after, or in competition with, the 2D "pseudo-oxide".

At low temperatures no ordered oxide structures are observed just as there are no ordered chemisorbed structures. Since there is little evidence to suggest any bonding differences and since the saturation coverage is very similar to that at 300 K it is likely that the only structural difference is the loss of long-range order. The one exception may be Ni{110} where the saturation coverage at 77 K is apparently less than at 300 K [243, 297]. We have suggested that this indicates a stabilization of the 2D "pseudo-oxide" against 3D island growth, but no significant work has been done to investigate this possibility.

3.4.3 Electronic structure of chemisorbed and oxide states

Most of the work related to electronic structure studies is on the {100} surface but the main features can probably be generalized to all three surfaces. Electronic structure calculations simulating O chemisorbed on Ni{100} indicate that the charge transfer is about unity and the bonding involves primarily O 2*p* and Ni 4*s*4*p* electrons with much less involvement of Ni(3*d*) [239, 240, 257–261, 263, 264]. This is confirmed by angle-integrated UPS data for Ni{111}/O [250], in addition to Ni{100}/O [286] where no shift in the Ni(3*d*) band was observed. This is quite different from the UPS in the oxidation stage where a 2 eV shifted Ni(3d) band is observed [250] in agreement with predictions of electronic cluster calculations [318] which simulate oxidation rather than chemisorption. The charge transfer in the oxide is higher than for the chemisorption, but is not equal to 2.

A claim based on $Ni_{25}O$ cluster calculations [239] that two electronically different chemisorption ground states existed for the p(2 × 2) O and c(2 × 2) O structures on Ni{100}, which implied a large difference in $d_\perp$ for the two structures and also accounted for differences in the vibrational frequencies, has been disproved. Good quality $Ni_{25}O$ and $Ni_{25}O_5$ cluster calculations predict that p(2 × 2) and c(2 × 2) O structures have the same ground state, nearly the same $d_\perp$, and show that the differences in vibrational frequencies are partly due to O atom lateral interactions [240, 264]. The remaining differences can be ascribed to different couplings to Ni substrate phonons [226].

Those calculations which include band structure dispersion effects, and are also sophisticated enough to represent properly the Ni–O bonding [255, 256], indicate that a simple decomposition of oxygen induced UPS features into O $2p_z$- and O $2p_{xy}$-like components (using angle-resolved and polarization information) is not possible. The strong Ni4sp and O 2p mixing results in O $2p_z$ and O $2p_{xy}$ character spread throughout the band. Though some experimental UPS work has claimed separations of O $2p_z$ and O $2p_{xy}$ components, the arguments are all very shaky and in addition there have often been complications from oxide, OH, or even bulk Ni features which have been incorrectly interpreted [253, 254].

The Ni{110}/O situation may be different from Ni{100}/O or Ni{111}/O. Since severe Ni reconstruction is involved in the chemisorption process it might be expected that this would show up as a difference in the UPS compared with the other two surfaces. Detailed comparisons have, however, not yet been made.

3.4.4 Kinetics and mechanism of chemisorption and initial oxidation

Sticking probability variations as a function of coverage and temperature have been studied in detail for Ni{100} and rather less detail for the other surfaces. From this data it has been established that, at 300 K and above for Ni{100}, the surface lifetime of any diffusing molecular oxygen (precursor state) is too small to influence the kinetics [244, 245, 247]. S is thus simply proportional to the number of vacant "dissociating sites" on the surface. From measurement of $S(\theta)$ and Monte Carlo simulations of the availability of different types of empty sites as a function of θ it has been suggested that the 8-site empty cluster is required for O_2 to dissociatively adsorb [245]. This indicates that the dissociation process cannot be considered as three separate steps: O_2 accommodation; dissociation; adatom interactions and re-ordering. The O_2 molecule takes account of the repulsive interactions with nearby O adatoms when approaching the surface and if the situation is unfavorable, which in this case means any nearest-neighbor O adatoms, it

References pp. 381–388

will reflect rather than dissociatively adsorb. A similar process appears to take place on the Ni{111} surface but because of the hexagonal geometry the site-cluster arrangements are different. As yet there are no Monte Carlo simulations to enable determination of the minimum cluster size for dissociative adsorption. On Ni{110} things again appear to be different. S apparently remains high for several tenths of a monolayer coverage (the measurements are sketchy), which indicates a strong precursor effect [241]. We suggest that the difference is due to the roughness of the Ni{110} surface (particularly after reconstruction). An incoming O_2 molecule may then both see a deeper potential well and also have its normal momentum accommodated more easily with the surface.

At low T all the faces behave similarly in that the geometry requirements for dissociation are overwhelmed by the increased effectiveness of a diffusing molecular precursor. In addition the structurally blocking ordered chemisorption structures are not formed. The result is that S remains high and the oxidation stage is entered at much lower exposure (1–3 L as opposed to 10–50 L).

The mechanisms operating during the oxide nucleation and growth stage also vary from face to face. Ni{100} oxidizes the slowest, Ni{110} the fastest. The original proposal [231] of lateral island growth of oxide is basically correct but some of the conclusions concerning the mechanism are highly questionable even though the oxidation kinetics can always be fit to the Holloway and Hudson formula [231]. The reason is that it is basically a fitting procedure where all the information is contained in the interpretation of the fitting parameter. The actual value of the parameter carries little information without the interpretation. On its own an increase in the parameter, $K_i N_0$, tells us only that either N_0, the nucleation site density is larger, or the reaction rate, K_i, at a fixed number of nucleation sites is greater. The exponential form of the equation incorporating $K_i N_0$ guarantees that a reasonable fit to the functional form of the experimental data will always be achieved. The Holloway and Hudson assumptions that went into their model, and which they believe are proven by the fit obtained, are that N_0 is fixed at time zero and is independent of T, that the nucleated islands are round and have their terminal thickness at time zero and that these islands grow because O_2 diffuses across the surface and is incorporated into the oxide at the perimeters. The diffusion and incorporation steps should have activation energies associated with them, which is where any T dependence of the oxidation comes from in the Holloway and Hudson model. We believe that, despite the good fits to the Holloway and Hudson formula, there is strong evidence that diffusing molecular oxygen plays no role in the oxidation of Ni{100} and {111} above 300 K and that the kinetics can be accounted for by direct impingement of O_2 from the gas phase at the island perimeters with no activation step. N_0 becomes fixed during the chemisorption process and is greater at lower T because nucleation sites are caused by the defects in the overlayer which otherwise blocks further dissociative adsorption. Both interpretations suggest a much larger nucleation site density on Ni{111} compared with Ni{100}. We have suggested that this is directly related to the geometric requirements for O_2 to dissociate. The final $\sqrt{3}$ O chemisorbed structure on Ni{111} is geometrically less blocking to further adsorption than is the c(2 × 2) O structure on Ni{100}. This allows a critical local O coverage, required for oxide nucleation, to be reached more easily on the {111} surface.

On Ni{110}, oxidation at 300 K is faster than on either Ni{100} or {111} because the molecular precursor state is effective (rougher surface), i.e. the Holloway and Hudson model of diffusing O_2 contributing to the kinetics becomes valid. At low T this is, of course, true for all three surfaces and accounts for why the oxidation reaches saturation coverage so quickly.

In our opinion the comparative aspects of the kinetics and mechanism of the dissociative adsorption and oxidation are the most interesting remaining aspects of the Ni/O system. More detailed measurements on Ni{110} and {111} of $S(\theta)$ are required to substantiate the suggestions put forward in this review.

3.5 SECONDARY ION MASS SPECTROSCOPY (SIMS) OF Ni/O SURFACES

In Sect. 2.10 we briefly reviewed the use of SIMS for studying chemisorption on metal surfaces in general and discussed the work on W/O surfaces in particular. As was the case for W/O we have separated out the Ni/O SIMS studies from the rest of the Ni/O work because we do not think that the understanding of SIMS is yet at a point where, on its own, it can provide much insight into the nature of the adsorption system. The reasoning behind this can be found in Sect. 2.10. Now that we have reviewed all the other existing work on Ni/O, we are in a position to comment on the interpretation of the SIMS data.

The volume of work is not extensive. There are several papers by Benninghoven and co-workers [265, 268] on polycrystalline surfaces where the positive and negative ion intensities were followed as a function of oxygen exposure at 300 K. Since coverages were not determined independently, correlations with structural or chemical changes are difficult. A couple of features emerge, however, which are common to the subsequent single-crystal work. Figure 80(a) shows the variation of Ni_2^+, NiO^+, O^- and NiO^- intensities as a function of exposure. From comparison with the single crystal results discussed below, where the surface characterization is in better shape, we can be sure of the following.

(i) The peak in the Ni_2^+ intensity represents the end of the chemisorption stage; i.e. the intensity grows rapidly during chemisorption (up to ~10 L here) and then decreases as oxide islands nucleate and grow laterally (10–40 L here). A roughly similar type of behavior is observed for Ni^+ and Ni_2O^+ except that the behavior during the oxide growth is variable and a plateau

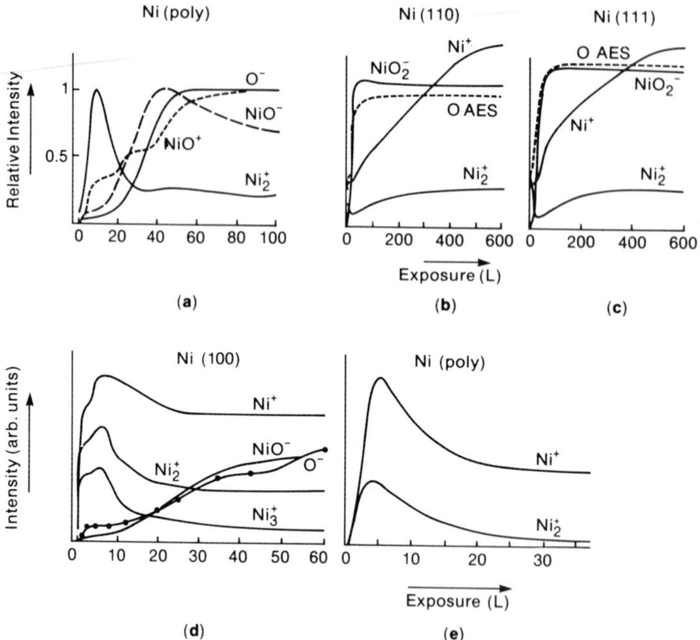

Fig. 80. Variation of SIMS cluster ion intensities as a function of exposure or coverage for Ni/O surfaces (300 K). (a) Ni(poly) [265, 268]; (b) Ni{110} [269]; (c) Ni{111} [269]; (d) Ni{100} [266]; (e) Ni(poly) [267].

rather than a decrease occurs in some cases (see later for a possible explanation).

(ii) Above 40 L there is insignificant increase in total oxygen coverage and so any growth in positive ion intensities above ~ 40 L (observed here for NiO^+) should not be associated with increasing coverage of a majority species, i.e. it is not related to an increase in oxide thickness.

(iii) The negative ion growth during the chemisorption region is weak and proportional to oxygen coverage, and at the end of the chemisorption stage intensities are still very low. There is then a very rapid increase during the oxide nucleation and island growth stage which correlates with oxygen coverage. Thus the negative ion intensities can be semi-quantitatively related to oxide species concentration.

We note that the property that is most likely to explain these observations is the local work function change. As discussed earlier in this chapter, $\Delta\phi$ is positive for O chemisorption on nickel and negative for oxide nuclei. Strong correlations of ion neutralization rates with work function are in general expected theoretically, as discussed by Yu and Lang [319]. In the present case it is not possible to decide whether the switch from positive ion to negative ion growth at the nucleation point is only related to a work function change, since there are also strong geometric and bonding changes occurring which could also affect SIMS ion intensities independently of work function changes.

Having made these general statements for polycrystalline Ni surfaces we will now review the work on Ni{100}, {110} and {111} surfaces. There are several studies on the {100} surface which exhibit some variations. Only one study exists for the {110} and {111} surfaces [269] so we will consider these surfaces first. Figure 80(b) and (c) show similar data to Fig. 80(a), but for the {110} and {111} surfaces. In addition, AES-monitored coverage changes are marked. One can clearly see that the general statements made above are also justified for the {110} and {111} surfaces. In the discussion of these results Rieder [269] noted the correlation of the negative ion SIMS with the oxide nucleation stage. He associated it with the formation of ionic bonds rather than the surface dipole effects we have suggested above. Rieder also suggests that the decrease in Ni_2^+ yield beyond the chemisorption plateau can be ascribed to a shielding effect of the Ni by O. We do not think this is a likely explanation since there is actually a rapid decrease in the region where the total oxygen coverage does not increase significantly. The correlation is rather with the conversion of chemisorbed O to oxide. The reason, then, is likely to be a combination of the fact that Ni_2 dimers are less likely to be produced from an NiO surface than a Ni surface, because there are no Ni–Ni neighbors on an NiO surface, plus the effect of work function change emphasizing negative ions and suppressing positive ions. Finally, we note that the Ni^+ and Ni_2^+ yields start to grow again only after the 2–3 NiO layer passivation, which was also the suggested situation for NiO^+ for polycrystalline Ni discussed above. Rieder [269] correlates this behavior with a gradual elimination of defects in the NiO over the 100–600 L range. This is a plausible explanation but an alternative one could be that O chemisorbing on top of completed NiO regions increases Ni^+ and Ni_2^+ yields just as O chemisorbed does on bare Ni. Since the total O coverage does not increase significantly in this region the implication would be that such chemisorbed O is a minority species causing a high SIMS yield.

Two studies were independently carried out on the Ni{100}/O system around the same time [236, 266]. The results are summarized in Figs. 80(d) and 81. The Fleisch et al. study [266] used XPS O(1*s*) intensities as a relative coverage calibration, but only 5 XPS data points are reported over the whole exposure range, which leads to ambiguity since $S(\theta)$ is known to oscillate so dramatically. The main thrust of this paper was to try and establish a correlation between cluster intensity changes and geometric changes [p(2 $\times$ 2) O, c(2 $\times$ 2) O, oxide] using classical trajectory calculations as a guide to predict the changes. The lack of simultaneous LEED data and the inadequacy of the XPS coverage calibration severely limited this objective, however. The observed Ni_2^+ and Ni_3^+ cluster dependence beyond the chemisorption stage were confirmed by trajectory calculations on NiO. In fact Ni_3^+ is predicted to have zero intensity for NiO whereas Ni_2^+ survives at a low concentration. This is in agreement with the situation for heavily oxidized

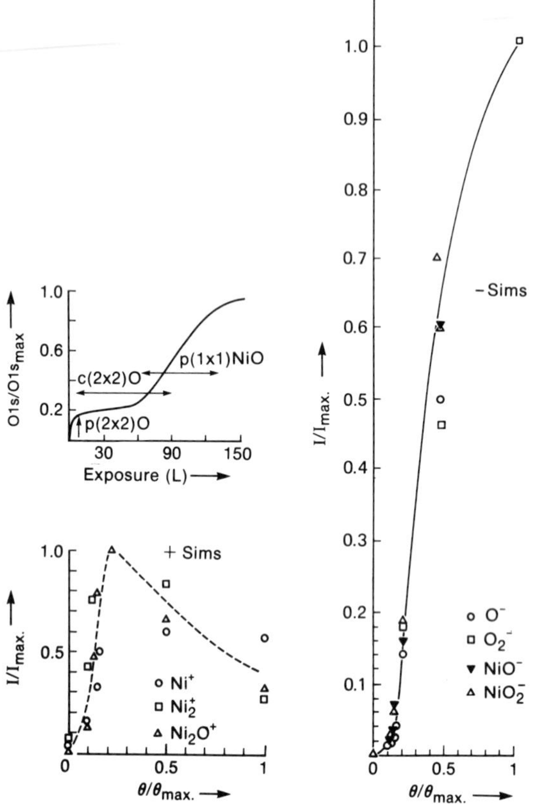

Fig. 81. Variation of positive and negative SIMS ion intensities as a function of coverage, relative to saturation, for Ni{100} at 300 K. The relationship between relative coverage, as determined from XPS O(1s) intensities, exposure, and the regions of chemisorption and oxidation are shown in the upper left part of the figure (Hopster and Brundle [236]).

Ni{100}. An attempt was made to correlate changes in O^- and O_2^- intensity with exposure with p(2 × 2) O and c(2 × 2) O formation. The small intensity increase up to ~5–10 L exposure was correlated with p(2 × 2) O formation and the large increase in the 10–35 L range with c(2 × 2) O formation. The former correlation is probably with both p(2 × 2) and c(2 × 2) regions (see the Ni{100} and {111} results), and we suggest that the latter is completely incorrect and in fact it represents the oxidation growth stage in a like manner to the {110} and {111} results. This is supported by the fact that the Ni^+ and Ni_2^+ intensities are strongly decreasing above 5 L, which as we have discussed above signifies the onset of conversion of O_{ad} to oxide. The NiO^- and NiO_2^- curves (not shown) rise smoothly to saturation coverage over the oxide nucleation and growth stage. The O^- behavior is actually very similar to these curves except for one data point at ~50 L exposure [see Fig. 80(d)]. The authors arguments [266] relating the O^- "plateau" between 35 and 50 L exposure to c(2 × 2) O are entirely dependent on the accuracy of this one point. We therefore believe that the Fleisch et al. [266] results on {100} are largely compatible with the {110}, {111} and polycrystalline results, i.e. the ion cluster intensities can distinguish chemisorbed O from oxide. We do not believe any special sensitivity to p(2 × 2) O or c(2 × 2) O has been demonstrated. One difference from the Rieder results on {110} and {111} is that, in those studies, the positive ion signal rose again after completion of the oxide layer. The rise was associated either with the gradual removal of defects (Rieder's explanation) or with minority chemisorption of O on NiO (our explanation). This rise is not observed in the Ni{100} work. The discrepancy may reflect a difference in the adsorption behavior or it may be due to differences in the KE of Ar^+ used (2 kV Rieder, 3 kV Fleisch et al.) or the KE of the ions detected. That the impact ion energy can have a significant effect is shown in the work of Dawson and Tam [267] on polycrystalline Ni. They showed (Fig. 8 of ref. 267) that, using low-KE Ar^+, no increase in Ni^+ yield occurred over the very high exposure range whereas at higher KE, yields grew over this range by a factor of 2–3. Figure 80(e) shows the positive ion data of Dawson and Tam over the low-exposure region. One can see that the general trends are in accord with the statements already made. Dawson and Tam [267] also rightly warn about the danger of correlating intensities with specific features because of the possibility that minority species are being preferentially detected.

The second study on Ni{100}/O is by Hopster and Brundle [236]. The SIMS data were correlated with XPS and LEED over the zero-to-saturation exposure range. The results are replotted in Fig. 81. One can see that the intensities of all the positive ions track the chemisorbed O coverage, while all the strong negative ion intensities track the oxide growth. In this work some care was taken to form well-defined p(2 × 2) O and c(2 × 2) O LEED patterns and to examine their ion cluster yields carefully for any evidence of anything specific to geometry. Nothing obvious was found and it was concluded that, in the chemisorption coverage range, the intensities were purely a function of coverage and independent of overlayer geometry.

Summarizing this last section, it seems obvious that the practical usefulness of SIMS for the Ni/O system is in establishing when oxide nucleation starts. Indeed it might well be the most sensitive way of doing this. In addition it is sensitive to minority changes occurring at the surface after the majority uptake has ceased. Whether these changes relate to defect sites or to chemisorption on NiO remains to be determined.

4. Conclusions

After critically reviewing and intercomparing such an enormous body of data it would be nice if one could arrive at a definitive set of conclusions which provide a simple framework of "understanding" for oxygen reaction with these metals. Unfortunately reality is too complicated for this, which may account for the dearth of critical reviews in this area. Some comments can be made, however.

First, as always in surface science, more intercomparative work is required, not only between techniques, but also between faces and between metals. Relying on individual techniques can easily be misleading. The ESD, SIMS, and TDS data reviewed here have provided examples. LEED has proved misleading in some areas where there are a combination of states present, such as large amounts of subsurface oxygen in the presence of an ordered overlayer. If one does not intercompare faces and metals one is unable to establish whether there are trends which lend themselves to simple interpretation. In the present case, even with the vast amount of data reviewed, there is barely enough direct comparison between faces and metals to make general conclusions. As stated in Sect. 2.11 the metal–metal bond strength is the decisive factor in establishing the relative potential for initial oxidation. Thus the hierarchy is Ni $>$ Cr $>$ Mo $>$ W and, while it is hard to stop oxide-island nucleation on single-crystal Ni surfaces even at low coverage ($\theta < 0.5$), it is very hard under normal UHV conditions at 300 K to get W to pass beyond the overlayer adsorption stage. The hierarchy for faces is generally related to the openness, i.e. in inverse proportion to the number of metal–metal bonds at the surface. For the b.c.c. series Cr, Mo, W, the order is $\{111\} > \{100\} > \{110\}$. For f.c.c. Ni, the series should be $\{110\} > \{100\} > \{111\}$, but, in fact oxide nucleation on the flattest surface, the $\{111\}$, is faster than on the $\{100\}$ surface. This inconsistency is a good example of a simple qualtitative understanding being thwarted because of the intrusion of other factors. In this case we believe that the protecting role of the ordered chemisorbed overlayer is crucial. The overlayer O structures on Ni$\{111\}$ are much less effective in inhibiting further oxygen dissociation and subsequent nucleation than is the c(2×2) O structure on Ni(100).

The role of steps and defects can also be critical, both for controlling the nature and orientation of adsorbed overlayers, and in determining oxide growth. A number of such examples are reported here for the W$\{110\}$/O system. We believe that, as direct observation of oxide initiation becomes possible using the enormous resolving power of the STM, it will become more apparent just how important surface imperfections are in controlling initial oxidation behavior.

The dynamics and mechanism of initial dissociative oxygen chemisorption need much more extensive and systemic studies of S as a function of θ, T_s and T_g before great faith is put in detailed models of the process invoking precursor or other effects. It can become apparent from the comparison of flat and stepped W surfaces that the traditional model of a precursor-induced dissociative adsorption, where incoming molecules are instantaneously accommodated to T_s in a precursor state and subsequently diffuse to a dissociation site, cannot be correct. Partial accommodation and rapid travel by the "hot" molecule over long distances to dissociation sites must be considered.

For the mechanistic interpretation of kinetics in the oxide nucleation stages for Ni it is equally clear that parameter fitting to non-unique models does not necessarily give much insight. For example, a good fit can be obtained to the Holloway and Hudson [231] island-growth model on Ni{100} which involves diffusion of molecular oxygen at 300 K, yet we know that precursor behavior for Ni{100}/O_2 is negligible at this temperature. Subsequently it is found that a model involving direct impingement of oxygen at the oxide nuclei perimeter will fit the data just as well.

Even establishing whether a particular chemisorbed phase consists of atomic or molecular oxygen on these metals has proven difficult. A number of earlier claims of detection of molecular oxygen phases have proven false and in fact only for W{110} below 45 K has a chemisorbed molecular phase been positively identified. As should be expected when dissociation is essentially unactivated, this molecular state fills in only after the completion of the dissociated state, even at 20 K.

Finally, we note that, though mundane, the influence of impurities on oxygen adsorption cannot be ignored. There have been numerous cases where the interaction of H_2O, CO_2, and possibly CO has affected both the kinetics and the resultant products during oxygen adsorption. On various occasions OH, CO, CO_2, and CO_3 species at the surface have been mistaken for atomic or molecular states of oxygen. Though pointed out on many occasions, these types of mistake continue to proliferate. A recent typical example of this is where the O(1*s*) XPS signature of surface carbonates and hydroxides on the $Y_1Ba_2Cu_3O_7$ superconductor compound have been mistaken for an elusive "special state" of oxygen in the material that might be responsible for the superconductivity [320].

References

1 D. Fowler, Ph.D. Thesis, Cornell University, Ithaca, NY, 1983.
2 W. Unertl and J.M. Blakely, Surf. Sci., 69 (1977) 23.
3 R.W. Joyner, K. Kishi and M.W. Roberts, Proc. R. Soc. London Ser. A, 358 (1977) 223.
4 K.O. Legg, F.P. Jona, P.W. Jepsen and P.M. Marcus, J. Phys. C, 8 (1975) 4492.
5 R.G. Smeenk, Thesis, University of Utrecht, 1982.
6 K. Griffiths, C. Kendon, D.A. King and J.B. Pendry, Phys. Rev. Lett., 46 (1981) 1584.
7 H.F. Winters, P. Morgen, S. Tougaard and J. Onsgaard, Proc. 4th Int. Conf. Solid Surf., Vide, 201 (1981) 303.
8 M.A. Barteau and R.J. Madix, in D.A. King and D.P. Woodruff (Eds.), The Chemical Physics of Solid Surfaces and Heterogeneous Catalysis, Vol. 4, Elsevier, Amsterdam, 1982.
9 M.K. Debe and D.A. King, J. Phys. C, 10 (1977) L303.
10 T.E. Felter, R.A. Barker and P.J. Estrup, Phys. Rev. Lett., 38 (1977) 1138.
11 G.B. Fisher, B.A. Sexton and J.L. Gland, J. Vac. Sci. Technol., 17 (1980) 144. J.L. Gland, B.A. Sexton and G.B. Fisher, Surf. Sci., 95 (1980) 587.
12 M.A. Barteau and R.J. Madix, Surf. Sci., 97 (1980) 101. B.A. Sexton and R.J. Madix, Chem. Phys. Lett., 76 (1980) 294.
13 R. Opila and R. Gomer, Surf. Sci., 105 (1981) 41.
14 D.A. King and M.G. Wells, Proc. R. Soc. London Ser. A, 339 (1974) 245.
15 N. Cabrera and N.F. Mott, Rep. Prog. Phys., 12 (1948–49) 163.

16 F.P. Fehlner and N.F. Mott, Oxid. Met., 2 (1970) 59.
17 C.R. Brundle, in A. Benninghoven, C.A. Evans, R.A. Powell, R. Shimuza and H.A. Storms (Eds.), Secondary Ion Mass Spectrometry, SIMS II, Springer Series in Chemical Physics, Vol. 9, Springer Verlag, Berlin, 1979.
18 R.C. Weast (Ed.), Handbook of Physics and Chemistry, Vol. 60, CRC Press, Boca Raton, FL, 1980.
19 D. Brennan, D.O. Hayward and B.M.W. Trapnell, Proc. R. Soc. London Ser. A, 256 (1960) 87.
20 E.G. King, W.W. Weller and A.U. Christensen, U.S. Dep. Inter. Bur. Mines, R.I. 5664 (1960).
21 H.M. Kennett, A.E. Lee and J.M. Wilson, Proc. R. Soc. London Ser. A, 331 (1972) 429.
22 W.F. Egelhoff, D.L. Perry and J.W. Linnett, J. Electron Spectrosc. Relat. Phenom., 5 (1974) 339.
23 L.H. Germer and J.W. May, Surf. Sci., 4 (1966) 452.
24 N.J. Taylor, Surf. Sci., 2 (1964) 544.
25 T.W. Haas and A.G. Jackson, J. Chem. Phys., 44 (1966) 2921.
26 J. Ferante and G.C. Barton, NASA Tech. Note D-4733 (1968).
27 H.M. Kennett and A.E. Lee, Surf. Sci., 33 (1972) 377.
28 J. Buchholz, G.C. Wang and M.G. Lagally, Surf. Sci., 49 (1975) 508.
29 M.A. Van Hove and S.Y. Tong, Phys. Rev. Lett., 35 (1975) 1892.
30 J.T. Grant and T.W. Haas, Surf. Sci., 17 (1969) 484.
31 S. Eklund and C. Leygraf, Surf. Sci., 40 (1973) 179.
32 A. Ignatiev, F. Jona, H.D. Shih, D.W. Jepsen and P.M. Marcus, Phys. Rev. B, 11 (1975) 4787.
33 L.J. Clark, Surf. Sci., 91 (1980) 131.
34 M.K. Debe and D.A. King, Surf. Sci., 81 (1979) 193.
35 T.E. Felter, R.A. Barker and P.J. Estrup, Phys. Rev. Lett., 38 (1977) 1138.
36 I. Stensgaard, L.C. Feldman and P.J. Silverman, Phys. Rev. Lett., 42 (1979) 247.
37 G. Gewinner, J.C. Peruchetti, A. Jaegle and R. Riedinger, Phys. Rev. Lett., 43 (1979) 935.
38 J.S. Foord, A.P.C. Reed and R.M. Lambert, Surf. Sci., 129 (1983) 79.
39 J.E. Inglesfield, J. Phys. C, 11 (1978) L69.
40 E. Tosatti, Solid State Commun., 25 (1978) 637.
41 M.C. Desjonquerés and F. Cyrot-Lackmann, J. Phys. F, 5 (1975) 1368.
42 G. Allen, Surf. Sci., 74 (1978) 79.
43 C.M. Bertoni, C. Cabandra and F. Mangli, Solid State Commun., 21 (1977) 121.
44 O. Bisi, C. Calandra, P. Flaviani and F. Mangli, Solid State Commun., 21 (1977) 121.
45 S. Weng and E.W. Plummer, Solid State Commun., 23 (1977) 518.
46 C. Noquera, D. Spanjaard, D. Jepsen, Y. Ballu, C. Guillot, J. Lecante, J. Paique, Y. Petroff, R. Pinchoux, P. Thiry and R. Cinti, Phys. Rev. Lett., 38 (1977) 1171.
47 J.C. Peruchetti, G. Gewinner and A. Jaegle, Surf. Sci., 88 (1979) 479.
48 B. Feuerbacher and R.F. Willis, Phys. Rev. Lett., 36 (1976) 1339.
49 B. McCarroll, J. Chem. Phys., 48 (1967) 863.
50 R.G. Musket, J. Less Common Met., 22 (1970) 175.
51 P. Michel and C. Jardin, Surf. Sci., 36 (1973) 478.
52 C.A. Haque and H.E. Farnsworth, Surf. Sci., 1 (1964) 378.
53 R.H. Sailors, G.L. Liedl and R.E. Grace, J. Appl. Phys., 38 (1967) 4928.
54 G. Gewinner, J.C. Peruchetti, A. Jaegle and A. Kalt, Surf. Sci., 78 (1978) 439.
55 D.L. Adams and L.H. Germer, Surf. Sci., 26 (1971) 109.
56 J.E. Hulse, K. Wandelt, J. Küppers and G. Ertl, Proceedings of the 4th International Conference on Solid Surfaces, Cannes, 1980, p. 108.
57 J. Oudar, Physics and Chemistry of Surfaces, Blockie, Edinburgh, 1975.
58 K.R. Lawless, Rep. Prog. Phys., 37 (1974) 231.
59 P.H. Holloway and J.B. Hudson, Surf. Sci., 43 (1974) 123.
60 P.H. Holloway and J.B. Hudson, Surf. Sci., 43 (1974) 141.
61 H.M. Kennett and A.E. Lee, Surf. Sci., 48 (1975) 591, 606, 617, 624.
62 K. Hayek, H.E. Farnsworth and R.L. Park, Surf. Sci., 10 (1968) 429.

63 B.M. Zukov, D.S. Ikonnikov and V.K. Tskhakaua, Sov. Phys. Solid State, 17 (1975) 163.
64 H.C.A. Kan and S. Feuerstein, J. Chem. Phys., 50 (1969) 3618.
65 E. Bauer and H. Poppa, Surf. Sci., 88 (1979) 31.
66 R. Riwan, C. Guillot and J. Paique, Surf. Sci., 47 (1973) 183.
67 J. Ferrante and G.C. Barton, NASA Tech. Note D-4735 (1963).
68 T. Engel, H. Niehus and E. Bauer, Surf. Sci., 52 (1975) 237.
69 E. Bauer and T. Engel, Surf. Sci., 71 (1981) 695.
70 J.C. Tracy and J.M. Blakely, Surf. Sci., 15 (1969) 257.
71 G.C. Wang, T.M. Lu and M.G. Lagally, J. Chem. Phys., 69 (1978) 479.
72 J.L. Desplat and C.A. Papageorgopoulos, Surf. Sci., 92 (1980) 97.
73 C.A. Papageorgopoulos and J.M. Chen, Surf. Sci., 39 (1973) 313.
74 J.L. Desplat, Proc. 2nd Int. Conf. Solid Surf., 1974.
75 P.J. Estrup and J. Anderson, Proc. 27th Phys. Electron. Conf., 1967, p. 47 (unpublished).
76 A.M. Bradshaw, D. Menzel and M. Steinkilberg, J. Appl. Phys. Suppl. 2, Pt. 2, (1974) 841; Discuss. Faraday Soc., 58 (1974) 46.
77 E. Bauer, H. Poppa and V. Viswanath, Surf. Sci., 58 (1976) 517.
78 D.A. Gorodelskii and Y.P. Mel'nik, Bull. Acad. Sci. USSR Phys. Ser., 35 (1971) 978.
79 P.E. Luscher and F.M. Propst, J. Vac. Sci. Technol., 14 (1977) 400.
80 J. Anderson and W.E. Danforth, J. Franklin Inst., 279 (1965) 160.
81 A.J. Melmed, H.P. Lauer and J. Kruger, Surf. Sci., 9 (1968) 476.
82 B.J. Hopkins, G.D. Watts and A.R. Jones, Surf. Sci., 52 (1975) 715.
83 H. Niehus, Surf. Sci., 80 (1979) 245; 87 (1979) 561.
84 N.J. Taylor, Surf. Sci., 2 (1964) 544.
85 T.E. Madey, J.J. Czyzewski and J.T. Yates, Jr., Surf. Sci., 49 (1975) 465.
86 G.J. Dooley, III and J.W. Haas, J. Vac. Sci. Technol., 7 (1970) 590; J. Chem. Phys., 52 (1970) 461.
87 D. Tabor and J.M. Wilson, J. Cryst. Growth, 9 (1971) 60.
88 A.E. Lee and K.E. Singer, Proc. R. Soc. London Ser. A, 323 (1971) 523.
89 N.R. Avery, Surf. Sci., 33 (1972) 107; 41 (1974) 533.
90 H.M. Kramer and E. Bauer, Surf. Sci., 92 (1980) 53; 93 (1980) 407.
91 E. Preuss, Surf. Sci., 94 (1980) 249.
92 J.C. Tracy and J.M. Blakely, Surf. Sci., 13 (1969) 313.
93 Y. Ballu, J. Lecante and H. Rousseau, J. Chem. Phys., 14 (1976) 3201.
94 J.M. Baker and D.E. Eastman, J. Vac. Sci. Technol., 10 (1973) 223.
95 T.E. Felter and P.J. Estrup, Appl. Surf. Sci., 1 (1977) 120.
96 M.L. Yu, J. Vac. Sci. Technol., 15 (1978) 668; Phys. Rev. B, 19 (1979) 5995.
97 B.J. Hopkins and M. Ibrahim, Vacuum, 23 (1973) 135.
98 C. Boizian, C. Garot, R. Nuvolone and J. Roussel, Surf. Sci., 91 (1980) 313.
99 R.M. Lambert, J.W. Linnett and J.A. Schwarz, Surf. Sci., 26 (1971) 572.
100 L.C. Walters and R.E. Grace, J. Appl. Phys., 26 (1965) 2331.
101 E. Bauer, Surf. Sci., 7 (1967) 351.
102 T. Engel, H. Niehus and E. Bauer, Vide, 164 (1973) 82.
103 H. Niehus and E. Bauer, Surf. Sci., 47 (1975) 222.
104 T. Engel, T. Van dem Hagen and E. Bauer, Surf. Sci., 62 (1977) 361.
105 S. Prigge, H. Niehus and E. Bauer, Surf. Sci., 65 (1977) 141.
106 S. Prigge, H. Niehus and E. Bauer, Surf. Sci., 75 (1978) 635.
107 C. Wang and R. Gomer, Surf. Sci., 84 (1979) 329.
108 C. Kohrt and R. Gomer, J. Chem. Phys., 52 (1969) 3283.
109 C. Leung and R. Gomer, Surf. Sci., 59 (1976) 638.
110 C. Steinbrüchel and R. Gomer, Surf. Sci., 67 (1977) 21.
111 C. Wang and R. Gomer, Surf. Sci., 74 (1978) 389.
112 J.R. Chen and R. Gomer, Surf. Sci., 79 (1979) 413.
113 H. Michel, R. Opila and R. Gomer, Surf. Sci., 105 (1981) 48.
114 T.E. Madey and J.T. Yates, Jr., Surf. Sci., 63 (1977) 203.

115 T.E. Madey, Surf. Sci., 94 (1980) 483.
116 T.E. Madey, Surf. Sci., 29 (1972) 571.
117 T.E. Madey, J.J. Czyzewski and J.T. Yates, Surf. Sci., 57 (1976) 580.
118 T.E. Madey, R. Stockbauer, J.F. Van der Veen and D.E. Eastman, Phys. Rev. Lett., 45 (1980) 187.
119 J. Buchholz and M.G. Lagally, Phys. Rev. Lett., 35 (1975) 442.
120 T.M. Lu, G.C. Wang and M.G. Lagally, Phys. Rev. Lett., 39 (1977) 411.
121 G.C. Wang, T.M. Lu and M.G. Lagally, J. Chem. Phys., 69 (1978) 479.
122 M.G. Lagally, G.C. Wang and T.M. Lu, in R. Vaneslow (Ed.), Chemistry and Physics of Solid Surfaces, Vol. II, C.R.C. Press, Cleveland, 1980.
123 W.Y. Ching, D.L. Huber, M.G. Lagally and G.C. Wang, Surf. Sci., 77 (1978) 550.
124 G.C. Wang and M.G. Lagally, Surf. Sci., 81 (1979) 69.
125 T.M. Lu, G.C. Wang and M.G. Lagally, Surf. Sci., 92 (1980) 133.
126 M.G. Lagally, T.M. Lu and D.G. Welkie, J. Vac. Sci. Technol., 17 (1980) 223.
127 T.M. Lu, G.C. Wang and M.G. Lagally, Surf. Sci., 107 (1981) 494.
128 M.A. Van Hove and S.Y. Tong, Phys. Rev. Lett., 35 (1975) 1092.
129 M.A. Van Hove, S.Y. Tong and M.H. Elconin, Surf. Sci., 64 (1977) 85.
130 J.I. Gersten, R. Janow and N. Tzoar, Phys. Rev. Lett., 36 (1976) 610.
131 R. Janow and N. Tzoar, Surf. Sci., 69 (1977) 253.
132 H. Niehus, Surf. Sci., 78 (1978) 667.
133 H. Froitzheim, H. Ibach and S. Lehwald, Phys. Rev. B, 14 (1976) 1362.
134 J.M. Baker and D.E. Eastman, J. Vac. Sci. Technol., 10 (1973) 223.
135 T.M. Duc, G. Guillott, Y. Lassailly, J. Lecante, Y. Jugnet and J.C. Vedrine, Phys. Rev. Lett., 43 (1979) 789.
136 J.C. Fuggle and D. Menzel, Surf. Sci., 79 (1979) 1.
137 G. Treglia, M.C. Desjonquéres, D. Spanjaard, Y. Lassailly, G. Guillot, Y. Jugnet, T.M. Duc and J. Lecante, J. Phys. C, 14 (1981) 3463.
138 B. Feuerbacher and M.R. Adriaens, Surf. Sci., 45 (1974) 553.
139 B. Feuerbacher, Surf. Sci., 47 (1975) 115.
140 J.F. Van der Veen, F.J. Himpsel and D.E. Eastman, Phys. Rev. B, 25 (1982) 7388.
141 J.T. Yates, Jr. and N.E. Erickson, Surf. Sci., 44 (1974) 489.
142 R. Butz and H. Wagner, Surf. Sci., 63 (1977) 448.
143 M. Bowker and D.A. King, Surf. Sci., 94 (1980) 564.
144 K.J. Rawlings, Surf. Sci., 99 (1980) 507.
145 M.G. Wells and D.A. King, J. Phys. C, 7 (1974) 4053.
146 Y.P. Zingermann and V.A. Ishchuk, Fiz. Tver. Tela, 9 (1967) 2529.
147 U. Banninger and E.B. Bas, Surf. Sci., 50 (1975) 279.
148 M.J. Grunze, J. Fuhler, M. Neumann, C.R. Brundle, D.J. Auerbach and R.J. Behm, Surf. Sci., 139 (1984) 109.
149 K. Besocke and S. Berger, Proc. 7th Int. Vac. Congr. and 3rd Int. Conf. Solid Surf., Vienna, 1977, Vol. 11, p. 893.
150 E. Bauer, Adsorption et Croissance Crystalline, C.C.N.R.S., Paris, 1965, p. 20.
151 K.J. Matysik, Surf. Sci., 38 (1973) 93.
152 M.G. Lagally, J.C. Buchholz and G.C. Wang, J. Vac. Sci. Technol., 12 (1975) 213.
153 G. Theodoreau, Surf. Sci., 81 (1979) 379.
154 G. Ertl and D. Schillinger, J. Chem. Phys., 66 (1977) 2569.
155 E.D. Williams, S.L. Cunningham and W.H. Weinberg, J. Chem. Phys., 68 (1978) 4688.
156 T.L. Einstein, Surf. Sci., 84 (1979) L497.
157 C.R. Brundle, unpublished results.
158 J.Q. Broughton and P.S. Bagus, J. Electron Spectrosc. Relat. Phenom., 20 (1980) 127, 261. P.S. Bagus and C.W. Bauschlicher, J. Electron Spectrosc. Relat. Phenom., 20 (1980) 183.
159 G. McGuire and T.A. Carlson, J. Electron Spectrosc. Relat. Phenom., 1 (1972) 161.
160 P. Alnot, C.R. Brundle, D.J. Auerbach, J. Behm and A. Viescas, Surf. Sci. 213 (1989) 1.
161 B.J. Hopkins, C.B. Williams and P.C. Wilmer, Surf. Sci., 25 (1971) 633.

162 N.R. Avery, Surf. Sci., 61 (1976) 391.
163 S.P. Singh-Boparai, M. Bowker and D.A. King, Surf. Sci., 53 (1975) 55.
164 E. Umbach, Thesis, Munich, 1978.
165 P. Kleban and R. Flagg, Surf. Sci., 103 (1981) 552.
166 D.A. King and G. Thomas, Surf. Sci., 92 (1980) 201.
167 R.A. Barker and P.J. Estrup, J. Chem. Phys., 74 (1981) 1442.
168 P. Luscher, Ph.D. Thesis, University of Illinois, Urbana, IL, 1977.
169 R. Gomer and J.K. Hulm, J. Chem. Phys., 27 (1957) 1363.
170 L.R. Clavenna and L.D. Schmidt, Surf. Sci., 33 (1972) 11.
171 M. Bacal, J.L. Desplat and T. Alleau, J. Vac. Sci. Technol., 9 (1972) 851.
172 J. Hölzl and J. Schäfer, Surf. Sci., 108 (1981) L387.
173 J.F. Wendelken and J. Kirschner, Surf. Sci., 110 (1981) 1.
174 A.M. Bradshaw and D. Menzel, Ber. Bunsenges. Phys. Chem., 78 (1974) 1140.
175 B.J. Waclawski, T.V. Vorburger and R.J. Stein, J. Vac. Sci. Technol., 12 (1975) 301.
176 J.T. Yates, T.E. Madey, E. Erickson and S.D. Worley, Chem. Phys. Lett., 39 (1976) 113.
177 N.P. Vasko, Y.G. Ptushinskii and A.A. Mitryaev, Sov. Phys. Solid State, 15 (1975) 1942.
178 M.L. Tarng and G.K. Wehner, J. Appl. Phys., 44 (1974) 1534.
179 M.L. Knotek and P.J. Feibelman, Phys. Rev. Lett., 40 (1978) 964; Phys. Rev. B, 18 (1978) 6531.
180 P.A. Redhead, Can. J. Phys., 42 (1964) 886.
181 D. Menzel and R. Gomer, J. Chem. Phys., 41 (1964) 3311.
182 D.P. Woodruff, M.M. Traum, H.H. Farrell, N.V. Smith, P.D. Johnson, D.A. King, R.L. Benbow and Z. Hurych, Phys. Rev. B, 21 (1980) 5642.
183 P.R. Antoniewicz, Phys. Rev. B, 21 (1980) 3811.
184 S.L. Weng, Phys. Rev. B, 23 (1981) 1699.
185 D.A. King, T.E. Madey and J.T. Yates, Jr., J. Chem. Soc. Faraday Trans. 1, 68 (1972) 1347.
186 S.W. Bellard and E.M. Williams, Surf. Sci., 80 (1979) 450.
187 D.A. King, T. Madey and J.C. Yates, J. Chem. Phys., 55 (1971) 3236.
188 Y.G. Ptushinskii and B.A. Chuikov, Sov. Phys. Solid State, 10 (1968) 565.
189 J.A. McHugh, in A.W. Czanderna (Ed.), Methods of Surface Analysis, Elsevier, Amsterdam, 1975.
190 A. Benninghoven, Surf. Sci., 53 (1973) 596.
191 R.J. Blattner and C.A. Evans, in Scanning Electron Microscopy, IV, 1980, SEM, AMF O'Hare, Chicago.
192 A. Benninghoven, E. Loebach, C. Plog and N. Trietz, Surf. Sci., 39 (1973) 397.
193 M. Yu, Surf. Sci., 71 (1978) 121.
194 S. Andersson, B. Kasemo, J.B. Pendry and M.A. Van Hove, Phys. Rev. Lett., 31 (1973) 595.
195 J.E. Demuth and T.N. Rhodin, Surf. Sci., 45 (1974) 249.
196 P.M. Marcus, J.E. Demuth and J.W. Jepsen, Surf. Sci., 53 (1975) 501. See also J.E. Demuth, D.W. Jepsen and P.M. Marcus, Phys. Rev. Lett., 31 (1973) 540.
197 G. Hanke, E. Lang, K. Heinz and K. Müller, Surf. Sci., 91 (1980) 551.
198 S.Y. Tong and K.H. Lau, Phys. Rev. B, 25 (1982) 7382.
199 J.E. Demuth, N.J. DiNardo and G.S. Cargill, III, Phys. Rev. Lett., 50 (1983) 1373.
200 J. Stöhr, R. Jaeger and T. Kendelewicz, Phys. Rev. Lett., 49 (1982) 142.
201 K.H. Rieder, IBM Internal Rep. RZ1166.
202 K.H. Rieder, IBM Internal Rep. RZ1173; Surf. Sci., 128 (1983) 325.
203 J.A. Barker and I.P. Batra, Phys. Rev. B, 27 (1983) 3138.
204 T. Engel and K.H. Rieder, IBM Internal Rep. RZ1181; Surf. Sci., 148 (1984) 321.
205 E. Taglauer and W. Heiland, Surf. Sci., 47 (1974) 234.
206 H.H. Brongersma and J.B. Theeten, Surf. Sci., 54 (1976) 519.
207 K.K. Verheij, J.A. Van den Berg and D.G. Armour, Surf. Sci., 84 (1979) 408.
208 J.A. Van den Berg, L.K. Verheij and D.G. Armour, Surf. Sci., 91 (1980) 208.
209 R.G. Smeenk, R.M. Tromp, J.W.M. Frenken and F.W. Saris, Surf. Sci., 112 (1981) 261.
210 R.G. Smeenk, Thesis, University of Utrecht, 1982.

211 J.F. Van der Veen, R.G. Smeenk, R.M. Tromp and F.W. Saris, Surf. Sci., 79 (1979) 212.
212 R.G. Smeenk, R.M. Tromp, J.F. Van der Veen and F.W. Saris, Surf. Sci., 95 (1980) 156.
213 R.G. Smeenk, R.M. Tromp and F.W. Saris, Surf. Sci., 107 (1981) 429.
214 J.W.M. Frenken, R.G. Smeenk and J.F. Van der Veen, Surf. Sci., 135 (1983) 147.
215 T. Narusawa, W.M. Gibson and E. Törnquist, Phys. Rev. Lett., 47 (1981) 417.
216 D.H. Rosenblatt, J.G. Tobin, M.G. Mason, R.F. Davis, S.D. Kevan, D.A. Shirley, C.H. Li and S.Y. Tong, Phys. Rev. B, 23 (1981) 828.
217 L.G. Petersson, S. Kono, N.F.T. Hall, S. Goldberg, J.T. Lloyd, C.S. Fadley and J.B. Pendry, Mater. Sci. Eng., 42 (1980) 111.
218 G. Armand and J.B. Theeten, Solid State Commun., 13 (1973) 563.
219 S. Andersson, Solid State Commun., 20 (1976) 229.
220 S. Andersson, Surf. Sci., 79 (1979) 385.
221 G. Dalmai-Imelik, J.C. Bertolini and J. Rousseau, Surf. Sci., 62 (1977) 67.
222 (a) S. Lehwald and H. Ibach in R. Caudano, J.M. Gilles and A.A. Lucas (Eds.), Vibrations at Surfaces, Plenum Press, New York, 1982. (b) T.S. Rahman, D.L. Mills, J.E. Black, J.M. Szeftel, S. Lehwald and H. Ibach, Phys. Rev. B, 30 (1984) 589.
223 G. Allan and J. Lopez, Surf. Sci., 95 (1980) 214.
224 T.S. Rahman, J.E. Black and D.L. Mills, Phys. Rev. Lett., 46 (1981) 1469.
225 S. Masuda, M. Nishijima, Y. Sakisaki and M. Onchi, Phys. Rev. B, 25 (1982) 863.
226 S. Andersson, P.A. Karlsson and M. Persson, Phys. Rev. Lett., 51 (1983) 2378.
227 D.F. Mitchell, P.B. Sewell and M. Cohen, Surf. Sci., 61 (1976) 355.
228 D.F. Mitchell, P.B. Sewell and M. Cohen, Surf. Sci., 69 (1977) 310.
229 T.W. Capehart and T.N. Rhodin, J. Vac. Sci. Technol., 16 (1979) 594.
230 H.D. Hagstrum and G.E. Becker, J. Chem. Phys., 54 (1971) 1015.
231 P.H. Holloway and J.B. Hudson, Surf. Sci., 43 (1974) 123.
232 P.H. Holloway and J.B. Hudson, Surf. Sci., 43 (1974) 141.
233 P.H. Holloway and R.A. Outlaw, Surf. Sci., 111 (1981) 300.
234 D.E. Taylor and R.L. Park, unpublished results.
235 A.R. Kortan and R.L. Park, Phys. Rev. B, 23 (1981) 6340.
236 H. Hopster and C.R. Brundle, J. Vac. Sci. Technol., 16 (1979) 548.
237 H. Hopster and C.R. Brundle, Vide, 201 (1981) 275.
238 R.J. Behm and C.R. Brundle, J. Vac. Sci. Technol. A, 1(2) (1983) 1223.
239 T.H. Upton and W.A. Goddard, Phys. Rev. Lett., 46 (1981) 1635; Critical Reviews in Solid State and Material Science, CRC Press, Boca Raton, FL, 1981.
240 C.W. Bauschlicher, S.P. Walch, P.S. Bagus and C.R. Brundle, Phys. Rev. Lett., 50 (1983) 864.
241 A. Winkler, K.D. Rendulic and K. Wendl, Appl. Surf. Sci., 14 (1982) 209.
242 H.T. Liu, A.F. Armitage and D.P. Woodruff, Surf. Sci., 114 (1982) 440.
243 P.R. Norton, R.L. Tapping and J.W. Goodale, Surf. Sci., 65 (1977) 13.
244 C.R. Brundle and H. Hopster, J. Vac. Sci. Technol., 18 (1981) 663.
245 C.R. Brundle, R.J. Behm and J.A. Barker, J. Vac. Sci. Technol. A, 2 (1984) 1038.
246 R. Miranda, J.M. Rojo and M. Salmeron, Solid State Commun., 35 (1980) 83.
247 C.R. Brundle, in U. Landman (Ed.), AIP Conf. Proc. No. 61, AIP, New York, 1980, p. 57.
248 G.B. Fisher, Surf. Sci., 62 (1977) 31.
249 D.E. Eastman and J.K. Cashion, Phys. Rev. Lett., 27 (1971) 1520.
250 H. Conrad, G. Ertl, J. Küppers and E.E. Latta, Solid State Commun., 17 (1975) 497.
251 S.P. Weeks and E.W. Plummer, Solid State Commun., 21 (1977) 695.
252 S.Y. Tong, C.H. Li and A.R. Lubinsky, Phys. Rev. Lett., 39 (1977) 498.
253 K. Jacobi, M. Scheffler, K. Kambe and F. Forstmann, Solid State Commun., 22 (1977) 17.
254 D. Jepsen, G. Noguera, D. Spanjaard, G. Guillot, Y. Ballu and P. Thiry, Solid State Commun., 28 (1978) 741.
255 A. Liebsch, Phys. Rev. B, 17 (1978) 1653.
256 C.S. Wang and A.J. Freeman, Phys. Rev. B, 19 (1979) 4930.
257 S.P. Walch and W.A. Goddard, Solid State Commun., 23 (1977) 907.

258 S.P. Walch and W.A. Goddard, III, Surf. Sci., 75 (1978) 609.
259 I.P. Batra and O. Robaux, Surf. Sci., 49 (1975) 653.
260 C.H. Li and J.W.D. Connolly, Surf. Sci., 65 (1977) 700.
261 D.E. Ellis, H. Adachi and F.W. Averill, Surf. Sci., 58 (1976) 497.
262 D.W. Bullett and M.L. Cohen, J. Phys. C, 10 (1977) 2101.
263 J.H. Gallagher, R. Haydock and V. Heine, J. Phys. C, 12 (1979) L13.
264 C.W. Bauschlicher and P.S. Bagus, Phys. Rev. Lett., 52 (1984) 200.
265 A. Müller and A. Benninghoven, Surf. Sci., 41 (1974) 493.
266 T. Fleisch, N. Winograd and W.N. Delgass, Surf. Sci., 78 (1978) 141.
267 P.H. Dawson and W.C. Tam, Surf. Sci., 81 (1979) 264.
268 K.H. Müller, P. Beckmann, M. Schemmer and A. Benninghoven, Surf. Sci., 80 (1979) 325.
269 K.H. Rieder, Appl. Surf. Sci., 2 (1978) 74.
270 A.M. Horgan and D.A. King, Trans. Faraday Soc., 67 (1971) 2145.
271 P.K. de Bokx, F. Labohm, O.L.J. Gijzeman, G.A. Bootsma and J.W. Geus, Appl. Surf. Sci., 5 (1980) 321.
272 T. Matsudaira and M. Onchi, Solid State Commun., 29 (1979) 549.
273 K. Akimoto, Y. Sakisaka, M. Nishijima and M. Onchi, Surf. Sci., 82 (1979) 349.
274 C.A. Papageorgopoulos and J.M. Chen., Surf. Sci., 52 (1975) 40.
275 N.G. Krishnan, W.N. Delgass and W.D. Robertson, Surf. Sci., 57 (1976) 1.
276 K.S. Kim and N. Winograd, Surf. Sci., 43 (1974) 625.
277 M.A. Van Hove and S.Y. Tong, J. Vac. Sci. Technol., 12 (1975) 230.
278 J.B. Pendry, referred to as a private communication in ref. 196.
279 J.E. Demuth, private communication, 1982.
280 H. Ibach, private communication, 1983.
281 J.E. Demuth, private communication, 1983.
282 S. Lehwald, J.M. Szeftel, H. Ibach, T.S. Rahman and D.L. Mills, Phys. Rev. Lett., 50 (1983) 518.
283 C.S. Fadley, private communication, 1984.
284 D. Norman, P.J. Durham, J.B. Pendry, J. Stöhr and R. Jaeger, Proc. Int. Conf. EXAFS, September 1982.
285 K.H. Rieder, private communication, 1983.
286 C.R. Brundle, unpublished data.
287 C. Guillot, Y. Ballu, J. Paigne, J. Lecante, J. Thiry, P. Pinchaux and Y. Petroff, Proc. 3rd Int. Conf. Solid Surf., Vienna, 1977, p. 1159.
288 J.G. Lapeyre, Surf. Sci., 89 (1979) 304.
289 A. Liebsch, Phys. Rev. Lett., 38 (1977) 248.
290 D. Brennan and M.J. Graham, Discuss. Faraday Soc., 41 (1966) 95.
291 P.S. Bagus, private communication, 1984.
292 J.C. Tracy, J. Chem. Phys., 56 (1972) 2736.
293 P.H. Holloway, J. Vac. Sci. Technol. 18 (1981) 653.
294 R.N. Bloomer, Br. J. Appl. Phys., 8 (1957) 321.
295 W.H. Orr, Thesis, Cornell University, Ithaca, NY, 1962.
296 W. Avrami, J. Chem. Phys., 7 (1939) 1103; 8 (1940) 212; 9 (1941) 177.
297 C. Benndorf, B. Egert, C. Nöbl, H. Seidel and F. Thieme, Surf. Sci., 92 (1980) 636.
298 K. Jacobi and H.H. Rotermund, Surf. Sci., 116 (1982) 435.
299 Y.-P. Hsu, K. Jacobi and H.H. Rotermund, Surf. Sci., 117 (1982) 581.
300 K. Jacobi and H.H. Rotermund, Surf. Sci., 133 (1983) 401.
301 L.H. Germer and A.U. MacRae, J. Appl. Phys., 33 (1962) 2923; A.U. MacRae, Surf. Sci., 1 (1964) 319.
302 R.L. Park and H.E. Farnsworth, J. Appl. Phys., 35 (1964) 2220.
303 E. Bauer, Surf. Sci., 5 (1966) 152.
304 J.E. Demuth, J. Colloid Interface Sci., 58 (1977) 184.
305 W. Heiland and E. Taglauer, J. Vac. Sci. Technol., 9 (1972) 620.
306 W. Heiland and E. Taglauer, Surf. Sci., 68 (1977) 96.

307 J.E. Demuth, P.M. Marcus and D.W. Jepsen, Phys. Rev. B, 11 (1975) 1460.
308 J.W. May and L.H. Germer, Surf. Sci., 11 (1968) 443.
309 P.J. Kisliuk, Phys. Chem. Solids, 3 (1957) 95; 5 (1958) 78.
310 (a) M. Schuster and C. Varelas, Surf. Sci., 134 (1983) 195. (b) A.M. Baro, Ch. Gerber, E. Stoll, A. Baratoff, G. Binnig, H. Rohrer and F. Salvon, Phys. Rev. Lett., 52 (1984) 1304.
311 C. Davisson and L.H. Germer, Phys. Rev., 30 (1927) 705.
312 A.U. MacRae, Appl. Phys. Lett., 2 (1963) 88.
313 L.D. Roelofs, A.R. Kortan, T.L. Einstein and R.L. Park, J. Vac. Sci. Technol., 18 (1981) 492.
314 R.L. Park and H.E. Farnsworth, Appl. Phys. Lett., 3 (1963) 167.
315 E. Taglauer and W. Heiland, Surf. Sci., 47 (1975) 234.
316 H. Ibach and D. Bruchman, Phys. Rev. Lett., 44 (1980) 36.
317 G.E. Becker and H.D. Hagstrum, Surf. Sci., 30 (1972) 505.
318 R.P. Messmer, C.W. Tucker and K.H. Johnson, Surf. Sci., 42 (1974) 341.
319 M.L. Yu and N.D. Lang, Phys. Rev. Lett., 50 (1983) 127.
320 D.C. Miller, D. Fowler, W.Y. Lee and C.R. Brundle, in J.M. Harper, R.J. Cotton and L.C. Feldman (Eds.), Thin Film Processing and Characterization of High-temperature Superconductors, AIP Conference Proceedings No. 165, AIP, New York, 1988.

Chapter 4

The Adsorption of Carbon Monoxide by the Transition Metals

J.C. CAMPUZANO

Department of Physics, University of Illinois at Chicago, Chicago, IL 60680 (U.S.A.)

1. Introduction

Carbon monoxide adsorbs on all of the transition metals with initial heats of chemisorption ranging from 6.4 kcal mol^{-1} in the case of silver surfaces to 150 kcal mol^{-1} in the case of titanium and zirconium surfaces. As one proceeds from left to right in the periodic table, the tendency for dissociative chemisorption of CO on the transition metals decreases. However, even for metals on the right-hand side of the transition series, activated dissociative adsorption of CO is observed. Thus, the surface chemistry of carbon monoxide will be determined by the properties of the adsorbed molecule itself, as well as by the dissociation products which may form.

Carbon monoxide chemistry on transition metal surfaces is of enormous technological importance. Synthetic hydrocarbon fuels of the future will be produced by heterogeneous catalytic chemistry using CO derived from coal and from other carbon sources [1, 2]. This will be done with transition metal catalysts employed for Fischer–Tropsch synthesis of liquid hydrocarbons [3] or for the catalytic production of methane [4]. In addition, the synthesis of aldehydes and ketones from CO, H_2, and olefins (oxo-synthesis or hydroformulation reaction) is carried out over transition metal catalysts [5].

Aside from its technological importance, the chemisorption of CO by the transition metals forms part of the core of surface chemistry knowledge. Probably no other adsorbate has been as well studied as CO, due to three basic factors:

(1) the simplicity of CO compared with more complex molecules,

(2) the technological importance of CO-based surface chemistry, and

(3) the rich chemistry of the transition metal carbonyls.

In Fig. 1 we show the level of contemporary research on the surface chemistry of CO. From a computer survey of *Chemical Abstracts* for 1977–1979, using the keywords shown in the figure, we found a current annual average rate of publication of 253 papers on CO adsorption on the transition metals. It is clear that, as one enters Group VIII, interest grows in comparison with the elements having a lower number of *d* electrons. The work on tungsten surfaces is counter to this trend, and probably stems from the

References pp. 460–469

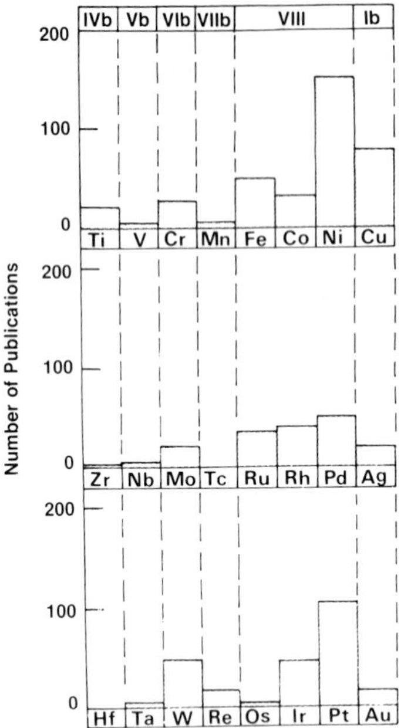

Fig. 1. Production of scientific publications concerned with CO adsorption on transition metals (1977–1979). Source: Chemical Abstracts. Keyboard combination: carbon monoxide, chemisor, adsorption, metal name. Average annual rate = 253 publications per year.

fact that Langmuir focused much of his attention on tungsten because of its use in incandescent bulbs [6]. Research workers have continued studying tungsten surfaces, adding to this large body of information.

The extensive work done with carbon monoxide adsorbed on the transition metals places us in a favorable position to make detailed comparisons from system to system. Generalizations derived in this manner are likely to be useful throughout the field of surface chemistry. Hence, the literature on CO surface chemistry is a valuable resource in the development of the field as a whole.

2. The carbon monoxide molecule

The result of an SCF-Xα–MSW calculation [7] of the upper valence orbitals of CO is shown in Fig. 2. The C atom is placed to the left. The highest filled orbital is the 5σ level which contains 2 electrons. The 2π orbital is antibonding and is empty in the free molecule, lying about 7 eV above the vacuum level. It is currently believed that, in the bonding of CO to transition metals either by chemisorption and surfaces or by the formation of ligands in metal carbonyls, only the 5σ and the 2π orbitals are involved. The conventionally accepted belief is that 5σ donation from CO is accompanied by donation of *sp* and *d* electrons from the metal into the 2π antibonding level,

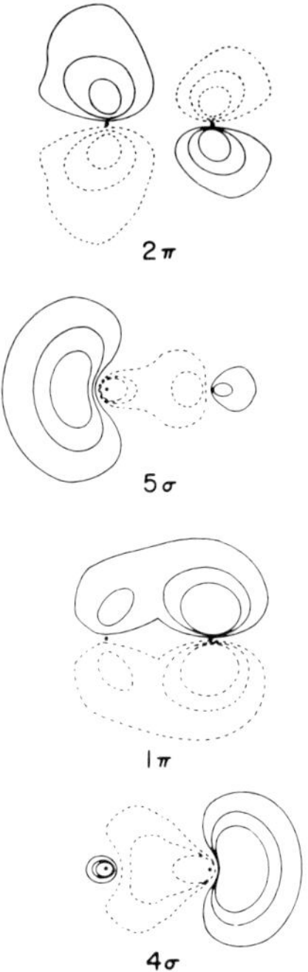

Fig. 2. Wave function contours of selected CO molecular orbitals. The solid and broken lines indicate contours of opposite sign having absolute values of 0.3, 0.2 and 0.1. The molecule is oriented with carbon on the left and oxygen on the right [10].

causing a weakening of the C–O bond [8]. On the other hand, the removal of charge from the 5σ level of CO results in a more uniform distribution in the 1π level, thus strengthening the C–O bond. This is evident in the gas phase: if one removes one electron from the 5σ level of CO, thereby converting it to CO^+, the C–O stretching frequency increases from $2143\,cm^{-1}$ in neutral CO to $2184\,cm^{-1}$ in CO^+, indicating higher bonding strength in CO^+. The gas-phase photoelectron spectrum for CO using 1253.7 eV X-radiation is shown in Fig. 3 with appropriate orbital labeling [8].

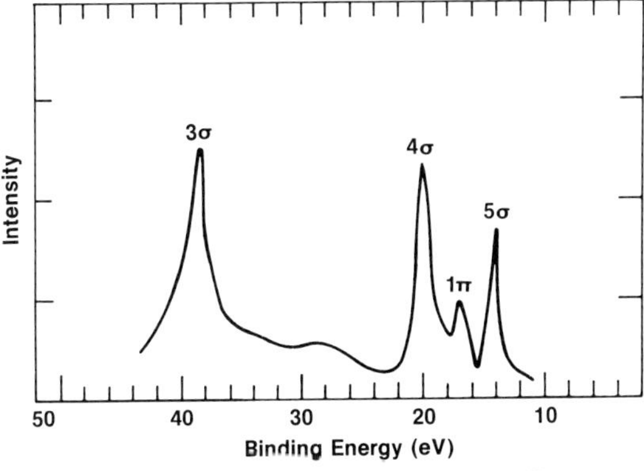

Fig. 3. Photoelectron spectrum of gaseous CO. X-ray energy = 1253.7 eV.

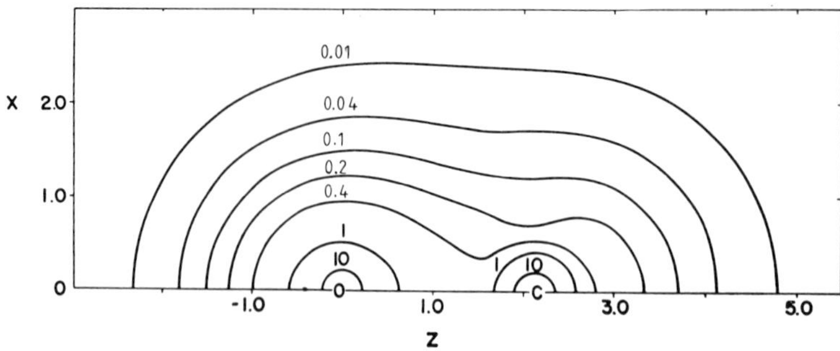

Fig. 4. Total molecular charge densities in gaseous CO. Dimensions on the x and z axes are in Å [10].

In a careful self-consistent field calculation, Huo [9] has determined the total molecular charge densities of CO, as shown in Fig. 4 where the dimensions on the two axes are in Å. However, the dipole moment calculated from her analysis is of the wrong sign. More recent calculations of the CO dipole moment by Grimaldi et al. [10] have shown that this difference may be explained by the coupling of single excitations to the SCF function through double excitations [11]. The accepted experimental dipole moment is 0.112 D with the charge distribution C^{-}–O^{+} [12, 13].

3. The bonding of CO in metal carbonyls

The bonding of CO to transition metal surfaces is thought to be similar to the bonding in metal carbonyls [14]. In carbonyls the bond is formed by transfer of electrons from the CO 5σ orbital to the metal, and transfer of charge from the metal atom into the 2π level of CO. As more electrons are transferred into the 2π molecular orbital, more electrons can be donated from the 5σ level of CO to the metal. This is called a synergic process. But the donation and back-donation of electrons have opposite effects on the strength of the C–O bond. As explained above, removal of charge from the 5σ level strengthens the bond. On the other hand, addition of charge to the antibonding 2π level weakens the bond. When the CO molecule is bonded to the metal, the C–O bond is weakened by 2π back-donation more than it is strengthened by 5σ donation. This effect can be seen from the spectrum of metal carbonyl complexes where the species with the highest metal–carbon stretching frequencies show the lowest carbon–oxygen stretching frequency [15]. (The relationship between charge transfer from the 5σ orbital and into the 2π orbital has been quantified by Baerends and Ros [16].) They have carried out a Hartree–Fock calculation of the orbital populations in carbonyl complexes, fitting them to the experimentally observed C–O stretching frequencies ($\omega_{C\text{–}O}$, in cm^{-1}) according to

$$\omega_{C\text{–}O} = -926.2P(5\sigma) - 707.2P(2\pi) + 4030.1$$

where $P(5\sigma)$ and $P(2\pi)$ refer to gross orbital electron populations.

Sheppard and Nguyen [17] have correlated the bonding of CO to multiple metal centers with frequency ranges for the C–O stretching motion. In general, as the CO coordination increases from one (terminally bonded CO) to two or three (bridge-bonded CO), the C–O frequency falls in both the ligands in carbonyls and in CO species chemisorbed on single crystal surfaces. This makes vibrational spectroscopy a useful tool in the determination of adsorption sites. The coordination is deduced from comparison of the C–O frequency in the adsorbed molecule to those of the equivalent carbonyl complex.

4. The bonding of CO to bulk metal surfaces

At room temperature, when CO adsorbs on transition metals in the left-hand region of the periodic table, rapid dissociation and high heats of chemisorption are observed. But for metals on the right-hand side of the table, molecular adsorption is often observed. Broden et al. [18] have divided the periodic table as shown in Fig. 5. The heavy line separates those elements exhibiting rapid dissociative chemisorption from those which tend to adsorb CO molecularly. But at sufficiently low temperatures, most transition metals on the left will adsorb CO molecularly, and at high temperatures, most of the transition metals will dissociate CO as the activation energy barrier for dissociation is surmounted. Benziger [19] has used thermodynamic arguments to show that molecular and dissociative adsorption are competing processes, and that the state of adsorption depends on pressure and temperature. Molecular adsorption predominates at low temperatures and high pressure, while dissociative adsorption is preferred at high temperature and low pressure. For molecular adsorption, the molecular orbital picture of Eischens and Pliskin [20] has been applied to metal surfaces in several molecular orbital calculations [21–26]. There have also been semi-empirical calculations based on the Newns–Anderson Hamiltonian [27]. Molecular orbital calculations provide an intuitive understanding of CO chemisorption, but for a more detailed understanding we must turn to more accurate

IIIB	IVB	VB	VIB	VIIB	VIII	VIII	VIII	IB
Sc	Ti D	V	Cr	Mn	Fe D 3.5	Co	Ni M 3.08	Cu
Y	Zr	Nb	Mo D 3.5	Tc	Ru M 3.15	Rh	Pd M 2.90	Ag
La	Hf	Ta	W D,M 3.2	Re	Os	Ir M 2.75	Pt M 2.60	Au

Fig. 5. Tendency of CO to dissociate on the transition metals at 300 K. Rapid dissociation occurs to the left of the heavy line on the clean metal [37].

calculations such as the one by Rosen et al. [28]. This is also a Hartree–Fock calculation of a Ni_5CO cluster, but one in which the exchange interaction has been approximated with a statistical exchange potential proportional to $\alpha\rho^{1/3}$ ($\alpha = 0.7$). This calculation shows that the CO (5σ) orbital interacts with Ni(3*d*) and Ni(4*sp*) levels. There is also some very localized bonding of the CO(1π) to the top Ni atom. Also, cluster states mix with the CO(2π) orbital, and it is interesting to note that the neighboring Ni atoms are more involved than the top Ni atom.

Since the original model of Eischens and Pliskin, it has been assumed that most of the bonding energy of CO to metals derives from the interaction of CO 5σ electrons with the metal. This belief has been reinforced by the observation of Eastman and Cashion [29] that the UPS spectrum of CO molecularly adsorbed on Ni(100) produces only two peaks, one at 8 eV and another at 11 eV below the Fermi level of the metal substrate. A very similar result is obtained for CO molecularly adsorbed on other metal surfaces. Fuggle et al. [24] had suggested that the peak at 8 eV consists of the overlapping 5σ- and 1π-derived levels. The 11 eV peak is the 4σ-derived level. This suggestion has been conclusively proven by Allyn et al. [25] using angular resolution, variable photon energy and Davenport's cross-section calculation [30]. As described above, the analogy with metal carbonyls becomes even more obvious. The similarity of the spectrum to that of gas-phase $Ni(CO_4)$ had, in fact, prompted Lloyd [31] to propose a similar interpretation to that of Fuggle et al. sometime earlier. The assignment, and the orientation of the molecule on the surface, has been confirmed by Smith et al. [32] and Apai et al. [33] using photoemission selection rules. The separation of the 4σ and 1π peaks is about 3 eV, similar to the gas-phase spectrum of CO. But the 5σ orbital is shifted by 3 eV. It was argued that, if the final state effects are the same for all levels, the shift reflects the fact that it is the 5σ level which interacts most with the substrate. The 1π peak of CO/Ni(100) is at a lower energy than the 5σ peak by 0.3–0.5 eV [25, 33], i.e. the order of the ionization energies is reversed relative to the free molecule. Allyn et al. [25] have shown that this is not the case for CO/Cu(100) where the 5σ level is centered at a few tenths of an eV below 1π. CO adsorbs much more weakly on Cu than on Ni, which would explain the smaller 5σ shift.

However, Bagus et al. [34] have proposed another interpretation. In a CSOV calculation they find that CO-to-metal σ donation only makes a small contribution to the adsorption energy. The metal-to-CO π donation (mostly from metal *d* electrons, although the *sp* contribution is not negligible) gives a contribution to the adsorption energy which is more than 2.5 times larger than the contribution from the ligand σ donation. If *d* electron correlation is included in the calculation, the π contribution is even larger. Hesketh et al. [35] have argued that the calculation of Bagus et al. [34] is not representative of CO adsorption on metallic Cu since there was no evidence of 2π levels in photoemission, and that when CO is weakly adsorbed, as in the case of Cu, there is no 2π participation at all. But McConville et al. [36] have now found clear evidence of photoemission from the 2π levels of CO/Cu(100). Other interesting conclusions of Bagus et al. are that the ligand and metal polarization also contribute significantly to their interaction energy, but the electrostatic interaction between ligand and metal is repulsive. They explain the large binding energy shift of the 5σ level by the fact that the occupied metal σ levels and ligand σ lone pair orbitals form bonding and antibonding combinations. The mixing of metal and ligand σ orbitals does not however imply that they contribute significantly to the adsorption energy. The lone pair binding energy shift shows that it is directed towards the metal.

Allyn et al. [25] were the first to show beyond reasonable doubt that the CO molecule at low coverages is bonded to Ni(100) through the carbon atom, with its molecular axis perpendicular to the surface. Horn et al. [37] have found the same results by using photoemission selection rules, and Petersson et al. [38] by using photoelectron diffraction. But there are cases in which CO adsorbs with its molecular axis tilted to the surface.

Hofmann et al. [39] have found that, when CO forms the (2×1)p1g1 structure on Pt(110), the axes of the CO molecules are tilted 26° away from the normal. Since then, other studies have found the axis tilted when CO is adsorbed on its own [40] or when co-adsorbed with Na [41].

Vibrational spectroscopy has shown that, as a CO overlayer compresses, the C–O stretching frequency can vary by more than 100 cm^{-1} [42, 43], partly due to adsorption site change and partly due to strong coupling between adsorbed CO molecules. Such coupling should induce some dispersion in the CO levels. The dispersion has indeed been observed by Horn et al. [37]. They find a dependence of the initial state energy on $k_{\parallel}$, the wavevector component parallel to the surface, as direct evidence for the partial wave-like character of the CO-derived levels.

5. Plan of this review

The chemisorption of CO on all of the transition metals is reviewed through mid-1986. Emphasis is placed on studies using the modern tools of surface science for investigation of CO adsorption on single-crystal surfaces, although some connections are also made with investigations of CO chemisorption on high-area, supported metal surfaces. No attempt was made to review the extensive literature dealing with the catalytic reactions of CO. While the emphasis of the review is on the properties of the transition metal interacting with CO, the author also hopes that he has dealt with some problems which have a significant impact on the field of surface science and catalysis in a more general way than strictly through the properties of CO itself. Indeed, CO should be considered as a molecule of significant historical importance to the field of surface science since many general principles have been discovered through studies of CO chemisorption.

References pp. 460–469

The following abbreviations are used in this article:

AES	Auger electron spectroscopy
EELS	electron energy loss spectroscopy (high resolution)
ESDIAD	electron-stimulated desorption ion angular distribution
LEED	low-energy electron diffraction
RAIRS	reflection–absorption infrared spectroscopy
S	sticking probability
SIMS	secondary ion mass spectrometry
TDS	thermal desorption spectroscopy
UPS	ultraviolet photoelectron spectroscopy
XPS	X-ray photoelectron spectroscopy
θ	coverage relative to the number of substrate atoms
ν	vibrational frequency

6. Titanium

The initial heat of adsorption of CO on Ti films at 273 K is very high (150 kcal mol^{-1}), suggesting that dissociative adsorption occurs [44]. The adsorption and thermal desorption of CO from polycrystalline Ti has been studied by Schwartz et al. [45]. The Ti was cleaned of S, C, and O using argon bombardment, yielding a surface containing small quantities of O following annealing. CO was found to yield two desorption states at ~1100 and ~400 K. Assuming first-order desorption with a standard pre-exponential factor of 10^{-13} s^{-1}, the 400 K state exhibited an activation energy for desorption of 20.3 kcal mol^{-1}. Further studies using UPS [46], soft X-ray appearance potential spectroscopy [47], AES [48], SIMS [49], and Auger electron appearance potential spectroscopy [50] have led to conclusions regarding CO chemisorption at room temperature ranging from molecular chemisorption to completely dissociative chemisorption. In a photoemission study for CO on Ti(1011)[51], the XPS values of the O(1*s*) and C(1*s*) binding energies each agree (within 0.1 eV) with values for elemental oxygen and carbon bound to Ti, supporting a model of dissociative chemisorption. The UPS results [51] also support dissociative CO chemisorption, since emission from ($5\sigma + 1\pi$)(~9 eV) and (4σ)(~11 eV) CO molecular orbitals is missing when CO adsorbs on Ti at room temperature. A comparison of the behavior of the O(1*s*) and C(1*s*) peaks on heating the monolayer indicate that only a small amount of adsorbate actually desorbs as CO below 573 K. In the temperature range 573–773 K, the majority of the C and O surface species diffuse into the bulk. Similar results were obtained by Fukuda et al. [52] for CO on Ti(0001) at 300 K. They find C(2*p*) and O(2*p*) levels at 3.4 and 6.3 eV below E_F. On the other hand, for adsorption of CO on Ti(0001) at 90 K, Stockbauer et al. [53] find that dissociation occurs at low coverages, but there is clear evidence for molecular adsorption at higher coverages in UPS.

The work function increase on Ti films has been found to be $\Delta\phi = 0.49$ eV for saturated CO chemisorption at room temperature [54]. Kasemo and Tornqvist [55] employed a sensitive quartz crystal microbalance to measure the O/Ti and CO/Ti ratios in adsorption experiments on Ti films. The results for CO suggest that C(ads) remains on the Ti surface, blocking CO adsorption at CO/Ti ratios of 0.5. A LEED I–V analysis of a p(2 × 2) chemisorbed layer on Ti(0001)[56] revealed that structure models with undissociated CO molecules perpendicular to the surface were unsatisfactory. The high sticking coefficient [55] of CO on Ti combined with the irreversible dissociative chemisorption are responsible for the well-known and widely used gettering property of this metal.

No published vibrational spectroscopic studies of CO on Ti are available.

7. Vanadium

The interaction of V with CO remains unstudied by either thin film calorimetric techniques [44], thermal desorption techniques, or any of the modern spectroscopic techniques currently available [57]. On the basis of correlations with heat of adsorption data for CO on Ta and Nb, the heat of adsorption would be expected to be near 120 kcal mol^{-1}, and dissociative chemisorption would therefore be expected [44]. There is one incomplete report of the chemisorption of CO on V studied by LEED [57]. Many studies of oxidized vanadium surfaces using various surface spectroscopies have been published [58–61].

On report of the infrared spectrum of CO on V [62] is available. CO adsorption bands at 1940 and 1890 cm^{-1} are observed on hydrocarbon-supported metal particles evaporated from a filament. This method of sample preparation renders these results suspect and probably unrepresentative of the properties of the clean metal.

8. Chromium

Unfortunately, the integral heat of adsorption of CO on Cr has not been measured by Brennan and Hayes [44]. However, Hayward [44] has suggested that close correlation with the heat of adsorption on Mo and W should be appropriate, since metals having the same number of outer electrons seem to exhibit similar heats of adsorption for CO. On this basis, the heat of adsorption of CO on Cr should be about 60–80 kcal mol^{-1}, the undissociated CO species might be stable on Cr. Some of the first studies of CO adsorption on a Cr(100) single crystal were carried out by Hague and Farnsworth [63] who reported that CO did not dissociate upon chemisorption at room temperature. Baker et al. [64] observed an infrared spectrum of CO on Cr films at 10^{-3} torr background pressure. The films were purposely deposited in the presence of CO to produce detectable spectra. A broad band at 1950 cm^{-1} is indicative of molecular CO adsorption at room temperature or below. Grant and Haas [65] report that, if CO/Cr(100) is heated to 1273 K, it decomposes. In contrast to this earlier work, which indicates molecular adsorption at room temperature or below, later work by Andersson and Nyberg [66] shows that it is likely that CO dissociates on Cr, either during adsorption or by electron impact during measurement. Kato et al. [67] found that the electron energy loss levels correspond entirely to those of chemisorbed O, and therefore arrived at the conclusion that CO/Cr(100) is dissociatively adsorbed at 300 K. Furthermore, they did not observe any LEED patterns due to ordered overlayers.

One might conclude from the work described so far that CO adsorbs molecularly at low temperatures, and that raising the temperature of the surface, or bombarding it with electrons, causes substantial CO dissociation. Indeed, the more recent EELS, ESDIAD, LEED and AES study by Shinn and Madey [68–70] of the adsorption of CO/Cr(110) clarifies the picture. They find that CO adsorbs molecularly at 120 K with its axis nearly parallel to the surface, which they denote by α_1 CO. This state gives rise to vibrational C–O bands at 1150–1330 cm^{-1}, the lowest reported to date for *any* clean or "promoted" single-crystal surface (Fig. 6). Shinn and Madey conclude that the tilted CO molecule facilitates substantial $2\pi^*$ back-donation, leading to such a low C–O stretching frequency, and indicating substantial bond weakening. In agreement with this conclusion, they find that the α_1O layer undergoes dissociation upon heating to $\geqslant 170$ K, indicating that α_1 CO serves as a precursor state to CO dissociation.

If the Cr(110) surface is kept at 120 K, CO adsorbs with an initial sticking probability of ~ 0.9, which drops smoothly to 0.4 with increasing coverage. At $\theta = 0.25$ the α_1 CO layer becomes fully ordered, giving rise to a c(4 × 2) LEED pattern. Further dosing causes the α_2 CO state to be populated with a sticking probability of 0.1, and EELS and ESDIAD show that the α_2 CO molecules are adsorbed on top sites with their axes perpendicular to the surface. Shinn and Madey observe that the CO adlayer disorders as the α_2 CO state forms, and that annealing causes dissociation to occur at higher temperatures ($\gtrsim 250$ K) than for the α_1 CO state alone.

A number of oxidation studies of Cr surfaces have been carried out [71–74]. O_2 and H_2O have been studied by X-ray photoelectron spectroscopy of Cr films [75]. The infrared spectrum of CO on hydrocarbon-oil supported Cr particles has been reported [62]. Absorption bands at 1940 and 1880 cm^{-1} are observed. The spectra are probably representative of a contaminated surface. Baker et al. [64] report a single CO adsorption band at 1930 cm^{-1} for an evaporated Cr film on CaF_2.

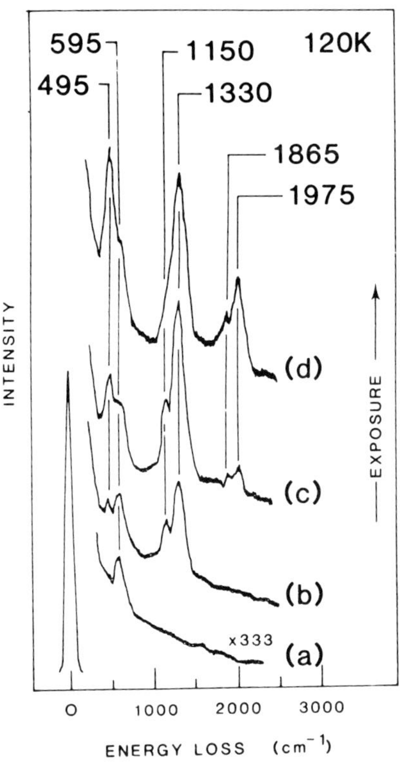

Fig. 6. High-resolution (90 cm^{-1} FWHM) EELS data for CO/Cr(110) at 120 K: Curve (a), clean; curve (b), 0.3 L CO; curve (c), 0.7 L CO; curve (d), 0.9 L CO. All spectra are normalized to the elastic peak height for the clean surface (shown)[68].

9. Manganese

Brennan and Hayes [44] measured the initial heat of chemisorption of CO on a Mn film (thought to be clean), obtaining a value of about 80 kcal mol^{-1}. The infrared spectrum of CO on Mn films was first observed by Baker et al. [64]. At 170 K, a Mn film deposited in 10^{-3} torr of CO exhibited two carbonyl absorption peaks, at 2080 and 2010 cm^{-1} (broad). Upon warming to room temperature, a feature at 1980 cm^{-1} remains. This evidence suggests that some adsorbed and undissociated CO species exist on Mn surfaces. Bickley et al. [76] have reported that the kinetics of CO interaction with a Mn film favors diffusion of adsorbate species into the Mn lattice. The chemisorption of CO on polycrystalline Mn has been studied by means UPS and XPS by Hu and Rabalais [77]. Up to 1 L exposure, when presumably 1 monolayer is formed, CO is dissociatively chemisorbed. On increasing exposure, immediately after one monolayer, molecular adsorption occurs. This is not too surprising a result considering that, even on the (111) face of Mn, there arc holes about 2 Å in diameter.

10. Iron

The initial heat of chemisorption of CO on an evaporated iron film is 46 kcal mol^{-1}, according to Brennan and Hayes [44]. Wedler et al. [78] have measured an initial heat of adsorption of 37.0 ± 1 kcal mol^{-1} for CO on an iron film, and report a stepwise decrease in the adsorption heat as the coverage increases. From work function and film resistance changes, they find evidence for slow CO decomposition at 273 K, which is absent at 77 K. Thermal desorption measurements indicate that CO desorbs over a wide temperature range from these films up to about 600 K.

The ultraviolet photoelectron spectrum of CO on polycrystalline Fe was first investigated by Yu et al. [79] and Kishi and Roberts [80]. Ertl et al. [81] studied the ultraviolet photoelectron spectrum of CO/Fe(110) at 220 K, finding strong spectroscopic evidence for undissociated CO. Characteristic photoemission features of the $5\sigma + 1\pi$ CO orbitals and the 4σ oxygen lone pair orbital were present in the spectra. By making corrections for the work function and "relaxation" effects, the spectral features correspond closely to those of $Ni(CO)_4$(g). CO dissociation near 400 K can be easily followed by characteristic changes in the photoemission spectrum [81]. Broden et al. [82] obtained similar spectra for CO on Fe(110) at 320 K, and observed marked changes in the spectrum at 385 K due to CO dissociation. They observe that approximately 20% of the molecules are dissociated upon heating the crystal during thermal desorption. Broden et al. [83] also studied the influence of co-adsorbed potassium on the thermal dissociation of CO/Fe(110), as well as the influence of K on the CO UPS spectrum. Although K increases the probability of CO dissociation, it does not appreciably decrease the temperature of dissociation. The 4σ molecular orbital (oxygen lone pair) feature is shifted by 0.8 eV, to higher binding energy, while the $1\pi + 5\sigma$ feature is split into a double peak by co-adsorption of K. There is thermal desorption evidence that K slightly increases the activation energy for CO desorption. Gonzalez et al. [84] find, from a TDS study, that the activation barrier for dissociation not only depends on temperature, but also on coverage. For adsorption below 400 K, at low coverages ($\theta = 0.05$), about 75% of the molecules desorb from the β state (dissociated), while at saturation coverage, only 25% desorb from the β state and 75% from the α state (molecular state). For adsorption above 450 K, all molecules desorb from the β state, whatever the initial coverage. They also conclude that a rough surface is 1.6 times more efficient at CO dissociation than an annealed surface. Rhodin and Brucker [85] have demonstrated that CO dissociates at 300 K on Fe(100). Varying degrees of pre-adsorption of C, O, P, or S reduces the tendency of this surface to dissociate CO. The influence of S poisoning on the CO UPS spectrum is rationalized on the basis of an enhanced 1π–5σ splitting for CO + S/Fe(100) compared with CO/Fe(100).

Jona et al. [86] have performed a LEED intensity analysis of CO/Fe(100) at room temperature. They find that no model involving undissociated CO fits their data, and conclude that the observed c(2 × 2) pattern is due to random occupancy of only 1/2 of the 4-fold hollow sites. There is 1/4 of a monolayer of C and 1/4 of a monolayer of O atoms randomly mixed together. Yoshida and Somorjai [87] have studied CO chemisorption on Fe(100) and Fe(111) using LEED, without finding any ordered structures. More recent XPS and TDS studies by Moon et al. [88] and EELS and TDS studies by Benndorf et al. [89] clarify the earlier results, and reveal that the behavior of CO/Fe(100) is rather similar to the behavior of CO/Cr(110) seen by Shinn and Madey [68, 70].

The TDS results of Moon et al. and Benndorf et al. show that adsorption of CO at 110–150 K leads to the sequential population of three states. The three states were observed previously, also by TDS, by Benziger and Madix [90] who ascribed them to molecularly adsorbed CO. The first state to be filled, designated α_3, is the precursor to CO dissociation, which occurs if the sample is heated to 440 K. Benndorf et al. [89] find that this state is due to CO adsorbed nearly parallel to the surface, and shows a very low C–O stretching frequency of 1180–1245 cm^{-1}, as expected. After the α_3 state is filled, the α_1 states are populated, which are due to terminally bonded CO. Moon et al. [88] find that this behavior is also reflected in the XPS spectrum, giving rise to distinguishable peaks, and assuming first-order desorption kinetics with a frequency factor of $10^{13}\,s^{-1}$, they find activation energies for desorption of 26.2, 18.0 and 12.8 kcal mol^{-1} for the α_3, α_2, and α_1 states respectively. These values are lower than the calorimetric values [44, 78], indicating that extensive dissociation occurred in the early experiments. In agreement with the results of Gonzales et al. [84], they find that the fraction of CO(α_3) molecules that undergo dissociation upon heating depends on the coverage. However, not all the α_3CO molecules dissociated on the Fe(100) surface even at the lowest coverage, in contrast to results on the (111) surface to be described below. Also, on the (110) and (111) surfaces, the low-frequency mode was not detectable. Erley [91] found that, on the (110) surface at 120 K, CO adsorbs on top sites. Moon et al. [88] pointed out that all these effects indicate a marked influence of the surface structure on CO dissociation, as found on many metal surfaces.

Textor et al. [92] studied CO adsorption on Fe(111) using both UPS and XPS methods. Their results at 300 K are consistent with dissociative CO chemisorption at low coverages, followed by molecular chemisorption as the coverage increases. Dissociative chemisorption is impeded by pre-adsorption of hydrogen on this surface. Seip et al. [93] studied CO/Fe(111) by means of EELS, LEED, TDS, and work function measurements. In agreement with Textor et al. [92] they find that, during the initial stages of adsorption, CO dissociates at 350 K. As the exposure is increased further to 1.5 L, CO adsorbs

References pp. 460–469

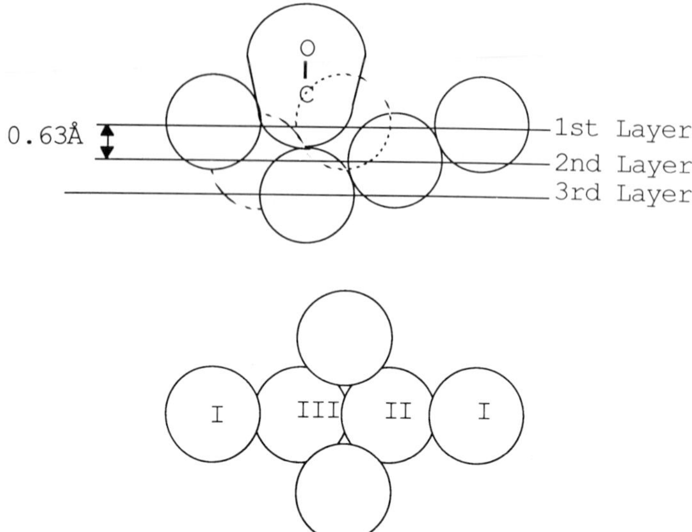

Fig. 7. Top view (lower part) and cross-section along the broken line (upper part) of the Fe(111) surface showing the three types of exposed Fe atoms. The upper part also shows a CO molecule (with appropriate dimensions) adsorbed in the "deep hollow" site [93].

in the shallow hollows of the (111) surface, as shown in Fig. 7, bridge-bonded between the first- and second-layer atoms. More interestingly, if the coverage is increased still further, some CO adsorbs on the third-layer atoms in the deep hollows with a very low C–O stretching frequency of 1530 cm^{-1}. Another interesting observation of Seip et al. [93] is the increase in work function as molecularly adsorbed CO dissociates when the sample temperature is raised.

Thirteen papers* have been written about the infrared spectrum of CO on evaporated Fe films and on supported metal particles. In each case, CO stretching frequencies in the range 2040–1915 cm^{-1} were observed, proving that undissociated CO exists on these surfaces, at least at full coverage. Bradshaw and Pritchard [94] observed that the initial adsorption of CO did not appear to produce CO adsorption bands. This may be due to dissociation at low coverages.

11. Cobalt

According to Brennan and Hayes [44], the initial heat of chemisorption of CO on a Co film is 48 kcal mol^{-1}. The activation energy for CO desorption

* An excellent review of the chemisorption of CO on supported metals and on single crystals has been published by Sheppard and Nguyen [17]. The reader is referred to this comprehensive review for information derived by infrared spectroscopy for CO chemisorption on the transition metals.

from a Co film was estimated to be only 18 kcal mol^{-1} by Benndorf and Thieme [95], although their work was done on Co films having detectable C and O impurity levels by Auger spectroscopy.

11.1 Co(0001)

A study of CO/Co(0001) at room temperature was carried out by Bridge et al. [96]. They also showed that the bulk phase transition between h.c.p. and f.c.c. forms of Co could be observed by LEED: a sixfold symmetric pattern becomes threefold above $\sim$700 K as the f.c.c. structure is produced. The thermal desorption of CO exhibits a peak maximum at 430 K for low coverages, shifting to about 410 K at the highest coverage. For oxygen-contaminated surfaces, there is a second CO state near 650 K. LEED studies indicate that a well-ordered ($\sqrt{3} \times \sqrt{3}$)R 30° phase is produced at intermediate coverages ($\theta = 0.33$). Beyond this coverage, continuous overlayer compression occurs, and new thermal desorption features become visible, overlapping the main CO desorption peak. An estimate of the CO desorption energy was made from an analysis of the thermal desorption curves at $\theta = 0.022$, giving an activation energy of 24.6 $\pm$ 2 kcal mol^{-1}. However, Papp [97] points out that this value includes a large error due to the assumption made by Bridge et al. [96] of a constant frequency factor of 10^{13} s^{-1}. Papp's results agree with those of Bridge et al., but Papp found that, when CO is adsorbed on Co at 100 K, a ($\sqrt{7}/2 \times \sqrt{7}/2$)R 19° structure is formed, as shown in Fig. 8, when the work function has changed by 1.35 V. This corresponds to a coverage of 3/7. Therefore, the assumption of Bridge et al. that the compressed ($\sqrt{3} \times \sqrt{3}$)R 30° structure constitutes the onset of a ($\sqrt{7} \times \sqrt{7}$)R 19.2° structure with $\theta = 4/7$ is incorrect. Papp finds a linear relationship between work function change and coverage which is completely reversible with coverage, and no evidence of thermal dissociation over the temperature range 100–450 K. In that study, the initial heat of chemisorption of 128 kJ mol^{-1} remains constant until $\theta = 0.3$, and then drops suddenly to a

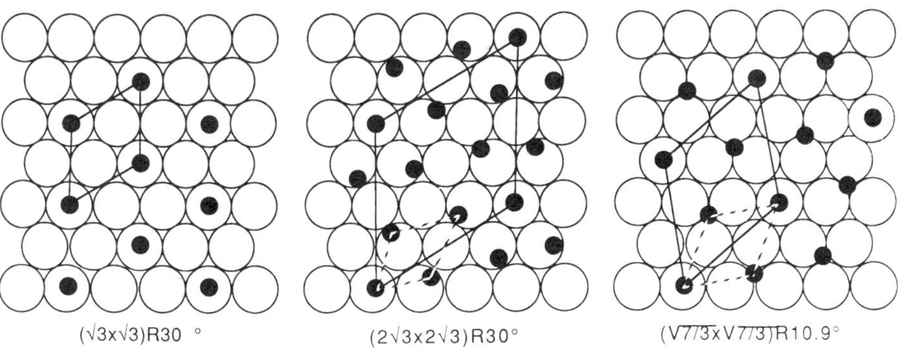

Fig. 8. Schematic structure models for the CO adsorption on Co(0001) at 100 K. Open circles, Co atoms; filled, small circles, CO molecules [97].

plateau at 96 kJ mol^{-1}. The $(\sqrt{3} \times \sqrt{3})$R 30° LEED pattern is formed below $\theta = 0.3$, and this, together with the constant heat of adsorption, is indicative of an attractive interaction between CO molecules leading to island formation. After the $(\sqrt{3} \times \sqrt{3})$ 30° structure is completed, a continuous splitting of the LEED spots is interpreted as a continuous compression of the CO overlayer above $\theta = 0.33$. However, Biberian and Van Hove [98] have pointed out that this interpretation, originally suggested by Tracy [99] is incorrect. A most likely cause of splitting of the LEED beams is the formation of antiphase domains in the overlayer. The antiphase domains can accommodate the excess CO molecules. Interestingly, despite the decrease in heat of adsorption and the onset of repulsive CO–CO interactions, the change in work function versus coverage remains constant. Papp [97] and Freund and Hollinger [100] studied CO/Co(0001) by UPS. Papp was able to distinguish between linear and bridge-bonded CO by the different amounts of $(5\sigma - 1\pi)$ shift. In the $(\sqrt{3} \times \sqrt{3})$R 30° structure all molecules can be arranged on one type of site, and the $(5\sigma - 1\pi)$ emission is observed at 8 eV below E_F. As the coverage increases, more than one type of site becomes occupied and another $(5\sigma–1\pi)$ photoemission peak is observed at 7.3 eV. The 4σ state does not interact with the metal, and so remains constant at 11 eV below E_F. But its decrease in intensity with increasing coverage remains unexplained. Papp [101] finds similar results for Co(1010), although the heat of adsorption is higher. It is surprising to find that the saturation coverage of CO/Co(1010) is the same as on Co(0001): 1.1×10^{19} molecules m^{-2}, despite the large difference in the number of Co atoms on the surface [Co(0001) has 1.83×10^{19} and Co(1010) has 0.97×10^{19} atoms m^{-2}]. Both Bridge et al. [96] and Papp [97] find that CO/Co(0001) exhibits a high cross-section ($\geqslant 10^{-16}$ cm^2) for electron-stimulated desorption and that this occurs mainly by desorption of CO rather than by dissociation. Greuter et al. [102] studied adsorption of CO/Co(0001) by angle-resolved UPS, LEED and X-ray photoelectron diffraction. They found the 4σ level at 10.7 eV and the $5\sigma + 1\pi$ level at 7.9 eV, and determined that CO is bound perpendicularly to the surface. As in the $(10\bar{1}2)$ and $(11\bar{2}0)$ surfaces, Greuter et al. find some CO dissociation after extended LEED and UPS measurements at high pressures. Furthermore, as in the case of CO adsorption on Ni(100), Pd(100), Fe(110) and Ir(111), they find that the CO levels interact sufficiently strongly to form two-dimensional bands. In the saturated room-temperature phase the 4σ bandwidth is 0.15 eV and the $5\sigma + \pi$ bandwidth is 0.35 eV. Upon cooling the surface to 170 K, Greuter et al. obtain a $(2\sqrt{2} \times 2\sqrt{3})$R 30° structure, in agreement with the results of Papp. In this incommensurate overlayer the CO–CO distance is 3.29 Å at a coverage of $\theta = 7/12$. Greuter et al. point out that it is puzzling that CO/Ni(111) does not show this incommensurate overlayer, considering that the Ni lattice spacing in the (111) surface is only 0.8% smaller than that of Co. The saturation overlayer of CO/Ni(111) has $\theta = 4/7$, only 2% less than on the Co(0001) surface. Greuter et al. find that, in this incommensurate overlayer, the 1π levels are no longer degenerate but split, possibly due to the lack of symmetry. Also, due to the much tighter packing, the bandwidths increase by approximately a factor of 3. Their data and calculations suggest that the dispersion is primarily due to direct CO–CO interactions, and not due to indirect interactions through the metal.

11.2 Co(1012)

Prior et al. [103] have studied the (1012) plane of Co which, in contrast to Co(0001), exhibits a channelled surface in two possible forms (unreconstructed). Here however, CO adsorption–desorption is non-reversible, and there is evidence for carbon formation following cycles of adsorption and desorption. It was found that the rate of temperature increase was the factor determining the extent of CO dissociation, leading to the conclusions that

(a) the activation energy for desorption ($\sim 25\,\text{kcal mol}^{-1}$) is very similar to the activation energy for CO dissociation on this surface and

(b) the transition state to desorption is mobile, whereas the transition state to dissociation is localized.

These results agree with the early work of Kehrer and Leidheiser [104] who heated a spherical single crystal of Co in 1 atm of CO(g). Carbon deposition did not visibly occur at the (0001) pole, but did occur around the 1012 regions. On Co(1012), it is postulated that a surface carbide is formed from CO; this carbide *prevents* further CO dissociation, but does not impede the adsorption of undissociated CO. CO chemisorption on the Co(11$\bar{2}$0) surface has been studied by Papp [105] whose LEED results show that no ordered structures are formed during adsorption at 100 K. Warming the overlayer above 300 K, or adsorption at 340 K, leads to a (3×1) structure at 2/3 of a monolayer. The adsorption is molecular in nature at low temperatures, but above 300 K the molecules in the trough tend to dissociate, leaving C and O atoms. The calculated initial activation energy barrier for dissociation is $85\,\text{kcal mol}^{-1}$. But surface sites are still available for adsorption. Papp finds that the dissociation is not linked to the heat of adsorption, which is identical to that on Co(10$\bar{1}$0), where no dissociation takes place. Also, the UPS spectrum of CO/Co(1$\bar{1}$20) is identical to the one of CO/Co(10$\bar{1}$0) and CO/Co(0001), as is the electron energy loss spectrum, which shows transitions between Co(d) $\rightarrow$ CO($2\pi^*$) levels and CO($5\sigma + 1\pi$) $\rightarrow$ CO($2\pi^*$) levels. These results point to the influence of structure on dissociation, in analogy to the Cr surfaces.

The interaction of hydrogen and CO on Co(0001) has been studied [106] using thermal desorption techniques. No chemical interaction between the two adsorbates is detected, although CO will displace H_2 at 300 K with an efficiency of ~ 0.7.

The infrared spectrum of CO on Co has been reported in 11 studies summarized by Sheppard and Nguyen [17]. Both terminal and bridged CO adsorption bands are observed on oxide-supported Co particles and on evaporated films of Co.

12. Nickel

The heat of adsorption of CO on a Ni film has been measured by calorimetric means at 273 K by Wedler et al. [107]. The films, which consist mainly of mixtures of (100) and (111) planes, show an initial heat of adsorption of 30 kcal mol^{-1}. The differential heat of adsorption drops to 25 kcal mol^{-1} at θ = 0.5. Earlier, Brennan and Hayes [44] had reported an initial heat of adsorption of 42 kcal mol^{-1}.

The isoteric heat of adsorption of CO on Ni(100) was first measured by Tracy in an elegant and pioneering LEED-work function study [99] as $\simeq$ 31 kcal mol^{-1}. Madden et al. [108] have investigated CO chemisorption on Ni(110) using a method similar to Tracy's, and also obtain an initial heat of about 30 kcal mol^{-1}, with evidence of a fall-off to a plateau at intermediate coverages. Taylor and Estrup [109] have measured the isoteric heat for the equilibrium condition involving the (4 $\times$ 2) $\leftrightarrow$ c(2 $\times$ 1) phase transition on Ni(110), to be 26 kcal mol^{-1}, in agreement with the result of Madden et al. [108]. The isoteric heat of CO/Ni(111) has been measured by Christmann et al. [110], again using a method similar to Tracy's. A summary of these results is shown in Fig. 9. Major effects of surface structure on the initial heats of chemisorption seem to be absent for CO on Ni. Similar findings are reported for CO on five planes of Pd [111].

12.1 Ni(100)

The adsorption of CO on Ni(100) proceeds through three structural regimes [99]: at low coverage or high temperatures a disordered phase is

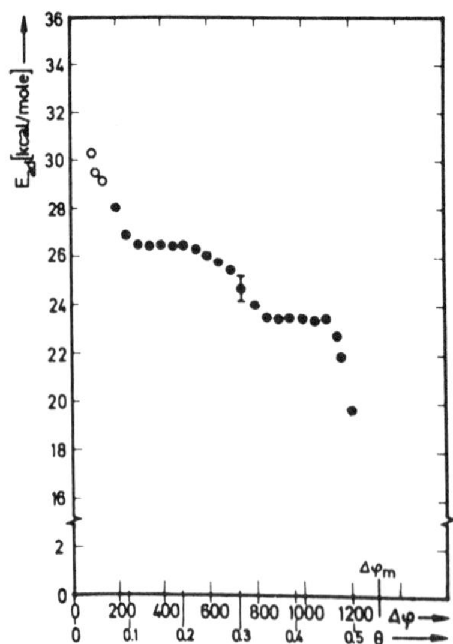

Fig. 9. The isosteric heat of adsorption, E_{ad}, as a function of coverage for CO/Ni(111)[110].

observed. A c(2 × 2) phase exists near $\theta = 0.5$, and a compressible hexagonal phase exists between $\theta = 0.61$ and $\theta = 0.69$. Tracy's postulate that the hexagonal phase is incommensurate with the underlying Ni lattice is of considerable interest, suggesting that, at high coverage, the repulsive forces between CO molecules dominate over directed chemical bonding. The highest coverage, $\theta = 0.69$, corresponds to 1.10×10^{15} CO cm^{-2}. This coverage is in remarkable agreement with absolute coverage measurements of Klier et al. [112] who found using ^{14}CO and radioactive counting techniques, that the saturation coverage is 1.11×10^{15} CO cm^{-2}.

The development of the vibrational spectrum of CO/Ni(100) has been thoroughly studied by Andersson using electron energy loss spectroscopy, EELS [113, 114]. At low coverages, linear and bridged CO coexist in a disordered layer. At higher coverages, the dominant C–O stretching feature of the c(2 × 2) overlayer at 175 K is a loss of 2070 cm^{-1} (256.5 meV), indicating that CO is linearly bonded. For mixtures of the c(2 × 2) + hexagonal structures, comparable loss intensities are observed at 2070 and 1930 cm^{-1}, indicating that both bridge-bonded and linear CO coexist. At $\theta = 0.61$, where the pure hexagonal phase exists, a *broad* C–O loss feature is observed which is almost the envelope of the 2070 and 1930 cm^{-1} features. This was thought to be a good indication that, in the incoherent hexagonal phase, the CO molecules adopt all types of sites, ranging from approximately linear to bridged, in accordance with the model of Tracy [99]. Bertolini and Tardy [115] found similar results for the C–O stretch, although at slightly lower wave numbers. They were able to observe two carbon–metal stretch losses, one due to bridge-bonded CO at 365 meV and another, due to linearly bonded CO, at 460 meV. Dignam [116] also studied the CO vibrations on Ni(100) with infrared ellipsometric spectroscopy. He also found two bands, at 1900 and 2068 cm^{-1} but at high coverages he sees a series of bands, unlike the broad band of Andersson [113, 114], although Dignam does not see the compression structure. Unfortunately, the question of whether CO does indeed adsorb on a range of sites during the compression stage has not been answered. In light of the many bands seen by Dignam [116] with higher resolution, the broad loss peak seen by Andersson could be a collection of peaks corresponding to discrete adsorption sites. Biberian and Van Hove [98] point out that Tracy's original suggestion of the occupation of a continuous range of adsorption sites as the overlayer compresses is not consistent with the observed LEED patterns and high-resolution vibrational spectra. More likely, as the coverage increases, antiphase domains are formed, splitting the LEED beams and accommodating the excess molecules. According to Tobin et al. [117], possibly all previous vibrational spectroscopy measurements of CO/Ni(100) could suffer from contamination problems. Tobin et al. find that they can obtain bridge-bonded CO at room temperature only if the sample is ever so slightly contaminated. Otherwise, only linearly bonded CO is observed. Mitchell et al. [118] observe that, when adsorbing CO at 90 K, only bridge bonding is observed, although over a range of sites. As the sample is warmed

References pp. 460–469

up to 265 K, a large amount of CO desorbs. At 265 K, the band sharpens considerably. Mitchell et al. have not interpreted their results in light of previous results. A report of the observation of vibrational bands of CO/Ni(100) by unenhanced Raman spectroscopy by Marzouk et al. has been published [119]. Unfortunately, the signal-to-noise ratio displayed is rather poor, and they do not present any evidence of sample cleanliness. Their conclusions, which run contrary to all other experimental evidence, are therefore suspect.

According to Andersson [114], CO adsorption on a hydrogen-covered Ni(100) surface yields only the 1930 cm^{-1} loss feature, implying that chemisorbed hydrogen hinders the adsorption of terminal CO. No evidence of C–H or O–H vibrations is found with H + CO mixtures on Ni(100). Mitchell et al. [118] find the opposite: only linearly adsorbed CO is present on a H-saturated Ni(100) surface. Only as the sample is warmed, and half the hydrogen is lost, do the CO molecules shift to bridge bonding. But most of these effects are not completely reproducible, presumably because at low temperatures it takes a long time to reach thermodynamic equilibrium. Yates et al. [120] have observed that unusual thermal desorption states are present for H + CO mixtures on Ni(100). The mixed layer yields Σ-H_2 and Σ-CO desorbing at ~230 K in states which are unique only to the mixed layer. These states are not produced by either H_2CO or CH_3OH adsorbates, and are therefore not believed to originate from a chemical complex such as HCO(ads) or CH_3O(ads) on the surface [120]. Instead, a surface complex involving multiple H(ads) species interacting with an adsorbed CO species is postulated [120]. UPS studies of this same adsorption system by Koel et al. [121] indicate that CO in the Σ-complex is strongly involved in bonding through its 5σ orbital and that final state screening effects due to the metal are less than in pure chemisorbed CO/Ni(111).

The EELS results for pure CO chemisorption on Ni(100) are consistent with early infrared studies of CO chemisorption on supported Ni particles. In particular, Yates and Garland [122] observed major C–O stretching features centered at 2050 and 1940 cm^{-1} at full coverage on large Ni crystallites supported on Al_2O_3. These features were assigned to linear and bridging CO, respectively, following the work of Eischens et al. [123]. These band positions are within ~10 cm^{-1} of the EELS observations on Ni(100)[113–117]. Sheppard and Nguyen [17] review 32 references treating CO chemisorption on supported Ni and on Ni films from the viewpoint of infrared spectroscopy.

The structure of CO/Ni(100) has been investigated by LEED I–V analysis [124, 125], with conclusions that CO is bonded atop Ni atoms when in the c(2 × 2) structure. Terminally bonded CO is reported to have Ni–C distances of 1.72–1.80 Å and C–O distances of 1.15–1.10 Å [124, 125]. The results were obtained in spite of major difficulties with electron-stimulated desorption [124]. Angle-resolved UPS studies [32, 33, 37, 126] and calculations [127] are consistent with this model of bonding. The angle-resolved XPS study of Petersson et al. [128] also comes to the same conclusion.

The ultraviolet photoemission spectrum of CO chemisorbed on Ni(100) exhibits two features related to CO molecular orbitals [126]. A feature at -11 eV below the Fermi level is related to the 4σ molecular orbital (oxygen lone pair) and a feature at -8 eV is caused by the overlap of 5σ- and 1π-derived states. Himpsel and Fauster [129] measured the $2\pi^*$ level of CO/Ni(111) by inverse photoemission at about 3 eV above the Fermi level. The $2\pi^*$ level of CO/Cu(110) is found at the same energy.

SIMS has also been used to investigate the system CO/Ni(100)[130]. Only clusters of ions containing CO were observed (not C or O clusters), supporting a model for non-dissociative chemisorption. Upon electron bombardment of the CO layer, dissociation occurs and Ni_2O^+ ions are observed by SIMS. As the bridged-to-linear CO ratio is changed, little change in the $Ni_2CO^+/NiCO^+$ ratio is observed, suggesting that SIMS is not able to detect coverage effects. These results are in disagreement with those of Bordoli et al. [131] and Fleisch et al. [132] where a higher Ni_2CO^+ ratio is observed at low coverages, suggesting that bridged CO predominates at low coverages. Fleisch et al. [132], studied the O(1s) and C(1s) behavior of CO/Ni(100) using XPS, and found evidence supporting the role of surface carbon in dissociating CO to yield multilayers of surface carbon.

12.2 Ni(110)

The adsorption of CO/Ni(110) has been studied by Klier et al. [112] using a radiotracer technique. Only a single state of adsorption of 25.3 kcal mol^{-1} was detected with a maximum coverage of 1.1×10^{15} molecules cm^{-2}. Later, Taylor and Estrup [109] studied adsorption at surface temperatures as low as 128 K by LEED, AES, work function changes, and TDS. The LEED patterns showed considerable disorder, but could be described by the sequence $c(2 \times 2) \rightarrow (4 \times 2) \rightarrow c(2 \times 1)$. TDS revealed two binding states. These results were not reproduced in subsequent studies by Madden et al. [108]. Their study showed only one peak by TDS. The LEED pattern sequence was also different: a one-dimensionally incoherent structure is followed by a $c(2 \times 1)$ structure. Lambert [133] pointed out that the $c(2 \times 1)$ designation is not consistent with the coverage, and proposed that the structure is really (2×1)p1g1. The glide line causes systematic absences of the $(h/2,0)$ LEED beams for h odd, which can then be mistaken for a $c(2 \times 1)$ pattern. Lambert pointed out that the same structure is observed for CO adsorption on the (110) surfaces of Pd, Pt, Ir, and Ru. Infrared studies by Mahaffy and Dignam [134, 135] explained that part of the discrepancy between the works of Taylor and Estrup [109] and Madden et al. [108] were due to the different temperature regimes employed. Mahaffy and Dignam find five IR bands growing as a function of coverage. Subsequently, EELS studies were carried out by Bertolini and Tardy [136] and by Nishijima et al. [137]. An EELS study by Bandi et al. [138] finds general agreement with the previous vibrational spectroscopy studies [134–137]. However, Bandi et al. do not find the two highest C–O bands found by Mahaffy and Dignam [134, 135], and suggest that they may have arisen from incomplete removal of oxygen from the Ni surface. Bandi et al. find that the linear site is preferred at low coverages, but suggest that the simultaneous growth of linear and bridged species indicates that the binding energy difference between the two sites is small, and therefore their relative populations will be temperature- and coverage-dependent. At full coverage however, Bandi et al. [138] find that only the short bridge site is occupied, which coincides with the appearance of the (2 × 1)p1g1 structure. The appearance of this structure is also signaled by the appearance of a low-temperature shoulder in the thermal desorption spectrum, indicating that adsorbate–absorbate interactions are responsible for the change in site preference. EELS measurements by Gurney and Ho [139] and Madix et al. [140] yielded similar results to those of Bandi et al.

Lambert [133] suggested that the short CO–CO distance could be accommodated by tilting the molecular axis away from the surface normal. Some circumstantial evidence existed to support this suggestion for the case of CO/Ni(110)[138, 141] which has been shown to be the case for CO/Pt(110)[435]. But since then, the ESDIAD experiments of Riedl and Menzel [142] and Alvey et al. [143] have shown the tilting in a dramatic way, as shown in Fig. 10. Alvey et al. find a tilt of 19° along the [001] direction, which really makes the structure a (2 × 1)p2mg, as first pointed out by Mahaffy and Dignam [134].

The adsorption kinetics of CO/Ni(110) were studied by Falconer and

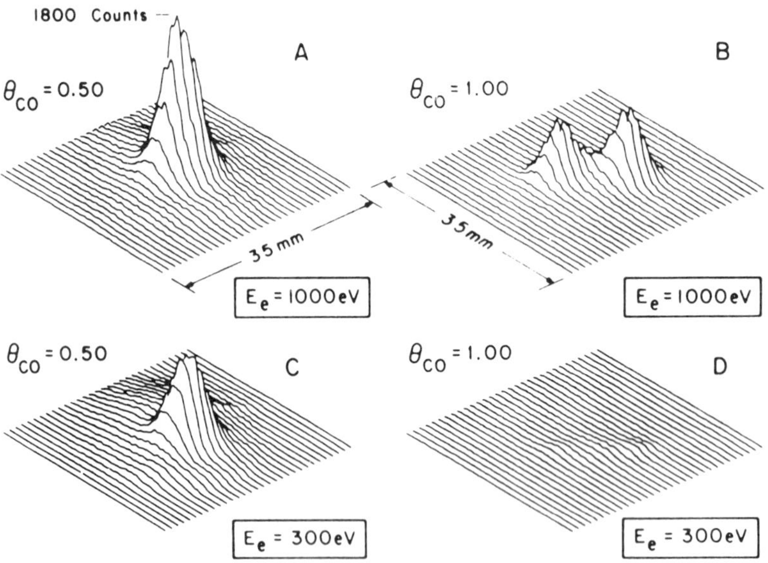

Fig. 10. ESDIAD patterns obtained at two coverages of CO/Ni(110) and at two electron energies. (A) and (B) are for θ_{CO} = 0.50 and 1.00 respectively, for electrons of 1000 eV energy; (C) and (D) are for the same CO coverages for electrons of 300 eV energy [143].

Madix [144], Lee et al. [141], Behm et al. [145], and Alvey et al. [143]. Adsorption is found to proceed via a precursor state, and quite accurately follows the first-order Kisliuk equation

$$S(\theta) = S_0(1 - K\theta)$$

Behm et al. determined K to be 0.95, which implies a very long-lived precursor molecule. The nature of this precursor will be discussed in more detail in Section 12.3. The initial sticking probability is approximately 0.9, and at low coverages, a single desorption peak is observed with an activation energy and preexponential for desorption in the ranges of 27–33 kcal·mol^{-1} and 8.5×10^{15} s^{-1}, respectively. There is a minor disagreement between Alvey et al. [143] and Behm et al. [145] as to the exact coverage at which a second desorption peak appears, but it is between 0.72 and 0.75. Behm et al. observe some intermediate LEED structures before (2×1)p2mg, and their results shed some light on the previous LEED observations. The intermediate structures show considerable disorder, which tends to support the proposal of Bandi et al. [138] that the relative coverage of the linear and bridge sites is temperature-dependent, as is found for CO/Ni(111)(see below).

Hsu et al. [146] have shown by means of UPS that CO chemisorbs on Ni(110) even at 20 K, implying that the activation energy for chemisorption is very small. This might be characteristic of very reactive metals, unlike the case of Cu, where the conversion to chemisorption does not start until 30 K.

12.3 Ni(111)

The interaction of CO with Ni(111) has been studied by angle-resolved UPS by Apai et al. [33]. By comparing the experimental intensity ratio of the 4σ to $5\sigma + 1\pi$ emission as a function of the angle of the detector to the surface normal with the theory of Davenport [30], they concluded that CO binds perpendicularly to the surface via the carbon end of the molecule. The results were not in exact accord with theory. This may be due to a perturbation of the 5σ orbital due to its interaction with the substrate, to mixture of angles of bonding, or to low-frequency vibrations in the terminal CO species. The results do not permit one to determine the adsorption sites of CO/Ni(111).

The adsorption sites as a function of coverage were studied by Campuzano and Greenler [43] by means of IRAS. They find that, at room temperature and $\theta < 0.20$, CO is adsorbed on threefold coordinated sites with a C–O stretching frequency of 1817 cm^{-1}. By separating the frequency shift due to the filling of the $2\pi^*$ level from the shift due to dipole–dipole coupling, they showed that the frequency remains constant while CO is adsorbed on threefold sites. For coverages greater than 0.25, the molecules suddenly shift to twofold sites, with an equally abrupt frequency shift to 1900 cm^{-1}, which again remains constant as the molecules remain in twofold sites with increasing coverage. A previous EELS results of Bertolini et al. [147] found the same frequency shift as that of Campuzano et al., but the poor frequency resolution did not permit a detailed assignment of the adsorption site. Another previous EELS study, by Erley et al. [148], carried out at 140 K showed the same general features as that of Bertolini et al., but in addition showed a small amount of linearly bonded CO at low coverages. Tang, et al. [149] explained the discrepancy by showing that the coverage-dependent binding energy difference between bridge and atop-bonded CO is very small, as shown in Fig. 11. Therefore, the relative equilibrium population of the two sites depends on temperature and coverage. Similar results were found earlier by Bandi et al. [138] on the (110) surface, by Hayden and Bradshaw [196] for CO/Pt(111), and by Hayden et al. [194] for CO/Cu(111). Hayden et al. suggested that, since the energy difference between the two states is small, their relative population should be given by the Boltzmann distribution, and the conversion from one state to the other should be aided by the increasing population of the frustrated translation modes as the temperature increases. At $\theta = 0.33$, a $(\sqrt{3} \times \sqrt{3})R\,30°$ LEED pattern is observed by Campuzano et al. [43, 150], in agreement with the previous LEED results of Christmann et al. [151] and Conrad et al. [152]. The pattern occurs over a narrow range of coverage and is rather diffuse, implying that CO/Ni(111) does not form islands, in contrast to CO/Co(0001)[96, 97, 102] or CO/Pd(111)[305–308]. At $\theta = 0.5$, CO/Ni(111) produces a $c(2 \times 2)$ LEED pattern, and at room temperature, this is the maximum irreversible coverage [43, 148–150]. The IRAS results of Campuzano and Greenler [43] and Trenary et al. [153, 154] and the previous EELS results of Erley et al. [148] show that, at this coverage, all the CO molecules are adsorbed on twofold sites. The structure models are shown in Fig. 12. Trenary et al. [153, 154] observed a strong temperature dependence of the bandwidths of both atop and bridge-bonded CO due to a dephasing, brought about by rapid energy exchange between low-frequency modes of the substrate and low-frequency modes of the molecule (such as the frustrated translation, which causes the bridge-to-atop site conversion), which are

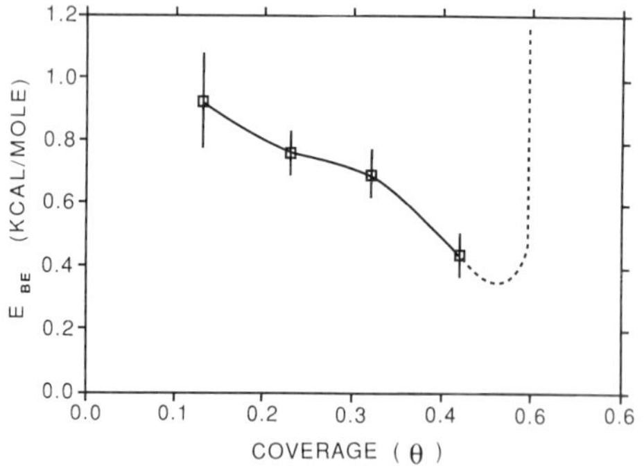

Fig. 11. The binding energy difference, $\Delta E_{BE} = E_B - E_A$, between bridge and atop-bonded CO as a function of coverage. A line is drawn through the data points to illustrate the trend in ΔE_{BE} with coverage. The bridge site is more stable than the atop site [149].

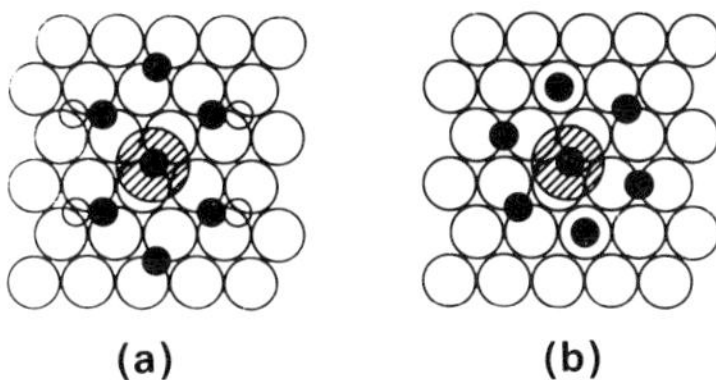

Fig. 12. (a) Structure model for the CO c(4 × 2) structure at $\theta = 0.5$ (dark circles). The open circles represent the positions of the adsorbed molecule at $\theta = 1/3$ [the ($\sqrt{3} \times \sqrt{3}$)R 30° structure]. Three equivalent domain orientations rotated by 60° to each other are possible. (b) Structure model for the ($\sqrt{7}/2 \times \sqrt{7}/2$)R 19.1° structure at $q = 0.57$. Three domain orientations are also possible. The shaded circles indicate the van der Waals size of the CO molecules.

anharmonically coupled to the high-frequency band being studied. Richardson and Bradshaw [155] calculated that the frustrated translation mode should have a frequency of 649 cm^{-1}, although this mode had not been observed by vibrational spectroscopy. Persson and Ryberg [156] repeated the experiments of Trenary et al., obtained the same results using the same theory, and came to the same conclusions. If the sample is cooled below 243 K, the CO coverage can be increased to 0.57, at which point a ($\sqrt{7}/2 \times \sqrt{7}/2$)R 19.2° LEED pattern is formed at $\theta = 0.57$ [43, 150, 152]. At this coverage, the IRAS results of Campuzano et al. [43] show that 3/4 of the molecules are adsorbed on twofold bridge sites, and 1/4 are adsorbed on atop sites, with a frequency of 2045 cm^{-1}. The ESDIAD, LEED, and TPD results of Netzer and Madey [159] are in agreement with previous results, and furthermore show that CO is adsorbed perpendicularly to the surface at room temperature. They find that adsorption below 260 K leads to a complex mixture of the c(4 × 2) and ($\sqrt{7}/2 \times \sqrt{7}/2$)R 19.1° patterns, as found by Erley et al., [148] and that gentle annealing restores the order of the overlayer.

The adsorption kinetics of CO/Ni(111) were thoroughly studied by Tang et al. [158] by means of AES, TDS, EELS, and supersonic molecular beam dosing. As did Campuzano et al. earlier [150], they find that adsorption of CO on the (111) surface proceeds via a precursor state for molecules incident on the surface with low energy. Tang et al. find that incident molecules with energies lower than 4 kcal mol^{-1} are adsorbed into an intrinsic precursor state (over empty Ni sites) with an initial sticking probability of 0.85. The binding energies of the precursor molecules are calculated to be 9–15 kcal mol^{-1}. By carrying out spatially resolved AES measurements, Tang et al. find that, once in the precursor state, the molecules are able to translate over macroscopic regions of the surface. This high mobility then leads to the independence of sticking probability as a function of coverage (for low coverages) observed in the earlier experiments [150]. Tang et al. [158] suggest that molecules having energies below 4 kcal mol^{-1} arriving at the surface in unfavorable configurations (not with the carbon end closest to the surface) do not chemisorb, i.e. there is an effective activation energy barrier. But the attraction between the carbon end and the metal surface applies a torque to the molecule, preventing it from leaving the surface and sending it tumbling along over long distances until they are finally chemisorbed at a particular site. This mechanism implies that molecules incident on the surface with large kinetic energies in unfavorable configurations will bounce back with sufficient kinetic energy to avoid being trapped. Indeed, Tang et al. [158] observe that molecules with energies larger than 7 kcal mol^{-1} are not trapped near the surface. The sticking probability then drops to about 1/5, as half the incident molecules will approach the surface with the carbon end closest to it, and chemisorb. Shayegan et al. [159] have also presented evidence for a precursor state trapped at 5.5 K on the (111) surface, which converts to chemisorbed CO at 60 K. This precursor state is of a different nature to the one described above, since Shayegan et al. obtain a pre-exponential factor of only 5×10^{-3} and an activation energy to chemisorption of only 0.01 kcal mol^{-1}. These results point to this precursor state being physisorbed CO, as found on low-temperature adsorption on Cu by Norton et al. [160].

Englert et al. [161] have employed ion scattering spectroscopy to confirm that CO bonds through the carbon end, and chemisorbs perpendicularly to a Ni(111) surface. Bertolini et al. [147] also observed that co-adsorption of C_2H_4 or C_6H_6 led to a lowering of the C–O stretching frequency, presumably by electron donation from the hydrocarbon into the $2\pi^*$ antibonding orbitals of CO. The magnitude of this effect seems to correlate qualitatively with the decrease in work function due to the hydrocarbon adsorption.

Erley and Wagner [162] have reported measurements of the effect of S poisoning on CO chemisorption on Ni(111). It was found by thermal desorption measurements that 1/3 of a monolayer of S blocks CO adsorption completely. The binding energy of CO is observed to decrease as S pre-coverage increases.

Rubloff and Freeouf [163] have carried out a combined surface reflectance spectroscopy–EELS study of CO/Ni(111). Correlations between the spectral measurement of chemisorbed CO, solid CO, and CO adsorbed on transition metal carbonyls have suggested possible identification of the observed optical transitions. These include charge transfer excitations from the metal to the antibonding $2\pi^*$ molecular orbital, $5\sigma \rightarrow 2\pi^*$ transitions, and Rydberg excitations to CO orbitals near and above 13 eV. The authors caution that other factors could contribute to the surface optical transitions in the case of CO/Ni(111), such as CO-induced changes in the optical response of the metal itself.

The question of surface structure on the dissociation of chemisorbed CO on Ni surfaces is of current interest since the hydrogenation of CO to produce hydrocarbon products has been shown to proceed via a CO dissociation mechanism [164–168]. Thermal desorption spectroscopy studies have shown that a *stepped* Ni(111) surface yields a high-temperature form of CO desorption, designated β_2-CO, at about 850 K, whereas Ni(111) yields only an α-CO state. The solution into the bulk of carbon produced form CO dissocia-

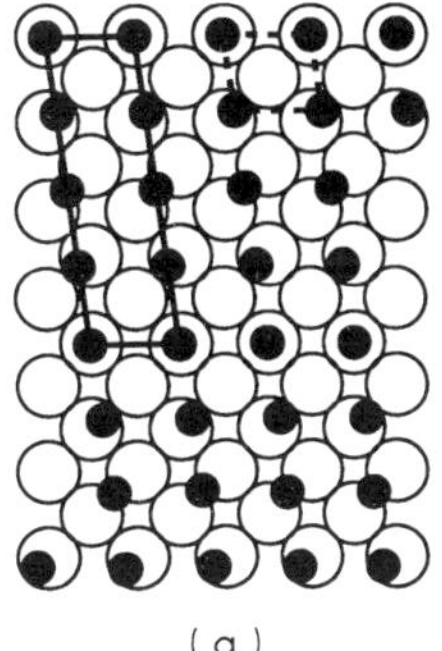

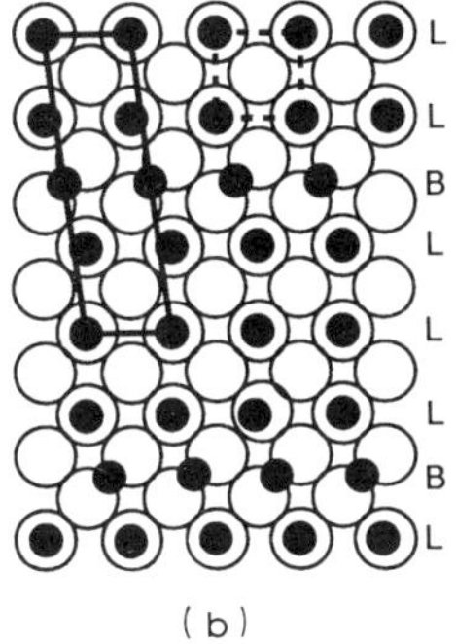

Fig. 13. (a) A coincidence net, c($7\sqrt{2} \times \sqrt{2}$)R 45° for CO/Cu(100), based on uniform compression of the overlayer, and (b) after relaxation into linear (L) and bridged (B) rows [175].

tion is an important process which competes with β_2-CO formation. It is postulated that CO dissociation occurs on stepped Ni(111) near 450 K [169]. These results are confirmed by the FEM results of Murayama et al. [170]. Studies of the isotopic mixing reaction

$$^{12}C^{18}O + {}^{13}C^{16}O \rightarrow {}^{12}C^{16}O + {}^{13}C^{18}O$$

on Ni(100)[171] are in accord with this result, since the reaction between C and O to produce isotopically mixed CO is not observed until ~800 K, suggesting that the Ni surface is protected by carbidic and graphitic overlayers which prevent the isotopic mixing process. Free Ni sites become available above ~800 K and the continuous reaction begins.

In summary, it appears that surface structure has an influence on CO dissociation kinetics. The catalytic activity of Ni (for CO-related reactions) is determined by a combination of factors involving dissociative chemisorption and the behavior of C(ads) and O(ads) on the Ni surface. Interactions between CO species and between CO and other species such as H(ads) or S(ads) can lead to profound structural effects and hence to kinetic effects in the reactions of chemisorbed CO on the surface of Ni.

13. Copper

Carbon monoxide chemisorption on Cu is weak and reversible. Using LEED and work function methods to measure isoteric heats, Tracy [172] obtained an initial heat of chemisorption of 17 kcal mol^{-1} for CO/Cu(100). LEED studies by Chesters and Pritchard [173] and Tracy [172] indicate that the first ordered structure to form gives a c(2×2) pattern. At 77 K, for exposures greater than 1.7 L, the pattern continuously changes, and this was interpreted as a smooth compression of the c(2×2) structure along the $\langle 110 \rangle$ direction. At the point where compression begins to occur ($\theta = 0.5$), a sharp drop in isoteric heat of adsorption is observed. Chesters and Pritchard [173] suggested that repulsive CO–CO interactions at all coverages prevent the molecules from sticking to high coordination sites.

Horn and Pritchard [174] obtained the vibrational spectrum of CO/Cu(100) at 77 K, the first of a molecule on a single crystal, using reflection IR spectroscopy. A single sharp ($\nu_{1/2} = 10\,cm^{-1}$) C–O band grows at 2078 cm^{-1} with negligible frequency shift until just before a maximum in surface potential (minimum work function) is achieved. At this point, a sharp 9 cm^{-1} upward shift occurs. The shift is associated with the establishment of good long-range order in the c(2 × 2) structure rather than in the compression of the overlayer. These results did not support the model which proposed a continuous compression of the overlayer [172, 173].

Pritchard [175] then proposed a different model to explain the continuous compression of the LEED pattern without invoking a continuous compression of the overlayer. The model suggests that the CO molecules adsorb only on top sites until the c(2 × 2) pattern is fully developed. These molecules have their dipoles similarly aligned such as to decrease the work function. As the coverage increases beyond $\theta = 0.5$, CO begins to adsorb on bridge sites with opposite dipole moment, causing the work function to increase again. The resulting unit cell then accounts for the LEED observations, as shown in Fig. 13. The only problem with this model was that the predicted bridge-bonded CO was not observed, either by the earlier infrared results of Horn and Pritchard [174] or subsequent EELS results of Sexton [176] or Andersson [177], although Andersson did see a second C–Cu stretching frequency consistent with bridge-bonded CO. Andersson suggested that only one mode was visible, perhaps because there was a close coupling between the linear and bridge-bonded molecules. Subsequently, Ryberg [178] used CO with isotopes of C and O to decouple the two modes and showed that Pritchard's suggestion [175] was correct. Indeed, there are bridge and linearly bonded molecules in the ratio 1:3. Furthermore, he showed that the small frequency shift observed for one isotopic species contains a large upward shift of 40 cm^{-1} due to dipole–dipole coupling and an opposite chemical shift of 30 cm^{-1}, in accordance with observations of other CO–metal systems. Other possibilities, such as some CO molecules lying flat on the surface, are ruled out by photoemission experiments.

Allyn et al. [179] employed angle-resolved photoelectron spectroscopy to study CO adsorption on Cu(100) at 80 K. Three CO-related photoemission features have been resolved and assigned as shown in Fig. 14. For polarized radiation in the y–z plane, P_1 is assigned as a combination of 1π and 5σ states, P_2 as a 4σ orbital, and P_3 as a shake-up of the parent 4σ orbital. A comparison with the UPS spectrum (curve 3) of $Cr(CO_6)$ is illuminating.

When s-polarized radiation is used (curve 2), P_2 and P_3 are missing. On the basis of the energy and angle dependence of the photoemission intensity, this work concludes that binding is primarily through the 5σ-carbon-centered orbital, and that the axis of the molecule is within 20° of the

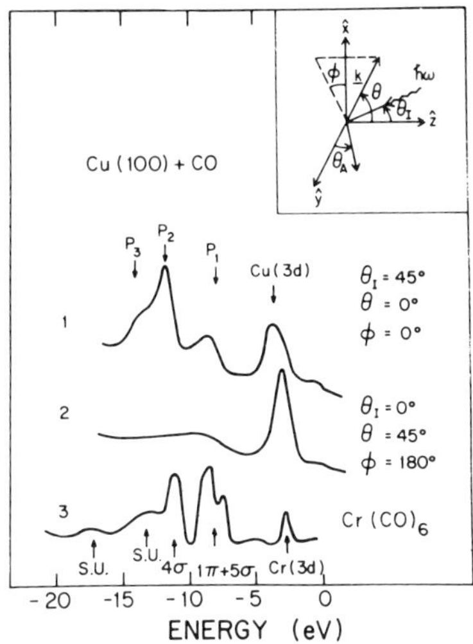

Fig. 14. Angle-resolved photoemission spectrum of CO chemisorbed on Cu(100) at 80 K. (1) Radiation polarized in the y–z plane; (2) s-polarized radiation; (3) $Cr(CO)_6$(gas)[179].

normal. Similar UPS spectra have been observed by Küppers et al. [180] for CO adsorbed on Cu(100) and Cu(110).

One of the more interesting developments in photoemission from Cu is the observation by McConville et al. [36] of the filling of the $2\pi^*$ level as it hybridizes with thermal *d*- (and to a lesser degree *sp*-) levels. Previous to this work, Bozso et al. [181] also found evidence for $2\pi^*$ emission by electron emission spectroscopy excited by incident metastable helium atoms. However, Conrad et al. [182], with the same technique, did not find $2\pi^*$ emission. This group does now find $2\pi^*$ emission from CO on other metals, so on balance, we conclude that there is indeed $2\pi^*$ filling. This indicates that, even though CO only adsorbs on Cu(100) below room temperature, it is chemisorbed, as shown by Norton et al. [183]. They have obtained UPS spectra from physisorbed CO on polycrystalline Cu at 20 K. The spectral features in this layer correspond closely to gas-phase CO features broadened to ~ 1 eV by interaction with the surface. Upon warming to 30 K, conversion to chemisorbed CO results in considerable change in the valence band photoemission line shape to that reported in refs. 179 and 180. The experiment illustrates that the activation energy for the process CO-(physisorbed) → CO(chemisorbed) is low. In distinct contrast, Isa et al. [184] reported the observation of irreversible CO adsorption on Cu(100) at 295 K, with UPS spectra different to those obtained by others [179, 180, 183], and this remains unexplained. Spitzer and Lüth [185] studied the electronic transitions of CO on the (100), (110) and (111) faces of Cu. They observe the $5\sigma \rightarrow 2\pi^*$ transition on all surfaces at 9 eV.

Horn et al. [186] have studied the heat of adsorption, LEED IRAS behavior of CO/Cu(110) adsorption on Cu(110). The isoteric heat of adsorption at zero coverage is 14 kcal mol^{-1}. This value drops slightly to 13 kcal mol^{-1} as the coverage increases up to where the work function drops to a minimum (− 0.3 eV); here a (2 × 1) LEED structure forms, and the isoteric heat falls rapidly. The adsorption kinetics were studied in greater detail by Harendt et al. [187] by means of LEEDS and TDS. The LEED patterns are in good agreement with those of Horn et al. [186]. Harendt et al. measured the initial sticking probability to be 0.46, which remains constant at this value until $\theta = 0.5$, indicating the presence of a precursor state. The behavior of the TDS peaks of CO/Cu(110) is most unusual.

In contrast to the monotonic shift of the TDS peak towards higher temperatures with increasing heating rate, as expected for first-order desorption, Harendt et al. [187] find that the trend reverses at higher heating rates. They attribute this effect to a very low mobility of adsorbed CO. The binding energy of CO/Cu(110) is measured to be 51–53 kJ mol^{-1} in the (2 × 1) structure, and decreases at higher coverages. Two IR bands at $\nu_{CO} = 2088\ cm^{-1}$ and $\nu_{CO} = 2014\ cm^{-1}$ exist at low coverages in temperature range 77–295 K. As coverage increases, the IR bands shift and merge into a single band at 2094 cm^{-1} accompanied by a compression of the (2 × 1) overlayer in the (110) direction. Model structures for the (2 × 1) and compressed overlayer are shown in Fig. 15. Horn et al. [186] postulated that the two IR bands correspond to two kinds of CO domains of similar energy. But in a more recent work, Hollins et al. [188] concluded that the high-frequency band corresponds to adsorption of CO on step or defect sites. Surprisingly, as the coverage increases, and therefore the ratio of CO adsorbed on top sites to CO adsorbed on defect sites also increases, so does the intensity of the high-

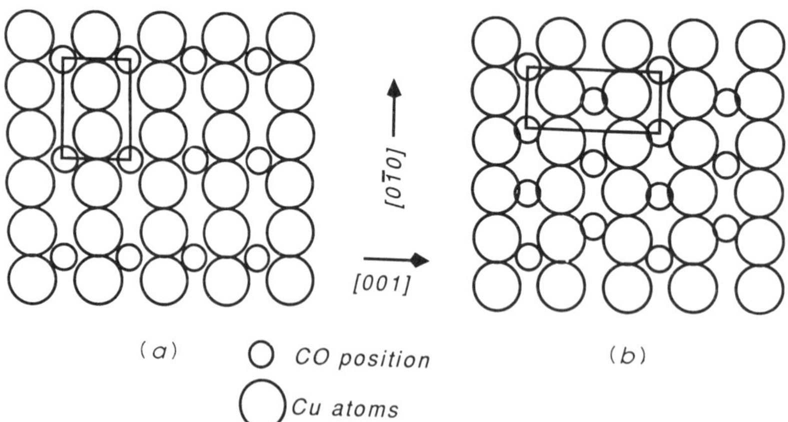

Fig. 15. Structure models of CO chemisorbed on Cu(110) at 80 K. (a)(2 × 1) overlayer; (b) compressed overlayer [186].

frequency band. Hollins et al. [188] showed that this is due to coupling between the two types of molecules, which causes the high-frequency mode to steal intensity from the low frequency mode. Woodruff et al. [189] showed that the small frequency shift observed for the CO molecules adsorbed on top sites is due to a cancellation effect, as in CO/Cu(100). There is a large upwards shift due to dipole–dipole interactions, partially offset by a large downward chemical shift. In this respect, the adsorption of CO on Cu is quite unlike that on Pt, which shows no chemical shift, or CO on Pd, which shows a large upward chemical shift. Woodruff et al. [189] proposed that there is a through-space interaction between adsorbed molecules due to the overlap of $2\pi^*$ orbitals. In a qualitative fashion, this model accounts for the downward chemical shift in noble metals and upward chemical shift in transition metals.

Wendelken and Ulehla [190] have employed EELS to study CO adsorption on Cu(110) at 80–200 K. A CO feature at 258 meV (2080 cm^{-1}) shifts to 2095 cm^{-1} with increasing coverage. Heating the sample to 200 K results in removal of CO responsible for this loss feature. Rogozik et al. [191] were able to observe the empty $2\pi^*$ level of CO/Cu(110) by Bremsstrahlung Isochromat spectroscopy. They find the level at 3.4 eV above the Fermi energy, and the width varies from 1.9 eV at low coverage to 2.6 eV at high coverage. Rogozik et al. suggest that the increase in width with increasing coverage is due to delocalization of captured electrons in the final state.

13.2 Cu(111)

Pritchard et al. [192] studied the adsorption of CO/Cu(111), and other high-index planes of Cu. They reported that a $(\sqrt{3} \times \sqrt{3})R\ 30°$ structure forms first, with a surface potential increase of 0.45 V and a C–O stretching frequency of 2080 cm^{-1}. With increasing coverage, a $c(4 \times 2)$ pattern appears, and the surface potential decreases by 0.1 V, suggesting that a different type of adsorption site is populated. The only CO band decreases in frequency with increasing coverage. Further increase in coverage yields a hexagonal (1.4×1.4) diffraction pattern. Hollins and Pritchard [193], in a clever use of isotopes, accurately determined the coverage of the intermediate LEED pattern and found it to be a hexagonal $(1.5 \times 1.5)R\ 18°$ structure with coverage 0.44 ± 0.01, and not a $c(4 \times 2)$. The sequence of LEED patterns was interpreted as a continuous compression of the overlayer which is out of registry with the substrate. But subsequently, Pritchard [175] departed from this interpretation, suggesting instead that the CO molecules occupy linear- and bridge-bonded sites. In general, very little distortion of the out-of-registry pattern is required to achieve linear and bridge bonding, and the LEED pattern remains unchanged. The model explains the surface potential measurements if bridge-bonded molecules have the opposite dipole moment of linearly bonded ones. It also explains the sharp IR band, as out-of-registry molecules would considerably broaden the bands. Although Pritchard suggested that the reason for the absence of IR bands due to bridge-bonded molecules is the same as in the case of CO/Cu(110), Hayden et al. [194] have since found the bridge-bonded molecules by means of IRAS for coverages greater than 1/3. At saturation, Hayden et al. [194] find almost equal numbers of linear- and bridge-bonded CO. Pritchard's model [175] predicts a ratio of 13:12 of linear to bridging CO. Therefore, the results of Hayden et al. give strong support to Pritchard's model of how apparently incommensurate structures as observed by LEED can be produced by models of adsorption on high-symmetry sites with dislocations to accommodate the excess molecules. Their results are confirmed by Chesters et al. [195], also by IRAS. There are two- and threefold bridging molecules, as also found on Pt(111)[196].

Furthermore, they find that the ratio of twofold-to-threefold bridged molecules is a strong function of temperature. This result implies that the binding energy difference between the two states is small, therefore their relative populations are given by the Boltzmann distribution. Hayden et al. [196] suggest that the conversion of twofold to threefold bridging is aided by increasing the population of the frustrated translation modes as the temperature is increased. The same effect was found earlier by Hayden and Bradshaw for CO/Pt(111)[196], and has been found since by Tang et al. for Cu/Ni(111)[149]. Hayden et al. [194] detected a physisorbed state which is populated only *after* chemisorption of the linear species is completed (near saturation). Furthermore, they show that the physisorbed state is *not* a precursor for chemisorption, as also found on Ni(111).

Jugnet and Minh Duc [197] have studied the sequence of UPS spectra obtained at various coverages of CO on Cu(111). They observe changes in relative intensity of the various features with coverage. Thus, their tentative assignment differs significantly from that of Allyn et al. [179] who invoked a shake-up process for total assignments. In agreement with the more recent interpretation of Pritchard [175], Jugnet and Minh Duc [197] assign two pairs of photoemission features to 4σ and $5\sigma + 1\pi$ orbitals associated with CO in two different sites. The variation in orbital energy differences for the two sites range from 0.9 to 2.0 eV. This difference seems beyond that expected for normal variations in CO structure.

UPS studies of CO on Cu(111) carried out by Kanski et al. [198] are also in disagreement with the results of Allyn [179]. By pre-adsorbing O on Cu(111), Kanski et al. found that the feature assigned as the 4σ orbital was strongly attenuated and suggested it was therefore due to the 5σ orbital. This difference compared with Ni is interpreted as being due to lower-lying metal states in Cu than in Ni. Also, the 4σ level is found to be shifted towards higher binding energy. The angular-resolved data suggest that CO on Cu(111) is tilted about 35° with respect to the normal.

The vibrational spectrum of CO on Cu surfaces has been thoroughly reviewed by Sheppard and Ngugen [17]. They point out that the low-index planes of Cu seem to yield lower C–O frequencies than do the high-index planes. They also list 10 evaporated film studies and 20 supported metal studies. It is apparent from the frequencies of CO adsorbed on supported Cu surfaces that low-index faces are seldom found. In addition, CO adsorption on partially oxidized Cu yields frequencies near 2120 cm^{-1} in many cases. Moskovits and Hulse [199] have studied CO ligation on Cu_n clusters in inert gas matrices at low temperatures. As n increases from 2, the CO band in Cu_n CO is observed to decrease in frequency.

14. Zirconium

The initial heat of adsorption of CO on a Zr film has been determined to be very high (148 kcal mol^{-1})[44], suggesting that dissociative adsorption must occur, as in the case of Ti, its homolog in Group IVB [1]. Quinn and Roberts [200], using a heated Zr ribbon, did find evidence for dissociative chemisorption of CO, which is followed by diffusion of O and C atoms into the bulk at 900–1000 K. The estimated activation energy of diffusion is 70 kcal mol^{-1}. A more detailed study of the adsorption of CO on polycrystalline zirconium has been carried out by Foord [201]. They find that CO adsorbs at 300 K with unit sticking probability to a density of one molecule per surface atom. The adsorption kinetics follows the ordinary Langmuir model for random one-site adsorption, but modified by the effect of a physically adsorbed precursor state with an appreciable surface lifetime. Even though these results indicate that only one end of the molecule is involved in the bonding, other measurements indicate that adsorption is dissociative. For example, the cross-section for electron-stimulated desorption is very small. Also, Foord et al. measured the diffusion coefficient of C and O into the bulk as the sample was heated, and found the values to be independent of the presence of the other species. They related the fact that both C and O diffuse into the bulk to the existence of both interstitial Zr carbides and oxides. Although most of the adsorbed C and O diffuse into the bulk upon heating the sample, a small amount of CO (< 0.05 monolayer) desorbs with first-order kinetics, with an activation energy of 92 kJ. Field ion microscopy has been carried out on Zr tips [202]. Haas et al. [203] have reported studies of impurity segregation to the surface after annealing the tip in vacuum.

The use of Zr centers for the reduction of CO is an area of current interest in organometallic chemistry. Stoichiometric reduction of coordinated CO by H_2 has been reported by Manriquez et al. [204] in a Zr complex. The postulated mechanism for the first reduction step is

$$(\eta_5\text{-}C_5](CH_3)_5)_2Zr(CO)_2 + H_2(g) \rightarrow (\eta_5\text{-}C_5](CH_3)_5)_2Zr\langle{}^{H}_{H}\rangle CO - (\eta_5\text{-}C_5(CH_3)_5)_2Zr{}^{H}_{H} - O$$

This reaction mechanism is important in Fisher–Tropsh-type reactions since it represents the first characterized example in which a coordinated carbonyl is reduced by molecular hydrogen. Analogous Ti complexes yield CH_4 stoichiometrically but not catalytically [205].

15. Niobium

Klein and Little [206] measured the work function changes due to adsorption of CO on a Nb field emission tip to be 0.7 eV at 100 K. They also deduced that there are two binding states, the β state being filled before the α state, plus a weakly bonded physisorbed state. The α state was found to desorb completely by heating the sample to 650 K, the remaining β population being less than 1/5th of a monolayer. The remaining β phase is found to undergo surface diffusion at 675 K, with an activation energy of 30 kcal. Haas et al. [207, 208] have observed that adsorption of CO/Nb(110) and Nb(112) does not produce ordered LEED patterns. But if the sample is warmed slightly, ordered patterns result which are characteristic of an oxide layer. Haas et al. postulate that CO dissociates upon sample warming, and that the C atoms do not contribute to the diffraction pattern, possibly due to migration into the bulk. An ESD study by Davis at al. [209] of CO/Nb(111) also shows the presence of at least two binding states which are characterized by the emission of O^+ and CO^+ ions, respectively. The O^+-producing state is filled first and removed after sample heating, and therefore represents the most strongly bonded species, in analogy to CO adsorption on Mo and W. But unlike the case of Mo and W, no thermal desorption is observed during heating to 700 K, after which the O^+ current has disappeared, suggesting conversion to an even more strongly bonded species, presumably dissociated CO, as suggested by previous work [206, 207]. Synchrotron radiation was used to study the adsorption of CO/Nb(100) at 300 and 100 K [210]. The study states that CO is dissociatively adsorbed over the entire coverage range at 300 K. Molecular CO was found at 100 K, which dissociates upon heating.

16. Molybdenum

The initial heat of adsorption of CO on polycrystalline Mo films was measured calorimetrically by Brennan and Hayes to be 75 kcal mol^{-1} [44]. Little et al. [211] found that the isosteric heat of adsorption in the coverage range 0.75–0.95 follows the relationship $\Delta H = 50\,\theta–50.5$ kcal mol^{-1}, suggesting that a low binding energy form of chemisorbed CO is reversibly adsorbed in this coverage range.

16.1 Mo(100)

Felter and Estrup [212] have studied the adsorption of CO/Mo(100) using TDS, ESD, LEED, work function changes and AES. Mainly on the basis of

Auger lineshape analysis, they conclude that the two β states which occur in thermal desorption near 1000 and 1300 K are dissociatively adsorbed, and exhibit an activation energy for 2nd-order desorption in the range 87–40 kcal/mol. Yates and King [213, 214] and Viswanath and Schmidt [215] report three β desorption states (at 1100, 1440 and 1650 K). In addition, for adsorption at room temperature, a molecular α state is detected near 490 K during TDS from high coverages. At 200 K, CO chemisorbs into a molecular state called virgin CO (v-CO), which does not dissociate into β-CO if the sample is kept at 200 K [212]. v-CO adsorbs with constant sticking probability, suggesting the presence of a precursor state. The discrimination between the three states, β-CO (dissociated into atomic C and O), α-CO (room-temperature molecular state), and v-CO (low-temperature molecular state) can be made by careful lineshape analysis of the carbon KLL Auger transition, taking pains to avoid extensive electron beam damage.

The ESD studies of Felter and Estrup [212] have detected two ionic products after CO adsorption on Mo(100): CO^+ and O^+. Just as in the case of CO/W(100), both of these seem to originate from undissociated CO. As the CO^+-yielding state desorbs, either thermally or by electron stimulation, the O^+-yielding state increases in coverage. Heating Mo(100) to $\sim$600 K causes the loss of the O^+-yielding state as molecular CO desorbs from the surface. Lecante et al. [216] have studied the LEED, TDS, and work function changes of CO/Mo(100), finding that the three β states can be distinguished from each other by their different work function behavior. Lecante et al. did not address the question of whether β-CO is dissociated. In a later paper, Guillot et al. [217] carried out LEED and quantitative AES measurements of the coverage of pure carbon layers, compared with carbon layers produced by CO adsorption. They concluded that β-CO on Mo(100) is probably due to dissociative CO adsorption. Ko and Madix [218] presented further evidence for the dissociation of CO, and also clarified the discrepancy between the work of Felter and Estrup [212] and Viswanath and Schmidt [215] by showing that Felter and Estrup did not resolve all three β states. Ko and Madix [219] showed that the presence of C or O on the Mo(100) surface severely hinders the dissociation of CO. Semancik and Estrup [220] studied the adsorption of CO/Mo(100) by UPS and XPS. Their results, which are similar to previous ones using polycrystalline samples [221], can distinguish V, β and α states. The $5\sigma + 1\pi$ levels of undissociated CO give a broad feature between 7.6 and 8.7 eV below the Fermi energy, while dissociated CO gives the $2p$ levels characteristic of atomic C and O. Furthermore, the binding energy of the O(1s) level is the same as that of $Mo(CO)_6$, indicating that α-CO is carbonyl-like. Semancik and Estrup suggest that the β substates observed in TDS could be due to adsorbate–adsorbate interactions, and not due to adsorbed molecules with different binding energies. They also show how UPS can be used to follow the dissociation kinetics of CO. In a subsequent paper, Semancik and Estrup [222] show that the dissociation of molecularly adsorbed CO does not obey simple first-order kinetics, but can be explained by an activa-

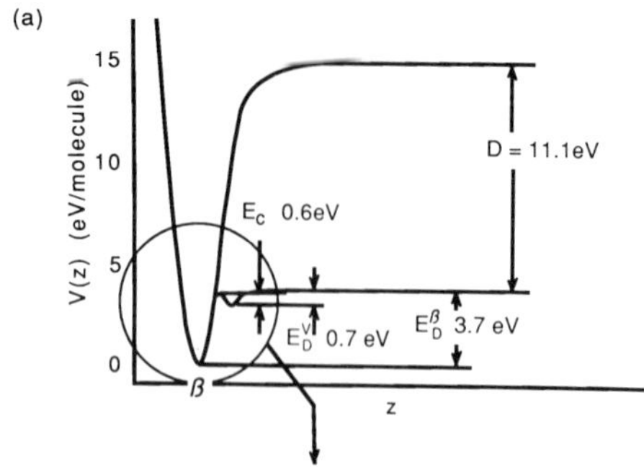

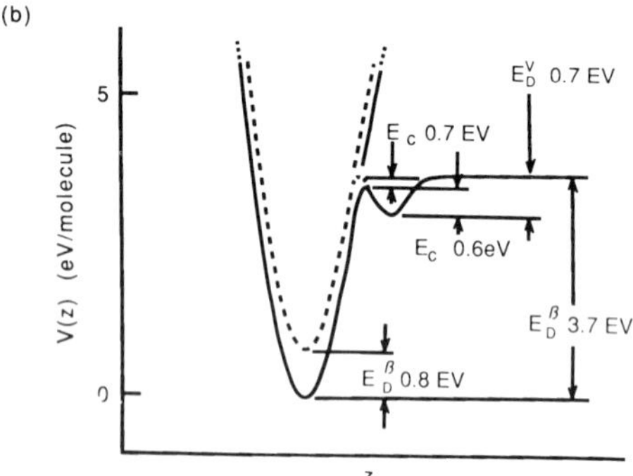

Fig. 16. (a) Potential energy diagram for the CO/Mo(100) system at low coverage, including the activation energy of conversion, E_c, the gas-phase dissociation energy, D, and the desorption energies of V-CO and β-CO, E_D^V and E_D^b, respectively; z is the distance from the surface. (b) Enlarged view of the circled region in (a); the solid curve again represents the low coverage potential energy, while the broken curve represents the β potential energy at higher coverage [222].

tion energy for dissociation which has an initial value of 0.6 eV, but increases with coverage, as shown in Fig. 16. Kellogg [223] also measured a dissociation barrier of 0.7 eV, although assuming a first-order process. He also showed that the dissociation barrier on the (110) surface is higher by 0.12 eV. The ESDIAD measurements of Niehus [224] showed that α-CO must adsorb either in twofold bridge sites or fourfold hollow sites, and that there is some tilting of the molecules at low α-CO coverages.

16.2 Mo(110)

The thermal desorption of CO/Mo(110) has been extensively studied by Gillet et al. [225] using TDS, LEED, and AES. Two thermal desorption states evolve, one at 1050 K and another at 1150 K, via first-order kinetics. No ordered layer was observed by LEED at room temperature. They did not consider whether CO is dissociatively chemisorbed on Mo(110). Boiziau et al. [226] studied the adsorption of CO/Mo(110) by metastable helium de-excitation spectroscopy, and found dissociated CO at low coverages, together with molecular adsorption at higher coverages.

Only two studies of the infrared spectrum of CO on evaporated Mo films have been reported [17]. The observation of C–O stretching frequencies near and above 2000 cm^{-1} suggests that, to a limited extent, non-dissociative chemisorption of CO occurs at full coverage on Mo surfaces.

It is of interest to note that the concepts associated with CO adsorption on Mo parallel the developments in the understanding of CO chemisorption on W.

17. Ruthenium

The adsorption of CO on Ru field emission tips has been studied by three groups of workers [227–229]. All of the field emission work suggests that a monolayer of CO on Ru desorbs completely at 700 K without leaving a detectable residue of C or O. The field emission data also suggest that there are no appreciable differences in the behavior of CO upon chemisorption on different crystal planes of Ru. Klein [227] observed that migration of CO on Ru occurred with an activation energy of 2.8 kcal mol^{-1} at temperatures as low as 105 K. He also observed work function increases of 1.3 eV for monolayer coverage at 60 K, whereas Kraemer and Menzel [229] observed an increase of 1.8 eV for full coverage at 230 K.

17.1 Ru(0001)

CO/Ru(0001) has been studied by almost every available technique, each of which made a vital contribution to the understanding of the process. Thus, Madey and Menzel [230] using LEED, thermal desorption spectroscopy and work function measurements found that CO chemisorption at 300 K leads to a $\sqrt{3} \times \sqrt{3}$ R30° structure at $\theta = 1/3$. A more detailed LEED study by Pfnür and Menzel [231] finds that the $\sqrt{3}$ structure exists over a wide range of coverage, due to the repulsive interaction between nearest-neighbor CO molecules and the attractive interaction between next-nearest-neighbor molecules. Furthermore, Pfnür and Menzel [231] have determined the phase diagram of CO/Ru(0001) as a function of coverage. Madey and Menzel [230] also found two thermal desorption peaks at 450 and 400 K as coverage increases. Williams and Weinberg [232] found that, if CO is adsorbed below room temperature, at 110 K, the $\sqrt{3}$ structure is followed by a $(2\sqrt{3} \times 2\sqrt{3})$R 30° structure at $\theta = 7/12$. They also found a complex structure at higher coverages that has not been identified.

Although the earlier results were interpreted to mean that CO adsorbs in two states, the vibrational spectroscopy results of Thomas and Weinberg [233] and Pfnür et al. [234] show only the vibrational band due to CO linearly adsorbed on top of Ru atoms. No evidence of bridging CO is found at any

References pp. 460–469

coverage. Earlier XPS and UPS results [235, 236] also showed evidence of only one adsorption state. The extra peak in the thermal desorption spectra has been explained by the onset of repulsive lateral interactions in the adsorbed layer by Pfnür et al. [234], and a calculation of Wearie [237] based on a random configuration of adsorbed molecules with only nearest-neighbor interactions convincingly reproduces not only the frequency shift of the observed vibrational bands, but also their widths. The variation of the widths of the vibrational bands is attributed to the lack of translational symmetry in a randomly adsorbed layer, and not to the multitude of adsorption sites. The work of Brown and Vickerman [238, 239] shows the unsuspected sensitivity of SIMS to the adsorbate geometry and the variation of the lateral interactions between adsorbed molecules. The results are consistent with only top-bonded CO below $\theta = 1/3$. ESDIAD is also in agreement with vibrational spectroscopy. The angular distributions obtained by Madey [240] for O^+ and CO^+ ions produced by electron impact show that the CO ligand is oriented perpendicularly to the surface. The dependence of the angular distribution of the ionic beams on temperature is consistent with estimates of vibrational amplitudes made using reasonable force constants for bending modes.

Recent detailed UPS studies of CO/Ru(0001) by Hofmann et al. [241] and Heskett et al. [242] confirms the results described above, showing that CO is adsorbed upright, even at high coverages. Hofmann et al. show that CO in the $\sqrt{3}$ structure is not tilted by more than 10° from the surface normal, and therefore tends to discount the high-density packing model of Biberian and Van Hove [243] for this structure. The Biberian–Van Hove model would require tilted CO in analogy to CO/Pt(110). Riedl and Menzel [244] have used ESDIAD to show that as the coverage increases beyond 1/3 there is a small tilting of $\sim 5°$ along all azimuths. This tilting accommodates the strain of an overcrowded CO layer, while still allowing for bonding of CO on atop sites.

The most dramatic tilting of CO/Ru(0001) is obtained by co-adsorbing a saturation coverage of CO at 80 K, with $\lesssim 15\%$ of a monolayer of Na, as shown by ESDIAD by Netzer et al. [245] (Fig. 17). The maximum emission is seen at $\sim 50°$ to the normal. Netzer et al. [245] also observed that CO tilts if a molecule of CO interacts only with one atom of Na. If the coverage of Na is increased beyond 0.15 monolayer, the tilting disappears. Hoffmann and de Paola [246] claim that CO also tilts when co-adsorbed with potassium. During co-adsorption, a single adsorption state is observed in which the C–O stretching frequency decreases from 2009 cm^{-1} to 1460 cm^{-1}, a very low value for CO adsorbed on a metal surface. Hoffmann and de Paola interpret this dramatic reduction of the stretching frequency to mean that the C–O bond strength is greatly reduced, and the CO molecule has its axis parallel to or inclined towards the metal surface. Unfortunately, in this case the ESDIAD study of Madey and Benndorf [247] does not show direct evidence of tilted CO. The ESDIAD data for low coverages ($\theta_K < 0.1$) of CO + K on Ru(0001) show that the O^+ ESD signal from CO is suppressed by co-adsorption with

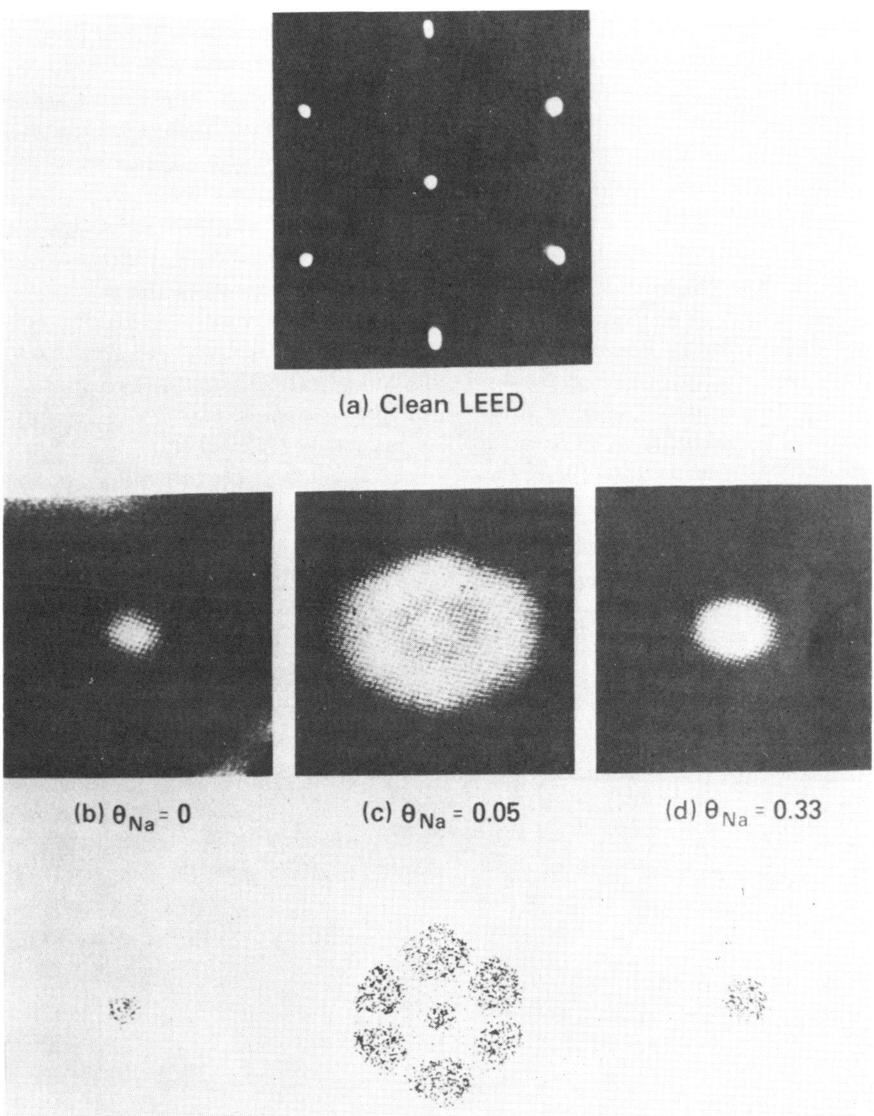

Fig. 17. LEED and ESDIAD patterns illustrating the interaction of CO with Na on Ru(001) at 80 K. (a) LEED pattern for clean Ru(001); (b) ESDIAD pattern for saturation coverage of CO on the clean surface; (c) ESDIAD pattern for CO co-adsorbed with $\theta_{Na} = 0.10$ mL, in equilibrium with $P_{CO} = 3 \times 10^{-9}$ Torr; (d) ESDIAD pattern for CO co-adsorbed with $\theta_{Na} = 0.33$ mL. Beneath patterns (b), (c) and (d) are sketches to indicate the detailed structures which are difficult to photograph. For all ESDIAD patterns, the electron energy was 300 eV; crystal bias was 160 eV. The distorted shape of the hexagonal halo in pattern (c) is due tc the bias field [245].

K. This can arise from a reorientation of the molecular axis, from perpendicular to inclined, or from increased reneutralization rates due to co-adsorption with K. Weimer et al. [248] have carried out an extensive study of the CO + K on Ru(0001) system using angle-resolved UPS and AES, and propose that CO remains perpendicularly bonded to the surface at all K coverages, in an sp^2-hybridized configuration. The co-adsorption experiments were carried out at $\theta_{CO} = \theta_{Na} = 1/3$, but Weimar et al. cited the similarity with less extensive data sets at other coverages to state that the sp^2 configuration applies at all coverages. More recent experiments by Eberhardt et al. [249] (carried out at different coverage than those of Weimer et al.) find evidence for a π-bonded configuration of CO, but no structural model is proposed. At this point the controversy concerning the orientation of CO in the presence of K or Ru(0001) is not settled.

The works of Hofmann et al. [241] and Heskett et al. [242] show the effects of the lateral interactions between adsorbed molecules on the energy of the CO-derived bands, which disperse as a function of parallel momentum by as much as 0.5 eV for the 4σ and 0.9 eV for the 5σ levels. Heskett et al. [242] point out that the value of the bandwidth of the 4σ level of CO/Ru(0001) follows the linear trend between the natural log of the bandwidth and the CO–CO nearest-neighbor distance of all systems studied so far [see Fig. 18(a)]. On the other hand, the bandwidth of the 5σ level is not correlated to the CO spacing, as shown in Fig. 18(b). Heskett et al. [242] attribute the breakdown of

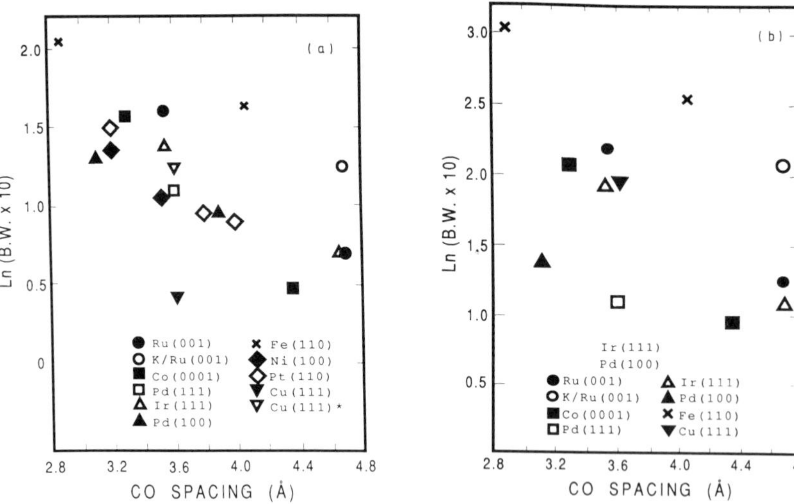

Fig. 18. (a) Plot of ln (bandwidth × 10) versus nearest-neighbor distances of the CO 4σ for different systems. The CO/Cu(111) value has been adjusted with respect to the intensity and binding energy associated shake-up peak (see ref. 242). The structure of the CO + K/Ru(001) overlayer is unknown; the CO molecules were assumed to be isotropically distributed on the surface at $\theta_{CO} = 1/3$. (b) Plot of ln (bandwidth × 10) versus nearest-neighbor distances of the CO $5s$ for different systems. (From ref. 242.)

correlation of the 5σ level shift with coverage to the fact that this level plays a major role in the bonding of the CO molecule to the metal surface, mixing quite strongly with metal levels. Therefore, only a calculation which properly takes into account the electronic structure of chemisorbed CO might be able to explain the 5σ level bandwidths. On the other hand, the 4σ levels do not participate much in the bonding of the molecules to the surface, so their dispertion is determined mostly by the lateral interaction between adsorbed CO molecules.

In contrast, Shincho et al. [250], using angle-resolved UPS have found that, when CO is adsorbed on a stepped surface such as Ru(1,1,10), it does not adsorb perpendicularly to the surface on the steps, although it does adsorb perpendicularly to the terraces. On the steps, the CO axis forms an angle of 30° with the (001) direction, tilted towards the $(5\bar{5}1)$ direction. Also, the steps tend to dissociate the CO molecule when the stepped surface at 500 K is exposed to gas-phase CO, with an activation energy of 20 kcal mol^{-1} [251]. There is no evidence that flat surfaces dissociate CO.

The $(\sqrt{3} \times \sqrt{3})$ structure of CO/Ru(0001) is one of the few cases where an accurate structure determination of adsorbed CO has been carried out by LEED IV analysis by Michalk et al. [252]. The results clearly rule out the two threefold hollow sites. However, the authors find that the differences in r factors between the bridge and on-top sites is not as large, particularly if only the Zanazzi–Jona r factor is employed. The Pendry r factor gives a decisive preference to the atop site, and therefore Michalk et al. conclude that CO is adsorbed on atop sites in the $\sqrt{3}$ structure, with a C–O distance 1.1 ± 0.1 Å and a Ru–C distance of 2.0 ± 0.1 Å. This fits in the trend 1.72, 1.90, 1.93 and 1.95 Å for Ni, Cu, Pd and Rh, respectively.

The adsorption and desorption kinetics of CO/Ru(0001) have also been extensively studied. The behavior of the sticking coefficient with coverage and sample temperature has been studied in detail by Pfnür and Menzel [253] who show that the initial sticking coefficient $S = 0.7$ remains constant up to $\theta = 0.2$, independent of sample temperature T_S between $T_S = 100$ and $T_S = 400$ K. The sticking coefficient then drops proportionally to $(1 - \theta)$ for $T_S > 200$ K, and more gradually below 200 K. Above $\theta = 0.23$, S drops sharply for $T_S > 200$ K up to $\theta = 0.36$, where the decrease becomes weaker. Around $\theta = 0.45$ the sticking probably again drops sharply until it becomes $< 10^{-2}$ above $\theta = 0.55$. For sample temperatures below 200 K the strong variations in the sticking probability become smoother, as S decreases continually with coverage. Pfnür and Menzel have explained these results in terms of the role played by intrinsic and extrinsic precursor states in the adsorption kinetics. Up to $\theta = 0.2$ and $T_S > 200$ K, gas-phase molecules adsorb directly on any uncovered site. The fact that next-nearest-neighbor sites are not blocked by CO–CO interactions, but are available for adsorption, has been explained by Pfnür and Menzel [253] as due to the path that the gas-phase molecules take during adsorption. However, they cannot rule out that adsorption takes place via molecules temporarily being trapped above the surface by selective adsorption (the so-called dynamical precursor state). Pfnür and Menzel [253] find that, at higher coverages, desorption proceeds via true intrinsic precursor states, i.e. oriented molecules weakly chemisorbed above empty sites not accessible for chemisorption, and separated from available sites by adsorbed molecules. The properties of the precursor state change strongly with coverage, leading to a change in S. The extrinsic precursor states (CO molecules physisorbed on top of the chemisorbed layer) make a significant contribution only for sample temperatures below 200 K, leading to a weak linear dependence of S on the coverage up to $\theta = 0.5$. This work is one of the few cases where the role of the different precursor states proposed by Cassuto and King [254] has been elucidated. Yamada and Tamaru [255] have used isotopes to measure the absolute rate of adsorption and desorption of CO/Ru(0001) and CO/Ru(2$\bar{1}$22). Although their data are somewhat inconsistent (with sticking probabilities greater than one) and they analyzed the data considering only one type of precursor state, the results generally agree with those of Pfnür and Menzel [253]. They also find that the rate of desorption increases with increasing pressure of gas-phase CO, a process they call "adsorption-assisted desorption". However, this conclusion has been questioned by Zhdanov [256]. The equilibrium adsorption energy, the activation energy for desorption and the pre-exponentials for desorption have been measured by Pfnür [257] and Menzel et al. [258]. They find very high values for the pre-exponential, up to $10^{19}\,s^{-1}$, which they attribute to the decreased phase space of the adsorbed layer in the range where the strong CO–CO repulsion leads to localization of the CO molecules even at high temperatures. Interestingly, both the pre-exponential and the adsorption energy values drop abruptly above $\theta = 1/3$, as the nearest-neighbor repulsion still prevents occupation of the nearest-neighbor sites. As adsorption proceeds, the adsorbed molecules must shift from the optimal on-top position, which leads to the strong drop in the adsorption energy. The delocalization of the adsorbed molecules also increases the entropy of the adsorbed layer, leading to the abrupt drop in the pre-exponential factor for desorption.

In distinct contrast to the non-dissociative behavior of CO/Ru(0001), it has been reported by McCarty and Wise [259] that the reaction

$$^{12}C^{18}O + {}^{13}C^{16}O \rightarrow {}^{12}C^{16}O + {}^{13}C^{18}O$$

occurs readily above 375 K for Ru supported on Al_2O_3. A model involving dissociative CO chemisorption is proposed. Above about 550 K, extra ^{16}O appears in the CO desorption products although no explanation is suggested for this effect, it seems likely that it is due to the presence of partially oxidized Ru, or to the involvement of the Al_2O_3 in oxygen exchange reactions with CO when Ru is present. Control experiments with pure Al_2O_3 did not yield the effect.

Although much less work has been carried out on other crystal faces of Ru, a comparison of the thermal desorption spectra of CO from three single-crystal planes is possible: on Ru(0001), two binding states are observed at high coverages [253, 255–258]; on Ru(101), two less clearly resolved states are reported [260]. On this surface, which contains kinked furrows, the states are not resolved but a broadening of the single peak on the low-temperature side occurred as the initial CO coverage was increased [261]. Two CO desorption peaks were resolved on Ru(10$\bar{1}$0)[262]. The temperature range for desorption of CO from all these crystallographic orientations is essentially identical (340–540 K).

18. Rhodium

The initial heat of chemisorption of CO on evaporated Rh films was measured by Brennan and Hayes to be 44 kcal mol^{-1} [44]. However, careful thermal desorption studies from CO/Rh(111) surfaces [263, 264] indicate that this value might be too high. In the limit of zero coverage [264], CO desorbs according to a first-order rate expression

$$-\frac{\mathrm{d}N}{\mathrm{d}t} = N \times 10^{13.6 \pm 0.3} \times \mathrm{e}^{-31600/Rt}$$

Similar results are obtained for CO/Rh(100)[265]. These activation energies are in accordance with values obtained on other close-packed faces of the Group VIII transition metals, all of which fall in the range 27–35 kcal mol^{-1}. The coverage versus exposure data do not fit a simple Langmuir adsorption isotherm. A mobile precursor model is strongly suggested by a long linear region to $\theta = 0.56$ for CO/Rh(111)[264] and $\theta = 0.4$ for CO/Rh(100)[265]. TDS of CO/Rh(110) also yields a desorption energy $E_\mathrm{d} = 32$ kcal mol^{-1}, assuming a pre-exponential factor of 10^{13} s^{-1} [266]. Similar studies on polycrystalline Rh yielded $E_\mathrm{d} = 32$ kcal mol^{-1} [267], assuming the same pre-exponential factor. In order to reconcile the calorimetric heat of adsorption on Rh films [44] with the kinetic measurements for the activation energy of desorption from the various Rh surfaces, a pre-exponential factor of $\sim 10^{18}$ s^{-1} would have to be postulated. Measurements of CO desorption from Ru(0001)[257, 258] suggest that large values of the pre-exponential factor may occur, in some cases, due to a mobile precursor state in desorption. But it is also possible that errors have occurred in the calorimetric measurement of the heat of adsorption on films, leading to too high a value for the adsorption energy. A possible error might arise from oxygen impurities on the surface, which has been shown to increase the binding energy of CO/Rh(110) to 44 kcal mol^{-1} [266], in fortuitous agreement with the early calorimetric value [44].

18.1 Rh(111)

The Rh(111) is the most studied surface. Grant and Haas [268], Castner et al. [263] and Thiel et al. [264] studied the LEED patterns occurring during adsorption of CO/Rh(111). A primitive (2 × 2) pattern evolves first, followed by a ($\sqrt{3} \times \sqrt{3}$)R 30° pattern at $\theta = 1/3$. As the coverage increases further, a non-primitive (2 × 2) LEED pattern appears at $\theta = 0.75$, with three CO molecules per unit cell. A dynamical LEED analysis by Koestner et al. [269] of the $\sqrt{3}$ structure shows the CO molecules to stand on end, carbon end down, on top of Rh atoms. These results confirm the previous conclusion of Dubois and Somorjai [270] whose EELS study showed only one C–O stretching frequency at 2070 cm^{-1} for $\theta = 1/3$, assigned to terminal CO. EELS shows that, for $\theta > 1/3$, CO is adsorbed on top and bridge sites, giving rise to a new C–O stretching frequency at 1870 cm^{-1}. In the first dynamical LEED analysis of molecular adsorption on multiple sites, Van Hove et al. [271, 272] studied the structure of the non-primitive (2 × 2) pattern. They find the $\theta = 0.75$ overlayer to contain one bridge-bonded and two near-top-bonded molecules in each unit cell, as shown in Fig. 19. The near top molecule is bonded asymmetrically, displaced from the symmetry position by ~0.53 Å, with the CO axis close to the surface normal. The nearest-neighbor distance to a bridge-bonded molecule is found to be 2.85 Å, one of the shortest CO–CO distances determined so far. The CO overlayer is also buckled by 0.36 Å due to strong metal–CO bonding. The top-bonded molecules are found to be within 5° of the surface normal, in distinct contrast to other cases such as CO/Pt(110)[435, 440], which achieve close-packing by tilting the CO axis away from the surface normal. A continuous overpressure of gaseous CO is required to maintain this high-density structure.

CO chemisorption on the Rh(111) and Rh(331) surfaces has also been studied by SIMS and XPS by DeLouise et al. [273, 274]. XPS shows two O 1*s* peaks, clearly distinguishing atop and bridge-bonded CO. In this case, as well as in the case of CO/Pt(111)[33, 411, 427], it is possible to distinguish the two adsorption sites because the XPS lineshape is not complicated by shake-up features. For Rh(111), the XPS results are in agreement with previous TDS, LEED and EELS results. Furthermore, the ratio of XPS intensities for atop- and bridge-bonded CO are in agreement with EELS intensity ratios, showing that, in some cases, EELS yields quantitative intensity information. The behavior of CO/Rh(331) is slightly different. The top and bridge sites are populated simultaneously from the beginning, in a bridged-to-atop ratio of 0.26. UPS spectra of CO/Rh(111)[274] show that chemisorption results in a 2.7 eV shift of the CO 5σ orbital.

The interaction of chemisorbed CO with chemisorbed hydrogen on the Rh(111) surface [275], and with chemisorbed deuterium on the Rh(100) surface [264] have been studied using LEED and thermal desorption spectroscopy. The adsorption of CO onto the clean or hydrogen/deuterium-covered surface proceeds via a mobile precursor state. There is a strong repulsive

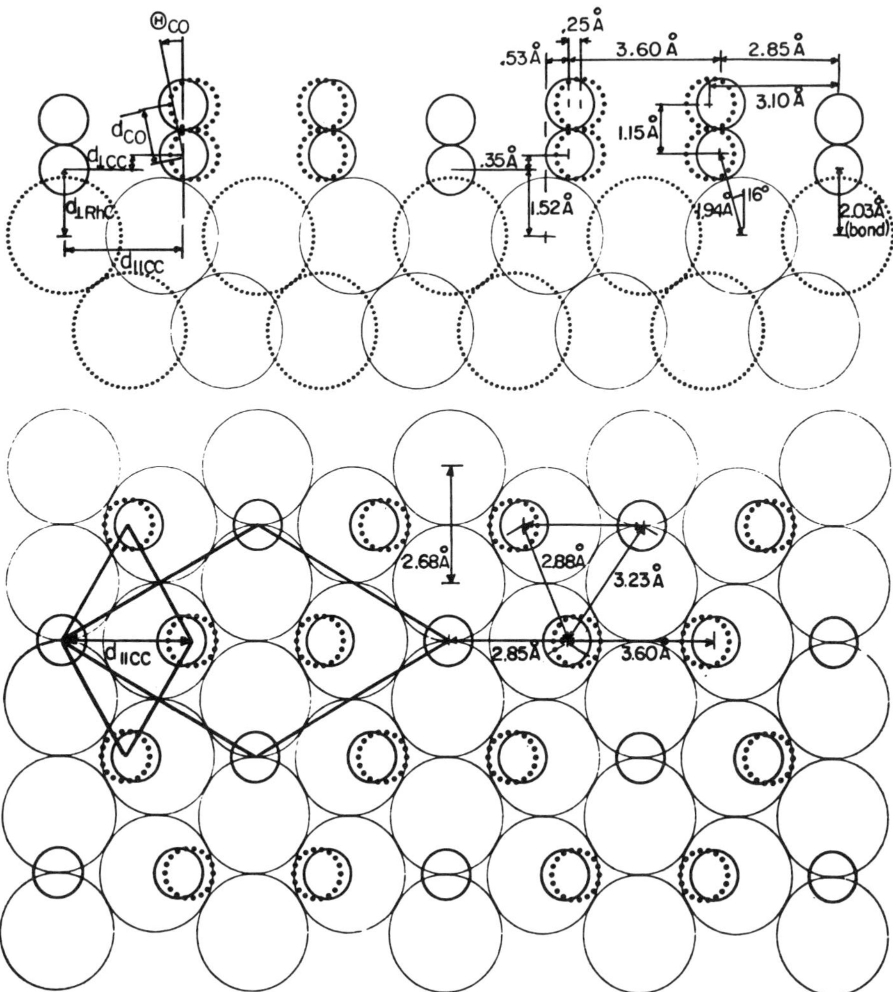

Fig. 19. The Rh(111)(2 × 2) − 3CO surface structure, showing its projections on the surface plane (at bottom) and on a mirror plane of the structure (at top); a (2 × 2) unit cell is outlined. Large circles represent Rh atoms, the dotted circles being outside the plane of the figure. Small dotted circles indicate C and O positions satisfying a hexagonal CO overlayer of coverage 3/4 (one unit cell of which is outlined), while small full circles indicate the optimum positions found in this study (but leaving all CO axes perpendicular to the surface). The independent structural parameters are labelled and some of the optimum interatomic and interlayer distances are indicated [271, 272].

interaction between adsorbed CO and adsorbed H, which is thought to be a "through-metal" effect [275]. LEED evidence suggests that island formation of CO occurs on both Rh(100) and Rh(111) in the presence of pre-adsorbed hydrogen.

CO/Rh(100) has been studied briefly by Castner et al. [263] using TPD and LEED. They report a c(2 × 2) pattern followed by a split (2 × 1) pattern at high coverages. In contrast to results in other laboratories [276], they also interpret their LEED results to mean that ESD does not occur for CO/Rh(100). Kim et al. [265], who also studied CO/Rh(100) by LEED and TDS, do report electron beam degradation of the c(2 × 2) overlayer. Kim et al. assign $\theta = 0.5$ to the c(2 × 2) structure. They show that further exposure to CO leads to a compression of the CO overlayer, coupled with the appearance of a second peak in the thermal desorption spectrum. They did not analyze the LEED patterns in detail, nor did they interpret the nature of the second peak in the thermal desorption spectrum.

18.2 Rh(110)

The behavior of CO on the (110) surface of Rh differs markedly from the behavior on the (111) and (100) surfaces, and is common to all the (110) faces of cubic metals in the platinum group. Marbrow and Lambert [266] found in their LEED, work function change, AES, and TPD study that CO adsorbs on near-top sites, displaced towards the troughs along the [11] direction. In this manner it can adsorb to a coverage of $\theta = 1$. The resulting zig-zag structure has p1g1 symmetry, and is therefore denoted (2 × 1)p1g1. In a pioneering infrared reflection–absorption study of (110)-oriented Rh films, Eckstrom et al. [278] found two bands, at 2078 and 2060 cm^{-1}, which they ascribed to a single adsorption state with p1g1 symmetry. Marbrow and Lambert find that the dipole moment per molecule is only constant at low coverage, resulting in a total work function change of 0.97 eV at $\theta = 1$. They also find that the layer undergoes dissociation and desorption with equal cross-sections ($10^{-22} m^2$) upon electron impact. Furthermore, Marbrow and Lambert [266] find that CO undergoes slow thermal dissociation at high temperatures and pressures, leaving deposits of carbon and oxygen on the surface. This behavior is also different from that on (111) surfaces. Although Sexton and Somorjai [279] and Castner et al. [263, 280] find thermal dissociation of CO on Rh(111), and suggest it is a route to the catalytic production of CH_4 by Rh, many subsequent attempts by other investigators [280–284] have found no evidence of thermal dissociation of CO on Rh surfaces. Baird et al. [284] have also carried out an XPS, UPS, and TPD study of CO/Rh(110). They find the CO 4σ and $1\pi + 5\sigma$ molecular orbital energies at 7.6 and 10.6 eV, respectively. Their conclusions as to the structure of adsorbed CO are the same as those of Marbrow and Lambert [266].

Rhodium surfaces in evaporated films and oxide-supported particles have been well studied by infrared spectroscopy [17]. Three types of chemisorbed CO exist on Al_2O_3-supported Rh surfaces [285, 288], viz.

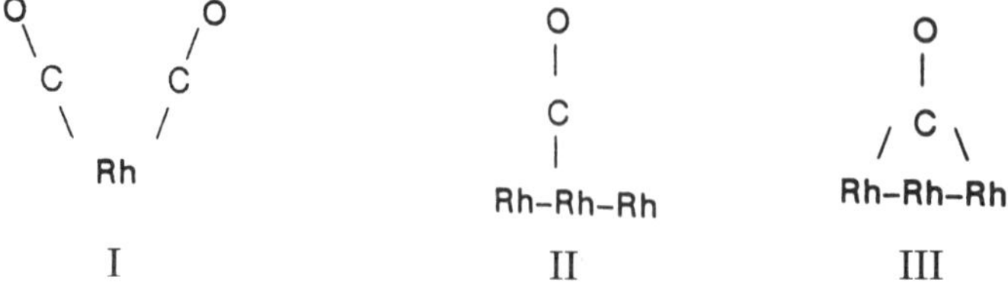

The $Rh(CO)_2$ species exhibit a doublet in the infrared spectrum at 2100 and 2030 cm^{-1}, which does not shift in frequency as the intensity changes due to increased exposure to CO. The lack of shift is attributed to species I being an atomically dispersed site. It is possible that the Rh site is in a higher oxidation state, $Rh^{\delta+}$, which would be consistent with its special properties. Species II and III are thought to exist on crystalline Rh sites, possibly two-dimensional rafts of Rh [289, 290]. The stoichiometry of I has been proven by volumetric uptake measurement [288] and isotopic studies [285]. The carbonyl frequencies for II and III are identical to those observed on Rh(111) [269]. Both species may be eliminated by treatment with O_2(g) at 300 K (as judged by the disappearance of their IR bands). Species I persists after the treatment. It has been found that isotopic exchange of ^{12}CO(ads) ligands with gas phase ^{13}CO occurs preferentially at 200 K for species I at higher temperatures, rapid exchange of II and III occur with ^{13}CO. ^{13}C-NMR spectroscopic studies have discriminated species I from a mixture of species II and III on the basis of ^{13}C-spin-lattice relaxation times [287]. The interpretation of species I as an isolated site is somewhat controversial [289].

Inelastic electron tunneling spectroscopy has also been used as a probe of the vibrational character of CO chemisorbed on evaporated Rh particles [290, 291]. Evidence for both bridged and linear CO species was found in both the C–O and metal–C stretching regions.

19. Palladium

The initial heat of chemisorption of CO on Pd films has been measured by calorimetric methods to be about 40 kcal mol^{-1} [44]. This may be compared with the isoteric heat of adsorption (in the limit of zero coverage) of CO on various single crystal planes, shown in Table 1 [292]. These results are in approximate agreement with the calorimetric results [44].

19.1 Pd(100)

In a pioneering study using LEED, AES, and work function changes, Tracy and Palmberg [296] found that the adsorption of CO on the Pd(100) surface is reversible, and characterized by four phases: (1) a lattice gas

TABLE 1

Initial heats of chemisorption for CO on various Pd surfaces (reversible isoteric heat measurements [292])

Pd plane	(111)	(100)	(110)	(210)	(311)
E_{ad} (kcal mol^{-1})	34	36.5	40	35.5	35

structure at low coverages, (2) a liquid-like short-range order phase just below 0.5 monolayers, (3) a two-dimensional crystalline phase at $\theta = 0.5$ with a c($2\sqrt{2} \times 2\sqrt{2}$)R 45° structure [called c(4×2) by Tracy and Palmberg], and (4) a compressible solid-like phase between $\theta = 0.5$ and $\theta = 0.8$. From the symmetry of the LEED pattern at $\theta = 0.5$, Tracy and Palmberg [294] correctly deduced that the CO molecules must be bridge-bonded. This conclusion has since been confirmed by Bradshaw and Hoffmann using IRAS [295], by Behm et al. using EELS and dynamical LEED analysis [296, 297], and by Ortega et al. [298], also using IRAS. The structure models are shown in Fig. 20. An unusual property of the c($2\sqrt{2} \times \sqrt{2}$)R 45° structure is that the surface lattice is a "coincidence lattice", different from both the substrate lattice and the overlayer lattice. In a coincidence lattice some of the diffraction spots are purely due to multiple scattering, and therefore cannot be predicted by kinematic theory [296]. Above $\theta = 0.5$, Tracy and Palmberg [294] proposed that the CO overlayer continuously compresses with increasing coverage along the $\langle 100 \rangle$ direction, while in the direction perpendicular to it, the dimension of the unit cell remains unchanged. Another possible structure is the fault model proposed by Pritchard [175] and later by Avery [299] where the excess CO molecules are accommodated along adjacent bridge sites, forming a dislocation line. This model requires the CO–CO distance to be as small as 2.75 Å. In the past, such a small separation between CO molecules was thought unlikely until Van Hove et al. [271, 272] showed

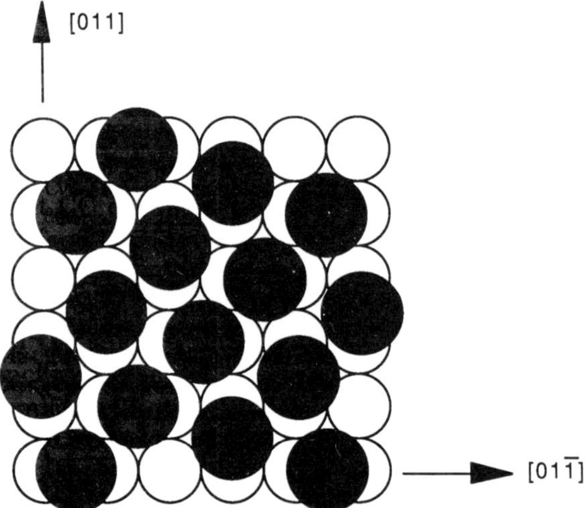

Fig. 20. Surface geometry for one domain of Pd(100) + ($2\sqrt{2} \times \sqrt{2}$)R 45° 2CO (left) and corresponding reciprocal lattice (right). A 90° rotation gives the second possible domain orientation. At left, crosses and circles indicate substrate atoms and adsorbates, respectively. At right, crosses correspond to clean surface diffraction spots, while large filled circles are kinematically predicted extra spots, and small filled circles are multiple-scattering spots. At normal incidence, the multiple-scattering spots emitted at azimuths $\phi = \pm 45°$ vanish [$\phi = 0$ being the (10) spot azimuth][296, 297].

it to occur. Van Hove's result notwithstanding, Ortega et al. [298] state that their infrared results favor the continuous compression model. They postulate that if the metal–carbon distance is kept constant for increasing coverage, the hybridization of the carbon atoms will remain the same as in the ordered structure. In this case no new features will appear in the IR spectrum during the compression stage, and indeed none are observed. A more recent EELS and SIMS study by Brown and Vickerman [300] finds linearly adsorbed CO even at $\theta = 0.48$, which is not observed in any other study. Brown and Vickerman's result could be explained by a non-smooth surface, since they do not show LEED evidence of a smooth surface. Behm et al. [297] and Ortega et al. [298] also find that the overlayer remains disordered if CO is adsorbed onto a sample below 300 K. The sample needs to be warmed above 340 K if adsorbed CO is to have enough mobility to produce an ordered overlayer. This result again points to the lattice-gas nature of CO/Pd(100).

Tracy and Palmberg found a linear decrease of the isoteric heat of adsorption versus coverage up to $\theta = 0.5$ [293], but a subsequent study by Behm et al. [297] found that the isoteric heat of adsorption remains constant with coverage at 38.6 kcal mol^{-1} up to $\theta = 0.5$. Behm et al. attribute the difference to a very slight C contamination of the sample used by Tracy and Palmberg. After $\theta = 0.5$, Behm et al. [297] find a steep decrease in the isoteric heat with increasing coverage. They also find a constant sticking probability up to $\theta = 0.2$, which is indicative of a mobile precursor state for adsorption. TDS shows first-order desorption with an activation energy of 36.8 kcal mol^{-1}, and an almost constant pre-exponential factor of $2 \times 10^{16}\,s^{-1}$. These, and the isoteric heat results, are expected from a lattice gas with small repulsive interactions. The UPS spectrum of adsorbed CO/Pd(100) was first obtained by Lloyd et al. [301] who assigned emission at 7.9 eV below the Fermi energy to the $1\pi + 5\sigma$ orbital and emission at 11.2 eV to the 4σ orbital. Subsequently, Horn et al. [302] found that, at sufficiently high coverages, the overlap of the CO orbitals form Bloch states, leading to dispersion of the photoemission features as a function of momentum parallel to the surface. They found the 4σ state to disperse the most, by as much as 0.5 eV. Furthermore, Horn et al. observed the change of the Brillouin zone size as the CO overlayer undergoes compression. The $2\pi^*$ state of CO/Pd(100) has been observed by inverse photoemission (or Bremsstrahlung Isochromat spectroscopy) by Rogozik et al. [303] who put it at 4.8 eV above the Fermi energy. Although this state has a large width of 2 eV, one does not expect it to be occupied. This effect is also observed with zero-valent carbonyl complexes [303]. The $5\sigma + 1\pi$ to $2\pi^*$ electronic transitions have been observed by Bader et al. [305]. Bader et al. also studied stepped Pd surfaces, and although they did not identify the patterns, they did find that CO adsorbs in different structures on a stepped surface compared with the flat surface. This result is significant in explaining why Brown and Vickerman alone [300] obtain linearly bonded CO on Pd(100).

References pp. 460–469

19.2 Pd(110)

The adsorption of CO/Pd(110) was studied by Ertl et al. [306], who observed three distinct structures with increasing coverage. At low coverages, a c(2 × 2) pattern appears, which converts rapidly into an incomplete p(4 × 2) pattern at intermediate coverages. The p(4 × 2) pattern persists until near saturation, when a (2 × 1) pattern appears. Ertl et al. placed the molecules in high-coordination B_5 sites, but Chesters et al. [307] showed by EELS that CO adsorbs preferentially in B_2 bridging sites. Chesters et al. also showed that the (2 × 1) structure has p2mg symmetry, as it is now known to have on several (110) transition metal surfaces. They find that adsorption at 110 K results in a disordered layer, until near saturation, where the (2 × 1)p2mg structure appears. The vibrational spectra are dominated by a band due to twofold bridged CO until θ approximately 0.8, where linear CO is also observed. The LEED observation of p2mg symmetry suggests that the molecules in the (2 × 1) structure are also tilted. The appearance of linearly bonded CO is somewhat surprising, but Chesters et al. point out that the density of CO/Pd(110) is smaller ($0.91 \times 10^{15}\,cm^{-2}$) than that of the maximum density attainable ($1 \times 10^{15}\,cm^{-2}$). Therefore, if the (2 × 1) structure is compressed further, there will be adsorption on some linear sites. During adsorption at 300 K, Chesters et al. also observe the sequence of LEED pattern observed by Ertl et al. [306]. At low exposures, where the c(2 × 2) structure forms, Chesters et al. [307] find only bridged CO. At intermediate coverages they find two bands separated by 60 cm^{-1}, which indicate that two types of bridged CO coexist. Also, a small amount of linearly adsorbed CO is found. Chesters et al. [307] attribute the additional band to bridging CO which is not adsorbed on high-symmetry positions, but rather displaced along the (110) rows. The linear species at high coverages is common to the Pd (111) and (100) surfaces, and is attributed to molecules adsorbed on atop sites at the boundaries of domains of bridged species. Structure models are shown in Fig. 21.

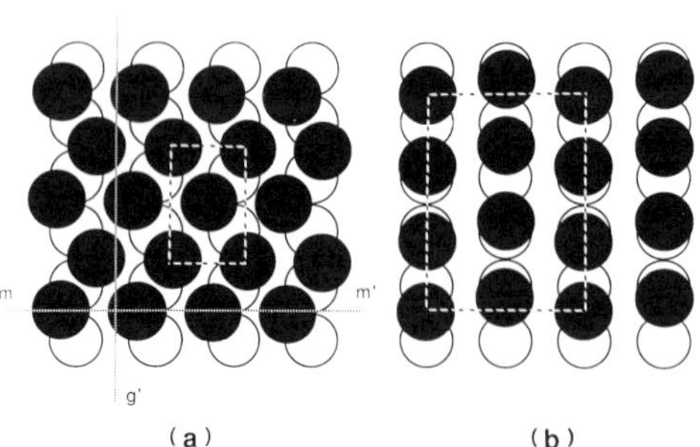

Fig. 21. A schematic representation of the p2mg overlayer formed via the adsorption of CO/Pd(110) at 300 K based on an out-of-register structure. CO molecules are represented by spheres 3 Å in diameter taken from calculations based upon CO adsorption on Ni and Pd surfaces [307].

Ertl et al. [306, 308, 309] find that, unlike on the (100) surface, CO adsorbed on Pd(111) shows attractive interactions between neighboring molecules. As a consequence, sharp LEED spots start to appear even at low coverages, indicating that CO forms islands with a $(\sqrt{3} \times \sqrt{3})$R 30° structure. The structure completes at $\theta = 1/3$. In agreement with this result, the C–O IR band obtained by Bradshaw and Hoffmann [310] remains stationary at 1823 cm^{-1} until $\theta = 0.32$, showing that CO is adsorbed on threefold sites. The isoteric heat also remains constant at 34 kcal mol^{-1} [305, 307, 308]. In the coverage range 0.32–0.37 the $\sqrt{3}$ LEED pattern is sharp, but IR shows that some molecules have shifted to twofold sites. Compression of the overlayer above $\theta = 0.37$ is accompanied by a drop in the isoteric heat of adsorption [305], as well as a shift of the adsorbate to two-bridge sites [309]. At $\theta = 0.5$, a c(4 × 2) overlayer is formed, with all the molecules adsorbed on twofold sites [309]. Above $\theta = 0.5$, a band corresponding to linear CO starts to appear [309]. Bradshaw and Hoffmann find that the appearance of linearly adsorbed CO does not depend on the kinetics of adsorption, and attribute it to adsorption on linear sites left vacant by a poor ordering of the c(4 × 2) overlayer. Interestingly, Matolin et al. [311] find by means of SIMS that bridge-bonded CO changes to linearly bonded CO with increasing surface temperature. The change occurs rather sharply at 430 K, and correlates with the start of catalytic activity (oxidation of adsorbed CO by adsorbed O_2). Both Matolin et al. [311] and Brown and Vickerman [312] find that the ion ratios in SIMS reflect not only the adsorption site of CO but also the lateral interactions between adsorbed molecules.

The kinetics of adsorption of CO/Pd(111) were studied by molecular beam scattering by Engel [313], who found the initial sticking probability to be unity and independent of sample temperature below 650 K. The decrease of the sticking coefficient with coverage is less than linear, indicating the presence of a precursor state for adsorption. The angular distribution of scattered CO is independent of coverage for $\theta < 0.5$ and independent of sample temperature below 1200 K, indicating that all CO molecules are initially adsorbed and subsequently desorbed. In order to explain the modulated CO scattering results, Engel postulates that there exist three states of CO/Pd(111) at high coverages. One of these states is a precursor state, but the nature of the other two is not elucidated. Kiskinova and Bliznakov also find three states at high coverage in their TDS and ESD experiments [314]. In an angle-resolved photoemission study, Miranda et al. [315] determined that the $5\sigma + 1\pi$ level of CO/Pd(111) has a binding energy of 8.02 eV and the 4σ level has a binding energy of 10.95 eV. In the high-coverage c(4 × 2) structure they also observe dispersion of both levels by 0.3 eV. Miranda et al., also point out that there is an exponential decay of the bandwidth with increasing CO–CO spacing, which suggests that direct "through space" interactions are responsible for the dispersion. The valence levels of CO are also observed by means of Penning ionization electron spectroscopy by Conrad et al. [316] for CO/Pd(110) and Sesselmann et al. [317] for CO/Pd(111), who point out that emission near the Fermi energy can be due to σ (bonding) or σ^* (antibonding) levels as well as from $2\pi^*$ antibonding orbitals (as concluded by Boszo et al. [181]), and at present it is not possible to decide which level contributes. Netzer and El Gomati [318] observe an electron energy loss at 13.5 eV, which they attribute to $5\sigma + 1\pi$ to $2\pi^*$ electronic transition, as observed from CO adsorbed on most transition metals.

In complete contradiction to all the results described above, Vankar et al. [319] and Poppa and Soria [320] report that CO adsorption is greatly reduced on epitaxially grown Pd films which have a (111) orientation. Epitaxial films are reported to have a much lower defect density than conventionally prepared single-crystal surfaces [318], but Nieuwenhuys and Kok [321] point out that surfaces in field emission tips are also defect-free, and still show CO adsorption, contradicting the epitaxial film results.

The interaction of chemisorbed CO with chemisorbed oxygen on Pd(111) has been studied by Conrad et al. [322] using LEED, UPS, and TDS. CO adsorption at 200 K was shown to produce additional ordered compression structures, and to lead to additional adsorbed CO binding states which in the limit extend down to 200 K. The chemical reaction between CO(ads) and O(ads) leads to CO_2, which desorbs. Control experiments with CO_2 suggests that CO(ads) and O(ads) on Pd(111) undergo repulsive interactions leading to mixed ordered phases. The chemical reaction CO and O appears to proceed via a Langmuir–Hinshelwood mechanism on Pd(111).

19.4 Pd(210)

The chemisorption of CO on the rough Pd(210) surface has been studied by LEED [305], IRAS [294], and ESDIAD [322]. IR measurements indicate that CO occupies a twofold bridging site on Pd(210). The structure of this surface is shown in Fig. 22: it may be seen that the interatomic distance between atop Pd atoms is 3.88 Å. This spacing is too large for a bridging CO to occupy. However, a number of Pd bridge sites (labeled B, C, and E) exist on the surface when bonding between Pd atoms in the first two layers is included. For adsorption of a symmetrical bridged CO on site C, the CO axis would be inclined 18° from the normal, as shown in the sectional view. ESDIAD is able to image the direction of the C–O bond, because ESD yields O^+ ions which are ejected along the bond direction. The results indicate that, at low CO coverages for sample temperatures of 80 K, sites C are filled first, giving non-normal O^+ ejection. On increasing the coverage, site B fills next, giving another non-normal ESDIAD beam. On increasing the coverage still further, the ESDIAD pattern switches to normal emission. It is not currently known whether the switching represents the formation of top-bonded CO, perhaps at site A, or to other effects. IR data at this low temperature are not available to elucidate the point.

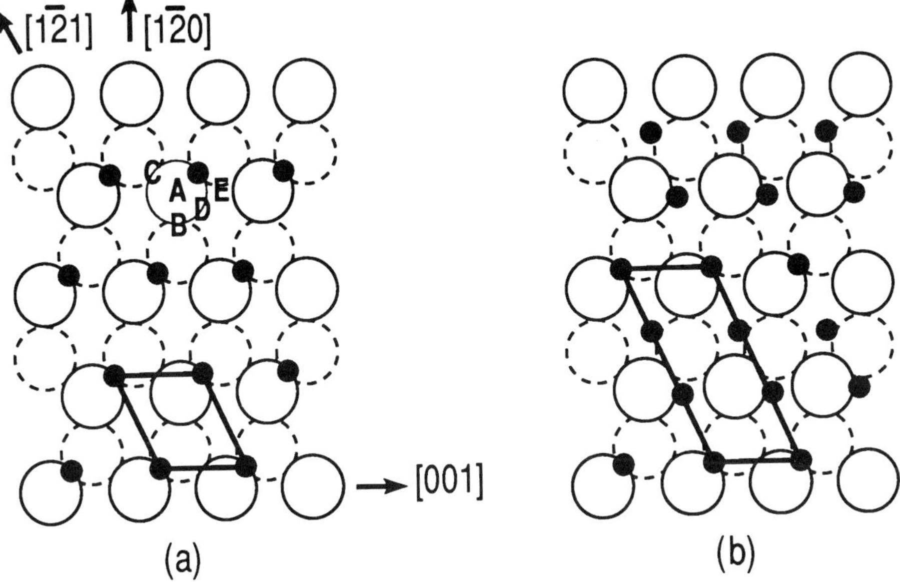

Fig. 22. (a) Surface model of the assumed (1 × 1) structure for CO/Pd(210). (b) Surface model of the (1 × 2) structure.

19.5 OTHER SURFACES

Davies and Lambert [324] studied even rougher surfaces, the (1 × 1)-Pd(331) and a thermally faceted surface [312]. The (331) surface may be regarded as a staggered array of (111) planes with 40% of the surface atoms in stepped positions, while the faceted surface was shown to be a mixture of 37% (111) planes and 31% of both (320) and (230) planes. Davies and Lambert [324] conclude from TDS and LEED that CO adsorbs first on the step sites, which have higher binding energy to CO. Only when the step sites are fully occupied does CO adsorb on the close-packed regions of the surface.

UPS has also been applied to highly dispersed Pd deposits on SiO_2 and on C substrates [325, 326]. At low Pd coverages, where only isolated Pd atoms are thought to exist, the *d* level lies approximately 3 eV below the Fermi level. As the Pd coverage increases, *d*-band formation is observed, and in the limit of 1.4×10^{16} Pd cm^{-2}, the photoemission spectra closely resemble that from bulk Pd. CO chemisorption occurs on these deposits giving new photoemission features corresponding to CO adsorption experiments on Pd single-crystal surfaces [325]. However, for highly dispersed Pd deposits, CO-induced photoemission features are entirely different, since adsorption is postulated to occur on single Pd atom sites or possibly on $Pd^{\delta+}$ sites [325, 326], depending on whether the electronic configuration of an isolated atom

References pp. 460–469

is d^9s^0 or d^9s^1. The free Pd atom has an electronic configuration $d^{10}s^0$, and would not be expected to receive 5σ donation from a CO molecule [326].

The study of CO adsorption on supported Pd surfaces has been extensively reviewed by Sheppard and Nguyen [17]. The review points out a number of correlations between the results of work on single crystals and supported samples. It seems that CO bonds preferentially as a bridged species to bulk Pd surfaces, while in highly dispersed samples where isolated Pd sites are available, terminal bonding becomes dominant [327]. These conclusions are in agreement with the results of Davies and Lambert [325].

20. Silver

McElhiney et al. [329] found CO to adsorb reversibly on Ag(111) at temperatures above 77 K. By measuring the surface potential changes over the range 77–123 K and in the pressure range 10^{-1} to 10^{-8} Torr, the isoteric heat of adsorption was found to vary linearly with surface potential (and coverage) from 6.4 to 4.0 kcal mol^{-1}, in agreement with Temkin-like behavior. Ford [330] was able to make surface potential measurements up to $\sim 10^{-2}$ Torr on an Ag film, observing a work function decrease of 0.5 eV. McElhiney et al. [329] classified this behavior as weak chemisorption. They did not observe any LEED patterns due to adsorbed CO, indicating that ordered structures are not formed. Measurements of the electronic excitations of CO/Ag(111) indicated losses at 13.5 and 2 eV, which are assigned to transitions of chemisorption levels to an unoccupied level.

Marbrow and Lambert [331] found no detectable adsorption of CO on the (331) face of Ag at 300 K and in the pressure range 10^{-9} to 10^{-7} Torr. Engelhard et al. [332] and Bowker et al. [333] also found no trace of adsorption on the (110) face at 300 K and pressures up to 10^{-6} Torr. At a surface temperature of 160 K, Bowker et al. [333] find that, if the temperature of the surface is lowered to 50 K, one monolayer of CO adsorbs on the surface. No LEED patterns were observed, indicating that the adsorbed layer is disordered. Krause et al. [334] show that the UPS spectrum of CO/Ag(110) is similar to that of gaseous CO, but there is a relaxation shift of 0.7 eV. These results indicate that CO is physisorbed, and not weakly chemisorbed. Furthermore, experiments with polarized light show that the molecular axis is not oriented perpendicularly to the surface. Krause et al. [334] observe a slight shift of the 1π peak relative to the 4π and 5σ levels of 0.2 eV. They interpret this to mean that the molecular axis is parallel to the surface, as the proximity of the 1π level to the metal would result in a higher image charge screening of the hole left by the photoemitted electron, and therefore a higher relaxation energy. Krause et al. also carried out XPS experiments, and propose that the process causing the satellites in XPS and the large relaxation shifts are the same in the core and valence regions.

21. Tantalum

The initial heat of adsorption of CO on Ta films at 300 K is 135 kcal mol^{-1} [44], higher than any of the metals except Ti and Zr. This result, along with a variety of other measurements, indicates that CO is dissociatively adsorbed on Ta at 300 K. Direct evidence for CO dissociation is seen in AES by Haas et al. [335] and Ducros et al. [336]. These authors find that the carbon Auger line shape following adsorption of a monolayer of CO at room temperature is very similar to that of carbides of Ta, W and Mo. In contrast, the carbon Auger line shapes for molecular CO adsorbed on W(112)[334] and Ni(110)[336] are quite similar to each other and different from a carbide line shape. Thus Auger spectroscopy is a useful fingerprinting method for distinguishing between carbon in molecular CO and atomic carbon adsorbed on a metal surface.

Haas et al. [335] and Ko and Schmidt [337] observed that heating a CO covered Ta(100) surface above 300 K results in a decrease of the carbon Auger signal due to diffusion of C into the bulk. The dissociation of carbon is essentially complete after 100 s at 800 K (negligible Auger signal). The oxygen remains on the surface until the sample is heated above 1100 K. The dissociation kinetics are rather complex, and these measurements suggest that layers of intermediate binding energy may exist between the outermost surface layer and bulk Ta.

In a LEED study of CO/Ta(100), Chesters et al. [339] reported that exposure to 25 L of CO at 300 K resulted in an increase in background intensity, but that no new beams were seen. Heating to 1200 K led to almost complete disorder, so that the Ta diffraction beams nearly disappear. Surface order was restored by heating to 1400 K, when a c(2 × 1) pattern appeared. This pattern was indistinguishable from that obtained by heating an oxygen-covered surface.

Further evidence of extensive disordering of the heated CO-covered Ta surfaces is provided in the field emission experiments of Klein and Leder [340] and Belov et al. [341] where highly "mottled" patterns were observed. Klein and Leder [340] also reported that surface diffusion of the CO monolayer occurred at a measurable rate for $T > 650$ K, with an activation energy of 38 kcal mol^{-1}.

The kinetics of adsorption and dissolution of CO by polycrystalline Ta have been examined by Gasser and Thwaites [342], who found the sticking probability of CO to be 0.14 and independent of temperature between −70 and 20°C. The adsorption of CO by Ta is reversible between 790 and 1960 K for CO pressures in the range 2×10^{-8} to 1×10^{-6} Torr. The maximum uptake observed (at ~1500 K) corresponds to 300 monolayers of CO! The dissolution of CO into Ta(100) has been studied by Belov et al. [341].

Tantalum temperatures above 2000 K in an atmosphere of 10^{-6} Torr of CO reacts irreversibly to form volatile TaO and carbon, which apparently diffuses into the bulk [342]. An Arrhenius plot of the rate of TaO formation yields an activation energy of 5.12 eV molecule^{-1}.

The desorption of CO from Ta is complicated by the dissolution process. There is a small α state seen with a peak temperature of 500 K [340]. The onset of desorption of the β states is about 1500 K [340] and heating to 2500 K is necessary to remove dissolved C + O, yielding a clean surface [339].

Evidence for a weakly bound molecular CO state on Ta is seen in the field emission studies of Klein and Leder [340] who observed desorption of CO at 170 K from a Ta tip saturated by spreading CO at 70 K. Finally, there is a single measurement [343] of the infrared spectrum for Ta evaporated in 10^{-2} Torr of CO, revealing a linear-CO band at 2096 cm^{-1}.

22. Tungsten

Over the years, the chemisorption of CO on tungsten has been one of the most studied systems. Because tungsten is one of the few metals which can be reliably cleaned in vacuum by thermal means, even older studies have a high degree of credibility. Despite the attention given to this chemisorption system, there are so many complexities that a number of questions regarding the structure and kinetics still remain.

Several excellent reviews concerning various aspects of CO chemisorption on tungsten have been published in the literature: Ford [345] surveyed the data through 1970, and provided a comprehensive discussion of the binding states; Gomer [346] and Schmidt [347] concentrated on the kinetics of CO adsorption and desorption on W single crystals and ESD; Plummer et al. [348] provided a detailed discussion of the electronic properties derived from UPS measurements.

Therefore, in the present review we shall not attempt a detailed historical survey. We shall instead briefly review the general features of the CO/W system, and then discuss the particulars of chemisorption on several crystallographic faces.

Brennan and Hayes [44] measured the initial heat of adsorption of CO/W films to be about 125 kcal mol^{-1}. The heat of adsorption drops rapidly with increasing coverage, and agrees well with thermal desorption energies of CO from polycrystalline W over a wide range of coverages [44].

Redhead [349] and Ehrlich [350] were the first to show that there are multiple binding states of CO on polycrystalline W, identifying (by means of TDS) at least three high-temperature β states and several low-temperature α states. Based on studies using field emission, field desorption, work function changes and ESD, Gomer [346] also demonstrated that CO must exist in several forms: linearly bonded, bridge-bonded, and perhaps dissociated CO.

At temperatures below 100 K, CO adsorbs on W in what Swanson and Gomer [351] termed the virgin state, which is molecular in nature and the molecules stand upright with the C atoms nearest the surface. During adsorption of the virgin state the work function increases by 0.6 eV. Upon heating the virgin-CO layer, a fraction of it desorbs in the temperature range 200–400 K, and the remainder dissociates into the β states, which desorb at temperatures above 900 K. The β states are mostly electronegative and have low ESD cross-sections. The saturation coverage of β-CO corresponds to one CO molecule for every two surface W atoms. There is extensive evidence (including isotopic mixing experiments [351]) that, in the β state, CO is dissociated into W–C and W–O, and that, upon heating, the C and O atoms recombine to desorb as CO. If adsorption proceeds beyond the saturation of the β layer, CO adsorbs in a molecular form called α-CO, which is electropositive (it decreases the work function). If adsorption is carried out at 300 K, the β states are occupied first, followed by the α states.

22.1 W(110)

The adsorption of CO/W(110) can be summarized as follows [352]. For $T < 200$ K, CO adsorbs molecularly upright (virgin state), with the carbon end of the molecule closest to the surface [347]. The maximum coverage is 1.1×10^{15} molecules cm^{-2}, corresponding to $\theta = 0.72$ to 0.8 [353]. LEED measurements yield a series of ordered structures which can be understood [354]. If virgin-CO (v-CO) is heated to 350–400 K, it partly desorbs and partly dissociates into β_1 CO. The dissociated nature of the β_1 state can be seen by XPS [355] and UPS [347, 356]. LEED shows a (2×1) pattern, which is best explained by C and O atoms lying along alternate close-packed rows. Heating β_1-CO to 900 K results in additional desorption, and the formation of complicated LEED patterns which may be due to the reconstruction of the W substrate [354]. Re-adsorption of CO on the β_1 state leads to the formation of α-CO, which desorbs at 350 K and may consist of molecules adsorbed in the empty rows, between those occupied by the C and O atoms of the β_1 layer [354].

Adsorption above 200 K on clean W leads to partial dissociation (β-CO). The ratio of virgin CO to β-CO depends strongly on temperature and pressure [355]. For instance, at room temperature the ratio can vary between 5 and 0.2 if the pressure during exposure to 30 L of CO is varied between 2×10^{-6} and 10^{-8} Torr [357]. The kinetics of dissociation have been followed in detail by Umbach and Menzel [358] with XPS, and the activation energy for dissociation measured to be ~ 21 kJ mol^{-1}, with a pre-exponential of $\sim 2 \times 10^{2}$ s^{-1}. Umbach and Menzel [358], together with the data provided by Kohrt and Gomer [359, 360], have put forward a detailed potential energy diagram of the CO/W(110) system, which is shown in Fig. 23.

In an interesting measurement of the time autocorrelation function of the field emission current fluctuations of CO/W(110), Chen and Gomer [353] found that there was *no* evidence for surface diffusion of either v-CO or α-CO prior to desorption. For $T > 650$ K, diffusion of β_1-CO was observed, with an activation energy of 23 kcal mol^{-1}. Madey et al. [361] used ESDIAD to

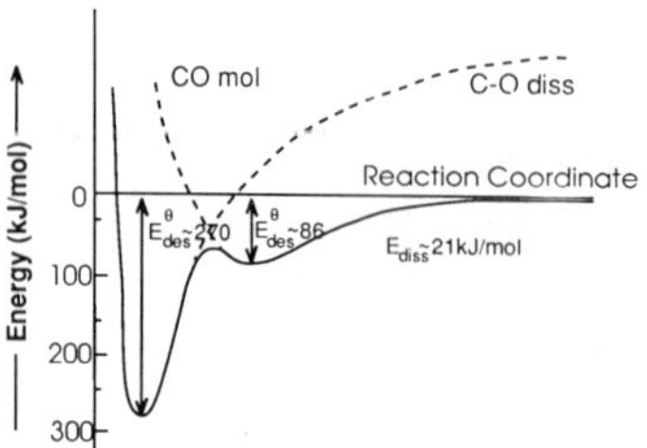

Fig. 23. Potential energy diagram of CO on a clean W(110) surface [358–360].

provide a direct view of the structures of v-CO on a planar W(110) surface, as well as on stepped surfaces containing (110) terraces. For CO adsorbed on the close-packed (110) surface, the dominant bonding mode is via the C atom, with the molecular axis perpendicular to the plane of the surface. For CO adsorbed at stepped sites on 4 different surfaces vicinal to the (110), the axis of the molecule is tilted away from the surface normal. These experiments were performed on a 7 mm diameter W crystal cut in the form of a truncated pyramid to expose 5 separate facets. The ESDIAD patterns of a monolayer of CO adsorbed on the multifaceted sample at 273 K are shown in Fig. 24. The central (110) facet yields a single ESDIAD beam which desorbs perpendicularly to the surface, giving a single spot in the center of the photograph. Each of the stepped surfaces yields an ESD beam which also desorbs perpendicularly to the (110) terraces. The images of these beams appear in the center of the photos for each of the 4 outer facets in Fig. 24. Each of the stepped surfaces also yields an ESD beam which desorbs in a down step direction, along an azimuth perpendicular to the step edge. Thus, we have a direct determination of "perpendicular" and "inclined" v-CO on planar and stepped W(110) surfaces, respectively.

22.2 W(100)

Adsorption of CO/W(100) at 80 K leads mostly to v-CO, but also produces some dissociated β-CO, as indicated by the XPS results of Yates et al. [362]. Figure 25 shows the O(1s) XPS spectrum due to a monolayer at 80 K. Upon heating to $\sim$275 K, and then to 550 K, dramatic changes in the band shapes occur. The principal changes in Fig. 25 are due to the coversion of v-CO to β-CO.

Thermal desorption of CO layers formed at $\sim$100 K shows peaks at 220 and 350 K [362], corresponding to the partial desorption (and concomittant dissociation) of v-CO and to the desorption of α-CO. Further heating leads to a β spectrum [346, 361, 362] showing three peaks at $\sim$1000, 1150, and 1450 K, labeled β_1, β_2, and β_3, respectively. Horton and Masel [364] have measured the angular distribution of the TDS products of CO/W(110) and found that the flux of α-CO varies as $\cos\theta$, where θ is the angle between the surface normal and the detector. They find that the flux of β_1-CO varies as $\cos^4\theta$, the

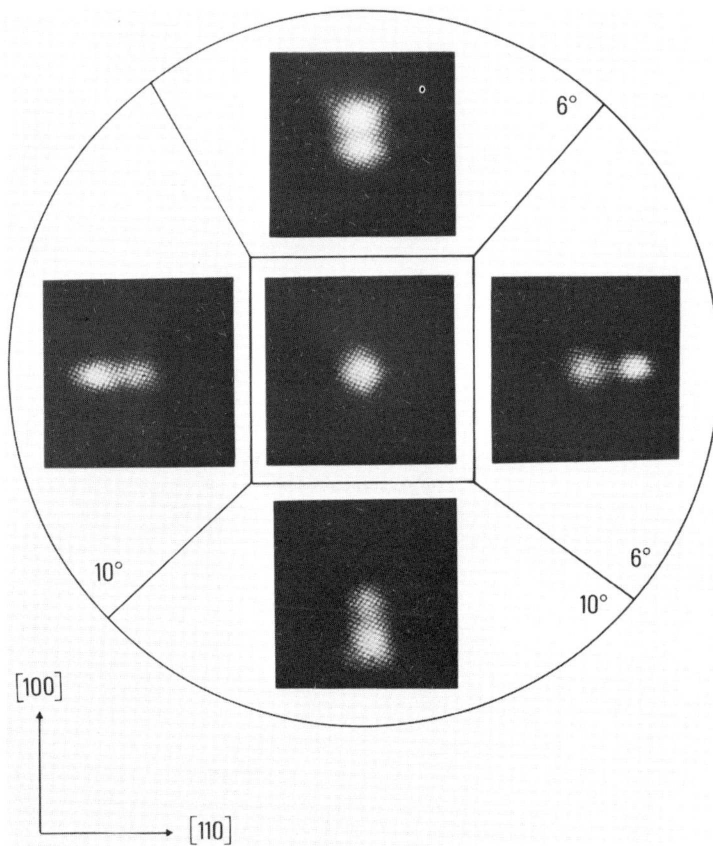

Fig. 24. ESDIAD patterns due to desorption of ions from CO adsorbed on a multifaceted W crystal at 273 K. The central facet is oriented with its surface parallel to the (110) plane, and the four surrounding facets are stepped surfaces of different step densities [6° and 10° off the (110) plane] and with steps parallel to [100] and [110] directions. In all of the patterns, the spot in the center of the picture corresponds to a beam of O^+ and CO^+ ions desorbing normal to the surface. The off-normal beams from each of the stepped facets provide evidence for "inclined" CO [361].

flux of β_2-CO varies as $\cos^3\theta$, and the flux of β_3-CO varies as $\cos\theta$. Horton and Masel interpret their results as an indication of the reconstruction of the W surface during desorption of β-CO, as is the case on the (110) surface [354]. The results also indicate that adsorption and desorption are not necessarily related by microscopic reversibility when the adsorption and desorption temperatures are different. In this light, one must consider the agreement between adsorption and desorption energies obtained by Brennan and Hayes [44] with caution.

LEED indicates that the β_3 layer forms a c(2×2) structure, while the full β layer shows a (1×1) pattern [362]. Absolute coverage measurements by Wang and Gomer [363] show that full coverage of the β layer corresponds to $\theta = 0.5$ and that the β_3 layer has a coverage of $\theta = 0.25$. Thus, the LEED

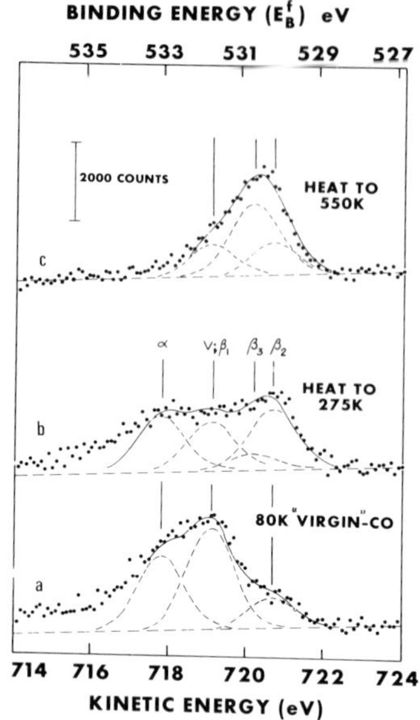

Fig. 25. O(1*s*) XPS spectra of CO/W(100). (a) Co exposure of 2.4×10^{-4} Torrs at ~80 K; (b) virgin-CO to β-CO conversion by heating to 550 K. All Gaussian curves should have FWHM = 1.50 eV [362].

results are consistent with a structure in which a full β layer has either a C or O atom in every other unit cell.

While adsorption at 80 K leads to simultaneous formation of v-CO and β-CO on W(100), adsorption at 300 K leads to a nearly saturated β-CO layer before adsorption into the α state occurs. This follows from the XPS results of Yates et al. [362], the UPS results of Plummer et al. [348] and the EELS data of Froitzheim et al. [365] and Franchy and Ibach [366]. In the EELS work, loss features at 360, 545, 625, and 2050 cm^{-1} are seen. The 545 and 625 cm^{-1} losses are due to C and O atoms, and the other two losses are due to the M–CO and C–O vibrations of molecular CO. These results also agree with the ESD results of Yates and King [367]. Adnot and Carette [368] report details in their EELS data which reveal an additional α state, as well as complexities in the β states.

There has been much discussion [362] concerning the origin of the three β states on W(100): do they exist simultaneously, or do they evolve during desorption as a consequence of lateral interaction? Wang and Gomer [363] have concluded that three distinct β states do *not* exist simultaneously ab initio. It appears that an initially homogeneous layer evolves, by heating and partial desorption, into new β states. Yates et al. [362] have shown that there are chemical shifts of the XPS O(1*s*) peak of the β-layer, and therefore the changes in the β layer during desorption may involve substrate atom rearrangements.

Direct evidence of the bonding structures of CO/W(100) have been provided by ion scattering [368] and ESDIAD [369]. Heiland et al. [369] have found that the β states are either lying down with the CO axis parallel to the surface or dissociated (they cannot distinguish between these configurations). In contrast, in the α state the molecules are standing up, bonded via the C atom. Jaeger and Menzel [370] have used ESDIAD to find both perpendicular and inclined bonding configurations of the α states.

Finally we note that the extensive ESD studies of CO/W(100) have confirmed the concepts described above [366]. Also, Franchy and Menzel [371] found a new mechanism for O^+ desorption from v-CO: creation of an O(1*s*) core hole opens a new channel for ion desorption, perhaps via a "coulomb explosion". A similar effect has been found in the case of CO/W(110) by Houston and Madey [372].

22.3 W(111)

Niehus [373] has used LEED and ESDIAD to examine CO/W(111), and draws conclusions consistent with the previously discussed concepts of CO chemisorption. Of particular note, however, are his observations of "off-normal" O^+ ion emission from the β-CO layer, indicating that O^+ originates from a configuration which is the same whether oxygen or CO is initially adsorbed! The molecular virgin and α states are bonded more-or-less perpendicularly to the surface [373, 374].

23. Rhenium

There appears to be no calorimetric measurements of CO adsorption on rhenium films reported in the literature. However, thermal desorption measurements on both polycrystalline [375–377] and single crystal [377] Re surfaces demonstrate that strong chemisorption occurs. Carbon monoxide adsorbs largely in molecular form at 300 K. A fraction of the molecular CO desorbs at temperatures below 500 K, and the remainder dissociates into carbide and oxide forms which recombine and desorb as CO at temperatures as high as 1300 K.

Direct evidence for molecular adsorption at 300 K of CO on polycrystalline Re [379] and Re(0001)[379] is provided by UPS. Adsorbed CO exhibits two distinct peaks in the UPS spectra at 7.4 and 11.1 eV below the Fermi level [379]. A comparison of the ionization potentials of the molecular orbitals of gaseous CO and $Re_2(CO)_{10}$ with CO on a Re surface [377, 379] indicates that the adsorbed CO is molecular, and probably bonded to single Re atoms via the C atom. Heating the CO-covered polycrystalline surface to 800 K yields a UPS spectrum which indicates that only dissociated species remain on the surface above that temperature. The angular dependence of UPS photoemission from CO on Re(0001) exhibits a non-uniform attenuation of photoemission from the Re 5*d* bands. Based on this evidence, the authors [379] have attempted to assign the specific metal d states involved in the chemisorption bond, suggesting that e'' and a_1' states of Re participate in bonding to CO.

CO adsorbs on both polycrystalline [375–377] and single-crystal [378] Re surfaces with a sticking probability of approximately ~0.1–0.5 at 300 K. TDS shows desorption from the molecular α states in the range 300–500 K and desorption from the (dissociated) β states at temperatures above 700 K [375–378]. The thermal desorption data of Housley et al. [378] of CO/Re(0001) shows desorption energies of 24 and 27 kcal mol^{-1} for the α states and 50 kcal-mol^{-1} for the β states. Multiple β states have been reported on polycrystalline surfaces [375, 376], as well as an ion-bombarded and stepped Re(0001) surface [378].

In a LEED study of CO on planar and stepped Re(0001) surfaces, Housley et al. [378] found evidence from the LEED patterns of a rather pronounced structural specificity. No ordered structures were seen on the planar surface at 300 K, whereas a poorly resolved (2 × 2) pattern was seen on the stepped (0001) surface [designated Re(14(0001) × (10$\bar{1}$1))]. Adsorption at a higher temperature (550 K), produced a $(2 \times \sqrt{3})R\,30°$ structure on the planar (0001) surface, and a (2 × 1) structure on the stepped surface. On both planar and stepped surfaces, the LEED results are dominated by scattering from (0001) surfaces, yet molecular arrangements on the two surfaces are different. It appears that the presence of steps plays a key role in the structures of both molecular and dissociated forms of CO. Furthermore, the relative coverages of α and β states are different on the two surfaces, providing further evidence for the influence of surface structure on adsorption: The α-state coverage is more-or-less insensitive to the surface structure whereas the β-state coverage depends on structure. Defects produced by Ar^+ ion bombardment of Re(0001) produced a surface on which the β states were similar to polycrystalline Re. Tatarenko et al. [380] studied the adsorption of CO/Re(0001) by means of TDS, AES, and LEED, finding three α states and one β state which desorb with first-order kinetics. As in adsorption of CO on Mo and W, the α states correspond to undissociated CO and the β state to dissociated CO. But unlike Mo or W, CO/Re(0001) is adsorbed in molecular form at room temperature or below, and only dissociates as the surface is heated above 650 K. Tatarenko et al. determined that the initial sticking probability is near unity, and the coverages of the α and β states are 4.8×10^{14} and 10^{14} molecules cm^{-2}, respectively. They observe a $(\sqrt{3} \times 4)$ LEED pattern near saturation. Ducros et al. [381] confirmed the proposition that CO is molecularly adsorbed at room temperature by observing the vibrational modes of CO with EELS. They observe one band at 1990 cm^{-1} which shifts to 2050 cm^{-1} at saturation. Ducros et al. assign the band to linearly bonded CO, in contradiction to the earlier suggestions of Braun et al. [379] who deduced from their UPS and work function measurements that α-CO is adsorbed on bridge or hollow sites. Ducros et al. [381] observe that the EELS spectrum of CO changes to that of atomic C and O as the sample is heated. Subsequently, Tatarenko et al. [382] used TDS, XPS, UPS and work function changes to show that only molecularly chemisorbed CO is present at 300 K. Upon heating, this state partly desorbs as the α_2 state and partly dissociates. However, if CO adsorption takes place at 100 K, after the chemisorbed layer is nearly saturated, a physisorbed layer is populated. The physisorbed layer partly converts to a chemisorbed layer and partly desorbs as the α_1 state.

The work function of a Re field emission tip [382] increases by 0.8 eV upon adsorption of CO at 125 K. There is no measurable diffusion of α-CO over the clean regions of a "shadowed" field emission tip upon heating. An ill-defined, boundary-free diffusion is seen only upon heating above 600 K, near the onset of desorption of β-CO. CO desorbs cleanly from Re [382, 383]. The dynamics of the CO–Re interactions have been measured using isotopic mixing [374, 384]. Only the β states undergo mixing, and the reaction is poisoned by pre-adsorbed oxygen.

24. Iridium

There appears to be no calorimetric measurement of the heat of adsorption of CO on Ir films. However, isosteric heat measurements of CO on several single-crystal surfaces indicate that the binding energy of CO varies between 37 and 39 kcal mol^{-1} [385–387]. As will be discussed in detail below, CO adsorbs molecularly on Ir at 300 K and desorbs rapidly at about 475 K. At temperatures in the range 500–600 K, heating an Ir surface in the presence of gaseous CO will result in dissociation, with the formation of a carbonaceous overlayer [386, 388].

CO adsorbs on the (110), (111), and (100) surfaces of Ir with a sticking probability near unity [385–387, 389] at low coverages. The desorption of CO from Ir(110) proceeds via first-order kinetics at low coverages, consistent with molecular adsorption [385, 390]. The multiple peaks seen in the thermal desorption spectra of CO from Ir(110) and (111) have been associated with lateral interactions and structural changes in the CO overlayer [385, 386, 389].

The clean Ir(100) and (110) surfaces are reconstructed [388, 389, 391], whereas the (111) surface exhibits the unreconstructed bulk structure [386]. There are conflicting reports concerning the LEED patterns observed during CO adsorption in Ir(100) and (110). Broden and Rhodin [390] report that the clean Ir(100)–(5 × 1) LEED pattern changes to (1 × 1) at an exposure of 10 L of CO at 300 K. Grant [393] had earlier reported (2 × 2) patterns for CO on Ir(100). Rhodin and Broden [394] further observed that there was no difference in reactivity toward CO between the (1 × 1) and (5 × 1) surfaces. On the clean reconstructed Ir(110)–(1 × 2) surface, Taylor et al. [392] reported that two patterns, a p(2 × 2) and a (4 × 2), appear near 90 K, but are poorly ordered at 300 K. These structures are different from those observed in an earlier study by Christman and Ertl [389], and a subsequent study by Nieuwenhuys and Somorjai [391], who observe a (2 × 1) structure. The (2×1) structure is common to all the $\{110\}$ faces of cubic metals in the platinum group, and have been explained by Lambert [133]. Furthermore, Nieuwenhuys and Somorjai find that, if the sample is contaminated with atomic carbon, a $c(2 \times 2)$ CO structure forms, which suggests that the structures obtained by Taylor et al. [392] might be due to contaminated samples. The saturation coverage of CO on Ir(110) is 9.6×10^{14} molecules cm^{-2}, corresponding to $\theta = 1$.

Comrie and Weinberg [387] report that CO adsorbed on the clean, unreconstructed (111) surface of Ir forms a $(\sqrt{3} \times \sqrt{3})R\ 30°$ structure at $\theta = 1/3$, and a $(2\sqrt{3} \times 2\sqrt{3})R\ 30°$ coincidence lattice at $\theta = 7/12$. These structures have been reported for CO adsorbed on many transition metal substrates having threefold rotational symmetry [395]. Using thermal desorption spectroscopy, two states are seen which correlate with the LEED structures [386]. The energy difference between the states indicates a change in the CO–CO interaction energy of about 5 kcal mol^{-1} as CO shifts from one configuration to another.

A saturated layer of CO/Ir(110) changes the work function at 90 and 300 K by 0.28 and 0.23 eV, respectively [389]. The differences are not due to coverage changes. Broden and Rhodin interpret them to be due to rotational randomization of the adsorbed molecules at temperatures greater than 300 K. They also find that the maximum work function change of CO/Ir(100) at 300 K is 0.2 eV.

The vibrational spectra of CO adsorbed on Ir films evaporated under UHV conditions were studied by Reinalda and Ponec [397]. Only a single adsorption band was observed at 300 K, which shifts continuously from 2010 cm^{-1} at low coverages to 2093 cm^{-1} at saturation. The band is attributed to molecular CO adsorbed via the carbon end of the molecule on atop sites. A decrease in the adsorption intensity was observed at coverages greater than $\theta = 0.4$. Reinalda and Ponec attributed the decrease in intensity to the formation of a compressed overlayer in which part of the molecules are adsorbed in low symmetry sites, and therefore have lower adsorption. However, none of the low-frequency bands attributed to bridging CO observed during CO adsorption on supported Ir particles [76] were seen by Reinalda and Ponec.

The electron spectroscopy of CO/Ir supports the assignment of molecular CO bound to Ir via the C atom. Taylor et al. [392] carried out AES lineshape analysis of CO/Ir(110), and observed spectra that were similar to both gaseous CO and $Ir_4(CO)_{12}$ [397, 398], indicating molecular adsorption. The adsorption of CO/Ir(100) and (111) was characterized by UPS [389, 393, 399, 400]. Single crystals of Ir appear to be the only Group VIII metal surfaces where the three photoemission bands due to the 4σ, 1π, and 5σ orbitals of adsorbed CO can be partially resolved. Seabury et al. [401] found that, on Ir(111) and Ir(110), 4σ emission occurs at -11.7 eV, 5σ emission at -9.2 eV, and 1π emission at -8.6 eV. They also observe final state resonances for the σ orbitals which are significantly shifted with respect to the gas-phase values.

Seabury et al. used symmetry-adapted selection rules to determine that the molecules are adsorbed perpendicularly to the surface. Zhdan et al. [402] observed only two features in the UPS of CO on polycrystalline Ir, unlike the case of single crystal surfaces.

Shek et al. [403] and Nieuwenhuys and Somorjai [391] studied the interaction of electron beams with CO/Ir(111) and CO/Ir(110), respectively. Shek et al. [403] find that, for a beam energy of 86 eV and $\theta \sim 1/3$, the total interaction cross-section for ESD and dissociation is in the range 0.8–$1.7 \times 10^{-17}\,cm^2$. Electron-stimulated dissociation occurs at only 1–2% of the rate of ESD. Nieuwenhuys and Somorjai [391] find that the electron beam favors the formation of the c(2 × 2) structure on Ir(110), presumably because it dissociates CO to create atomic carbon.

25. Platinum

The adsorption properties of platinum have received a great deal of attention due to its extensive use as a catalyst. CO adsorbs molecularly and reversibly at room temperature on all faces of Pt studied so far, with heats of adsorption of the order of 30 kcal mol^{-1}. Desorption takes place with negligible dissociation around 400–500 K. Thermal decomposition of CO/Pt(111) has been reported for $T > 600$ K [408].

25.1 Pt(111)

Adsorption of CO/Pt(111) at 170 K leads to a $(\sqrt{3} \times \sqrt{3})$R 30° structure at $\theta = 1/3$, which was thought to continuously compress along the $[1\bar{1}0]$ direction with increasing coverage, leading to a c(4 × 2) structure at $\theta = 1/2$ and finally to a hexagonal close-packed layer at saturation coverage ($\theta = 0.68$)[403–416]. The earlier vibrational spectroscopy studies [403–406, 410, 412, 413] showed that C adsorbs only on top sites until the $\sqrt{3}$ structure is completed, while at higher coverages twofold bridge sites begin to fill. At $\theta = 1/2$, half the molecules are adsorbed on top sites and the other half on twofold bridge sites. Furthermore, RAIRS showed that there is an attractive interaction between adsorbed molecules below $\theta = 1/3$, leading to island formation [412]. In an EELS experiment with 48 cm^{-1} resolution, Avery [418] did not observe any shift of the bands or changes in their widths as the coverage increased above $\theta = 1/2$, and therefore proposed that the excess CO molecules are accommodated along packing fault lines, and not by a continuous compression of the overlayer, as originally suggested by Pritchard for CO/Cu(100) and Cu(111)[175]. The CO–CO distance along the fault lines is only 2.76 Å, closer than the Van der Waals radius of CO of 1.5 Å. However, Van Hove et al. have found that such small CO–CO distances exist [271]. The IRAS study of Hayden and Bradshaw [419] also favors the model of Avery for the high-coverage structure, based on the relative intensities of the bands in the bridging and linear regions. They also find that at higher temperatures the molecules adsorbed in twofold bridge sites shift over to threefold bridge sites, which accounts for the disorder seen in LEED at higher temperatures. The twofold and threefold bridge-bonded sites cannot be differentiated by EELS due to the limited resolution. Hayden and Bradshaw also see no shift in the bands above $\theta = 0.5$, again suggesting that Avery's model might be correct. Norton et al. [420] used nuclear microanalysis to determine the absolute coverage of CO, confirming that the c(4 × 2) structure occurs at $\theta = 0.5$. A subsequent study by Steininger et al. [421] has found that, when adsorption takes place at 100 K, the fuzzy $\sqrt{3}$ structure at 150 K converts to a series of distinct structures at $\theta = 0.17$, 0.22, and 0.33. They did not find any satisfactory real space models for the structures below $\theta = 1/3$. Furthermore, Steininger et al. find that CO adsorbs only on top sites until the structure at $\theta = 0.17$ is completed. For greater coverages, the bridging sites begin to fill. Steininger et al. [421] observed that after $\theta \approx 0.35$ at 100 K, LEED shows a mixture of the $\sqrt{3}$ and c(4 × 2) patterns, which indicates that even at this low temperature CO is mobile, and the attractive interactions cause islands with c(4 × 2) structure to form below a coverage of 1/2. Ogletree et al. [422] have carried out a dynamical LEED analysis of the c(4 × 2) structure, and find that indeed half the molecules are adsorbed on top sites and half on bridge sites, in agreement with vibrational spectroscopy. They further find a remarkable correlation between the stretching frequencies determined by vibrational spectroscopy and the bond lengths determined by LEED, as seen in Fig. 26

$$d_{\mathrm{M\text{-}C}}\ (\text{Å}) \ = \ -\ 6.25 \times 10^{-4} \quad \nu_{\mathrm{CO}}\ (\mathrm{cm}^{-1}) \ = \ +\ 3.18$$

$$d_{\mathrm{C\text{-}O}}\ (\text{Å}) \ = \ -\ 2 \times 10^{-4} \quad \nu_{\mathrm{CO}}\ (\mathrm{cm}^{-1}) \ = \ +\ 1.56$$

The M–C bond length shows a much stronger correlation than does the C–O bond length.

Ertl et al. [409] found the work function behavior of CO/Pt(111) to be unlike the behavior on most transition metals, but similar to the behavior on Cu. It initially *decreases* with increasing coverage, passes through a temperature-dependent minimum at 1/3 coverage, and rises to 0 at $\theta = 1/2$, with a further minimum at saturation coverage. Steininger et al. [421] and Poelsema et al. [423] have established that the minimum in the work function occurs when the first LEED structure is formed at $\theta = 0.17$. Therefore, the top-bonded molecules decrease the work function while the bridge-bonded molecules increase it. The work function results imply there is charge transfer *to* the substrate from terminally bonded CO and *from* the substrate to bridge-bonded CO, consistent with the chemical shifts of the O 1*s* and C 1*s* binding energies observed by Norton et al. [412] and Dückers et al. [424].

The adsorption and desorption kinetics of CO/Pt(111) have been extensively studied, and a fairly complete understanding has emerged. An important realization came with the work of Collins and Spicer [410], who found

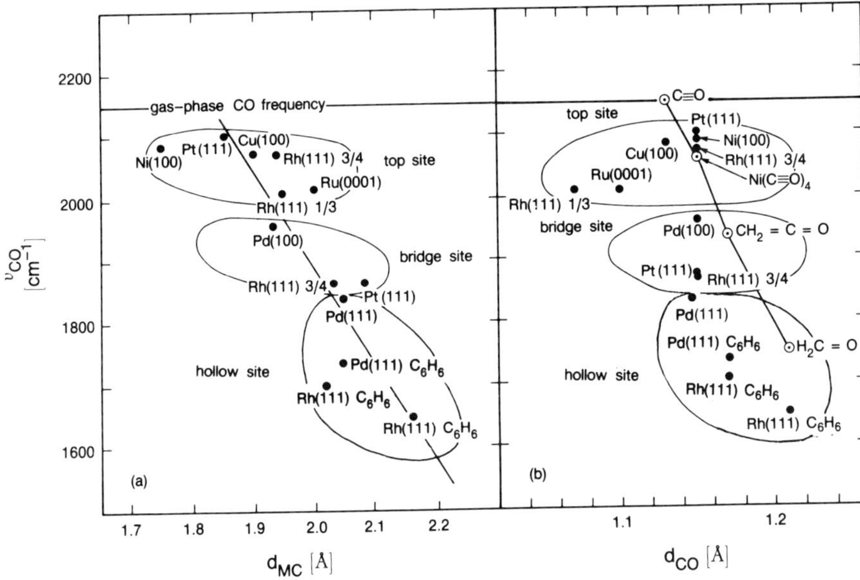

Fig. 26. (a) Metal–carbon and (b) carbon–oxygen bond lengths versus measured C–O stretch frequency for CO adsorbed on surfaces and for a few CO-containing molecules. Circles, squares, and triangles indicate top, bridge, hollow sites, i.e. 1-, 2-, and 3-fold coordinated sites, respectively. The labels 1/3 and 3/4 indicate different coverages of CO in units of monolayers [422].

that, of the two peaks observed in thermal desorption, one at 450 and another at 550 K, the high-temperature one can be attributed to adsorption of CO on defect sites. Poelsema et al. [425] have confirmed that conclusion by preparing a nearly defect-free surface on which the high-temperature peak is totally absent. Poelsema et al. also find that the minimum in the work function depends on the sample defect density. In what follows we will discuss the properties of defect-free surfaces. At low temperatures, Steininger et al. [421] find that the sticking probability passes through a maximum. Prutton [426] has postulated that this effect is the result of island growth. Suppose that the rim of a nucleus consisting of several atoms contains adsorption sites with a high sticking probability. As the size of the nucleus increases, so does its circumference, and therefore the measured sticking probability also increases. Finally, when the islands contact each other and coalesce, the measured sticking probability decreases, as observed. At room temperature, Steininger et al. [421] find a weak sticking probability (0.63), but between $\theta = 0.01$ and $\theta = 0.17$ the sticking probability remains constant at 0.8. Island formation seems still to play a role, but the range of coverage over which the sticking probability remains constant indicates the presence of a precursor state. The same results were found by Campbell et al. [415], by Lin and Somorjai [416], and by Poelsema et al. [427]. In a molecular beam scattering experiment, Winicur et al. [417] found that the adsorption energy decreases linearly with coverage, with an average value of 31.1 kcal mol^{-1}

over the range $0 < \theta < 0.5$, and a pre-exponential factor of $2.67 \times 10^{13}\,s^{-1}$. Poelsema et al. [425] also find that the activation energy for desorption decreases linearly with coverage up to $\theta = 0.13$ according to the formula $E_d = E_{d0} - \varepsilon\theta$, where $E_{d0} = 133\,kJ\,mol^{-1}$ and $\varepsilon = 68.5\,kJ\,mol^{-1}$, and the pre-exponential factor for desorption is $1.4 \times 10^{14}\,s^{-1}$ at 400 K. It is quite remarkable that this set of three parameters allows a fully consistent description of CO adsorption, desorption, and equilibrium measurements [424]. Poelsema et al. [425] find that, even though the TDS curves shift to lower temperatures with coverage, that shift is only due to the linear decrease of the binding energy of CO with increasing coverage. The desorption process is strictly first order.

XPS and UPS measurements [33, 411, 427] indicate that the sequence of CO levels of the adsorbed molecule on Pt(111) is 1π, 5σ, and 4σ, in order of increasing binding energy. 1π and 5σ are inverted from gaseous CO. The results also indicate that the $1\pi - 4\sigma$ splitting in bridge-bonded CO is smaller (by ~0.5 eV) than in linearly bonded CO. Miller et al. [429] also reported observing photoemission from the 3σ level, and that the 5σ and 1π levels have the same order as in the gas phase, but all of the extensive subsequent work [411, 427, 429, 430] disagrees. Apai et al. [33] and subsequent workers [427, 429, 430] measured the angular distribution of the peak intensity emission from the levels of adsorbed CO. By comparing their results with the calculations of Davenport [30], they deduced that the molecule stands perpendicularly to the surface with the carbon end down, in agreement with vibrational spectroscopy. Apai et al. [33] also found that the Pt $5d(t_{2g})$ orbitals are strongly involved in the bonding of CO to the surface.

25.2 Pt(110)

For CO/Pt(110), Bonzel and Ku [432–434] and Comrie and Lambert [435] resolved two desorption states at ~420 and 500 K, whereas McCabe and Schmidt [436] reported a single broad thermal desorption peak at ~450 K. Bonzel and Ku [432–434] suggested that the two peaks might be due to interactions between adsorbed molecules, rather than to the presence of two states. Exposure of the clean reconstructed (2 × 1) Pt(110) surface to > 0.9 L of CO at 300 K results in a (1 × 1) LEED pattern, indicating that CO removes the reconstruction. Because the activation energy of formation of the ordered layer on this surface is comparable with the desorption energy, one must heat to ~500 K in 10^{-7} Torr of CO to form a highly ordered LEED structure. This structure is characterized by a glide plane and was indexed as p1g1 by Comrie and Lambert [435], and has a binding energy of $133\,kJ\,mol^{-1}$. In their model for the overlayer, Comrie and Lambert propose that the CO molecules form zig-zag chains down the rows of Pt atoms along the [11] direction, in a fashion similar to CO/Pd(110)(Fig. 21), which would allow for the required packing density. Hofmann et al. [437] have found by angle-resolved UPS, that the zig-zag comes from the CO molecules being tilted, rather than adsorbed away from the atop position. The UPS data of

Hofmann et al. shows a clear tilt of 26° away from the surface normal, which they interpret to be in between the [211] and the [433] directions. This is the first clear case of observation of CO adsorbed on top sites which is tilted to allow for a high packing density. Further detailed UPS studies by Bare et al. [438] confirm the structure, and determine that the $5\sigma/1\pi$ ordering of adsorbed CO is reversed from that of gas-phase CO.

The same group studied the adsorption of CO at all coverages by means of TDS, UPS, LEED, RAIRS and EELS [438, 439], and found several coverage-dependent phases. Infrared spectroscopy shows that, at low coverages ($\theta \approx 0.1$), there are isolated CO species adsorbed on top sites on the reconstructed Pt(110)–(1 × 2) surface, with a C–O stretching band at 2080 cm^{-1}. UPS shows that these molecules are adsorbed with their axes perpendicular to the surface. For $0.1 < \theta < 0.3$, the isolated CO molecules coexist with islands of CO adsorbed on top sites on the (1 × 1) surface, which have a C–O frequency of 2094 cm^{-1}. As the coverage increases, the vibrational spectroscopy results show that the islands grow in size. UPS shows that the molecules with the islands are adsorbed with their axes tilted, due to the high density of molecules in the island. At coverages greater than half a monolayer, the islands begin to coalesce, and the (1 × 2) to (1 × 1) transition is completed. When adsorption takes place at room temperature, the region where the islands coalesce contains many imperfections, and the overlayer cannot fully rearrange itself to give long-range order. However, if the sample is annealed to higher temperatures, long-range order forms and the p1g1 structure is visible in LEED. Bare et al. [440] also observe that if a full-coverage overlayer is heated so that partial desorption of CO occurs, then the CO adsorbed on top sites next to a vacancy will convert to upright, bridge-bonded CO to reduce the packing density. This process breaks the glide line, leading to the disappearance of the p1g1 LEED pattern, although a large amount of CO adsorbed is still tilted on top sites. Bare et al. [440] interpret the two peaks in the thermal desorption spectra to arise from (a) CO desorbing from the (1 × 1) surface, but which leaves the surface unreconstructed, and (b) CO desorbing from the (1 × 1) surface but which leaves the surface in the reconstructed (1 × 2) configuration. Although CO is thought to have the same binding energy in both cases, the extra energy required to reconstruct the surface generates the second TDS peak. That there are CO molecules adsorbed on the (1 × 1) and (1 × 2) surface has been confirmed by the inverse photoemission results of Ferrer et al. [441], who find that the empty 2π level of CO adsorbed on the (1 × 1) surface occurs at 5 eV above the Fermi energy, while the 2π level of CO adsorbed on the (1 × 2) surface occurs at 3.4 eV above the Fermi level. Ferrer et al. attribute the 1.6 eV shift to the strong interaction between the closely packed CO molecules on the (1 × 2) surface. The extensive UPS results of Rieger et al. [442] lead them to propose that the CO axis in the (1 × 2) structure is tilted by 20° along the [001] direction, and not in between the [211] and [433] directions as proposed by Hofmann et al. [437].

References pp. 460–469

25.3 Pt(100)

The clean Pt(100) surface can be prepared in two different structures [442]: (1) the (1 × 1) surface, which is a termination of the bulk [443] but is metastable and transforms into (2) a quasi-haxagonal structure ("hex") above 400 K [444, 445] through an intermediate (1 × 5) structure [446]. The hex structure is often denoted by (5 × 20), but Heilmann et al. [448] argued that the "hex" designation is more appropriate. The reconstructed hex surface is the stable structure obtained after annealing the sample to 1400 K. Bonzel et al. [449] showed that the unreconstructed (1 × 1) surface can be formed by adsorbing various molecules, such as CO and NO, on the hex surface, and subsequently removing the adsorbate by a low-temperature catalytic reaction. The electronic structure and chemical reactivity of the two Pt(100) surfaces is quite different. For example, the sticking probabilities of oxygen and hydrogen on the (1 × 1) surface are orders of magnitude higher than on the hex surface [449, 450], and although the initial sticking probability of CO on both surfaces is 0.75, the adsorption of CO proceeds in a different manner on the two surfaces.

Tucker [452] was the first investigator to report the LEED patterns formed upon the adsorption of CO on hex Pt(100). Morgan and Somorjai [453], Kneringer and Netzer [451], Barteau et al. [450], and Norton et al. [445] also reported that, after CO adsorption on the hex surface at room temperature, the LEED patterns progress from the (5 × 20) to a p(1 × 1) to a c(4 × 2). Detailed structural models were not proposed by the authors. Helms et al. [454], Broden et al. [443], Thiel et al. [459], and Behm et al. [455] studied the LEED patterns obtained after CO adsorption on the (1 × 1) surface, and reported that the first pattern obtained is a c(2 × 2) at $\theta = 0.5$, $(2 \times 3\sqrt{2})$R 45° at $\theta = 4/6$, and finally a c(4 × 2) at a saturation coverage of 0.75. Behm et al. [455], find that, if CO is adsorbed on a hex surface at temperatures higher than 395 K, the c(2 × 2) pattern is also observed.

The EELS results of Pirug et al. [457] and Behm et al. [455] indicate that CO adsorbs initially on both bridge and atop sites on both the hex and the (1 × 1) surface. The RAIRS results of Crossley and King [456] and Banholzer and Masel [458] only show the atop site to be populated, presumably due to the lower sensitivity of RAIRS, although studies to determine if the ratio of linear-to-bridged CO is temperature-dependent have not been carried out to date. On the (1 × 1) surface there is a much higher ratio of bridged CO, which decreases dramatically at half a monolayer, indicating that a perfect c(2 × 2) structure would consist of linearly adsorbed CO. On the hex surface the behavior is quite different. The bridge-to-linear ratio grows continuously until it reaches 0.2 at $\theta = 0.3$, and remains fairly constant at that value until saturation [454].

Behm et al. [455] find that, for coverages less than 0.4, they can effectively describe the adsorption kinetics of CO on the hex and the (1 × 1) surface by the mobile precursor model of Kisliuk. The value of K in the Kisliuk formula

$$S(\theta) = S_0\left[\frac{(1 - \theta)}{(1 - \theta + K\theta)}\right]$$

is best fitted with 0.05 for adsorption on the (1 × 1) surface and 0.35 for adsorption on the hex surface, indicating that the lifetime of the precursor is much larger on the (1 × 1) surface. The measured initial sticking probability on both surfaces is 0.75 [454]. From the IR frequency shifts, Crossley and King proposed that CO adsorbs in islands of constant size, each containing approximately 11 molecules. Behm et al. [455] also calculated the island size from the width of the c(2 × 2) LEED beams for CO on the hex surface, and their results agree with those of Crossley and King. However, Behm et al. find that the islands on the (1 × 1) surface vary in size as a function of coverage, the largest size being at $\theta = 0.5$ with 110 molecules per island. The LEED I–V curves of Thiel et al. [459] indicate that the structure of CO when in the c(2 × 2) islands is the same on the (1 × 1) and the hex surface. Behm et al. point out that the results described above do not imply that there is an attractive interaction between the adsorbed CO molecules in the islands. On the contrary, the molecule–molecule interaction is repulsive. The fact that adsorbed CO forms islands is related to the mechanism of the (1 × 1)–hex phase transition. Thiel et al. [459] show that, on the hex surface, the adsorption of CO and the phase transition occur in separate steps: CO adsorbs initially on (1 × 1) patches that might occur on the surface due to imperfections, with a small effect on the hex areas. As discussed above, the mobility of CO on the hex areas is sufficiently large to allow adsorbed molecules to migrate to the (1 × 1) patches, which act as traps for the adsorbed CO, causing islands to be formed. Only when the c(2 × 2) structure is fully formed in the (1 × 1) patches does CO begin to attack the hex areas, removing the reconstruction. Earlier, Rutherford backscattering experiments of Norton et al. [461] had shown that the reconstruction is completely removed at $\theta = 0.5$, when the c(2 × 2) structure is fully formed, a result which fits nicely in the model of Thiel et al. The heats of adsorption of CO are 27.5 kcal mol^{-1} on the hex surface at low coverage, and 33 kcal mol^{-1} on the (1 × 1) surface at $\theta = 0.5$ [459]. It is this difference in energies (minus the energy of the hex–(1 × 1) transition) which causes the trapping of CO in islands at low coverages. The energetics and kinetics of the process are discussed in detail in ref. 459.

Acknowledgements

This review article is an update of a previous, unpublished version written together with John Yates and Theodore Madey. The author is indebted to them for their extensive contributions.

References pp. 460–469

References

1 W.L. Lom and A.F. Williams, Substitute Natural Gas, Manufacture and Properties, Wiley, New York, 1976.
2 L.L. Anderson and D.A. Tillman, Synthetic Fuels from Coal, Wiley-Interscience, New York, 1979.
3 P.H. Emmett, Catalysis Then and Now, Part I, Franklin, Englewood, NJ, 1965, p. 154.
4 G.A. Mills and F.W. Steffgen, Catal. Rev., 8 (1973) 189.
5 P.H. Emmett, Catalysis Then and Now, Part I, Franklin, 1965, p. 186.
6 C.G. Suits and H.E. Way (Eds.), The Collected Works of Irving Langmuir, Vols. 2, 3 and 9, Pergamon Press, Oxford, 1960.
7 J.B. Johnson and W.G. Klemperer, J. Am. Chem. Soc., 99 (1977) 7132.
8 E.W. Plummer, W.R. Salaneck and J.S. Miller, Phys. Rev. B, 18 (1978) 1673.
9 W.M. Huo, J. Chem. Phys., 43 (1965) 624.
10 F. Grimaldi, A. Lecourt and C. Moser, Int. J. Quantum Chem., 1S (1967) 313.
11 A.K.Q. Siu and E.R. Davidson, Int. J. Quantum Chem., 4 (1970) 223.
12 C.A. Burrus, J. Chem. Phys., 28 (1958) 427.
13 B. Rosenblum, A.H. Nethercot and C.H. Townes, Phys. Rev., 109 (1958) 400.
14 A.M. Bradshaw and J. Pritchard, Proc. R. Soc. London Ser. A, 316 (1970) 169.
15 D.L. Adams, Metals Ligands and Related Vibrations, Arnold, London, 1967.
16 E.J. Baerends and P. Ros, Mol. Phys., 30 (1975) 1735.
17 N. Sheppard and T.T. Nguyen, in R.J. Clarke and R.E. Hester (Eds.), Advances in Infrared and Raman Spectroscopy, Vol. 5, Heyden, London, 1978.
18 G. Broden, T.N. Rhodin, C. Brucker, R. Benbow and Z. Hurych, Surf. Sci., 59 (1976) 593.
19 J.B. Benziger, Appl. Surf. Sci., 6 (1980) 105.
20 R.P. Eischens and W.A. Pliskin, Adv. Catal., 10 (1958) 1.
21 G. Blyholder, J. Phys. Chem., 68 (1968) 2772; J. Vac. Sci. Technol., 9 (1972) 901.
22 A.B. Anderson and R. Hoffman, J. Chem. Phys., 61 (1974) 4545.
23 J.C. Robertson and C.W. Wilmsen, J. Vac. Sci. Technol., 9 (1972) 901.
24 J.C. Fuggle, M. Steinkelberg and D. Menzel, Chem. Phys. 11 (1975) 307.
25 C.L. Allyn, T. Gustafsson and E.W. Plummer, Solid State Commun., 24 (1977) 531.
26 N.K. Ray and A.B. Anderson, Surf. Sci., 119 (1982) 35. H. Kobayashi, S. Yoshida and M. Yamaguchi, Surf. Sci., 107 (1981) 321.
27 G. Doyen and G. Ertl, Surf. Sci., 43 (1974) 197. G. Doyen, Surf. Sci., 59 (1976) 461.
28 A. Rosen, E.J. Baerends and D.E. Ellis, Surf. Sci., 82 (1979) 139.
29 D.E. Eastman and K. Cashion, Phys. Rev. Lett., 27 (1971) 1520.
30 J.W. Davenport, Phys. Rev. Lett., 36 (1976) 945.
31 D.R. Lloyd, Faraday Trans. Chem. Soc., 58 (1974) 136.
32 R.J. Smith, J.A. Anderson and G.J. Lepeyre, Phys. Rev. Lett., 37 (1976) 1081.
33 G. Apai, P.S. Wehner, R.S. Williams, J. Stohr and D.A. Shirley, Phys. Rev. Lett., 37 (1976) 1497.
34 P.S. Bagus, C.J. Nelin and C.W. Bauschlicher, Jr., Phys. Rev. B, 28 (1983) 5423. P.S. Bagus, K. Hermann and C.W. Bauschlicher, Jr., IBM Res. Rep. RJ4173 (46101)(1984).
35 Hesketh, E.W. Plummer and R.P. Messmer, Surf. Sci., 139 (1984) 558.
36 C.F. McConville, C.S. Somerton and P. Woodruff, Surf. Sci., 139 (1984) 75.
37 K. Horn, A. Bradshaw and K. Jacobi, Surf. Sci., 72 (1978) 719.
38 L.G. Petersson, S. Kono, N.F.T. Hall, C.S. Fadley and J.B. Pendry, Phys. Rev. Lett., 42 (1979) 1545.
39 P. Hofmann, S. Bare, N.V. Richardson and D.A. King, Solid State Commun., 42 (1982) 645. S. Bare, K. Griffiths, P. Hofmann, D.A. King, G.L. Nyberg and N.V. Richardson, Surf. Sci., 120 (1982) 367.
40 D. Rieger, R.D. Schnell and W. Steinmann, Surf. Sci., 143 (1984) 157.
41 F. Netzer, D.L. Doering and T.E. Madey, Surf. Sci., 143 (1984) L363.

42 A.M. Bradshaw and F.M. Hoffman, Surf. Sci., 72 (1978) 513.
43 J.C. Campuzano and R.G. Greenler, Surf. Sci., 72 (1978) 513.
44 D. Brennan and F.H. Hayes, Philos. Trans. R. Soc. London Ser. A, 258 (1965) 347. See also D. Hayward in J.R. Anderson (Ed.), Chemisorption and Reaction on Metal Films, Vol. 1, Academic Press, New York, 1971, p. 299.
45 J.W. Schwarz, R.S. Polizzotti and J.J. Burton, Surf. Sci., 67 (1977) 10.
46 D.W. Eastman, Solid State Commun., 10 (1972) 933.
47 S. Anderson and C. Nyberg, Surf. Sci., 52 (1975) 489. R.L. Park and J.E. Houston, Phys. Rev. B, 6 (1972) 1073.
48 M.P. Hooker and J.T. Grant, Surf. Sci., 62 (1977) 21.
49 P.H. Dawson, Surf. Sci., 65 (1977) 41.
50 Y. Fukuda, W.T. Elam and R.L. Park, Appl. Surf. Sci., 1 (1978) 278.
51 Y. Fukuda, G.M. Lancaster, F. Honda and J.W. Rabalais, J. Chem. Phys., 69 (1978) 3447.
52 Y. Fukuda, F. Honda and J.W. Rabalais, Surf. Sci., 91 (1980) 165.
53 D.M. Hanson, R. Stockbauer and T.E. Madey, Phys. Rev. B, 24 (1981) 5513.
54 N.A. Surplice and W. Brearley, Surf. Sci., 72 (1978) 84.
55 B. Kasemo and E. Tornqvist, Surf. Sci., 77 (1978) 209.
56 H.D. Shih, F. Jona, D.W. Jepsen and P.M. Marcus, J. Vac. Sci. Technol., 15 (1978) 596.
57 T.W. Haas, A.G. Jackson and M.P. Hooker, J. Chem. Phys., 46 (1967) 3025.
58 J. Kirschner, D. Menzel and P. Staib, Surf. Sci., 87 (1979) L267.
59 C.R. Brundle, Surf. Sci., 52 (1975) 426.
60 F.J. Szalkowski and G.A. Somorjai, Surf. Sci., 52 (1975) 431.
61 A. Benninghoven, Surf. Sci., 53 (1975) 596.
62 G. Blyholder and M.C. Allen, J. Chem. Soc., 91 (1969) 3158.
63 C.A. Hague and H.E. Farnsworth, Surf. Sci., 1 (1964) 378.
64 F.S. Baker, A.M. Bradshaw, J. Pritchard and K.W. Sykes, Surf. Sci., 12 (1968) 426.
65 J.T. Grant and T.W. Haas, Surf. Sci., 17 (1969) 484.
66 S. Andersson and C. Nyberg, Surf. Sci., 52 (1975) 489.
67 H. Kato, Y. Sakisaka, T. Miyano, K. Kamel, N. Nishijima and M. Onchi, Surf. Sci., 114 (1982) 96.
68 N.D. Shinn and T.E. Madey, Phys. Rev. Lett., 53 (1984) 2481.
69 N.D. Shinn and T.E. Madey, Phys. Rev. B, 33 (1986) 1469.
70 N.D. Shinn and T.E. Madey, J. Chem. Phys., 83 (1985) 5928.
71 H.M. Kennett and A.E. Lee, Surf. Sci., 33 (1972) 377.
72 P. Michel and Ch. Jardin, Surf. Sci., 36 (1973) 478.
73 A. Benninghoven and A. Muller, Surf. Sci., 39 (1973) 416.
74 S. Ekelund and C. Leygraf, Surf. Sci., 40 (1973) 179.
75 J.C. Fuggle, L.M. Watson, D.J. Fabian and S. Affrossman, Surf. Sci., 49 (1975) 61.
76 R.I. Bickley, M.W. Roberts and W.C. Storey, J. Chem. Soc. A, (1971) 2774.
77 H.K. Hu and J.W. Rabalais, Surf. Sci., 107 (1981) 376.
78 G. Wedler, K.G. Colb, G. McElhiney and W. Heinrich, Appl. Surf. Sci., 2 (1978) 30.
79 K.Y. Yu, W.E. Spicer, I. Lindau, P. Pianetta and S.F. Lin, Surf. Sci., 57 (1976) 157.
80 K. Kishi and M.W. Roberts, J. Chem. Soc. Faraday Trans. 1, 71 (1975) 1715.
81 G. Ertl, J. Kuppers, F. Nitschke and M. Weiss, Chem. Phys. Lett., 52 (1977) 309.
82 G. Broden, G. Gafner and H.P. Bonzel, Appl. Phys., 13 (1977) 333.
83 G. Broden, G. Gafner and H.P. Bonzel, Surf. Sci., 84 (1979) 295.
84 L. Gonzalez, R. Miranda and S. Ferrer, Surf. Sci., 119 (1982) 61.
85 T.N. Rhodin and C.F. Brucker, Solid State Commun. 23 (1977) 275.
86 F. Jona, K.O. Legg, H.D. Shih, D.W. Jepson and P.M. Marcus, Phys. Rev. Lett., 40 (1978) 1466.
87 K. Yoshida and G.A. Somorjai, Surf. Sci., 75 (1978) 46.
88 D.W. Moon, D.J. Dwyer and S.L. Bernaser, Surf. Sci., 163 (1985) 215.
89 C. Benndorf, B. Krüger and F. Thieme, Surf. Sci., 163 (1985) L675.
90 J. Benziger and R.J. Madix, Surf. Sci., 94 (1980) 119.

91 W. Erley, J. Vac. Sci. Technol., 18 (1981) 472.
92 M. Textor, I.D. Gay and R. Mason, Proc. R. Soc. London Ser. A, 356 (1977) 37.
93 U. Seip, M.-C. Tsai, K. Christmann, J. Kuppers and G. Ertl, Surf. Sci., 39 (1984) 29.
94 A.M. Bradshaw and J. Pritchard, Proc. R. Soc. London Ser. A, 316 (1970) 169.
95 C. Benndorf and F. Thieme, Z. Phys. Chem. NF, 102 (1976) 231.
96 M.E. Bridge, C.M. Comrie and R.M. Lambert, Surf. Sci., 67 (1977) 393.
97 H. Papp, Surf. Sci., 129 (1983) 205.
98 J.P. Biberian and M.A. Van Hove, Surf. Sci., 118 (1982) 443.
99 J.C. Tracy, J. Chem. Phys., 56 (1972) 2736.
100 H.J. Freund and G. Hollinger, Ber. Bunsenges. Phys. Chem., 83 (1979) 100.
101 H. Papp, Ber. Bunsenges. Phys. Chem., 86 (1982) 555.
102 F. Greuter, D. Heskett, E.W. Plummer and H.J. Freund, Phys. Rev. B, 27 (1983) 7117.
103 K.A. Prior, K. Schwaha and R.M. Lambert, Surf. Sci., 77 (1978) 193.
104 V.J. Kehrer and J. Leidheiser, J. Phys. Chem., 58 (1954) 550.
105 H. Papp, Surf. Sci., 149 (1985) 460.
106 M.E. Bridge, C.M. Comrie and R.M. Lambert, J. Catal., 58 (1979) 23.
107 G. Wedler, H. Papp and G. Schroll, Surf. Sci., 44 (1974) 463.
108 H.H. Madden, J. Kuppers and G. Ertl, J. Chem. Phys., 58 (1973) 3401.
109 T.N. Taylor and P.J. Estrup, J. Vac. Sci. Technol., 10 (1973) 26.
110 K. Christmann, O. Schober and G. Ertl, J. Chem. Phys., 60 (1974) 4719.
111 G. Ertl and J. Kuppers, Low Energy Electrons and Surface Chemistry, Verlag Chemie, Weinheim, 1974, p. 227.
112 K. Klier, A.C. Zettlemoyer and H. Leidheiser, Jr., J. Chem. Phys., 52 (1970) 589.
113 S. Andersson, Solid State Commun., 21 (1977) 75.
114 S. Andersson, in R. Dobrozemsky, F. Rüdenauer, F.P. Viehböck and A. Breth (Eds.), Proc. 7th Int. Vac. Congr. and 3rd Int. Conf. Solid Surf., Berger, Vienna, 1977, p. 1019.
115 J.C. Bertolini and B. Tardy, Surf. Sci., 102 (1981) 131.
116 M.J. Dignam, in R. Caudano, J.M. Giles and A.A. Lucas (Eds.), Vibrations at Surfaces, Plenum Press, New York, 1982.
117 R.G. Tobin, S. Chiang, P.A. Thiel and P.L. Richards, Surf. Sci., 140 (1984) 393. S. Chiang, R.G. Tobin and P.L. Richards, Phys. Rev. Lett., 52 (1984) 648.
118 G.E. Mitchell, J.L. Gland and J.M. White, Surf. Sci., 131 (1983) 167.
119 H.A. Marzouk, E. Bradley and K.A. Arunkumar, Surf. Sci., 161 (1985) 462.
120 J.T. Yates, Jr., D.W. Goodman and T.E. Madey, in R. Dobrozemsky, F. Rüdenauer, F.P. Vleböck and A. Breth (Eds.), Proc. 7th Int. Vac. Congr. and 3rd Int. Conf. Solid Surf., Berger, Vienna 1977, p. 1133. D.W. Goodman, J.T. Yates, Jr. and T.E. Madey, Surf. Sci., 93 (1980) L135.
121 B. Koel, D.E. Peebles and J.M. White, Surf. Sci., 107 (1981) L367.
122 J.T. Yates, Jr. and C.W. Garland, J. Phys. Chem., 65 (1961) 617.
123 R.P. Eischens, S.A. Francis and W.A. Pliskin, J. Phys. Chem., 60 (1956) 194.
124 M. Passler, A. Ignatiev, F. Jona, D.W. Jepsen and P.M. Marcus, Phys. Rev. Lett., 43 (1979) 360.
125 S. Andersson and J.B. Pendry, Phys. Rev. Lett., 43 (1979) 363.
126 C.L. Allyn, T. Gustafsson, E.W. Plummer, D.E. Eastman and J.L. Freeouf, Solid State Commun., 17 (1975) 391.
127 C.H. Li and S.Y. Tong, Phys. Rev. Lett., 40 (1978) 46.
128 L.G. Petersson, J. Kono, N.F.T. Hall, C.S. Fadley and J.B. Pendry, Phys. Rev. Lett., 42 (1979) 1545.
129 F.J. Himpsel and Th. Fauster, Phys. Rev. Lett., 49 (1982) 1583. Th. Fauster and F.J. Himpsel, Phys. Rev. B, 27 (1983) 1390.
130 H. Hopster and C.R. Brundle, J. Vac. Sci. Technol., 16 (1979) 548.
131 R.S. Bordoli, J.C. Vickerman and J. Wolstenholm, Surf. Sci., 85 (1979) 241.
132 T. Fleisch, G.L. Ott, W.N. Delgass and N. Winograd, Surf. Sci., 81 (1979) 1.
133 R. Lambert, Surf. Sci., 49 (1975) 325.

134 P.R. Mahaffy and M.J. Dignam, Surf. Sci., 97 (1980) 377.
135 P.R. Mahaffy and M.J. Dignam, J. Chem. Phys., 84 (1980) 2683.
136 J.C. Bertolini and B. Tardy, Surf. Sci., 102 (1981) 131.
137 N. Nishijima, S. Masuda, Y. Sukisaka and M. Onchi, Surf. Sci., 107 (1981) 31.
138 B.J. Bandi, M.A. Chesters, P. Hollins, J. Pritchard and N. Shepperd, J. Mol. Struct., 80 (1982) 203.
139 B.A. Gurney and W. Ho, J. Vac. Sci. Technol. A, 3 (1985) 1541.
140 R.J. Madix, J.L. Gland, G.E. Mitchell and B.A. Sexton, Surf. Sci., 125 (1981) 481.
141 J. Lee, J. Arias, C. Hanrahan, R. Martin, H. Metiu, C. Klauber, M.D. Alvey and J.T. Yates, Jr., Surf. Sci., 159 (1985) L460.
142 W. Riedl and D. Menzel, in W. Brenig and D. Menzel (Eds.), Proc. DIET II Conf., Springer, Berlin, 1985.
143 M.D. Alvey, M.J. Dresser and J.T. Yates, Jr., Surf. Sci., 165 (1986) 447.
144 J. Falconer and R.J. Madix, Surf. Sci., 48 (1975) 393.
145 R.J. Behm, G. Ertl and V. Penka, Surf. Sci., 160 (1985) 387.
146 Y. Hsu, K. Jacobi and H. Rotermund, Surf. Sci., 117 (1982) 581.
147 J.C. Bertolini, G. Dalmai-Imelik and J. Rousseau, Surf. Sci., 68 (1977) 539.
148 J. Erley, H. Wagner and H. Ibach, Surf. Sci., 80 (1979) 612.
149 S.L. Tang, M.B. Lee, Q.Y. Yang, J.D. Beckerle and S.T. Ceyer, J. Chem. Phys., 84 (1986) 1876.
150 J.C. Campuzano, R. Dus and R.G. Greenler, Surf. Sci., 102 (1981) 172.
151 K. Christmann, O. Schober and G. Ertl, J. Chem. Phys., 60 (1974) 4719.
152 H. Conrad, G. Ertl, J. Kuppers and E.E. Latta, Surf. Sci., 57 (1976) 475.
153 M. Trenary, K.J. Uram, F. Bozso and J.T. Yates, Jr., Surf. Sci., 157 (1985) 512.
154 M. Trenary, K.J. Uram, F. Bozso and J.T. Yates, Jr., Surf. Sci., 147 (1984) 269.
155 N.V. Richardson and A.M. Bradshaw, Surf. Sci., 88 (1979) 255.
156 B.N.J. Persson and R. Ryberg, Phys. Rev. Lett., 54 (1985) 2119.
157 F.P. Netzer and T.E. Madey, J. Chem. Phys., 76 (1982) 710.
158 S.L. Tang, J.D. Beckerle, M.B. Lee and S.T. Ceyer, J. Chem. Phys., 84 (1986) 1876.
159 M. Shayegan, E.D. Williams, R.E. Glover and R.L. Park, Surf. Sci. 154 (1985) L239.
160 P.R. Norton, R.L. Tapping and J.W. Goodale, Surf. Sci., 72 (1978) 33.
161 W. Englert, W. Heiland, E. Taglauer and D. Menzel, Surf. Sci., 83 (1979) 243.
162 W. Erley and H. Wagner, J. Catal. 53 (1978) 287.
163 G.W. Rubloff and J.L. Freeouf, Phys. Rev. B, 17 (1978) 4680.
164 E.L. Muetterties and J. Stein, Chem. Rev., 79 (1979) 479.
165 P.R. Wentrcek, B.J. Wood and H. Wise, J. Catal., 43 (1976) 363.
166 M. Araki and V. Ponec, J. Catal., 44 (1976) 439.
167 D.W. Goodman, R.D. Kelley, T.E. Madey and J.T. Yates, Jr., J. Catal., 63 (1980) 26 and references cited therein.
168 D.W. Goodman, R.D. Kelley, T.E. Madey and J.M. White, J. Catal., 64 (1980) 479.
169 W. Erley and H. Wagner, Surf. Sci., 74 (1978) 333.
170 Z. Murayama, I. Kojima, E. Miyazake and I. Yasumori, Surf. Sci., 118 (1982) L281.
171 D.W. Goodman and J.T. Yates, Jr., J. Catal., 82 (1983) 255.
172 J.C. Tracy, J. Chem. Phys., 56 (1972) 2748.
173 M.A. Chesters and J. Pritchard, Surf. Sci., 28 (1976) 460.
174 K. Horn and J. Pritchard, Surf. Sci., 55 (1979) 701.
175 J. Pritchard, Surf. Sci., 79 (1979) 231.
176 B.A. Sexton, Chem. Phys. Lett., 63 (1979) 451.
177 S. Andersson, Surf. Sci., 89 (1979) 477.
178 R. Ryberg, Surf. Sci., 114 (1982) 627.
179 C.C. Allyn, T. Gustafsson and E.W. Plummer, Solid State Commun., 24 (1977) 531.
180 J. Küppers, F. Nitschke, K. Wandelt, G. Ertl and C. Brundle, J. Chem. Soc. Faraday Trans. 1, 75 (1979) 984.

181 F. Bozso, J. Arias, J.T. Yates, R.M. Martin and H.H. Metiu, Chem. Phys. Lett., 94 (1983) 243.
182 H. Conrad, G. Ertl, J. Kuppers, S.W. Wang, K. Gerard and H. Haberland, Phys. Rev. Lett., 42 (1979) 1082.
183 P.R. Norton, R.L. Tapping and J.W. Godale, Surf. Sci., 72 (1978) 33.
184 S.A. Isa, R.W. Joyner and M.W. Roberts, J. Chem. Soc. Chem. Commun., 11 (1977) 377; J. Chem. Soc. Faraday Trans. 1, 74 (1978) 546.
185 A. Spitzer and H. Lüth, Surf. Sci., 102 (1981) 29.
186 K. Horn, M. Hussain and J. Pritchard, Surf. Sci., 63 (1977) 244.
187 C. Harendt, J. Goschnick and W. Hirschwald, Surf. Sci., 152/153 (1985) 453.
188 P. Hollins, K.J. Davies and J. Pritchard, Surf. Sci., 138 (1984) 75.
189 D.P. Woodruff, B.E. Hayden, K. Prince and A.M. Bradshaw, Surf. Sci., 123 (1982) 397.
190 J.F. Wendelken and M.V. Ulehla, J. Vac. Sci., Technol., 16 (1979) 441.
191 J. Rogozik, H. Scheidt, V. Dose, K.C. Prince and A.M. Bradshaw, Surf. Sci., 145 (1984) L481.
192 J. Pritchard, T. Catterick and R.K. Gupta, Surf. Sci., 53 (1975) 1. P. Hollins and J. Pritchard, Surf. Sci., 89 (1979) 486.
193 P. Hollins and J. Pritchard, Surf. Sci., 99 (1980) L389.
194 B.E. Hayden, K. Kretzchmar and A.M. Bradshaw, Surf. Sci., 155 (1985) 553.
195 M.A. Chesters, S.F. Parker and R. Raval, Surf. Sci., 165 (1986) 179.
196 B.E. Hayden and A.M. Bradshaw, Surf. Sci., 125 (1983) 787.
197 Y. Jugnet and Tran Minh Duc, Chem. Phys. Lett., 58 (1978) 243.
198 J. Kanski, L. Ilver and P.O. Nilsson, Solid State Commun., 26 (1978) 339.
199 M. Moskovits and J.E. Hulse, Surf. Sci., 21 (1976) 302; J. Phys. Chem., 81 (1977) 2004.
200 C.M. Quinn and M.W. Roberts, Trans. Faraday Soc., 59 (1963) 985.
201 J.S. Foord, P.J. Goddard and R.M. Lambert, Surf. Sci., 94 (1980) 339.
202 J.J. Carroll and A.J. Melmed, Surf. Sci., 58 (1976) 601.
203 T.W. Haas, J.T. Grant and G.J. Dooley, J. Vac. Sci. Technol., 7 (1970) 43.
204 J.M. Manriquez, D.R. McAlister, R.D. Janner and J.E. Bercaw, J. Am. Chem. Soc., 98 (1976) 6733.
205 J.C. Huffman, J.G. Stone, W.C. Krusell and K.G. Caulton, J. Am. Chem. Soc., 99 (1977) 5829.
206 R. Klein and J.W. Little, Surf. Sci., 2 (1964) 167.
207 T.W. Haas, A.G. Jackson and M.P. Hooker, J. Chem. Phys., 46 (1967) 3025.
208 G.J. Dooley and T.W. Haas, J. Vac. Sci. Technol., 7 (1970) 590.
209 P.R. Davis, E.E. Donaldson and D.R. Sandstrom, Surf. Sci., 34 (1973) 177.
210 D.M. Hanson, R. Stockbauer and T.E. Madey, unpublished data.
211 J.G. Little, C.M. Quinn and M.W. Roberts, J. Catal., 3 (1964) 57.
212 T.W. Felter and P.J. Estrup, Surf. Sci., 76 (1978) 464.
213 J.T. Yates, Jr. and D.A. King, Surf. Sci., 30 (1972) 601.
214 J.T. Yates, Jr. and D.A. King, Surf. Sci., 32 (1972) 479.
215 Y. Viswanath and L.D. Schmidt, J. Chem. Phys., 59 (1973) 4184.
216 J. Lecante, R. Riwan and G. Guillot, Surf. Sci., 35 (1973) 271.
217 G. Guillot, R. Riwan and J. Lecante, Surf. Sci., 59 (1976) 581.
218 E.I. Ko and R.J. Madix, Surf. Sci., 100 (1980) L505.
219 E.I. Ko and R.J. Madix, Surf. Sci., 109 (1981) 221.
220 S. Semancik and P.J. Estrup, J. Vac. Sci. Technol., 17 (1980) 233.
221 J. Atkinson, C.R. Brundle and M.W. Roberts, Faraday Discuss. Chem. Soc., 58 (1974) 62.
222 S. Semancik and P.J. Estrup, Surf. Sci., 104 (1981) 26.
223 G.L. Kellogg, Surf. Sci., 111 (1981) 205.
224 N. Niehus, Surf. Sci., 92 (1980) 88.
225 E. Gillet, J.C. Chiarena and M. Gillet, Surf. Sci., 66 (1977) 596, 613, 622.
226 C. Boiziau, C. Garot, R. Nuvolone and J. Roussel, Surf. Sci., 91 (1980) 313.
227 R. Klein, Surf. Sci., 20 (1970) 1.
228 S. Chakrabortty and H. Grenga, J. Appl. Phys., 44 (1973) 5000.
229 K. Kraemer and D. Menzel, Ber. Bunsenges. Phys. Chem., 78 (1974) 591.

230 T.E. Madey and D. Menzel, Proc. 2nd Int. Conf. Solid. Surf., Japan, J. Appl. Phys. Suppl. 2, Pt. 2 (1974) 229.
231 H. Pfnür and D. Menzel, Surf. Sci., 148 (1984) 411.
232 E.D. Williams and W.H. Weinberg, Surf. Sci., 82 (1979) 93.
233 G.E. Thomas and W.H. Weinberg, J. Chem. Phys., 70 (1979) 954.
234 H. Pfnür, D. Menzel, F.M. Hoffmann, A. Ortega and A.M. Bradshaw, Surf. Sci., 19 (1980) 431.
235 J.C. Függle, T.E. Madey, M. Steinkilberg and D. Menzel, Surf. Sci., 52 (1975) 521.
236 J.C. Függle, M. Steinkilberg and D. Menzel, Chem. Phys., 11 (1975) 307.
237 D. Wearie, Surf. Sci., 103 (1981) L115.
238 A. Brown and J.C. Vickerman, Surf. Sci., 117 (1982) 154.
239 A. Brown and J.C. Vickerman, Surf. Sci., 124 (1983) 267.
240 T.E. Madey, Surf. Sci., 79 (1979) 575.
241 P. Hofmann, J. Gossler, A. Zartner, G. Glanz and D. Menzel, Surf. Sci., 161 (1985) 303.
242 D. Heskett, E.W. Plummer, R.A. de Paola, W. Eberhardt, F.M. Hoffmann and H.R. Moser, Surf. Sci., 164 (1985) 490.
243 J.P. Biberian and M. Van Hove, Surf. Sci., 138 (1984) 361.
244 W. Riedl and D. Menzel, Surf. Sci., 163 (1985) 39.
245 F.P. Netzer, D.L. Doering and T.E. Madey, Surf. Sci., 143 (1985) L363.
246 F.M. Hoffmann and R.A. de Paola, Phys. Rev. Lett., 52 (1984) 1697.
247 T.E. Madey and C. Benndorf, Surf. Sci., 164 (1985) 602.
248 J.J. Weimer, E. Umbach and D. Menzel, Surf. Sci., 155 (1985) 132; 159 (1985) 83.
249 W. Eberhardt, F.M. Hoffmann, R. de Paola, D. Heskett, I. Strethy, E.W. Plummer and H.R. Moser, Phys. Rev. Lett., 54 (1985) 1856.
250 E. Shincho, C. Egawa, S. Naito and K. Tamaru, Surf. Sci., 149 (1985) 1.
251 E. Shincho, C. Egawa, S. Naito and K. Tamaru, Surf. Sci., 155 (1985) 153.
252 G. Michalk, W. Moritz, H. Pfnür and D. Menzel, Surf. Sci., 129 (1983) 92.
253 H. Pfnür and D. Menzel, J. Chem. Phys., 79 (1983) 2400.
254 A. Cassuto and D.A. King, Surf. Sci., 102 (1981) 388.
255 T. Yamada and K. Tamaru, Surf. Sci., 146 (1984) 347.
256 V.P. Zhdanov, Surf. Sci., 157 (1985) L384.
257 H. Pfnür, P. Feulner and D. Menzel, J. Chem. Phys., 79 (1983) 4613.
258 D. Menzel, H. Pfnür and P. Feulner, Surf. Sci., 126 (1983) 374.
259 J.G. McCarty and H. Wise, Chem. Phys. Lett., 61 (1979) 323.
260 P.H. Reed, C.M. Comrie and R.M. Lambert, Surf. Sci., 59 (1976) 33.
261 D.W. Goodman, T.E. Madey, M. Ono and J.T. Yates, Jr., J. Catal., 50 (1977) 279.
262 R. Ku, N.A. Gjostein and H.P. Bonzel, Surf. Sci., 64 (1977) 465.
263 D.G. Castner, B.A. Sexton and G.A. Somorjai, Surf. Sci., 71 (1978) 519.
264 P.A. Thiel, E.D. Williams, J.T. Yates, Jr. and W.H. Weinberg, Surf. Sci., 84 (1979) 54.
265 Y. Kim, H.C. Peebles and J.M. White, Surf. Sci., 114 (1982) 363.
266 R.A. Marbrow and R.M. Lambert, Surf. Sci., 67 (1977) 489.
267 C.T. Campbell and J.M. White, J. Catal., 54 (1978) 289.
268 J.T. Grant and T.W. Haas, Surf. Sci., 21 (1970) 76.
269 R.J. Koestner, M.A. Van Hove and G.A. Somorjai, Surf. Sci., 107 (1981) 439.
270 L.H. Dubois and G.A. Somorjai, Surf. Sci., 91 (1980) 514.
271 M.A. Van Hove, R.J. Koestner and G.A. Somorjai, Phys. Rev. Lett., 50 (1983) 903.
272 M.A. Van Hove, R.J. Koestner, J.C. Frost and G.A. Somorjai, Surf. Sci., 129 (1983) 482.
273 L.A. DeLouise and N. Winograd, Surf. Sci., 138 (1984) 417.
274 L.A. DeLouise, E.J. White and N. Winograd, Surf. Sci., 147 (1984) 252.
275 W. Braun, M. Neumann, M. Iwan and E.E. Koch, Solid State Commun., 27 (1978) 155.
276 E.D. Williams, P.A. Thiel, W.H. Weinberg and J.T. Yates, Jr., J. Chem. Phys., 72 (1980) 3496.
277 J.T. Yates, Jr., E.D. Williams and W.H. Weinberg, Surf. Sci., 91 (1980) 562.
278 H.C. Eckstrom, G.G. Possley and S.E. Hannum, J. Chem. Phys., 52 (1970) 5435.

279 B.A. Sexton and G.A. Somorjai, J. Catal., 46 (1977) 167.
280 D.G. Castner, L.H. Dubois, B.A. Sexton and G.A. Somorjai, Surf. Sci., 103 (1981) L134.
281 J.T. Yates, Jr., E.D. Williams and W.H. Weinberg, Surf. Sci., 91 (1980) 562.
282 J.T. Yates, Jr., E.D. Williams and W.H. Weinberg, Surf. Sci., 115 (1982) L93.
283 V.V. Gorodetskii and B.E. Nieuwenhuys, Surf. Sci., 105 (1981) 299.
284 R.J. Baird, R.C. Ku and P. Wynblatt, Surf. Sci., 97 (1980) 346.
285 J.T. Yates, Jr., T.M. Duncan, S.D. Worley and R.W. Vaughan, J. Chem. Phys., 70 (1979) 1219.
286 J.T. Yates, Jr., T.M. Duncan and R.W. Vaughan, J. Chem. Phys., 71 (1979) 3908.
287 T.M. Duncan, J.T. Yates, Jr. and R.W. Vaughan, J. Chem. Phys., 73 (1980) 975.
288 R.R. Cavanagh and J.T. Yates, Jr., J. Chem. Phys., 74 (1981) 4150.
289 D.J.C. Yates, L.L. Murrell and E.B. Prestridge, J. Catal., 57 (1979) 41.
290 E.B. Prestridge and D.J.C. Yates, Nature (London), 234 (1971) 345.
291 R.M. Kroeker, W.C. Kaska and P.K. Hansma, J. Catal., 57 (1979) 72.
292 J. Klein, A. Leger, S. DeCheveigne, G. Guinet, M. Belin and D. Deforneau, Surf. Sci., 82 (1979) L288.
293 G. Ertle and J. Kuppers, Low Energy Electrons and Surface Chemistry, Verlag Chemie, Weinheim, 1974, 227.
294 J.C. Tracy and P.W. Palmberg, J. Chem. Phys., 51 (1969) 4852.
295 A.M. Bradshaw and F.M. Hoffmann, Surf. Sci., 72 (1978) 513.
296 R.J. Behm, K. Christmann, G. Ertl, M.A. Van Hove, P.A. Thiel and W.H. Weinberg, Surf. Sci., 88 (1979) L59.
297 R.J. Behm, K. Christmann, G. Ertle and M.A. Van Hove, J. Chem. Phys., 73 (1980) 2984.
298 A. Ortega, F.M. Hoffman and A.M. Bradshaw, Surf. Sci., 119 (1982) 79.
299 N.R. Avery, J. Chem. Phys., 74 (1981) 4202.
300 A. Brown and J.C. Vickerman, Surf. Sci., 151 (1985) 319.
301 D.R. Lloyd, C.M. Quinn and N.V. Richardson, Solid State Commun., 20 (1976) 412.
302 K. Horn, A.M. Bradshaw, K. Hermann and I.P. Batra, Solid State Commun., 31 (1979) 257.
303 J. Rogozik, J. Kuppers and V. Dose, Surf. Sci., 148 (1984) L653.
304 F.A. Cotton and G. Wilkinson, Advanced Inorganic Chemistry, Wiley, New York, 1980.
305 S.D. Bader, J.M. Blakely, M.B. Brodsky, R.J. Friddle and R.L. Panosh, Surf. Sci., 74 (1978) 405.
306 G. Ertl, H. Conrad, J. Koch and E.E. Latta, Surf. Sci., 43 (1974) 462.
307 M.A. Chesters, G.S. McDougall, M.E. Pemble and N. Sheppard, Surf. Sci., 164 (1985) 425.
308 G. Ertl and J. Koch, Z. Naturforsch. Teil A, 25 (1970) 1906.
309 G. Ertl and J. Koch, in F. Ricca (Ed.), Adsorption–Desorption Phenomena, Academic Press, London, 1972, p. 345.
310 A.M. Bradshaw and F.M. Hoffmann, Surf. Sci., 72 (2978) 513.
311 V. Matolin, S. Channakhone and E. Gillet, Surf. Sci., 164 (1985) 209.
312 A. Brown and J.C. Vickerman, Surf. Sci., 124 (1983) 267.
313 T. Engel, J. Chem. Phys., 69 (1978) 373.
314 M.P. Kiskinova and G.M. Bliznakov, Surf. Sci., 123 (1982) 61.
315 R. Miranda, K. Wandelt, D. Rieger and R.D. Schnell, Surf. Sci., 139 (1984) 430.
316 H. Conrad, G. Ertl, J. Kupper, W. Sesselmann and H. Haberland, Surf. Sci., 121 (1982) 161.
317 W. Sesselmann, B. Woratschek, G. Ertl, J. Kupper and H. Haberland, Surf. Sci., 146 (1984) 12.
318 F.P. Netzer and M.M. El Gomati, Surf. Sci., 124 (1983) 26.
319 V.C. Vankar, R.W. Vook and B.C. De Cooman, Thin Solid Films, 102 (1983) 313.
320 H. Poppa and F. Soria, Phys. Rev. B, 27 (1983) 5166.
321 B.E. Nieuwenhuys and G.A. Kok, Thin Solid Films, 106 (1983) L95.
322 H. Conrad, G. Ertl and J. Kuppers, Surf. Sci., 76 (1978) 323.
323 T.E. Madey, J.T. Yates, Jr., A.M. Bradshaw and F.M. Hoffmann, Surf. Sci., 89 (1979) 270.
324 P.W. Davies and R.M. Lambert, Surf. Sci., 111 (1981) L671.
325 P.W. Davies and R.M. Lambert, Surf. Sci., 110 (1981) 227.

326 R. Unwin and A.M. Bradshaw, Chem. Phys. Lett., 58 (1978) 58.
327 M. Grunze, Chem. Phys. Lett., 58 (1978) 409.
328 Y. Soma-Noto and W.M.H. Sachtler, J. Catal., 32 (1974) 315.
329 G. McElhiney, H. Papp and J. Pritchard, Surf. Sci., 54 (1976) 617.
330 R.R. Ford, Adv. Catal. 21 (1970) 51.
331 R.A. Marbrow and R.M. Lambert, Surf. Sci., 71 (1978) 107.
332 H.A. Engelhard, A.M. Bradshaw and D. Menzel, Surf. Sci., 40 (1973) 410.
333 M. Bowker, M.A. Barteau and R.J. Madix, Surf. Sci., 92 (1980) 528.
334 S. Krause, C. Mariani, K.C. Prince and K. Horn, Surf. Sci., 138 (1984) 305.
335 T.W. Haas, J.T. Grant and G.T. Dooley, in F. Ricca (Ed.), Adsorption–Desorption Phenomena, Academic Press, London, 1972, p. 359.
336 R. Ducros, G. Piquard, B. Weber and A. Cassuto, Surf. Sci., 54 (1976) 513.
337 M.P. Hooker and J.T. Grant, Surf. Sci., 55 (1976) 741.
338 S.M. Ko and L.D. Schmidt, Surf. Sci., 47 (1975) 557.
339 M.A. Chesters, B.J. Hopkins and M.R. Leggett, Surf. Sci., 43 (1974) 1.
340 R. Klein and L.B. Leder, J. Chem. Phys., 38 (1963) 1866.
341 V.D. Belov, Yu. K. Ustinov and A.P. Lomar, Surf. Sci., 72 (1978) 390.
342 R.P.H. Gasser and R. Thwaites, Trans. Faraday Soc., 61 (1965) 2036.
343 M.D. Scheer and J. Fine, Surf. Sci., 12 (1968) 102.
344 D.O. Hayward and J.C. McManus, cited by D.O. Hayward, in J.R. Anderson (Ed.), Chemisorption and Reactions on Metallic Films, Vol. I, Academic Press, London, 1971, p. 225.
345 R.R. Ford, Adv. Catal., 21 (1970) 51.
346 R. Gomer, Japan, J. Appl. Phys. Suppl., 2(2)(1974) 213.
347 L.D. Schmidt, in R. Gomer (Ed.), Interaction on Metal Surfaces, Springer-Verlag, New York, 1975.
348 E.W. Plummer, B.J. Waclawski, T.V. Vorburger and C.E. Kuyatt, Prog. Surf. Sci., 7 (1976) 149.
349 P.A. Redhead, Trans. Faraday Soc., 57 (1961) 641.
350 G. Ehrlich, J. Chem. Phys., 34 (1961) 39.
351 L. Swanson and R. Gomer, J. Chem. Phys., 39 (1963) 2813.
352 T.E. Madey, J.T. Yates, Jr. and R.C. Stern, J. Chem. Phys., 42 (1965) 1372.
353 J.R. Chen and R. Gomer, Surf. Sci., 81 (1979) 589.
354 C. Wang and R. Gomer, Surf. Sci., 84 (1979) 329.
355 Ch. Steibruchel and R. Gomer, Surf. Sci., 67 (1977) 21.
356 E. Umbach, J.C. Függle and D. Menzel, J. Electron Spectrosc. Relat. Phenom., 10 (1977) 15. D. Menzel, J. Vac. Sci. Technol., 12 (1975) 313.
357 R. Opila and R. Gomer, Surf. Sci., 129 (1983) 563.
358 E. Umbach and D. Menzel, Surf. Sci., 135 (1983) 199.
359 C. Kohrt and R. Gomer, Surf. Sci., 24 (1971) 77.
360 C. Kohrt and R. Gomer, Surf. Sci., 40 (1973) 71.
361 T.E. Madey, J.E. Houston and S.C. Dahlberg, Vide, 201 (1980) 205.
362 J.T. Yates, Jr., N.E. Erickson, S.D. Worley and T.E. Madey, in E. Drauglis and R.I. Jaffee (Eds.), The Physical Basis of Heterogeneous Catalysis, Plenum Press, New York, 1975.
363 C. Wang and R. Gomer, Surf. Sci., 90 (1979) 10.
364 D.R. Horton and R.I. Masel, Surf. Sci., 125 (1983) 699.
365 H. Froitzheim, H. Ibach and S. Lehwald, Surf. Sci., 63 (1972) 56.
366 R. Franchy and H. Ibach, Surf. Sci., 155 (1985) 15.
367 J.T. Yates, Jr. and D.A. King, Surf. Sci., 38 (1973) 114.
368 A. Adnot and J.D. Carette, Surf. Sci., 74 (1978) 109.
369 W. Heiland, W. Englert and E. Taglauer, J. Vac. Sci. Technol., 15 (1978) 419.
370 R. Jaeger and D. Menzel, Surf. Sci., 93 (1980) 71.
371 R. Franchy and D. Menzel, Phys. Rev. Lett., 73 (1979) 865.
372 J.E. Houston and T.E. Madey, Phys. Rev. B, 26 (1982) 554.

373 N. Niehus, Surf. Sci., 87 (1979) 561.
374 T.E. Madey, J. Czyzewski and J.T. Yates, Jr., Surf. Sci., 49 (1975) 465.
375 J.T. Yates, Jr. and T.E. Madey, J. Chem. Phys., 51 (1969) 334.
376 M. Alnot, J. Ehrhardt, J. Fusy and A. Cassuto, Surf. Sci., 46 (1974) 81.
377 Y. Fukuda, F. Honda and J.W. Rabalais, Surf. Sci., 93 (1980) 338.
378 M. Housley, R. Ducros, G. Piquard and A. Cassuto, Surf. Sci., 68 (1977) 277.
379 W. Braun, G. Meyer-Ehmsen, M. Neuman and E. Schwarz, Solid State Commun., 30 (1979) 605.
380 S. Tatarenko, R. Ducros and M. Alnot, Surf. Sci., 126 (1983) 422.
381 R. Ducros, B. Tardy and J.C. Bertolini, Surf. Sci., 128 (1983) L219.
382 S. Tatarenko, M. Alnot, J.D. Ehrhardt and R. Ducros, Surf. Sci., 152/153 (1985) 471.
383 R. Klein, Surf. Sci., 20 (1970) 1.
384 K. Ishizuka, J. Res. Inst. Catal. Hokkaido Univ., 15 (1967) 95.
385 R.P.H. Gasser and D.E. Holt, Surf. Sci., 64 (1977) 520.
386 J.L. Taylor and W.H. Weinberg, Surf. Sci., 78 (1978) 259.
387 C.M. Comrie and W.H. Weinberg, J. Chem. Phys., 64 (1976) 250.
388 D.I. Hagen, B.E. Nieuwenhuys, G. Rovida and G.A. Somorjai, Surf. Sci., 57 (1976) 632.
389 K. Christman and G. Ertl, Z. Naturforsch. Teil A, 28 (1973) 1144.
390 G. Broden and T. Rhodin, Solid State Commun., 18 (1976) 105.
391 B.E. Nieuwenhuys and G.A. Somorjai, Surf. Sci., 72 (1978) 8.
392 J.L. Taylor, D.E. Ibbotson and W.H. Weinberg, J. Chem. Phys., 69 (1978) 4298.
393 J.T. Grant, Surf. Sci., 18 (1969) 228.
394 T.N. Rhodin and G. Broden, Surf. Sci., 60 (1976) 466.
395 C.-M. Chang, M.A. Van Hove, W.H. Weinberg and E.D. Williams, Surf. Sci., 91 (1980) 440.
396 D.G. Castner and G.A. Somorjai, Chem. Rev., 79 (1979) 233.
397 D. Reinalda and V. Ponec, Surf. Sci., 91 (1980) 113.
398 F.P. Netzer, Appl. Surf. Sci., 7 (1981) 289.
399 P. Netzer, E. Bertel and J.A.D. Matthew, J. Electron Spectrosc. Relat. Phenom., 18 (1980) 199.
400 T. Rhodin and C. Brucker, in R. Drobrozemsky, F. Rüdenauer, F.P. Vleböck and A. Breth (Eds.), Proc. 7th Int. Vac. Congr. and 3rd Int. Conf. Solid Surf., Berger, Vienna, 1977.
401 C.W. Seabury, T.N. Rhodin, M.M. Traum, R. Benbow and Z. Hurych, Surf. Sci., 97 (1980) 363.
402 P.A. Zhdan, G.K. Boreskov, A.I. Boronin, A.P. Schepelin, W. Egelhoff and W.H. Weinberg, Surf. Sci., 71 (1978) 267.
403 M.-L. Shek, S.P. Withrow and W.H. Weinberg, Surf. Sci., 72 (1978) 678.
404 R.A. Shigeishi and D.A. King, Surf. Sci., 58 (1976) 379.
405 H. Froitzheim, H. Hopster, H. Ibach and S. Lehwald, Appl. Phys., 13 (1977) 147.
406 K. Horn and J. Pritchard, J. Phys. (Paris), 38 (1977) C4–164.
407 H.J. Krebs and H. Luth, Appl. Phys., 14 (1977) 337.
408 R.W. McCabe and L.D. Schmidt, Surf. Sci., 64 (1977) 189.
409 G. Ertl, M. Neumann and K.M. Streit, Surf. Sci., 64 (1977) 393.
410 D.M. Collins and W.E. Spicer, Surf. Sci., 69 (1977) 85.
411 H. Hopster and H. Ibach, Surf. Sci., 77 (1978) 109.
412 P.R. Norton, J.W. Goodale and E.B. Selkirk, Surf. Sci., 83 (1979) 189.
413 A. Crossley and D.A. King, Surf. Sci., 95 (1980) 131.
414 A.M. Baro and H. Ibach, Surf. Sci., 103 (1981) 248.
415 C.T. Campbell, G. Ertl, H. Kuipers and J. Segner, Surf. Sci., 107 (1981) 207.
416 T.H. Lin and G.A. Somorjai, Surf. Sci., 107 (1981) 573.
417 D.H. Winicur, J. Hurst, C.A. Becker and L. Wharton, Surf. Sci., 109 (1981) 263.
418 N.R. Avery, J. Chem. Phys., 74 (1981) 4202.
419 B.E. Hayden and A.M. Bradshaw, Surf. Sci., 125 (1983) 787.
420 P.R. Norton, J.A. Davies and T.E. Jackman, Surf. Sci., 122 (1982) L593.
421 H. Steininger, S. Lehwald and H. Ibach, Surf. Sci., 123 (1982) 264.

422 D.F. Ogletree, M.A. Van Hove and G.A. Somorjai, Surf. Sci., 173 (1986) 351.
423 B. Poelsema, R.L. Palmer and G. Comsa, Surf. Sci., 123 (1982) 152.
424 K. Dückers, H.P. Bonzel and D.A. Wesner, Surf. Sci., 166 (1986) 141.
425 B. Poelsema, R.L. Palmer and G. Comsa, Surf. Sci., 136 (1984) 1.
426 M. Prutton, Surface Physics, Clarendon Press, Oxford, 1975.
427 B. Poelsema, L.K. Verheij and G. Comsa, Surf. Sci., 152/153 (1985) 496.
428 M. Trenary, S.L. Tang, R.J. Simonson and F.R. McFeely, Surf. Sci., 124 (1983) 555.
429 J.N. Miller, D.T. Lin, I. Lindau, P.M. Stefan and W.E. Spicer, Phys. Rev. Lett., 38 (1977) 1419. J.N. Miller, I. Lindau and W.E. Spicer, Surf. Sci., 111 (1981) 595.
430 P. Hofmann, S.R. Bare, N.V. Richardson and D.A. King, Solid State Commun., 42 (1982) 645.
431 D. Rieger, R.D. Schnell and W. Steinmann, Surf. Sci., 143 (1984) 157.
432 H.P. Bonzel and R. Ku, J. Vac. Sci. Technol., 9 (1972) 663.
433 H.P. Bonzel and R. Ku, Surf. Sci., 33 (1972) 91.
434 H.P. Bonzel and R. Ku, J. Chem. Phys., 58 (1973) 4617.
435 C.M. Comrie and R.M. Lambert, J. Chem. Soc. Faraday Trans. 1, 72 (1976) 1659.
436 R.W. McCabe and L.D. Schmidt, Surf. Sci., 60 (1976) 85.
437 P. Hofmann, S.R. Bare, N.V. Richardson and D.A. King, Solid State Commun., 42 (1982) 645.
438 S.R. Bare, K. Griffiths, P. Hofmann, D.A. King, G.L. Nyberg and N.V. Richardson, Surf. Sci., 120 (1982) 367.
439 P. Hofmann, S.R. Bare and D.A. King, Surf. Sci., 117 (1982) 245.
440 S.R. Bare, P. Hofmann and D.A. King, Surf. Sci., 144 (1984) 347.
441 S. Ferrer, K.H. Frank and B. Reihl, Surf. Sci., 162 (1985) 264.
442 D. Rieger, R.D. Schnell and W. Steinmann, Surf. Sci., 143 (1984) 157.
443 G. Broden, G. Pirug and H.P. Bonzel, Surf. Sci., 72 (1978) 45.
444 J.A. Davies, T.E. Jackman, D.P. Jackson and P.R. Norton, Surf. Sci., 109 (1981) 20.
445 P.R. Norton, J.A. Davies, D.K. Creber, C.W. Sitter and T.E. Jackman, Surf. Sci., 108 (1981) 205.
446 K. Heinz, E. Lang, K. Strauss and K. Muller, Appl. Surf. Sci., 11/12 (1982) 611.
447 K. Heinz, E. Lang, K. Strauss and K. Muller, Surf. Sci., 120 (1982) L401.
448 P. Heilmann, K. Heinz and K. Muller, Surf. Sci., 83 (1979) 487.
449 H.P. Bonzel, C.R. Helms and S. Kelemen, Phys. Rev. Lett., 35 (1975) 1237.
450 M.A. Barteau, E.I. Ko and R.J. Madix, Surf. Sci., 102 (1981) 99.
451 G. Keringer and F.P. Netzer, Surf. Sci., 49 (1975) 125.
452 C.W. Tucker, Surf. Sci., 2 (1964) 516.
453 A.E. Morgan and G.A. Somorjai, Surf. Sci., 12 (1968) 405.
454 C.R. Helms, H.P. Bonzel and S. Kelemen, J. Chem. Phys., 65 (1976) 1773.
455 R.J. Behm, P.A. Thiel, P.R. Norton and G. Ertl, J. Chem. Phys., 78 (1983) 7437.
456 A. Crossley and D.A. King, Surf. Sci., 95 (1980) 131.
457 G. Pirug, H. Hopster and H. Ibach, Ned. Tijdschr. Vacuumtech., 16 (1978) 152.
458 W.F. Banholzer and R.I. Masel, Surf. Sci., 137 (1984) 339.
459 P.A. Thiel, R.J. Behm, P.R. Norton and G. Ertl, Surf. Sci., 121 (1982) L553.
460 P.A. Thiel, R.J. Behm, P.R. Norton and G. Ertl, J. Chem. Phys., 78 (1983) 7448.
461 P.R. Norton, J.A. Davies, D.K. Creber, C.W. Sitter and T.E. Jackman, Surf. Sci., 108 (1981) 205.

Index

D

E

F

G

H

I

J

K

R

S

T

U

V

W

X

Z

W

X

Z